Varanoid Lizards of the World

Varanoid Lizards of the World

Edited by Eric R. Pianka and Dennis R. King
with Ruth Allen King

Indiana University Press
Bloomington & Indianapolis

This book is a publication of

Indiana University Press
Office of Scholarly Publishing
Herman B. Wells Library 350
1320 East 10th Street
Bloomington, Indiana 47405 USA

iupress.indiana.edu

The paper used in this publication meets
the minimum requirements of American
National Standard for Information
Sciences—Permanence of Paper for
Printed Library Materials, ANSI Z39.48-
1984.

Manufactured in the United States of
America

**Library of Congress Cataloging-in-
Publication Data**

Varanoid lizards of the world / edited by
Eric R. Pianka and Dennis R. King with
Ruth Allen King.
        p. cm.
Includes bibliographical references (p.   ).
  ISBN 0-253-34366-6 (alk. paper)
  1. Monitor lizards. I. Pianka, Eric R. II.
King, Dennis, date III. King, Ruth Allen.
  QL666.L29V27 2004
  597.95'96—dc22
                        2003023722

2 3 4 5 6 21 20 19 18 16

Dedicated to the memory of the late
Dennis R. King,
consummate varanophile, for whom there were "only two kinds of lizards, goannas and goanna food."

Seit ich—vor nunmehr 30 Jahren—meinen ersten lebenden Nilwaran erworben habe und mit seiner Lebensweise im Terrarium vertraut geworden bin, haben mich die W a r a n e, diese "stolzesten, bestproportionierten, mächtigsten und intelligentesten" Eidechsen, wie sie Werner treffend nennt, ständig gefesselt.

Since I—about 30 years ago—got my first living Nile monitor and became acquainted with his life habits in a terrarium, the m o n i t o r   l i z a r d s have fascinated me all the time, these "proudest, best-proportioned, mightiest and most intelligent" lizards as Werner strikingly called them.

—R. Mertens, 1942

CONTENTS

Color photos follow pages 178 and 274.

Contributors

Daniel D. Beck, Department of Biological Sciences,
Central Washington University, Ellensburg, WA 98926, USA

Gavin Bedford, Faculty of Science, Charles Darwin University,
Darwin NT 0909 Australia

Daniel Bennett, 118 Sheffield Road, Glossop, SK13 8QU,
United Kingdom

Wolfgang Böhme, Zoologisches Forschungsinstitut und Museum
Alexander Koenig, Adenauerallee 160, D-53113 Bonn, Germany

Keith Christian, Faculty of Science, Charles Darwin University,
Darwin NT 0909, Australia

Claudio Ciofi, Department of Ecology and Evolutionary Biology, 165
Prospect Street, Yale University, P.O. Box 208106, New Haven,
CT 06520-8106, USA; present address, Dipartimento di Biologia
Animale e Genetica, Via Romana 17, Università de Firenze,
50125 Firenze, Italy

Sean Doody, Applied Ecology Research Group, University of
Canberra, ACT 2601, Australia

Gil Dryden, Dalton Cardiovascular Research Center, University of
Missouri, Columbia, MO 65211, USA

Bernd Eidenmüller, Griesheimer Ufer 53, 65933 Frankfurt am Main,
Germany

Maren Gaulke, Bodenseestrasse 300, D-81249 München, Germany

Harry W. Greene, Department of Ecology and Evolutionary Biology,
Cornell University, Corson Hall, Ithaca, NY 14853, USA

Hans-Georg Horn, Hasslinghauser Strasse 51, D-45549 Sprockhövel,
Germany

Grant Husband, Territory Wildlife Park, P.O. Box 771, Palmerston,
NT 0831, Australia

Steve Irwin, Queensland Reptile Park, Brisbane, Queensland,
Australia

Hans J. Jacobs, Siekenweg 8, D-33178 Borchen, Germany

W. Bryan Jennings, Department of Zoology, University of Washington, Seattle, WA 98195, USA; present address, Museum of Comparative Zoology, Harvard University, Cambridge, MA 02138

Dennis R. King (deceased), 9 Lawrence Close, Darlington, Western Australia 6070, Australia

Ruth Allen King, 9 Lawrence Close, Darlington, Western Australia 6070, Australia

Max King, Baupie, Balranald, NSW 2715, Australia

Jeffrey M. Lemm, Applied Conservation Division, Center for Reproduction of Endangered Species, Zoological Society of San Diego, P.O. Box 120551, San Diego, CA, USA 92112-0551

Sigrid Lenz, Am Wallgraben 8, 56751 Polch, Germany

Ralph E. Molnar, Geology, Museum of Northern Arizona, 3101 North Fort Valley Road, Flagstaff, AZ 86001, USA

Mark A. Norell, American Museum of Natural History, Central Park West at 79th Street, New York, NY 10024-5192, USA

Kai M. Philipp, Zoologische Staatssammlung, Münchhausenstraße 21, D-81247 München, Germany

John ("Andy") Phillips, San Diego Zoo, P.O. Box 120551, San Diego, CA 92101, USA

Eric R. Pianka, Integrative Biology C0930, University of Texas, Austin, TX 78712-1064, USA

Tim Schultz, Faculty of Science, Charles Darwin University, Darwin NT 0909, Australia

Lawrence A. Smith, Western Australian Museum, Perth, WA 6000, Australia

Michael Stanner, 67 Kehilat Warsha Street, 69702 Tel Aviv, Israel

Samuel S. Sweet, Department of Ecology, Evolution and Marine Biology, University of California, Santa Barbara, Santa Barbara, CA 93106, USA

Graham Thompson, Centre for Ecosystem Management, Edith Cowan University, Joondalup Drive, Joondalup WA 6027, Australia

Gerard Visser, Rotterdam Zoological and Botanical Gardens, P.O. Box 532, 3000 AM Rotterdam, The Netherlands

Brian Weavers, New South Wales Department of Infrastructure, Planning and Natural Resources, P.O. Box 39, Sydney, NSW 2001, Australia

Thomas Ziegler, Zoologisches Forschungsinstitut und Museum Alexander Koenig, Adenauerallee 160, D-53113 Bonn, Germany; Present address: Zoologischer Garten Köln, Riehler Straße 173, 50735 Köln, Germany

Acknowledgments

James O. Farlow helped us obtain an advance contract to publish this book with Indiana University Press. We thank the curators of the major Australian Museums for providing locality records from their collections and for allowing us to print these and show them as dot range maps. Ric How, of the Western Australian Museum, was especially helpful in preparing these maps. Richard Bartlett, Daniel D. Beck, Daniel Bennett, Gavin Bedford, Wolfgang Böhme, Alain Compost, Robert Browne-Cooper, Hal Cogger, Indraneil Das, Alex Dudley, Bernd Eidenmüller, Walter Erdelen, Michael Gruschwitz, Adel Ibrahim, Hans J. Jacobs, Ron Johnstone, David Knowles, Jeff Lemm, Brad Maryan, Bob Murphy, Jim Murphy, Kai M. Philipp, Louis Porras, Samuel S. Sweet, Steve Wilson, Frank Yuwono, and Thomas Ziegler kindly allowed us to use some of their photographs. Mark Bayless, Gavin Bedford, Daniel Bennett, Wolfgang Böhme, Robert Browne-Cooper, Greg Fyfe, Steve Irwin, Samuel S. Sweet, Graham Thompson, and Alexei Tsellarins contributed to species accounts by providing references or information as personal communications. We are grateful to Kathleen Kendall and Marianna Grenadier for redrawing most of the line art. Ruth Allen King was deeply involved in the preparation of this book from its very beginnings.

Part I

1. Introduction

ERIC R. PIANKA AND DENNIS R. KING

In Latin, "*monere*" means "to warn." The word "monitor" comes from the Latin noun "*monitio,*" which means someone who is a warner. According to ancient belief, these lizards were supposed to warn people that crocodiles were in an area. The generic name *Varanus* comes from the Arabic word "*waran,*" the Egyptian name for the Nile monitor *Varanus niloticus,* which basically means "monitor" in Arabic. Many large varanids commonly adopt a bipedal stance in which the lizard raises itself vertically and supports itself on its hind legs and tail. Because a lizard then has a clear and elevated view of its surroundings, this upright posture may have given rise to the name "monitor" (Fig. 1.1).

Monitor lizards (genus *Varanus*) have attracted a great deal of interest; these large and impressive lizards are often the centerpieces of reptile house exhibits. Around the world, dedicated varanophiles keep and breed these magnificent lizards in captivity. Monitors tend to be fairly wary and difficult to observe; therefore, they are not particularly tractable research subjects, but they have nevertheless received an extraordinary amount of attention from devoted students. Often, each scrap of information about these lizards requires perseverance and hard work or simply a substantial amount of luck. *Varanus* enthusiasts tend to cooperate freely and exchange data among themselves. This has been emphatically demonstrated again and again during production of this book, as we have received willing assistance from monitor enthusiasts worldwide. The founder of varanid systematics was Robert Mertens, most of whose work was published in German. He wrote in 1942, "Since I—about 30 years ago—got my first living Nile monitor and became acquainted with his life habits in a terrarium, the m o n i t o r l i z a r d s have fascinated me all the time, these 'proudest, best-proportioned, mightiest and most intelligent' lizards as Werner strik-

Figure 1.1. A Varanus gouldii *surveys its surroundings (photo by Jeff Lemm).*

ingly called them" (Mertens 1942, p. 4). Two varanid symposia have been held in Germany (Böhme and Horn 1991; Horn and Böhme 1999). Walter Auffenberg of the University of Florida spent many years in the field following monitor lizards, studying their ecology. This dedicated student of varanids published a monumental trilogy of extremely important books containing a vast amount of information about the ecology and behavior of three species: *V. komodoensis, V. olivaceus,* and *V. bengalensis* (Auffenberg 1981, 1988, 1994). Two other extremely useful compendia of varanid biology are the books by Bennett (1998) and King and Green (1999).

Fifty-three species of *Varanus* are now currently recognized worldwide. All occur in Africa, Asia, Southeast Asia, and Australia (the New World is sadly impoverished, although fossil varanids and the sister group to monitors, Helodermatids, are known from North America). A rash of new species from remote Indonesian and Philippine Islands in Southeast Asia has recently been described (*cerambonensis, caerulivirens, finschi, melinus, mabitang, macraei,* and *yuwonoi*), several others have been recently elevated to species status (*kordensis, spinulosus,* and *ornatus*), and others doubtlessly remain to be described. Some "species" (*gouldii, indicus, tristis,* and *scalaris*) actually represent species complexes that will require revision and recognition of new species. The largest adaptive radiations have occurred in Australia, where they are commonly known as "goannas," and where 24 endemic species have been named (several other new Australian species remain to be described). Several Asian species, including *V. indicus* and *V. prasinus,* have also reached northern Australia. One very interesting Australian clade, the subgenus *Odatria,* has evolved dwarfism.

"Goanna" is believed to be a corruption of the name "iguana," which belongs to a family of lizards found mainly in North and South America, and the name is used to describe varanids of all sizes in Australia. Some Australians often also use the word incorrectly for large members of another family, the Scincidae.

Similarly, in South Africa in the old Dutch language Afrikaans, monitors are called *"leguaan,"* which is related to the German *"Leguan"* for iguana. Because Australia was also colonized by Dutchmen ("New Holland"), the origin of the South African and the Australian names for monitors may have the same origins.

On the basis of morphological evidence, Mertens (1942, 1958, 1963) recognized 10 subgenera, 5 of which survive today. Ziegler and Böhme (1997) examined hemipeneal structure and revised and extended Mertens's classification, erecting a new subgenus *Soterosaurus* for *V. salvator* and its subspecies. Two subgenera, *Odatria* and *Varanus,* both of which are speciose, have undergone adaptive radiations in Australia. The nine currently recognized subgenera (Böhme 2003) and the species belonging in each are listed in Table 1.1.

Monitor lizards live in a wide variety of habitats, ranging from mangrove swamps to dense forests to savannas to arid deserts. Some species are aquatic, some semiaquatic, others terrestrial, whereas still others are saxicolous (rock dwelling), semiarboreal, or truly arboreal.

The varanid lizard body plan appears to have been exceedingly successful: it has been around since the late Cretaceous, 80 million years ago, as evidenced by the Mongolian fossil *Estesia. Varanus* are morphologically conservative but vary widely in size, which makes this genus ideal for comparative studies of the evolution of body size. No matter what size a monitor happens to be, it always looks like a monitor! These lizards range from the diminutive Australian pygmy monitor *Varanus brevicauda* (~17–20 cm in total length and 8–20 g in mass) to Indonesian Komodo dragons (*Varanus komodoensis*), which attain lengths of 3 m and weights of 150 kg. Komodo monitors, however, are themselves dwarfed by a closely related, extinct, gigantic varanid *Varanus priscus* (formerly *Megalania prisca*). This Australian Pleistocene species is estimated to have reached more than 6 m in total length and to have weighed over 600 kg. *Varanus priscus* fossils have been dated at 19,000 to 26,000 years BP.

Monitors may well be more closely related to snakes than to most other lizards. Many monitor lizards are active predatory species that raid vertebrate nests and eat large vertebrate prey (some smaller species also feed extensively on invertebrates, including centipedes, large insects, earthworms, crustaceans, and snails). Many monitor lizards are the top predators in the communities in which they live.

Varanids are by far the most intelligent of all lizards. At the National Zoo in Washington, D.C., individual Komodo monitors have

TABLE 1.1.
Recognized subgenera of *Varanus* and species currently assigned to them.[a]

Empagusia (4 species)
    bengalensis (2 subspecies)
    dumerilii
    flavescens
    rudicollis

Euprepriosaurus (14 species)
    caerulivirens
    cerambonensis
    doreanus
    finschi
    indicus
    jobiensis
    juxtindicus
    keithhornei
    kordensis
    macraei
    melinus
    prasinus (2 subspecies)
    spinulosus
    yuwonoi

Odatria (17 species)
    acanthurus (3 subspecies)
    baritji
    brevicauda
    caudolineatus
    eremius
    gilleni
    glauerti
    glebopalma
    kingorum
    mitchelli
    pilbarensis
    primordius
    scalaris (2 subspecies)
    semiremex
    storri (2 subspecies)
    timorensis
    tristis (2 subspecies)

Papusaurus
    salvadorii

Philippinosaurus
    mabitang
    olivaceus

Polydaedalus
    niloticus
    ornatus
    albigularis (3 subspecies)
    exanthematicus
    yemenensis

Psammosaurus
    griseus (3 subspecies)

Soterosaurus
    salvator (8 subspecies)

Varanus (8 species)
    giganteus
    gouldii (2 subspecies)
    komodoensis
    mertensi
    panoptes (3 subspecies)
    rosenbergi
    spenceri
    varius

[a] From Ziegler and Böhme (1997) and Böhme (2003).

their own personalities and recognize each keeper. These big lizards exhibit curiosity: one lizard walked up to its keeper and gently climbed up on him before taking a notebook from the keeper's shirt pocket in its mouth! The Komodo then dropped down, put the notebook on the ground, and tongue flicked it, as if to ask, "Why is this of such interest?" Recent experiments on captive *V. albigularis* by John Phillips at the San Diego Zoo suggest that some varanids can actually count. Lizards were conditioned by feeding them groups of four snails in separate compartments with movable partitions, which were opened one at a time to allow monitors to eat each batch of four snails. Upon finishing the fourth snail, lizards were allowed into another chamber containing four more snails. After such conditioning, one snail was removed from some snail groups. Lizards searched extensively for the missing fourth snail, even when they had access to the next group. Similar experiments with varying numbers of snails showed that these varanids can count up to six, but with groups of snails larger than six, the monitors seemed to stop counting and merely classified them as "lots," eating them all before moving on to the next chamber (King and Green, 1999, p. 43). Such an ability to count probably evolved as a consequence of raiding nests of reptiles, birds, and mammals, because average clutch or litter size would be around six.

Teeth of most varanids are serrated along the rear edge, which facilitates cutting and tearing the skin and flesh of prey as these big lizards pull back on their bite. *Varanus komodoensis* routinely kill deer and pigs in this way, and one Komodo monitor actually eviscerated a water buffalo (Auffenberg 1981). *Varanus komodoensis* and the extinct *V. priscus* are ecological equivalents of large saber-toothed cats, using their slashing bite to disembowel large mammals. *V. priscus*, which was contemporary with aboriginal humans, probably ate *Homo*, although strangely, Australian aborigines seem not to have any dreamtime stories about gigantic, fierce, man-eating lizards.

Scientists have recently discovered useful pharmacological agents in varanoid lizards. Venoms of *Heloderma* are complex mixtures of over a dozen small peptides, neurotransmitters, proteins, and other molecules, which have powerful effects on mammalian physiology (Raufman 1996). Natural selection has invented molecular analogs that mimic important mammalian hormones such as the neurotransmitter serotonin, secretin, and a variety of peptides and proteins. One lowers blood pressure, another regulates insulin release, and still another attacks certain cancers. Yet another, a peptide called gilatide, improves memory in rats and is a candidate for development of a drug to treat Alzheimer's disease. Such molecules could prove to be useful drugs to control hypertension, diabetes, and cancer. One such drug. Exendin-4, derived from Gila monster venom, is currently being evaluated for treatment of type 2 diabetes (Edwards et al. 2001; Seppa 2001). Some, but not all, of these molecules are also found in snake venoms, which tend to be much more toxic.

Although Komodo dragons are not actually considered to be venomous, they virtually are. Their saliva harbors over 50 different strains of bacteria, some of which are highly septic. If a Komodo does not kill its prey outright, its bite introduces germs potent enough to kill its quarry with a massive infection in a few days. When these big lizards fight each other, their bites do not become infected by the bacteria of other dragons, suggesting that Komodo dragons may possess a natural immunity to bacterial infection. These observations prompted an investigation of blood plasma of Komodo dragons. Recent preliminary work by Dr. Gill Diamond at a Medical School in New Jersey has identified a powerful antibacterial agent in the blood plasma of Komodo dragons. It could be developed as a new useful antibiotic in our ongoing world-wide battle against the evolution of antibiotic-resistant microbes (Diamond, personal communication).

To compile this reference book, we have assembled 35 experts on various aspects of the biology of varanoid lizards. These people are of many different nationalities and their writing styles vary (we tried to standardize their treatments as much as possible). We attempt to provide a comprehensive account of virtually everything important known about monitors. We begin with a review of fossil varanoids, followed by a discussion of biogeography and phylogeny. The heart of the book consists of 56 detailed species accounts, most written by people with hands-on field experience with a particular species, although a few had to be based on literature reviews. Species accounts are separated into three sections, one for African species, another for Asian and Southeast Asian species (a few of which reach tropical northern Australia), and finally a section for endemic Australian species, some of which have reached New Guinea. Following this, we proceed to various comparative analyses and a chapter on the art of keeping these lizards healthy and breeding them in captivity, and the book concludes with a new taxonomy for *Varanus*.

In the process of reporting what is known, we also identify what remains to be learned about these lizards. A great deal is known about some species, but barely anything is known about others. Such a diverse monophyletic group can be exploited both to identify and to understand the actual course of evolution.

References

Auffenberg, W. 1981. *The behavioral ecology of the Komodo monitor.* Gainesville: University Press of Florida.

———. 1988. *Gray's monitor lizard.* Gainesville: University Press of Florida.

———. 1994. *The Bengal monitor.* Gainesville: University Press of Florida.

Bennett, D. 1998. *Monitor lizards: Natural history, biology and husbandry.* Frankfurt am Main: Edition Chimaira.

Böhme, W. 2003. Checklist of the living monitor lizards of the world (family Varanidae). *Zool. Verhand.* 341:3–43.

Böhme, W., and H.-G. Horn. 1991. *Advances in monitor research. Mertensiella* 2:1–266.

Edwards, C. M., S. A. Stanley, R. Davis, A. E. Brynes, G. S. Frost, L. J. Seal, M. A. Ghafel, and S. R. Bloom. 2001. Exendin-4 reduces fasting and postprandial glucose and decreases energy intake in healthy volunteers. *Am. J. Physiol. Endocrinol. Metab.* 281:155–161.

Horn, H.-G., and W. Böhme. 1999. *Advances in monitor research II. Mertensiella* 11:1–366.

King, D., and B. Green. 1999. *Goannas: The biology of varanid lizards.* Sydney: University of New South Wales Press.

Mertens, R. 1942. Die Familie der Warane (Varanidae). *Abh. Senck. Naturf. Ges.* 462, 466 467:1–391.

————. 1958. Bemerkungen über die warane Australiens. *Senckenbergia Biol.* 39:229–264.

————. 1963. Liste der rezenten Amphibien und Reptilien, Helodermatidae, Varanidae, Lanthanodtidae. *Tierreich Lief.* 79:v–x, 1–26.

Raufman, J. P. 1996. Bioactive peptides from lizard venoms. *Regul. Pept.* 61:1–18.

Seppa, N. 2001. Reptilian drug may help treat diabetes. *Sci. News* 160(3):47.

Ziegler, T., and W. Böhme. 1997. Genitalstrukturen und Paarungsbiologie bei squamaten Reptilien, speziell der Platynota, mit Bemerkungen zur Systematik. *Mertensiella* 8:3–207.

2. The Long and Honorable History of Monitors and Their Kin

RALPH E. MOLNAR

Introduction

Monitors are the most impressive of lizards. Unlike smaller scincids, iguanids, and agamids that flee if they believe they are noticed, monitors have a deliberate, unafraid demeanor. They are large and they can be fierce—and no one who has had a picnic visitation from a monitor will easily forget their presence. Thus, we are naturally curious about their history and ancestry. Today they are found through most of Africa, southern Asia—including Arabia, Iran, India, Indochina, and Indonesia—New Guinea, and Australia. In the past, monitors and their relatives inhabited central Asia, Europe, and North America, and they ranged into the South Pacific as far as New Caledonia. Their fossil record begins about 90 mya (Turonian), but their history must be longer than that.

The modern varanoid lizards, *Varanus, Lanthanotus,* and *Heloderma,* give little clue to the number of forms that existed in the past. This diversity is manifested in the group known as platynotans, which includes varanoids together with several extinct groups. Unlike crocodilians, whose history reveals forms remarkably different from those surviving today, most fossil varanoids would probably be readily recognized as relatives of *Varanus* (or *Heloderma*). However, there were also marine forms rather different from the living monitors, although if a modern herpetologist could see these creatures alive, they would be recognized as kin of monitors. The traditional view, supported by some recent phylogenetic work (Lee 1997), is that snakes are derived from platynotan stock.

Here we consider both the land-dwelling and marine platynotans, and—briefly—what are currently taken to be ancestral and primeval snakes. The land-dwelling lizards will be looked at in more detail if only because, poorly known as they are, they are still better understood than marine forms. Considerable advances in the understanding of monitors and their kin have been made in recent years, but even so—just as for living monitors—much remains to be learned.

Most of the fossil material of the land-dwelling platynotans is too fragmentary for useful application of cladistic methods, as can be seen from even a superficial glance at the number of question marks (indicating unknown character states) of the character matrices of two recent phylogenetic studies, those of Lee (1997) and Norell and Gao (1997). Only the analysis of Lee included representatives of all forms usually thought to be related to monitors, i.e., snakes and mosasaurs.

Clear discussion of the evolution of land-dwelling platynotans is also hampered by a lack of terms. (The situation for marine platynotans is worse.) Of the suprageneric clades, only Varanidae, Helodermatidae, Monstersauria, Varanoidea, and Platynota are named, comprising only 5 of 12 clades of the cladogram of Norell and Gao (1997). Often, and traditionally, the last two terms have been used as synonyms.

For many extinct land-dwelling platynotans, especially the Mongolian, only the skull is known. In many others, only vertebrae are known. Thus, most of the illustrations here are of skulls or vertebrae. For these creatures, there is the problem that reasonable numbers of animals are known from material that cannot be compared. In light of this, here, I regard all current phylogenies as tentative.

Marine Platynotans

Three groups of marine platynotans, aigialosaurs, dolichosaurs, and mosasaurs, have been traditionally regarded as closely related to monitors. Unfortunately, rather little is known of the first two groups —it isn't even entirely clear that they are two groups (cf. Dal Sasso and Pinna 1997). In addition, there are animals of uncertain relationships but plausibly (and generally) considered platynotans. Although many specimens include at least partial skeletons, few have been studied recently, although Mike Caldwell and his students are now working on these animals. Thus, relationships of these important early platynotans remain largely unknown. The forms included in aigialosaurs and dolichosaurs might make up one, two, or more taxa, or—conceivably—they might be plesiomorphic mosasaurs.

Aigialosaurs (and dolichosaurs) lived in the early part of the Late Cretaceous. During this time, Europe was an archipelago, perhaps appearing somewhat as Indonesia does now. Aigialosaurs have recently been studied by Carroll and Debraga (1992). There seem to be only two species: *Aigialosaurus dalmaticus* and *Opetiosaurus buccichi,* which Carroll and Debraga found indistinguishable from *Aigialosaurus novaki.* Applying the rules of nomenclature, this species would properly be known as *Opetiosaurus novaki.* An alleged third species, *Carsosaurus marchesetti,* is known only from the postcranial skeleton and may be a

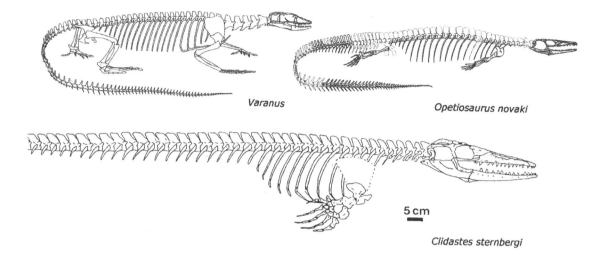

Varanus

Opetiosaurus novaki

5 cm

Clidastes sternbergi

Figure 2.1. Representative land-dwelling and aquatic platynotans. Varanus is, of course, a monitor, Opetiosaurus an aigialosaur, and Clidastes a (small) mosasaur. Only the anterior part of Clidastes is shown here. Images are approximately to scale. (Modified after Caldwell et al. 1995.)

dolichosaur or even a terrestrial platynotan. In the study of Lee and Caldwell (2000), aigialosaurs turned out to be the sister group of mosasaurs, which was also the conclusion of Carroll and Debraga.

Aigialosaurs may be characterized as having skulls similar to those of mosasaurs but with the postcranial features, at least in the skeleton, very similar to those of land-dwelling platynotans (Fig. 2.1). The limbs show no specializations for aquatic life but are relatively short compared to those of terrestrial platynotans, as is the neck. The transverse processes and zygapophyses are reduced in the tail, which was apparently laterally compressed, presumably for sculling.

On the other hand, the skulls resemble those of mosasaurs in their general proportions, as well as having a rounded, deeply notched quadrate, also like that of mosasaurs. Other apparent synapomorphies with mosasaurs include fusion of the two frontals, the mandibular component of the jaw joint being composed about equally of the articular and surangular bones, and a hinge joint between the angular and splenial of the mandible (probably enabling them to swallow larger prey).

O. novaki seems more plesiomorphic than A. dalmaticus in having relatively long ribs through much of the trunk. In A. dalmaticus, as in mosasaurs, the long ribs are found only at the front of the trunk.

Proaigialosaurus huenei was described as a Late Jurassic aigialosaur (Kuhn 1958). Known only from the impression of the dorsal surface of skull (Fig. 2.2), it was regarded by Russell (1967) as an aquatic varanoid. It would have been a small beast, well under a meter long. However, others have attributed the specimen to a pleurosaur, quite unrelated aquatic forms, and because the material could not be located for Carroll and Debraga (1992), it is best to leave the issue unresolved. This is a pity because P. huenei, if an aigialosaur, would be the oldest of these beasts, and possibly the oldest platynotan.

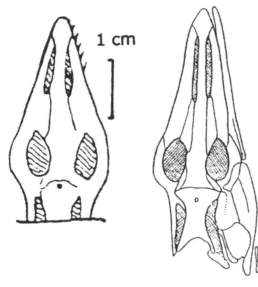

**Proaigialosaurus
huenei**

**Aigialosaurus
dalmaticus**

Figure 2.2. The skull of
Proaigialosaurus *compared with*
that of Aigialosaurus. *Not to*
scale (no scale given for
Aigialosaurus). *One can see why*
Kuhn thought Proaigialosaurus
was related to Aigialosaurus, *but*
whether or not this would stand
up to a phylogenetic analysis is
another matter. (Modified after
Kuhn 1958; McDowell and
Bogert 1954.)

Unlike aigialosaurs, dolichosaurs had a relatively small skull, apparently with a relatively short snout (again, unlike aigialosaurs), long neck, and reduced forelimbs. They also had a relatively long presacral vertebral column, with at least 34 vertebrae. The dolichosaurs include *Acteosaurus tommasini, Acteosaurus crassicostatus, Adriosaurus suessi,* and *Pontosaurus lesinensis* from Croatia and the Dalmatian Archipelago, *Aphanizocnemus libanensis* from Lebanon, and *Dolichosaurus longicollis* from England. They ranged as far east as Kazakhstan (Averianov 2001). Dolichosaurs have generally not been regarded as closely related to mosasaurs, and recent work by Lee and Caldwell (2000) suggests that they are more closely related to snakes. Lee and Caldwell also regard *A. suessi* as the sister group of snakes, rather than as a dolichosaur.

Of the dolichosaurs, *Aphanizocnemus libanensis* is the most remarkable (Dal Sasso and Pinna 1997). *A. libanensis* was a small lizard, known from a single specimen less than 30 cm long, of which over half is tail (Fig. 2.3). This animal lived during the beginning of the Late Cretaceous (Cenomanian). The forelimbs have enlarged "feet" that are almost as long as the humerus and antebrachial bones together. In the hind limb, the tibia is very reduced, to a greater degree even than in mosasaurs. Long chevrons suggest that the tail was laterally flattened and could have been used for swimming, and dorsoventral flattening of the phalanges suggests that the fore- and hind feet may have evolved into flippers. Dal Sasso and Pinna suggest that *A. libanensis* lived in shallow lagoons, foraging their bottoms for food.

Other Cretaceous platynotans have from time to time been included in the aigialosaurs or dolichosaurs, or even considered to be

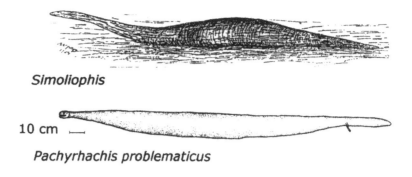

Simoliophis

10 cm

Pachyrhachis problematicus

1 cm

Aphanizocnemus libanensis

Figure 2.3. Some lesser known marine platynotans and platynotan derivatives. Simoliophis *and* Pachyrhachis *were primitive snakes,* Aphanizocnemus *a dolichosaur, and* Coniasaurus *a coniasaur. Not to scale. (After Nopcsa 1925; Scanlon et al. 1999; Caldwell and Cooper 1999.)*

5 cm

Coniasaurus crassidens

relatives of *Lanthanotus borneensis*. *Mesoleptos zendrini* and *Eidolosaurus trauthi* have been found in the Lower Cretaceous rocks of Croatia. *E. trauthi* had a relatively long tail and neck, and a spindleformed trunk, with the longest ribs in the middle. Both are thought to be marine forms because they have been found in marine beds and because they exhibit pachyostosis, an increase in bone density usually found in marine tetrapods.

Coniasaurus is a Cretaceous varanoid lizard, originally described from the Cenomanian chalks of England (Fig. 2.3). It has (at least) two species: *Coniasaurus crassidens* and *Coniasaurus gracilodens* (Fig. 2.4). Other material has been found in Cenomanian and Turonian beds in Texas (Bell et al. 1982), South Dakota, and, reportedly, Israel (Caldwell and Cooper 1999). Relationships of coniasaurs are currently no less tentative than those of the other forms here mentioned. Caldwell (1999a) regarded them as primitive lizards more closely related to mosasaurs than to terrestrial varanoids.

One of the unusual features of *C. crassidens* is the presence of a longitudinal groove on the distal edge of the first maxillary tooth (Fig. 2.5). Whether or not this groove might prove to be a homolog of those of monstersaurs is an interesting question. The more posterior maxillary teeth are bulbous but with a wrinkled (or crenulate) ridge on both

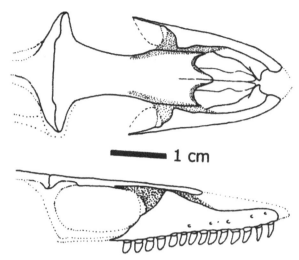

Coniasaurus gracilodens

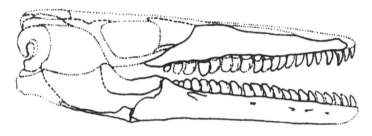

Coniasaurus crassidens

Figure 2.4. The skull of one of the smaller marine platynotans, Coniasaurus. *Differences in tooth size are clear, as well as in proportions of the snout and orbit. (Modified after Caldwell 1999a; Caldwell and Cooper 1999.)*

sides of the tip of each tooth (Caldwell and Cooper 1999). The dentary teeth are pleurodont, but set into sockets in the lateral wall of the mandible. In both upper and lower tooth rows, the anterior teeth are more slender than the posterior ones. The tips of both posterior maxillary and posterior dentary teeth overhang so that the wrinkled ridges near the tips of successive teeth form a single "edge" along the tooth row, thus creating a single cutting edge along the entire row (Fig. 2.5). Each posterior tooth is swollen, and the shaft of the crown, near the neck, extends lingually to form a shallow platform. Contact with this platform presumably accounts for the wear seen on some of the lower teeth (so far as I am aware, tooth wear is not seen in the land-dwelling varanoids). The teeth lack plicidentine. *C. crassidens* lacks the constricted centra seen in dolichosaurs, aigialosaurs, and other platynotans. This is a plesiomorphic feature and should not prohibit *C. crassidens* from being a member of one of these groups, if other features so indicate. *C. crassidens* probably didn't reach much over a meter in length (neither did *C. gracilodens*), but because it was found on both sides of the North Atlantic, it was presumably a marine or nearshore dweller. Interestingly, the English specimens are not found after the

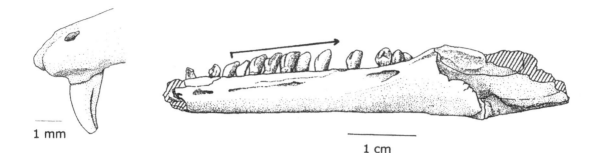

1 mm

1 cm

Figure 2.5. (left) The anteriormost maxillary tooth of Coniasaurus crassidens, *showing the groove in the crown.* (right) *The lower jaw of* C. crassidens, *showing how the teeth curve back to form a single cutting edge along the tooth row (bar). (Both modified after Caldwell and Cooper 1999.)*

Cenomanian (~91–97 mya), but those in North America survive well into the following Turonian (~88–91 mya).

Coniasaurus gracilodens shows significant similarities to *C. crassidens* in the maxilla and vertebrae but has a seemingly less derived dentition in some respects. The posterior teeth are basically similar to those found toward the front of the jaws but are more robust; also, they are not bulbous as in *C. crassidens.* On the other hand, the anterior teeth lack the groove found in that of the type specimen of *C. crassidens.*

Traditionally, i.e., before the widespread use of phylogenetic systematics, snakes were considered descendants of varanoids or varanoid-like lizards. *Lanthanotus borneensis* was a favored taxon in this context, regarded as what would now be called the sister group of snakes. Generally, snakes were thought to have arisen from burrowing ancestors, at least in part because the most plesiomorphic of modern snakes are burrowers (and *L. borneensis* is also thought to live in burrows). With the advent of phylogenetic methods, the position and ancestry of snakes became less clear. Two sets of phylogenetic analyses have been carried out. Neither of these has been as nearly definitive as may be hoped. Mike Caldwell and Mike Lee (Caldwell and Lee 1997; Lee and Caldwell 1997) based their analysis on the reexamination of a legged snake, *Pachyrhachis problematicus,* from the Cretaceous (Cenomanian) of Israel (Fig. 2.3). One specimen of this animal clearly shows that the pelvis, femur, tibia, fibula, and two tarsal bones were present (Fig. 2.6), although there is no sign of forelimbs. The fossils were found in marine beds and apparently deposited some considerable distance out to sea (Scanlon et al. 1999). *P. problematicus* was distinctly stout for a snake, with a massive trunk, short, thin neck, small head, and short tail (Scanlon et al. 1999). Their study led the team to suggest that snakes were derived from marine lepidosaurs. Lee (1997) went so far as to resurrect, following his own cladistic analysis, the old idea of Edward Drinker Cope that snakes were derived from mosasaurs. This would clearly place snakes in the platynotan clade. The dissident opinion, that mosasaurs and their aquatic kin were not platynotans (Caldwell 1999b), has now been abandoned (Lee and Caldwell 2000).

A second analysis, by Tchernov et al. (2000), on a second legged snake, *Haasiophis terrasanctus,* argued that this form and *P. problematicus* were actually quite advanced snakes. *H. terrasanctus* had more

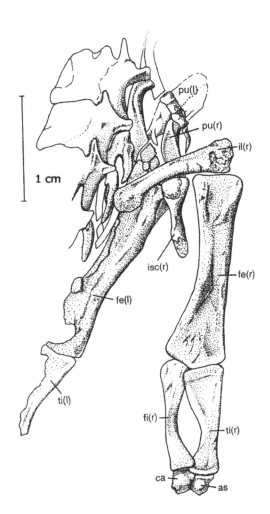

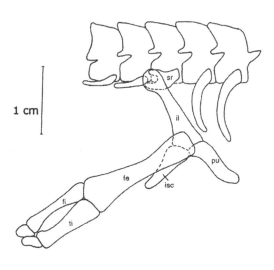

Figure 2.6. Snake's legs. The bones of the hind limb of Pachyrhachis problematicus, *as seen in the fossil on the left, as reconstructed on the right. Abbreviations: as, astragalus; ca, calcaneum; fe, femur; fi, fibula; il, ilium; isc, ischium; (l), left; pu, pubis; sr, sacral rib; ti, tibia; (r), right. (Modified after Lee and Caldwell 1997.)*

complex limbs than *P. problematicus,* with three tarsals, metatarsals, and phalanges present as well as the elements listed for *P. problematicus.* Tchernov and his team argued that *H. terrasanctus* and *P. problematicus* had reevolved limbs that had been previously lost, or that—possibly, as in biblical legend—all plesiomorphic snakes had limbs and these had been lost independently at least six times. Although several lineages of lepidosaurs have lost one or more pairs of limbs, at first sight, this scheme seems unparsimonious. Furthermore, Tchernov and his team's interpretation implies that at least five other lineages of snakes were present by the Cenomanian. Although we know that snakes predate the Cenomanian by about 20 million years (Rage and Richter 1994), the existence of so many ghost lineages (lineages not represented by fossils) also seems less than parsimonious. As Fraser (1997) pointed out, herpetologists working on the origins of snakes have tended to look for snake ancestors among lepidosaurs already showing the derived features of snakes, such as long, slim bodies and loss of limbs. But somewhere in their evolution, the ancestors must have had shorter bodies and four limbs, if only because early tetrapods were, after all, four-legged creatures.

I have always found it odd that popular and children's books on fossil animals uniformly and entirely ignore simoliophids. After all, simoliophids fit the criteria for "prehistoric monsters" in that they are large and unusual. In fact, they were literally sea serpents, large marine snakes (however, books on sea serpents also ignore them). Unfortunately, they are poorly known, a condition that doesn't always stop extinct animals from becoming well known to the general public. These snakes are generally represented by isolated vertebrae—rather like varanids. Only two species are known: *Simoliophis rochebrunei* from France, Portugal, and Egypt (Nopcsa 1925), and *C. lybicus* from Libya (Nessov et al. 1998). They are believed to have had relatively small heads and thin necks but broader bodies, somewhat like *Pachyrhachis* (Fig. 2.3).

The palaeophids, a group of large estuarine snakes, were once thought related to simoliophids but are now considered relatives of boas (Rage 1983). These Eocene forms included what is seemingly the largest known snake, *Palaeophis colossaeus*, possibly reaching as much as 9 m in length (Rage 1983), making this the largest platynotan derivative other than among the mosasaurs. Remains of this species were found in Mali, but six or seven other species occurred in England, France, Belgium, Egypt, Nigeria, and the United States—in other words, in the North Atlantic–Mediterranean region. Members of a second genus, *Pterosphenus,* also lived in these waters.

Palaeophids, like simoliophids, are largely known from their vertebrae. They are recognized as snakes by their possession of zygosphenes and zygantra, accessory articulatory structures of snake vertebrae, in addition to the zygapophyses. However, the fragmentary nature of the fossils of early snakes and platynotans can lead to difficulty. The vertebrae—all that is known—of *Pachyvaranus crassispondylus* from the terminal Cretaceous of Morocco are snakelike but lack the accessory articulations (Arambourg 1952). Thus, *P. crassispondylus* is looked upon as some other kind of platynotan, but just what kind isn't known. Like those of some primitive snakes and other platynotans, the vertebrae had tapered, flattened centra. They are also pachyostotic, and thus the animal, about 1 m long, is thought to have been aquatic. *P. crassispondylus* is also important in that it suggests that aquatic platynotans other than mosasaurs (and some snakes) survived until the end of the Mesozoic, at least in seas near north Africa, some 20 million years after the better-known dolichosaurs, aigialosaurs, and coniasaurs disappeared from the fossil record.

Mosasaurs were giant marine lizards of the Late Cretaceous (Fig. 2.1). They were apparently found throughout the world's oceans, and their fossils have been found on all continents, including Antarctica, and some islands, such as those of New Zealand and Indonesia. Some reached as much as 15 m in length (*Hainosaurus*), making them the largest lizards. The neck is short (with seven vertebrae) and the trunk and tail long. Expansion of the neural spines and chevrons at the tip of the tail suggests that some kinds had a tail fin. The feet had become modified into flippers—mosasaurs presumably spent very little, if any, time out of the water—and the number of phalanges in each digit was increased over the number usual in terrestrial lizards (hyperphalangy).

In most mosasaurs, the teeth were conical, basically similar to those of killer whales (*Orca*). But in others (*Leiodon*), they are sharp and recurved, but laterally flattened, reminiscent of those of the carnivorous dinosaurs. In some (*Globidens*), the teeth were bulbous and apparently suited for crushing shellfish, and those of one species (*Carinodens belgicus*) were longitudinally flattened and had three cusps per crown. In most mosasaurs, the skull is fairly slender, but some, such as *Prognathodon* and *Goronyosaurus,* developed more robust cranial bones. The differences in tooth form and in skull form and proportions led Theagarten Lingham-Soliar (1991) to suggest that mosasaurs employed a substantial variety of feeding methods.

Studies of the pathologies of mosasaur bones reveal interesting insights about their lives (Rothschild and Martin 1993). Healed injuries apparently inflicted by teeth may be found on the skulls and jaws. The tooth marks seem to be those of other mosasaurs, leading to the suggestion that mosasaurs may, like some modern lizards, have fought by grabbing one another on the head and attempting to roll the seized individual onto his back. Other mosasaurs, however, were not their only opponents, and one diseased caudal vertebra includes also part of a shark tooth. The shark bit the mosasaur and introduced microbes that resulted in the disease. Other mosasaur vertebrae show lesions (avascular necrosis) that are interpreted as having been caused by decompression syndrome (the bends). Thus, some forms (*Platecarpus* and *Tylosaurus*) were seemingly capable of deep diving, even if sometimes they returned to the surface too rapidly. A mosasaur braincase in Belgium (of *Mosasaurus*) suffered a blow to the head from a blunt instrument powerful enough to sever the brain and kill the beast (Lingham-Soliar 1998). This blow was plausibly the result of a ramming attack by a larger mosasaur (*Hainosaurus*). Modern monitors can engage in serious fighting: mosasaurs, it seems, were equally prone to such behavior, as well as to attacks from other predators in the seas.

Thus, mosasaurs are sufficiently common and sufficiently well known to provide some glimpse of their personal lives and deaths. Functional studies of mosasaur feeding and swimming (Lingham-Soliar and Nolf 1989) indicate that at the end of the Cretaceous, mosasaurs were becoming more diverse and adapted to more different modes of both feeding and swimming—in other words, more different modes of life. This suggests that whatever caused the Cretaceous-Tertiary extinction caught the mosasaurs as they were undergoing an adaptive radiation; their extinction, at least, seems not to have been the result of any gradually acting environmental factors (Molnar 1991).

The Oldest Claimed Varanoids

The oldest known land-dwelling platynotans—with a single exception—are Late Cretaceous, and so, given the appearance of dolichosaurs and aigialosaurs in the Early Cretaceous, platynotans must have arisen during (or before) that time. To my knowledge, only these potentially ancestral forms have ever been reported.

The formidably named *Nuthetes destructor,* from the latest Jurassic Purbeck beds of England (~145 million years old), was thought by

Sir Richard Owen (1854) to have been related to varanids. *N. destructor* is known from a small fragment of jaw, about 3.5 cm long, with six teeth. The tooth implantation was thought to be pleurodont, as in many lizards, among them monitors. Later illustrations (Owen 1884a) indicate that each tooth crown has a shallow trough extending up its side from the base, as in many theropod teeth. The teeth of varanoids show no such feature but instead have expanded, flattened bases for their pleurodont attachments to the jaw, quite different from the teeth of *N. destructor. N. destructor* was, as most paleontologists now believe, a small theropod dinosaur.

The elimination of *Nuthetes* leaves *Paikasisaurus indicus* (Yadagiri 1986) as a suggested ancestral platynotan. *P. indicus* comes from the Early Jurassic Kota Formation of central India. Two fragments of jaws (with teeth) and an ilium have been found, and the form of the teeth suggested platynotan relationships. Although the mandibular fragments are very small, only about 2 mm long, the crowns are laterally compressed (especially at the tips) and weakly recurved. They also have sharp edges and an expanded base, and they show basal wrinkles. However, a more recent study of the material (Evans et al. 2002) shows that the basal wrinkles are not plicidentine. Furthermore, there is no solid evidence for relating *P. indicus* to platynotans and no substantial indication that all of the material even belongs to a single species. This leaves platynotans without any plausible close ancestors.

But maybe not. As chance would have it, the same beds (the Purbeck Limestone Formation) that yielded fossils of *Nuthetes* have also produced several lizards. One of these, *Parviraptor estesi,* is not unlike what one would expect of an ancestral platynotan (Evans 1944). Fossils from this genus are found in even older deposits, the Middle Jurassic Forest Marble, about 170 million years old or even a little older. A second species, *Parviraptor gilmorei,* has been found in the Morrison Formation, of age intermediate between that of the Forest Marble and the Purbeck, in the western United States (Evans 1996). This lizard lived with such well-known dinosaurs as *Allosaurus, Diplodocus,* and *Stegosaurus.*

Parviraptor shows several character states—a palatine as wide as it is long; a broad interpterygoid vacuity; sharp, recurved teeth; and well-developed subolfactory processes on the frontals (among others)—that strongly suggest relationships with the platynotans. However, this lizard lacks plicidentine in the teeth, evidence of hypapophyses attaching at the rear of the cervicals, posteriorly placed ascending process of the maxilla, oblique condyles, and waisted centra. However, these are plesiomorphic states that should not bar *Parviraptor* from a position in, or near, the ancestry of the platynotans. Thus, monitor-like lizards may indeed date back well into the times of the dinosaurs.

Necrosaurs: Primitive Terrestrial Platynotans

In addition to the aquatic forms already mentioned, the Late Cretaceous saw lizards that looked like modern monitors because the body form—as far as we can tell—changed little from the Cretaceous to the present while the specific characteristics that distinguish modern varanids evolved. Forms such as *Estesia* and *Telmasaurus* would have looked familiar—at least to the readers of this book. They grew to a length of about 1 to 2 m.

The oldest group related to the varanids was the necrosaurids, a name that means "lizards of death." Lee's (1997) analysis indicates that these animals were primitive forms no more closely related to each other than to the lineages that went on to become mosasaurs, helodermatids, and monitors—in the jargon, they did not form a natural group. (In short, there is no good reason to lump these forms together but exclude lineages like the monitors or the mosasaurs.) Still, we can use "necrosaurs" as a term for basal terrestrial platynotans without implying that it is a clade (natural group) like monstersaurs and varanids.

Some necrosaurs lived in North America during Late Cretaceous to Middle Eocene times (~41–83 mya) and others in Europe during the Late Eocene (~41–43 mya). The oldest is from the Albian of Utah (~97–113 million years old). When alive, they doubtless looked much like monitors. They were also similar in the form of their teeth, with the basal wrinkles of infolded enamel—plicidentine—that still persist in the teeth of monitors today, and with similar relatively low, broad vertebral bodies (centra) in the vertebral column. When alive, the Eocene ones may have been as much as a meter long. Judging from their teeth, they were all predatory lizards.

The oldest land-dwelling platynotan is known from a single incomplete right maxilla from the Early Cretaceous of Utah (Cifelli and Nydam 1995). This element has now been referred to the monstersaur, *Primaderma,* and so will be discussed in more detail with those lizards. *Primaderma* does show that terrestrial platynotans must have had a previous history of some extent that is currently unknown, and that by almost 100 mya they were capable of feeding on relatively large prey.

Lev Nessov (1981) described a single isolated lizard maxilla from the Late Cretaceous of Uzbekistan (Fig. 2.7). The bone, the type of *Ekshmer bissektensis,* represented the first terrestrial lizard to be discovered from the Late Cretaceous of the USSR. Only the bases of four teeth remain. The nasal margin seems to rise abruptly, like that of *Parasaniwa wyomingensis,* and Nessov believed that *E. bissektensis* was related to *Parasaniwa.* The teeth, however, are clearly larger and more widely spaced in *E. bissektensis.* Nessov later (1997) referred it to the necrosaurs, and it appears to be the oldest one known, deriving (probably) from the Coniacian (~88 million years old).

A later reference to *E. bissektensis* by Zerova and Chkhikvadze (1984) refers to its occurrence in the Cretaceous, not to any later occurrence in the Cenozoic.

Paravaranus angustifrons was found later in the Cretaceous also in Asia, in Mongolia (Fig. 2.8). Its frontals are narrower (possibly an autapomorphy) and the orbits larger when viewed from above than in other platynotans. The size of the orbits is a consequence of the lateral projection of the jugal arches. This may indicate relatively large eyes or a relatively large pterygoideus muscle mass. Postorbital portions of the

Figure 2.7. The maxilla (and only known specimen) of Ekshmer bissektensis. (left) Medial view. (right) Lateral view. The teeth are badly worn, so only the bases remain. (Modified after Nessov 1981.)

Figure 2.8. (opposite page) Skulls, jaws, and parts thereof of advanced varanoids and varanids. The skull of Cherminotus in lateral view is taken from Borsuk-Białynicka (1984), but the dorsal view and the mandible are from Alifanov (2000). The jaw seems to be very undershot but was placed to fit as do those of modern monitors. (Modified after Alifanov 2000; Borsuk-Białynicka 1984; De Vis 1900; Gilmore 1928; Lydekker 1888; Levshakova, 1986; drawn from Balouet 1991; Clos 1995.)

skull also are broader relative to the width of snout than in other platynotans.

Borsuk-Białynicka (1984) considered this lineage as having diverged prior to the origin of the Platynota; however, Norell and Gao (1997) and Lee (1997) regard it as a basal platynotan (more closely related to *Varanus* than to anguids).

Parviderma inexacta is another Late Cretaceous Mongolian terrestrial platynotan. This animal is known from a single incomplete skull, lacking the snout and the cheek regions on both sides (Fig. 2.9).

Incomplete as it is, the skull apparently has relatively large orbits, thus giving uncommonly narrow frontals for a platynotan, though not quite as narrow as those of *P. angustifrons*. Borsuk-Białynicka (1984) thought the specimen might have come from a juvenile. The skull roof is encrusted with small platelike cranial osteoderms resembling those of the modern Central American *Xenosaurus grandis*. *P. inexacta* was probably a small lizard. The skull as preserved is about 1 cm long and, if complete, was probably no more than 2 cm in length. It had only seven or eight dentary teeth (as, incidentally, did *Paravaranus angustifrons*).

Norell and Gao (1997), like Borsuk-Białynicka (1984), regard *P. inexacta* as a platynotan of necrosaur grade; Lee (1997) considered it the sister group of *Gobiderma* and related to *Estesia* and helodermatids.

Colpodontosaurus cracens (Estes 1964), from the latest Cretaceous of Wyoming, is known from a single, slender (incomplete) dentary (Fig. 2.9) and two pieces of maxillae, both from the back of that bone. It lacks laterally compressed teeth and plicidentine. In life, the lizard was probably less than a meter long overall.

This material lacks the varanoid synapomorphies of Pregill et al. (1986), as does *Proplatynotia*, considered a varanoid by Norell and Gao (1997). In view of the very fragmentary material, one can't be sure that *C. cracens* is a platynotan, and Estes (1964) regarded it as "Diploglossa incertae sedis," which would now presumably be ascribed to an anguid or xenosaurid. But on the other hand, some features—teeth with very thin walls (shared with *Provaranosaurus acutus*) and almost complete absence of a ventral border to the intramandibular septum (Estes 1983)—suggest that it was a primitive member of this group related to *P. acutus*, and nothing contradicts this interpretation. In view

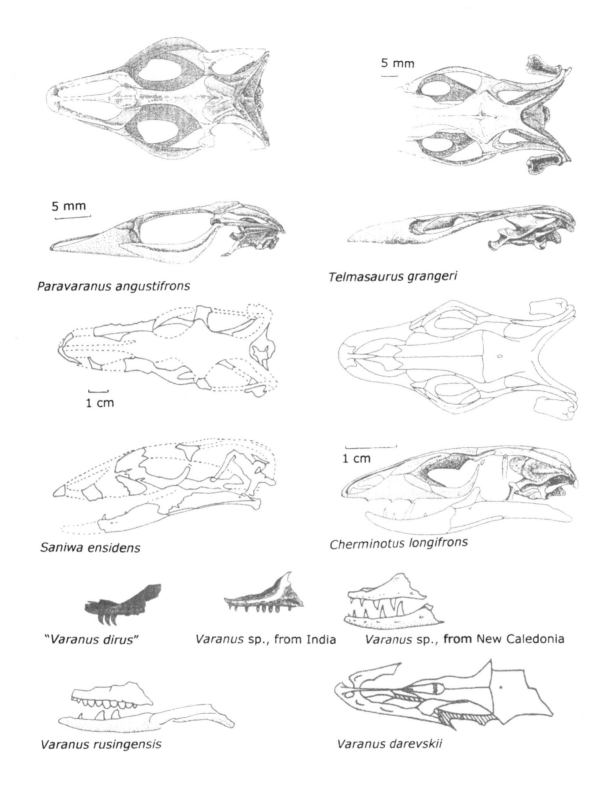

5 mm

5 mm

Paravaranus angustifrons

Telmasaurus grangeri

1 cm

Saniwa ensidens

1 cm

Cherminotus longifrons

"Varanus dirus"

Varanus sp., from India

Varanus sp., **from** New Caledonia

Varanus rusingensis

Varanus darevskii

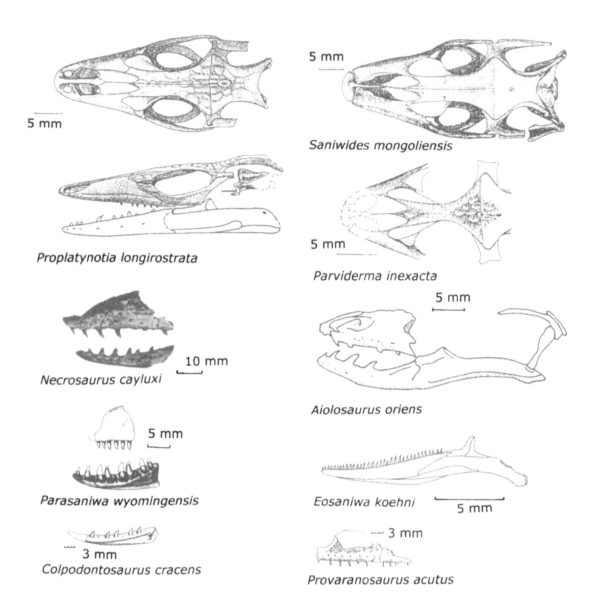

5 mm

Saniwides mongoliensis

5 mm

Proplatynotia longirostrata

5 mm

Parviderma inexacta

10 mm

Necrosaurus cayluxi

5 mm

Aiolosaurus oriens

5 mm

Parasaniwa wyomingensis

Eosaniwa koehni

5 mm

3 mm

Colpodontosaurus cracens

Provaranosaurus acutus

Figure 2.9. Skulls, jaws, and parts thereof of plesiomorphic platynotans ("necrosaurs"). The maxilla and dentary of Necrosaurus *and of* Parasaniwa *are not known to be from the same individuals. The dentaries of* Colpodontosaurus *and* Parasaniwa *are shown from the medial view; all other mandibles are shown from the lateral view. (Modified after Borsuk-Białynicka 1984; de Fejervary 1935; Estes 1964, 1983; drawn from Gao and Norell 2000.)*

of this, Estes (1983) suggested that *C. cracens* might be ancestral to *P. acutus*. If a platynotan, this species appears to be the most primitive known.

Parasaniwa wyomingensis seems to have been a moderately small platynotan, with a dentary about 2 cm long, from the latest Cretaceous of Wyoming and Montana (Fig. 2.9). It is known largely from fragmentary cranial bones, some only tentatively attributed (Estes 1964). The maxilla has a high dorsal process, with the posterior margin of the external naris rising abruptly. Perhaps this indicates a deep snout. Possible autapomorphies (following Estes 1983) include the nearly vertical narial border of the dorsal process of maxilla (but which is also found in *Provaranosaurus acutus*) and the intramandibular septum

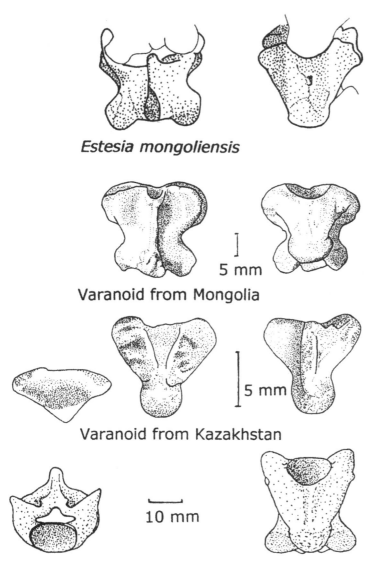

Estesia mongoliensis

Varanoid from Mongolia

5 mm

Varanoid from Kazakhstan

5 mm

10 mm

Palæosaniwa canadensis

Figure 2.10. Dorsal vertebrae of plesiomorphic varanoids and a monstersaur. (left) Anterior view. (middle) Dorsal view. (right) Ventral view. Estesia *was drawn from the photo of an articulated specimen (in Norell and Gao 1997). The Kazakhstani specimen is lacking the entire neural arch. (Modified after Estes 1964; drawn from Gao and Norell 2000; Kordikova et al. 2001.)*

with a notch for the surangular and fused to the floor of the Meckelian sulcus.

Parasaniwa obtusa was distinguished from *P. wyomingensis* by Gilmore (1928) on its tooth form, which was later shown by Estes (1964) to be due to postmortem erosion.

Palaeosaniwa canadensis (Gilmore 1928) was described from dorsal vertebrae (Fig. 2.10). More recently discovered material, including a skull (Estes 1983), remains undescribed. Of this Late Cretaceous lizard from Alberta, Montana, and Wyoming, Estes (1983, p. 183) writes that it was as large as "some of the largest species" of living monitors. The teeth are trenchant and (very slightly) recurved, and plicidentine is present (Estes 1964). The vertebrae seem to be less derived than the teeth, as in the tail the chevrons attach to the centrum closer to the condyle than in later forms, such as *Varanus*. This suggests that adaptations for feeding appeared before the derived vertebral features in monitor evolution. Examination of the character matrix of Norell and Gao (1997) suggests that *P. canadensis* is more derived than *Eosaniwa* and may be related to lizards such as *Necrosaurus* and *Parasaniwa*.

Provaranosaurus acutus, from the Paleocene of Montana and Wyoming, was a lizard of only moderate size, probably less than a meter long. The maxilla—such of it as is known—is characterized by a steeply rising narial margin (Fig. 2.9). Unfortunately, the specimen is not sufficiently complete to tell whether the maxilla was as high as in *Parasaniwa wyomingensis*. The teeth are high and slender—the slenderness being an apparent autapomorphy—but trenchant, and sometimes they lack basal striae. They share with those of *Colpodontosaurus cracens* very thin walls, and the form of intramandibular septum of the dentary is shared with *P. wyomingensis*.

Eosaniwa koehni was a relatively large lizard, with a skull 19 cm long, and so possibly the lizard was as long as 2 m. It is known from a single specimen from the Middle Eocene deposits at Geiseltal, Germany, and is the largest lizard from these beds. The skull and snout were elongate, and the latter was slightly laterally compressed. Palatal teeth were present on the vomer, palatines, and pterygoids, and there were more than 30 teeth on each mandible (Fig. 2.9). The number of these teeth and their slender construction, together with the elongation of the snout and the toothed palate, suggested to Estes (1983) that *E. koehni* was a fish-eater. Krumbiegel et al. (1983), on the other hand, felt that it was large enough to have preyed on mammals.

Relationships of *E. koehni* are unclear. It shows many plesiomorphic features, such as having teeth with plicidentine (Norell and Gao 1997) but weakly developed, and lacking the basal infolding of more advanced platynotans (Estes 1983). Some characteristics are shared with *Necrosaurus,* such as having oval keeled osteoderms with prominent central ridges. The narrow, slender, recurved teeth are somewhat similar to those of *Provaranosaurus acutus*. Possible autapomorphies include the posterior placement of the coronoid at approximately 70% of the total length of the mandible and the large number (30+) of dentary teeth. The posterior part of the jaw is more sharply downcurved than in other platynotans. Norell and Gao (1997) indicate that *E. koehni* was probably one of the most primitive platynotans known, considerably more plesiomorphic than lizards such as *Telmasaurus* and *Estesia* that lived some 40 million years earlier.

Necrosaurus, if not the best known of the plesiomorphic platynotans, is at least the longest known (Figs. 2.9, 2.11). It was initially described in 1873 from fossils from the Quercy phosphorite beds in France. These fossils make *Necrosaurus* one of the best represented of fossil platynotans, but much of the material has yet to be studied. Estes's (1983) discussion of this material is very helpful.

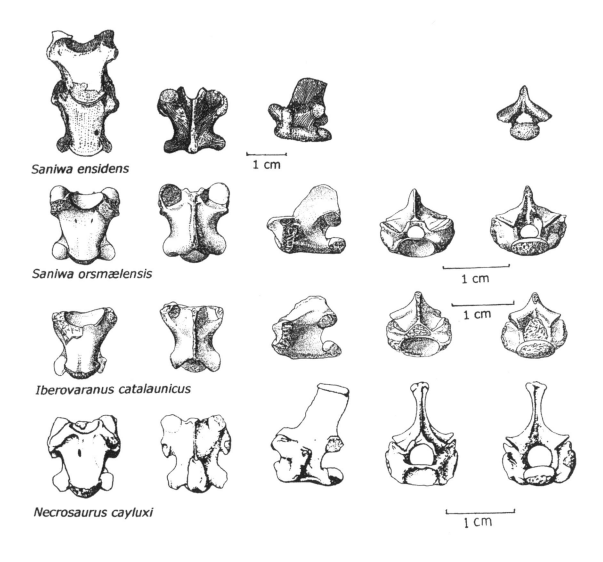

Saniwa ensidens

1 cm

Saniwa orsmælensis

1 cm

1 cm

Iberovaranus catalaunicus

Necrosaurus cayluxi

1 cm

Figure 2.11. Dorsal vertebrae of plesiomorphic varanids and Necrosaurus. *(left to right) Ventral, dorsal, lateral, anterior, posterior views. The diapophyses and transverse processes seen laterally in the posterior views of the other specimens are not drawn for S. ensidens. (Modified after Estes 1983; Gilmore 1928; Hoffstetter 1969.)*

Necrosaurus lived during the Paleocene and Eocene and apparently became extinct in the region of Quercy (and perhaps elsewhere) with the "Grande Coupure," a faunal turnover in Europe (and perhaps elsewhere) marking the end of the Eocene. Other material referred to *Necrosaurus* also dates from the Paleocene and Eocene, with the exception of material from the Early Oligocene of Belgium (Hecht and Hoffstetter 1962). These lizards were moderately large, with skulls to about 7 cm long.

Oval, keeled osteoderms having "a fine pit and ridge structure" (Estes 1983) are an obvious autapomorphy. Estes (1983) suggested that possession of many small oval cranial osteoderms may be another. The jaw adductor muscles were seemingly very extensive and extended onto the roof of the skull to attach on top of the parietals, which formed a sagittal crest. Two species are recognized, *Necrosaurus cayluxi* from

Quercy, in France, and *Necrosaurus eucarinatus* from the Geiseltal, in Germany. In the absence of a thorough study of *Necrosaurus*, the distinction between these is provisional: *N. eucarinatus* has more dentary teeth, about 18, as opposed to around 13 in *N. cayluxi*. *N. eucarinatus* also lacks the sagittal crest of the parietal, but this may indicate age rather than taxonomic distinction.

A considerable amount of other material has been referred to this genus (see list), suggesting that it was found over much of western and central Europe, with one report from Mongolia.

Monstersaurs

Norell and Gao (1997) described a clade of platynotan lizards more closely related to the beaded lizard and Gila monster than to living monitors, or more formally defined as the descendants of the common ancestor of *Gobiderma pulchrum* and *Heloderma suspectum*. Monstersaurs include six genera, of which one still survives. Their defining features include having the foramen ovale in front of the sphenooccipital tubercle of the braincase, having a parietal lappet on the quadrate, and having vertebral centra shorter than in varanids and with tall, narrow neural spines.

Monstersaurs date back at least into the Late Cretaceous, about 80 mya. They lived in Europe, Asia (where the oldest forms are found), and North America, but they survived later than the Eocene (~38 mya) only in North America. There were three genera during the Late Cretaceous and three during the Paleocene and Eocene, but *Heloderma* itself is apparently the only lineage to survive through the past 20 million years.

Primaderma nessovi is the oldest known monstersaur (Nydam 2000). Known from a few isolated cranial bones and two vertebrae, its remains were found in the Cedar Mountain Formation of central Utah. The age of this material is 98 to 99 million years. The bone, and the attached teeth, have several characteristics—e.g., plicidentine—suggesting that it belonged to a helodermatid-like lizard. The teeth were serrate but apparently lacked venom grooves. The bone suggests that the animal had a broad, blunt, helodermatid-like snout. Because it lacks some of the defining characteristics of the helodermatids, it was referred to the necrosaurs. The broad, blunt snout and widely spaced teeth suggest that the general form of the Gila monster and beaded lizard may predate the origin of the venom and that this form of lizard may date back some 100 million years.

Gobiderma pulchrum (Fig. 2.12) was a moderately large lizard from the Late Cretaceous of Mongolia (Borsuk-Białynicka 1984). Its skull is about 5 cm long, so the entire animal—even with a relatively long tail—was probably not more than a meter in length overall. But because only the skull and mandibles are known, this estimate of the length is speculative.

The skull is broader—especially at the back—when seen from above than those of many other varanoids. Unlike later varanids, the subolfactory processes of the frontals that partially enclose the olfac-

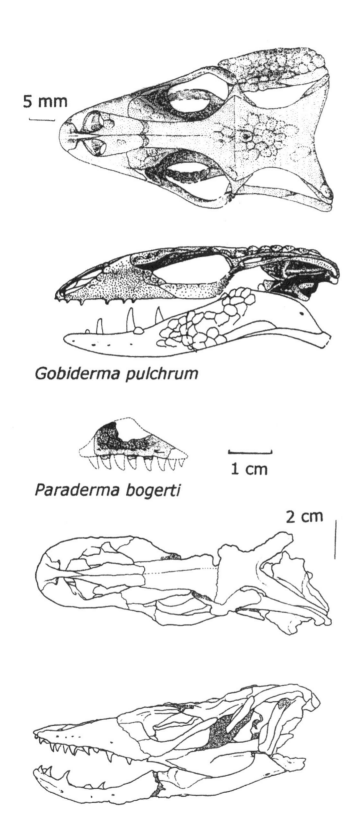

5 mm

Gobiderma pulchrum

Paraderma bogerti

1 cm

2 cm

Estesia mongoliensis

Figure 2.12. The skulls, jaws, and parts thereof of plesiomorphic monstersaurs. (Modified after Borsuk-Białynicka 1984; Estes 1964; Norell et al. 1992.)

tory tract are weak. In some skulls, the skull roof and cheeks retain a covering of osteoderms, much like that of helodermatids (hence the name "*Gobiderma*"). Also, the individual polygonal, pitted osteoderms resemble those of helodermatids. In general form, the mandible is reminiscent of those of helodermatids but with a weaker coronoid process. The teeth of *G. pulchrum* are trenchant and widely spaced in the jaws, and those of the lower jaw seem larger than those of the upper.

In their character matrix, Norell and Gao (1997) give no autapomorphies for *G. pulchrum,* which thus must be defined on the basis of a unique combination of character states. Its possession of a long, posterolaterally directed sphenooccipital tubercle (=basal tubera) in the braincase is shared only with the contemporaneous, but larger, *Estesia mongoliensis.* Lee (1997) regarded *G. pulchrum* as closely related to *Parviderma,* an allocation disputed by Norell and Gao (1997). Alifanov (2000) also disagreed with Norell and Gao and regarded *G. pulchrum* as closely related to *Necrosaurus* for reasons yet to be published.

Estesia mongoliensis is known from two skulls and associated jaws (Fig. 2.12) found in upper Cretaceous rocks in Mongolia (Norell et al. 1992; Norell and Gao 1997). The skull looks like that of a modern monitor in side view but has a low, blunt snout that was broader when seen from above.

Its autapomorphies were originally suggested to include the absence of osteoderms on the head and grooves that extend the lengths of the teeth anteriorly and posteriorly. The long grooves seemed to Norell and his colleagues similar to those of beaded lizards. This suggests that perhaps, like beaded lizards, *E. mongoliensis* had poison glands. Interestingly, oras (*Varanus komodoensis*) are effectively poisonous because of the septic, and rather virulent, bacteria they harbor in their mouths. Introduction of poison, either directly or as the result of bacterial infection, into wounds caused during attacks on prey may have had a long history in varanoids.

Although it probably would have appeared much like a modern monitor when alive, *E. mongoliensis* is regarded by Norell and Gao (1997; and Gao and Norell 1998) as a member of the Monstersauria and hence more closely related to *Heloderma* than to *Varanus.* Lee (1997) also independently classified it near helodermatids.

Paraderma bogerti (Estes 1964) is known from fragmentary cranial material from the latest Cretaceous (Maastrichtian) of Wyoming (Fig. 2.12). Estes (1983) was uncertain of its membership in the helodermatids, but this was supported by the analysis of Norell and Gao (1997). The skull apparently had the short, broad snout similar to, but not as developed as in, modern helodermatids (Estes 1964). A further similarity to helodermatids is the presence of relatively large, polygonal, pitted osteoderms fused to the maxilla. The presence of a venom groove in the teeth suggests that venomous lizards have inhabited the western United States since the time of the (last) dinosaurs. The osteoderms suggest that it may even have looked more or less like a beaded lizard. The size of the preserved cranial material suggests that *P. bogerti* was about the size of modern *Heloderma,* perhaps larger. Together, these three species represent the basal members of the monstersaurs.

Other forms developed during the Cenozoic. *Eurheloderma gallicum* (Hoffstetter 1957) demonstrates that helodermatids were not restricted to North America and Asia. Remains of this lizard were found in Eocene rocks in France. *E. gallicum* would have had a skull around 7 to 8 cm long, perhaps a little larger than that of a modern Gila monster. The parietal was longer than in *Heloderma,* and venom grooves of the teeth were less developed. This is consistent with having relatively larger jaw adductors, so perhaps they relied less on venom for subduing prey and more on simple bite force. This lizard bore polygonal sculptured osteoderms, resembling those of *Heloderma.*

A parietal similar to those of *E. gallicum* has been recovered from the Late Paleocene of Wyoming (Bartels 1983). This element derived from a subadult individual (osteoderms had become detached).

Lowesaurus matthewi (Gilmore 1928) lived from Middle Oligocene through Early Miocene times in Nebraska and Colorado. This was a moderately small lizard, with relatively larger cranial osteoderms than *Heloderma.* Pregill et al. (1986) distinguished this genus from other known helodermatids in its having triangular frontals (trapezoidal in the others). Estes (1983) remarked that *L. matthewi* had more primitive character states than *Heloderma.* Distinct grooves along the anterior edges of the teeth indicate that this genus, too, was venomous.

Fossils of *Heloderma* have been found in Nevada, New Mexico, and Texas. Those from Nevada, attributed to the living *Heloderma suspectum,* date back 8000 to 10,000 years (Estes 1983) and have yet to be described (as have those from New Mexico). They were recovered from Gypsum Cave in Clark County and represent skin and fragmentary osteoderms.

More informative are the fossils of *Heloderma texana,* a small species from the Early Miocene of Texas. The skull is very slightly flatter than those of the living species, and, because it seems to be of an adult, indicates a lizard significantly smaller than both modern forms (Stevens 1977). Pregill et al. (1986) conclude that *H. texana* was the sister group of *H. suspectum* and *Heloderma horridum.*

Estes (1983) referred previously described vertebrae and femora from the Thomas Farm Miocene of Florida to *Heloderma?* sp., thus suggesting that helodermatids ranged at least from the mountains of the Cordillera to the Atlantic coastal regions.

Monitors

Having surveyed monstersaurs and a variety of other platynotans, let us now consider those forms more closely related to living monitors than to any other platynotans. *Saniwides mongoliensis* (Borsuk-Białynicka 1984) lived during the Late Cretaceous (around 80 mya) in Mongolia (Fig. 2.9). Alive, it would probably have looked similar to living varanids, but it seems to have had a very low, flattened, and rather broad snout, perhaps giving the head a somewhat duck-billed appearance. The size of the skull, less than 5 cm long, suggests that the animal was less than a meter long overall. Features of the skull (absence of a broad contact of the supraoccipital with the parietal, the narrow supratemporal process of the parietal, and the reduced outer conch of the quadrate) indicate that *S. mongoliensis* was more derived than helodermatids.

Proplatynotia longirostrata was a platynotan lizard of moderate size, with a skull 4 cm long. This skull is long and slender (Fig. 2.9); it lacks the laterally projecting jugal arches seen in *Gobiderma pulchrum* or *Paravaranus angustifrons*. As in *S. mongoliensis,* the adductors of the jaws attach to the lateral sides of parietals. The parietals are also slightly elevated in this lizard, suggesting, perhaps, a larger adductor mass than in other monitors and hence a more powerful bite.

Borsuk-Białynicka (1984) and Lee (1997) regard *P. longirostrata* as a basal platynotan, but Norell and Gao (1997) found it to be more closely related to *Varanus* than to either helodermatids or necrosaurs. Alifanov (2000), on the other hand, regarded it as sharing significant (but unstated) similarities in dentary and tooth structure with *Colpodontosaurus cracens*.

Aiolosaurus oriens (Gao and Norell 2000) is the most recently described Cretaceous varanoid lizard (Fig. 2.9). The incomplete specimen indicates a rather small lizard, with a skull 4 to 5 cm in length, so it was probably less than a meter long overall. It is referred to the varanoids from its possession of precondylar constrictions of the vertebral centra. The seemingly relatively deep maxilla bore 9 or 10 sharply pointed, recurved teeth, with plicidentine weakly developed. Both *A. oriens* and *C. longifrons* have a single lacrimal foramen (as is usual in tetrapods), unlike *Telmasaurus, Saniwa, Lanthanotus,* and *Varanus,* which all have two. Thus, Gao and Norell (2000) suggest that *A. oriens* and *C. longifrons* are more primitive (basal) than those four taxa.

Telmasaurus grangeri (Gilmore 1943) is not as well known as *Estesia* because it is, unfortunately, represented by less complete specimens; for example, the snout is missing from both known skulls (Fig. 2.8). *T. grangeri* is regarded as closely related to *Saniwa, Lanthanotus,* and *Varanus* (Norell and Gao 1997). The skull was maybe 7 to 8 cm long, thus indicating a reasonably large lizard, but was quite low and flattened.

As for *Gobiderma*, Norell and Gao (1997) list no autapomorphies for *T. grangeri,* and neither does Lee (1997). However, Estes (1983) suggested as possible autapomorphies the parietal with a low median crest extending forward across the frontal, the parietal sharply constricted behind the skull deck, and the absence of the parietal foramen; however, a parietal foramen is present (Norell and Gao 1997).

The low crest along the midline of the top of the skull was missed by Gilmore, who saw the skull roof exposed from the bottom but mistook it for the top (Estes 1983). Estes also points out that this crest didn't result from the attachment of the adductor musculature because the adductors did not attach to the dorsum of the skull in this form. Presumably it was a display feature, possibly sexually dimorphic, clearly visible on the low, flattened skull.

Although some of the Cretaceous monitors, particularly those from Mongolia, are known from nice skulls, words like "fragmentary" and "frustrating" involuntarily spring to mind when considering the fossil record of varanids, particularly of *Varanus* itself. Mind you, such words are by no means always out of place describing the fossil record of the earlier terrestrial platynotans either. Of 28 specimens of fossil *Varanus* (excluding Australian material) recorded by Estes (1983), 23 included vertebrae, 7 had cranial elements, and only 3 had any limb material at all. The dentary was present in at least three of the specimens that included cranial elements. Thus, if any sensible phylogenetic scheme is to be found for these animals, it will have to be based on vertebral features. So far, all phylogenetic analyses conducted for terrestrial platynotans have strongly emphasized cranial features, which seem unable to provide much assistance here.

Cherminotus longifrons Borsuk-Białynicka, 1984, was given its (generic) name from its resemblance to *Lanthanotus borneensis,* a secretive and little-known Indonesian lizard without a fossil record. *C. longifrons* was a relatively small lizard, with a skull less than 3 cm long (Fig. 2.8). Both Lee (1997) and Norell and Gao (1997) regarded *C. longifrons* as the sister group of *L. borneensis,* but recently Alifanov (2000) and Gao and Norell (2000) noted that new material of *C. longifrons* casts doubt on this conclusion, a circumstance foreseen in the phylogenetic analysis of Carroll and Debraga (1992).

Borsuk-Białynicka (1984) reported that the upper jaw (maxilla) bears five widely spaced, conical teeth, probably with simple basal fluting. Gao and Norell (2000) found in new material that the teeth were laterally compressed and trenchant (Borsuk-Białynicka had reported that the teeth were not well preserved in the specimens available to her). Norell and Gao (1997) reported three synapomorphies with *L. borneensis*: a close approach or contact of the prefrontal with the postfrontal over the orbit, loss of the upper temporal arch, and a blunt, rounded muzzle (also independently developed in helodermatids). However, the new material shows that the upper arch was present (Alifanov 2000; Gao and Norell 2000). Borsuk-Białynicka (1984) believed that the orbits were relatively small, but Alifanov (2000) figures a skull with large orbits and laterally expanded jugal arches reminiscent of those of *Paravaranus*. Gao and Norell conclude that *C. longifrons* and *L. borneensis* do not seem to be closely related, leaving the latter with no obvious close relatives in the fossil record.

Saniwa was a large (to ~1 m long) lizard of the Eocene and Oligocene (~23–55 mya) in North America and Europe and one of the few extinct land-dwelling platynotans to have several named species. Like most land-dwelling platynotans, it would probably have looked much like a living monitor, but with a relatively larger head.

Although retaining a number of plesiomorphic states, *Saniwa* also possessed derived states relative to *Varanus* (Estes 1983). These include a relatively small premaxilla, with six premaxillary teeth, the dorsal lacrimal foramen completely enclosed by the lacrimal, rudimentary zygosphenes and zygantra, reduced cervical intercentra, and anteroposteriorly expanded neural spines in the (proximal) tail.

Saniwa ensidens is the only reasonably well known species; thus, it has provided most of the characteristics mentioned here. Estes (1983) points out that with only a single reasonably complete skeleton (*S. feisti* was unknown at the time he wrote), the range of individual—not mention sexual, geographical, etc.—variation is impossible to determine. Thus, the validity and relationships of the other species to *S. ensidens* and to each other cannot be determined. Some good comparative studies of the osteology of the various species of *Varanus* would not go amiss here, either.

Estes (1983), like Gilmore (1928) before him, was doubtful of the validity of the species of *Saniwa*—five from approximately contemporaneous times in North America and two from Europe. Unfortunately for simplicity, present-day Australia accommodates at least 23 species of *Varanus,* and possibly more. Future fossils from this fauna could easily provide five species from the same region at the same time. *Varanus* seems likely to be a good ecological model for *Saniwa,* so there is no reason to believe that *Saniwa* could not also have had a large number of species. However, this doesn't imply that the named species of *Saniwa* are actually all valid, and given the fragmentary nature of the fossils, prospects for clarifying the taxonomy could hardly be worse. Here, I simply assume that the species are valid.

Saniwa ensidens

One of the two species of *Saniwa* represented by a skeleton, *S. ensidens,* comes from the Eocene of Wyoming. Its skull was long and narrow, and an intramandibular joint was present (Figs. 2.8, 2.11). There were 6 premaxillary, 24 maxillary, at least 22 dentary teeth, each trenchant, sharp, and recurved, with dilated striate bases.

Saniwa agilis

Estes (1983; and Gilmore 1928) found no differences between the dentary (the lectotype) of this putative species and those of *S. ensidens.* Differences of the vertebrae—lower neural spine and variation in form of the V-shaped notch of the lamina of the posterior neural arch between the postzygapophysis and the neural spine—clearly are regional variation in the vertebral column. Estes had guessed this, and I have verified that he was correct, at least in Bosc's monitor, *Varanus exanthematicus.* Thus, *S. agilis* is probably actually *S. ensidens.* This species (or individual), from the Eocene of Wyoming, was rather small (vertebral length 10 mm).

Saniwa crassa

The precondylar constriction of the vertebral centra is less marked than in other species, and the ventral faces of caudals are flattened. This lizard (or individual) was more massively constructed than the others. The massiveness of construction may result from the greater size, and at least in *V. exanthematicus,* the degree of precondylar constriction of the centra varies along the dorsal column. So this species, too, seems without any distinguishing features. *S. crassa* also comes from the Wyoming Eocene.

Saniwa grandis

The largest species (or individual) from the Eocene of Wyoming, *S.*

grandis, is characterized by massive vertebrae, with rudimentary, medially inflected zygosphenes and strongly convex ventral surfaces of the centra. The first two of these features are to be expected in a larger or more massive individual (or species) and thus would seem not to be reliable indicators that this is a valid species: the convexity of the ventral faces of the centra, perhaps, may be.

Saniwa paucidens

The distinguishing character states (centra longitudinally concave in ventral view, ventral surfaces of centra convex) originally proposed for *S. paucidens* are no longer considered valid (Estes 1983). Because the holotype has been misplaced, there is currently no way to determine any other, possibly distinguishing states: the validity of this species is in limbo. It has been reported from the Eocene of Wyoming and Utah.

Saniwa brooksi

This species, from the Eocene of southern California, apparently did not inhabit the same region as other North American ones. It is the largest known species (or individuals?). Estes (1983) noted that vertebrae of *S. brooksi* differ more from those of *S. ensidens* in their proportions than those of other species, including *S. orsmaelensis* from western Europe. Known largely from vertebrae, *S. brooksi* differs from the others in that the height of the condylar "ball" is more than half its width, the ventral face of the centrum is even with that of the ball, there is no ridge around cotyle anteriorly, the ventral faces of centra are slightly convex laterally but straight anteroposteriorly, the neural spine is placed far posteriorly and projects beyond the postzygapophysis to the posterior extremity of centrum, and the neural arch is triangular across the postzygapophyses. A cursory look at modern varanid vertebrae suggests that these may be valid distinguishing characteristics, but they require confirmation. If any one North American species other than *S. ensidens* is valid, this one probably is.

Saniwa feisti

This is the second species reported from Europe, from the German Eocene, and the second represented by a skeleton (Fig. 2.13). Unusually for this genus (and terrestrial platynotans in general), it is an almost complete skeleton. The lizard was about 60 cm long, with a short neck. The maxilla held 20 teeth. *S. feisti* differs from the North American species in the number of maxillary teeth, having fewer presacral vertebrae and a smaller, thinner pterygoid. It has higher neural spines in on the posterior trunk vertebrae and proximal caudals than *S. orsmaelensis.* Rage (1993) suggested that this species may be a necrosaur and not *Saniwa.*

Saniwa orsmaelensis

S. orsmaelensis, from the Eocene of Belgium, was the first species described from Europe. Most of the material has yet to be figured, although Hoffstetter (1969) did illustrate a trunk vertebra (Fig. 2.11). Hecht and Hoffstetter (1962) were unable to distinguish this material

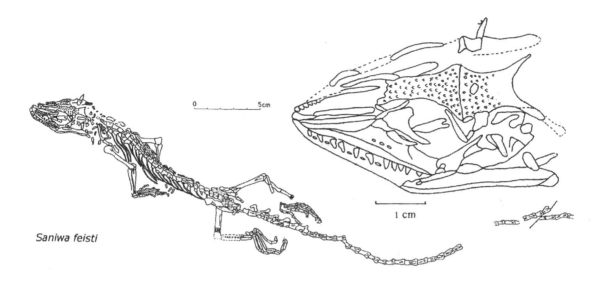

Figure 2.13. Saniwa feisti. *(left)* Skeleton as preserved. *(right)* Skull as preserved. *(Modified after Stritzke 1983.)*

from corresponding elements of *S. ensidens*. They also noted that two of the three femoral pieces indicated that they derived from juveniles.

If one had to guess which species of *Saniwa* might be valid, my guess would be *S. ensidens* in North America and Europe, *S. brooksi* in North America, and *S. feisti* in Europe, but this needs to be substantiated by detailed study. I think *S. agilis* and *S. crassa* are likely synonyms of *S. ensidens*. Both in Europe and North America, fossils of *Saniwa* are generally found in Eocene rocks. However, vertebrae have been found in Oligocene beds in Nebraska and Wyoming. Although it is not clear that all of this material actually pertains to *Saniwa* (Estes 1983), the material does confirm that monitors survived in North America at least as late as the Middle Oligocene, about 32 mya. Monitors survived, even thrived, in the Old World and monstersaurs in the New World (at least in the north, South America was an island continent at this time), although both lineages had originally inhabited both Eastern and Western Hemispheres.

Saniwa sp. has also been reported from the Late Cretaceous (Alifanov 1993) and Eocene (Averianov and Danilov 1997) of Mongolia. These reports are based on vertebrae.

The Miocene sees the last monitor, other than *Lanthanotus,* that has not been assigned to the genus *Varanus. Iberovaranus catalaunicus* (Fig. 2.11) was found in Spain and Portugal. Known only from its vertebrae, it differs from *Varanus* in having a more slender vertebral construction. The centra are constricted as in *Varanus,* but then taper slightly to the condyle; in *Varanus* (and *Saniwa*), there is little or no taper and the condyle expands abruptly from the waist of the constriction. In this respect, vertebrae of *I. catalaunicus* resemble those of helodermatids.

The Portuguese material was described as *Iberovaranus* cf. *I.*

catalaunicus, but the differences from the holotype may be due simply to individual variation (Telles Antunes and Rage 1974).

The fossil record of the genus *Varanus* begins before the Miocene, and in Europe, Africa, and Asia taken together, there is apparently a decrease in the occurrences with time from the Miocene to the present. At first sight, this might suggest that—as the scripts of some natural history videos proclaim—monitors are a prehistoric form that has been slowly dying out with the passing ages. In passing, we might note that this scenario applies better to the monstersaurs. Before leaping to such a conclusion, however, we need to see just how many occurrences of monitors there have been for equal durations of time—not for the periods into which the Cenozoic has been divided. Taking these equal periods into account, for the Miocene, there is (approximately) one occurrence of a monitor fossil per 2 million years, for the Pliocene one per every third of a million years, and for the Pleistocene about one for (a little more than) every half-million years. So there really is no indication that monitors were more common, say, 15 mya than they are now.

The oldest fossils attributed to *Varanus* sp. date from the Eocene (~38–55 mya) in Mongolia (Alifanov 1993), but these are very fragmentary. Those of the succeeding Oligocene are equally sparse. During the Miocene period (~5 to 25 mya), varanids were widespread, apparently living in most of the regions in which they still survive. Fossils have been found in Europe, central Asia (Kazakhstan), and east Africa. Bearing in mind that few, if any, fossil-bearing deposits of this age are known from India, Indochina, or Indonesia, and that Europe has been more extensively searched for fossils than central Asia or east Africa, we can plausibly conclude that varanids were already widespread and thus that they have reasonably well held their own for the past 20 million years.

Discussion of the Mongolian species of varanoids perforce consisted of comments about the skull because most of the taxa are represented by skulls, with postcranial elements few and far between. For *Varanus*, we must, as already mentioned, look at vertebrae because other elements are very largely absent. As for *Saniwa*, here we shall look at the individual species of *Varanus* arranged by region, starting with Africa, then Europe, across Asia to Australia, but the giant monitors will be treated separately.

Varanus rusingensis

V. rusingensis (Clos 1995) is the oldest named species of the genus *Varanus* (Fig. 2.8). It derives from the Early Miocene Hiwegi Formation of Rusinga Island, Kenya. This lizard reached a size of at least 2 m and possessed a differentiated dentition, with recurved teeth set anteriorly in the lower jaw, and stout, blunt ones posteriorly. Clos suggested that the lizard was aquatic—it had a relatively deep tail—and fed on mollusks or other shelled invertebrates.

Varanus hofmanni

V. hofmanni is based on vertebrae (Roger 1898) from the Middle and Upper Miocene beds of France and Germany (Fig. 2.14). These

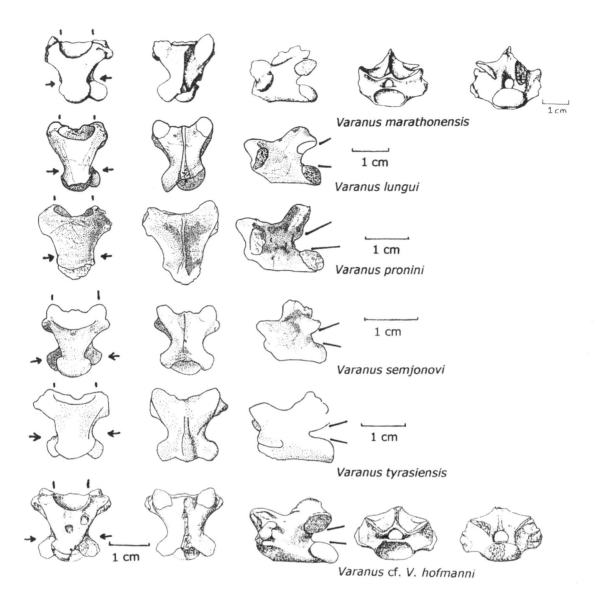

Varanus marathonensis

Varanus lungui

Varanus pronini

Varanus semjonovi

Varanus tyrasiensis

Varanus cf. V. hofmanni

vertebrae are characterized by depressions on the dorsal face of the lamina (of the neural arch) that are well developed anteriorly with an angulate bottom. This species is the oldest monitor (*Varanus*) known from Europe, dating from approximately 13 mya.

Varanus marathonensis

This is the most widespread fossil species of *Varanus* from Europe, having been found in the Pliocene of Hungary and Greece, and also Turkey. It has been reported from the Pleistocene of Italy (Morelli 1891), but Estes (1983) regarded this report as in need of confirmation.

In *V. marathonensis,* centra are longer than in the older *V. hofmanni,* with a larger posterior moiety of the neural arch; zygapophyses that are less transversely oriented; the condyle and cotyle are smaller;

and depressions on the dorsal face of the lamina (of the neural arch) are less deep anteriorly (Fig. 2.14). Even considering this, de Fejervary (1918) was unconvinced that this species was truly distinct from *V. hofmanni*. New and better material has been found at the type locality, Pikermi, in Greece (Estes 1983), but remains to be studied.

Varanus lungui

This species (Zerova and Chkhikvadze 1986) derives from the Miocene of Moldavia (Fig. 2.14).

The centrum is narrow and elongate, with only a slight precondylar constriction and a relatively small condyle and neural canal (Zerova and Chkhikvadze 1986). Presumed autapomorphies include narrow, deep, and groovelike anterior depressions on the dorsal face of the laminae (of the neural arch), more convex ventral face of centrum (than in *V. hofmanni* and *V. tyrasiensis*), a greater angle between the dorsal margin of the condyle and the ventral margin of postzygapophysis in lateral view than in those taxa, pre- and postzygapophyses that are equally broad, and less precondylar constriction than in *Varanus marathonensis*. However, the degree of precondylar constriction and the convexity of the ventral face of the centrum vary with position along the dorsal column in at least some modern varanids. In modern forms, the form of the angle between the dorsal margin of the condyle and the ventral margin of postzygapophysis in lateral view varies along the dorsal column, apparently being related to the height of the neural spine: the higher the spine, the greater the angle (at least in *V. exanthematicus*). Thus, these may not be valid diagnostic features.

Zerova and Chkhikvadze (1986) regard *V. lungui* as related to, but more primitive than, *Varanus indicus*. They estimate that the holotype individual was about 1 m long.

Varanus tyrasiensis

Another monitor from the Miocene beds of Moldavia, *V. tyrasiensis* (Lungu et al. 1983) was a relatively large form. It is based on a vertebra (Fig. 2.14) with a centrum approximately 170 mm long. (The enlargement given in the published figure caption disagrees with the measurements in the table; apparently the enlargement should be two times natural size, not one-half times.) The single autapomorphy initially given is that the parapophysis is larger than the diapophysis. However, Zerova and Chkhikvadze (1986) later reported that this feature, by which they originally distinguished this species, is in fact variable, and they implied that *V. tyrasiensis* may be *Varanus hofmanni*. They did not, however, synonymize the two.

Varanus pronini

This central Asian species (Zerova and Chkhikvadze 1986) comes from Middle Miocene sediments, about 13 million years old, from the northwest shore of the Aral Sea, in Kazakhstan. The centrum is narrow, with a strongly convex ventral surface, a relatively small condyle (as in *V. lungui*), but a large neural canal, in contrast to *V. lungui*. Autapomorphies may be depressions on the dorsal face of the lamina (of the

Figure 2.14. (opposite page) Dorsal vertebrae of European and central Asian species of Varanus. Arrangement as in Figure 2.11. The bars above each vertebra in the left column indicate the width of the cotyle. The centra are reproduced to approximately the same length so the relative proportions can be more easily seen. The arrows indicate the precondylar constructions. In the central column, the bars indicate the angle between the dorsal margin of the condyle and ventral margin of the postzygapophysis. (Modified after Hoffstetter 1969; redrawn from Lungu et al. 1983; Zerova and Chkhikvadze 1986.)

neural arch) that are anteriorly broad and well developed, and an angle between the dorsal margin of the condyle and ventral margin of the postzygapophysis that is greater in lateral view than that of *V. lungui* (Fig. 2.14).

The centrum seems to have a broader cotyle (relative to the length of the centrum) than in other species of *Varanus* (except *V. semjonovi*). In *V. pronini,* the cotyle is approximately 74% the length of the centrum, in *V. semjonovi* about 66%, in *V. lungui* about 60%, and, for comparison, in *V. hofmanni* approximately 53%. However, from the figure, the type specimen also seems to be rather poorly preserved, so the centrum may appear shorter than it was. Furthermore, this feature varies along the dorsal column in Bosc's monitor (*V. exanthematicus*), ranging at least from 52% to 60%. This is not as great as the range of values for the extinct forms; thus, this character, if used with care, may distinguish between different species.

Zerova and Chkhikvadze (1986) regard this species as closely related to both *V. marathonensis* and *V. lungui* and hence, by implication, to *V. indicus*. This is presumably the specimen mentioned by Estes (1983, p. 181) under "*Varanus* sp." from the "northwest shore of Aral Sea, Kazakhstan."

Varanus semjonovi

This is another of the Miocene monitors, from near Odessa, Ukraine (Zerova and Chkhikvadze 1986). In this form, the posterior portion of the centrum is ventrally flattened, and the centrum has a pronounced precondylar constriction, small neural canal, and short neural arch (relative to the length of the centrum) (Fig. 2.14). The condyle is directed more posteriorly than in *V. lungui* or *V. pronini,* and the cotyle is broader and deeper than in *V. lungui* or even *V. pronini,* Presumed autapomorphies are a centrum narrower than in other (fossil) species of *Varanus,* depressions on the dorsal face of the lamina (of the neural arch) that are narrow and deep anteriorly, and prezygapophyses broader than the postzygapophyses.

Zerova and Chkhikvadze (1986) estimate that the holotype individual was 50 to 70 cm long.

Varanus (Psammosaurus) darevskii

Varanus darevskii (Levshakova 1986), for a change, is known from a partial skull (Fig. 2.8), from the Early Pliocene (~4–5 mya) of Tadjikistan. The skull is similar to that of the living *Varanus* (*Psammosaurus*) *griseus,* which includes the gray, Caspian, and Indian Desert monitors (as subspecies). However, that of *V. darevskii* differs from the skull of the modern species in having a more pointed tip of the snout and a broader parietal region. The orbits may have been larger in *V. darevskii* than in *V. griseus.*

Levshakova attributes this species to the living subgenus *Psammosaurus,* which seems sensible on geographical grounds. However, the subgeneric status of the living species of *Varanus* is currently under review, and several putative subgenera have not stood the test of phylogenetic analysis. Because only the species *V. griseus* has previously been attributed to this subgenus, *Psammosaurus* so far remains valid—but whether phylogenetic analysis will support the attribution of *V. darevskii* to it remains to be seen.

Varanus bengalensis

Fossils of modern species of *Varanus* have been reported from India, the Malay Archipelago, and Australia. From India, there are specimens tentatively identified as of *V. bengalensis* (Fig. 2.8) from Pleistocene or Recent deposits in the Billa Surga (Lydekker 1886) and the nearby Karnool Caves (Prasad and Yadagiri 1986) near Madras in India. The Bengal monitor still inhabits this region; it is most common in dry, open forest (Bennett 1998). It is nice to have this fossil material, but it tells us little because we could have guessed that modern species of monitors have been around for the past few tens of thousands of years, at least. The material from Billa Surga was studied by Lydekker (1886, 1888) and de Fejervary (1918, 1935), both of whom figured the specimens (de Fejervary 1918) and concluded that the material was similar to that of *V. bengalensis*. Estes (1983) suggested that restudy of the material would not be out of place.

Varanus bolkayi

This species (de Fejervary 1935) is based on two vertebrae from the Pleistocene Kendeng Beds of Java (~70,000–160,000 years old), which have also yielded fossils of *Homo erectus*. The centra are long, slender, and low (Fig. 2.15). The larger is 2.7 cm long, indicating a large, but not very large, monitor.

Varanus hooijeri

Varanus hooijeri (Brongersma 1958), represented by parts of a skull from Recent cave deposits on Flores, is clearly not an ora. The skull has stout, blunt teeth, and Estes (1983) thought it was related to the Nile monitor (*Varanus niloticus*) or Gray's monitor (*Varanus olivaceus*). However, Auffenberg (1981) thought that both *V. bolkayi* and the Timor material may actually represent the Asiatic water monitor *V. salvator*, although to my knowledge, the reasons for this conclusion have not appeared. This lizard was about as big as modern *V. niloticus* (Estes 1983).

The swollen, blunt, molariform teeth are a possible autapomorphy, as is the supratemporal process that forms a broad roof over the parietal fossa.

Varanus giganteus

Australia has produced fossils attributed, more or less tentatively, to three species: *V. giganteus, V. gouldii,* and *V. varius.* Today *V. giganteus,* the perentie, is the largest Australian varanid (to ~1.9 m overall) and inhabits arid regions in a band from northwestern New South Wales to the west coast, at closest somewhat north and west of Wellington Caves. Bennett (1998) notes that *V. giganteus* lives in desert regions, as well as grasslands and shrublands. Its occurrence (Lydekker 1888; attributed by Nopcsa 1908) in the Wellington Caves, not far

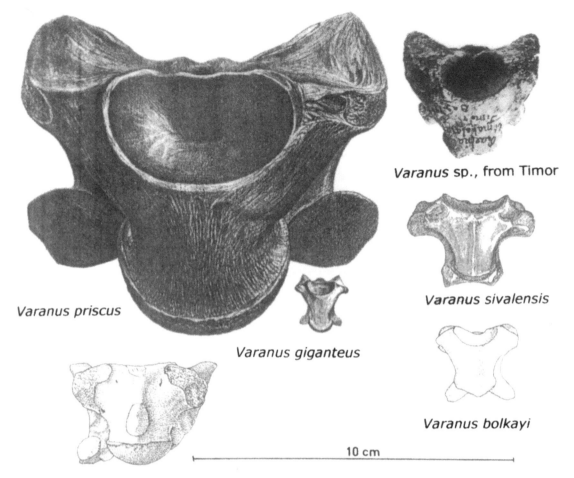

Varanus sp., from Timor

Varanus priscus

Varanus giganteus

Varanus sivalensis

Varanus bolkayi

10 cm

Palæophis colossæus

Figure 2.15. The dorsal vertebrae (ventral view) of the giant monitors and a giant snake. To scale. (Modified after Lydekker 1888; Owen 1859; Rage 1983; V. bolkayi redrawn from de Fejervary 1935; Timorese vertebra from Hooijer 1972.)

from the eastern coast of New South Wales, is well outside of the current range, although the area is now grassland. Perhaps they ranged further south and east during the interglacials or the warmer periods since the last ice age. Or perhaps their ranges have just contracted since people arrived in Australia. Estes (1983), like Lydekker, regarded this material as *Varanus* sp. pending restudy. The remains (referred to *Varanus* sp.) from Cathedral Cave (Wellington Caves) come from two levels, one older than about 33,000 years, before the last glacial maximum, and the other after it, probably between 3000 and 14,000 years (Dawson and Augee 1997).

Varanus gouldii

Bennett (1998) argues that *V. gouldii* (Gould's goanna) is poorly known because of taxonomic confusion with *V. flavirufus* and *V. panoptes*. Bennett's work follows the two reports (Smith 1976; Pledge

1990) of fossil *V. gouldii* from South Australian caves by 22 and 8 years, and thus it seems unprofitable to attempt to work out what taxa might be represented here without reexamining the original fossils. This material might pertain to *V. flavirufus* or *V. gouldii* (or both) or—just possibly—some extinct species.

Varanus varius

V. varius, the lace monitor, is the second largest Australian goanna and reaches about 2 m in total length, although specimens of that size are now rare (Bennett 1998). Victoria Fossil Cave, at Naracoorte, South Australia, lies just (less than 50 km) outside the current range of *V. varius* (Bennett 1998). If the southern limit of this range is set by ambient temperature, it suggests that fossils date from a time when the regional temperature was higher.

These Australian occurrences indicate that if the identifications are accurate, some Pleistocene goannas lived in regions where they are now absent. This is hardly a surprising conclusion, given the fluctuations of climate during that period, but is still significant.

Water Monitors

Wilkinson (1995) described two unusual caudal vertebrae from an unrecorded locality, probably on the Darling Downs, west of Brisbane in southeastern Queensland. The caudals were distinctive in having vertical and unusually tall neural spines (Fig. 2.16). They were similar to those of the so-called water monitors, such as Mertens' goanna, *Varanus mertensi.* They, however, also resembled the caudals of *V. panoptes.* At the time Wilkinson studied this material, this resemblance was problematic, but both Fuller et al. (1998) and Ast (2001) found that *V. mertensi* and *V. panoptes* are related, members of a clade (Ast's *gouldii* group). Thus, Wilkinson's material documents the existence of this clade during the Pleistocene (or Pliocene).

Fossil Monitors of the Southwestern Pacific Region

New Caledonia today lacks indigenous monitors. However, fossils of monitors (Fig. 2.8) have been found (Balouet 1991), some in deposits probably about 1750 years old. Unusually, this material contains cranial and limb elements in addition to vertebrae. Balouet found the bones similar to those of *Varanus indicus* (mangrove monitor), but with more elongate teeth in the dentary and a longer narial margin of the maxilla. The dentary is a little over 4.5 cm long, so representing a reasonably large lizard. The material may represent a new species. Monitors in this region today are restricted to Australia and islands on or near the Asian continental shelf. This lizard is not, and represents the farthest penetration of monitors into the South Pacific. It suggests that future research on southwestern Pacific islands may yet produce more evidence of now-extinct insular monitors.

Varanid fossils have been found in Kalimantan. In the Malayan provinces of Sarawak and Sabah, remains have been found that represent unidentified (and presumably small) monitors. Two sites in Sara-

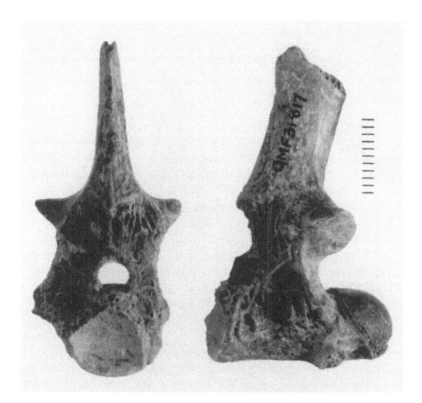

Figure 2.16. A caudal vertebra of a presumably Pleistocene monitor of the V. gouldii *"group" from the Darling Downs of Queensland, Australia, seen in anterior (left) and lateral (right) views. Scale in millimeters. (Courtesy of J. E. Wilkinson and the Queensland Museum.)*

wak have ages of 900 to 22,000 and 2000 to 40,000 years (Datan 1993; Bellwood 1985) and that in Sabah is 10,000 to 15,000 years old (Bellwood 1985). Bones of *Varanus* cf. *V. salvator* (Asiatic water monitor) have been found in cave deposits at two sites on Palawan, the Philippines (Reis and Garong 2001). These are probably less than 12,000 years old.

Giant Monitors

A number of species of extinct fossil monitors that were not just long and slender, but were big—over 2 m long—have been described from Australia: *Varanus priscus* (=*Megalania prisca*), *Varanus dirus,* and *Varanus emeritus. Notiosaurus dentatus* and *Varanus warburtonensis* are universally considered to be *Varanus priscus.* However, it is not so widely realized that giant monitors other than the ora (Komodo dragon) also existed in Asia. Here, we will travel backward from Australia into Asia.

Varanus priscus

V. priscus (Owen 1859), often termed *Megalania prisca,* is both the largest and best represented of the fossil giant monitors. This is, in fact, the largest land-dwelling lizard, probably reaching a length of 6 m— more if it had a relatively long tail like *V. varius.*

There is no recent comprehensive study of *V. priscus*, either of its anatomy or its phylogenetic relationships. It's not even clear that sufficient material exists for such a study. *V. priscus* certainly has some unusual features, such as the low sagittal crest on the frontals, but these are of unknown significance and may be simply the result, directly or indirectly, of its large size. Thus, here we shall revert to the opinion of Lydekker (1888) and use the name *Varanus priscus*.

This beast lived in eastern Australia, from the southern coast near Melbourne to the Cape York Peninsula. As for most other fossil monitors, most of the specimens of *Megalania* are vertebrae. The Queensland Museum, in Brisbane, holds the largest collection of *Megalania* fossils, which includes approximately 100 dorsals, about half as many caudals, five sacrals or parts of sacrals, and a single cervical. There are also 11 pieces from skulls or jaws, 1 piece of shoulder girdle, parts of 3 pelvic girdles, 4 bones (or parts thereof) from the forelimb, and also 4 from the hind limb. Limb bones, except for the ulna and fibula, are known from only a single specimen each, and so far as I can tell, there are no radii at all.

As the largest land-dwelling lizard, *V. priscus* is expected to show robust, massive bones. This expectation isn't disappointed: the humerus is stockier and more robust than in any other lizard (De Vis 1885). The vertebrae, particularly those of the back, are also massive, with inflated arches (Fig. 2.15). Bones of the skull roof, the frontals and parietal, are also massive, although, perhaps surprisingly, the jaw bones were not particularly robust (Molnar 1990). The frontals bore a low sagittal crest that I think was probably made more prominent in life by a scaly covering and functioned as a display device in sexual selection. Unexpected was that the upper surface of the frontals was also bore low ornamentation that looked as if small, damp spaghetti strands had been poured onto the bone and then ossified. In other species of *Varanus*, prefrontal and fused postfrontal–postorbital bones do not come into contact above the orbit (or anywhere else, for that matter). In *V. priscus*, however, they apparently did come into contact (Molnar 1990). For the genus *Varanus*, this is a derived feature apparently not possessed by any other species, but it is found in the monstersaurs, *Lanthanotus* and *Telmasaurus*. *V. priscus* also shows encroachment of the attachment areas of the jaw adductors well up the sides of the parietals, a feature also found in some Cretaceous platynotans. This does not, I feel, indicate that *V. priscus* shows any particular affinity to plesiomorphic forms, but that with the thickening of the skull roof following from the large size of the beast, the jaw muscles simply had more surface area on which to attach.

Work on the paleobiology of this animal is still in progress or in press, but a few comments won't be out of place here for those willing to take the reasoning on faith. Two individuals, one from near Springsure and the other from the eastern Darling Downs, both in southeastern Queensland, are known from partial skeletons. Both of these skeletons, although incomplete, preserved a single broken and healed rib. Perhaps the thickened skull roof was also useful in butting the flanks of opponents. Study of the histology of the fibular shaft of *V. priscus* by

Vivian de Buffrènil and Armand de Ricqlés shows localized but intense Haversian remodeling of the bone not seen in the other monitors. This implies that the large size of *V. priscus* resulted from both a prolongation of the relatively fast growth characteristic of juvenile large monitors into later life and a long life span. This notion finds independent support in the work of Gregory Erickson, who looked into the histological structure of small ossicles, presumably from the head (and resembling those found in the heads of oras). Erickson's work indicates that individuals of moderate size had lived over 15 years. The conspicuous cyclical growth marks in the long bone shaft often seen in monitors are absent, and together with the large size suggest that *V. priscus* maintained a higher core body temperature and metabolic rate than living varanids. These physiological features may well be the result of the increased body mass of this lizard, which would have conferred a substantial degree of inertial homeothermy on it (McNab and Auffenberg 1976).

Lee (1996) attempted to determine which living monitor is most closely related to *V. priscus*. He suggested the perentie, *V. giganteus*, one skull of which bears an incipient midline crest on the skull roof, similar to but smaller than that of *V. priscus*. I have seen such incipient crests on other perentie skulls but also in skulls of other Australian goannas. Lee suggested that the clade of *V. giganteus* and *V. priscus* had recently evolved this feature, which was thus a shared, derived character. However, low or incipient crests date back a long way among varanids—they are also found in the Cretaceous *Telmasaurus*—so the relationships of *V. priscus* still await a more thorough study.

The date of extinction of *V. priscus* is not known. Although found at Wellington Caves in east-central New South Wales, the fossils don't occur in the dated levels, most of which are less than 33,000 years old. Archeological excavations at Cuddie Springs, in northern New South Wales, also encountered fossils of this creature, but in undated units older than 30,000 years (Dodson et al. 1993; Furby et al. 1993). Extinctions in central Australia occurred about 50,000 years ago (Miller et al. 1999), but we don't know when they happened in coastal regions.

Although the Pleistocene remains of *V. priscus* are the most numerous and best known, bones from Early Pliocene deposits (~4.5 million years old) in southeastern Queensland indicate that it, or a very similar large monitor, also lived during that time (Hecht 1975). These earlier lizards were as large, or almost as large, as those of the Pleistocene.

Varanus dirus

V. dirus is based on an isolated tooth from the Darling Downs, presumably Pleistocene in age, and now generally accepted as from *V. priscus*, so *V. dirus* is a synonym of that species. The single specimen referred to *V. dirus* (properly *V. "dirus"*), a maxilla with three teeth, comes from the Early Pliocene deposits at Chinchilla, in southeastern Queensland, rather than the Pleistocene beds further east (Fig. 2.8). Charles De Vis thought the maxilla closely resembled that of Salvadori's monitor, *V. salvadorii*, but it was about twice as large as that of the approximately 2.1-m-long skeleton he had. It is considerably smaller and more lightly built than the maxillae of (adult) *V. priscus*. Further study is necessary to find out whether this is a juvenile of *V. priscus,* a new taxon, or a very large specimen of *V. salvadorii* (which does not now live in Australia, but may have when Australia and New Guinea were contiguous). One problem is the lack of studies of the changes of the skeletal form and proportions during the ontogeny—the growth to maturity—of any monitor.

Varanus emeritus

V. emeritus is based on half of a humerus and of a tibia from the Darling Downs of southeastern Queensland. I could not distinguish these bones from those of *V. salvadorii,* Salvadori's monitor. The humerus and tibia come from different individuals (the tibia indicates a lizard about 1.5 times as large as that yielding the humerus; Estes 1983), so it's possible that they derive from different kinds of monitors. Dunn (1927) remarked that *V. emeritus* seems to have been a slim lizard, similar to the lace monitor, and was perhaps 3 or 4 m long (presumably given a similarly proportioned tail).

Varanus komodoensis

A pair of vertebrae found in 1966 in Pleistocene gravels in (western) Timor may be those of the ora, *V. komodoensis* (Fig. 2.15). Hooijer (1972) noted that the vertebrae are similar to those of *V. bolkayi*: they are broader across the prezygapophyses and relatively smaller neural spine (in diameter) than those of *Varanus komodoensis*. Presumably regarding these differences as only variation, he suggested that these bones, together with the vertebrae from Trinil, Java (both those from *V. bolkayi,* and others), and from Kedungbrubus, probably represent fossil oras. Both Trinil and Kedungbrubus are probably about 160,000 to 70,000 years old (Watanabe and Kadar 1985).

Fossil ora teeth have been reported from beds 850,000 and 900,000 years old on the island of Flores (Morwood et al. 1999; Sondaar et al. 1994; van den Bergh et al. 2001). Thus, oras have been present on Flores for at least half of the Pleistocene and may once have inhabited Timor and Java. Clearly, both monitors of (presumably) moderate size and quite large ones—even if it isn't clear just how large—lived in Indonesia during much of the Pleistocene.

Because *Varanus komodoensis* is the largest living monitor—indeed, the largest living lizard—its history is of particular interest. Van den Bergh et al. (2001) note that *V. komodoensis* is the only land-dwelling tetrapod known to have survived the faunal turnover that occurred between the deposition of members A and B of the Ola Bula Formation in Flores. This turnover marked the replacement of dwarf stegodont elephants (*Stegodon sondaari*) by a large species (*Stegodon florensis*) and the loss of giant tortoises. Early last century, it was thought that the ora, or its ancestral lineage, originated in Australia and later migrated to Indonesia. This notion was based on anatomical similarities of the ora to Australian monitors, particularly *V. varius* (Dunn 1927), as well as on the large number of monitors found in Australia, including one shared with Indonesia. Although this conclusion was based on "prephylogenetic" studies, it has been supported by the most recent phylogenetic analysis (Ast 2001). This work, using mitochondrial DNA, indicates that *V. varius* is the sister group of *V. komodoensis,* with *V. salvadorii* as the sister group of both. The distribution of these species—*V. salvadorii* in southern New Guinea and *V. varius* in southeastern and eastern Australia—suggests that a Australasian lineage did reenter Indonesia. The Malayan Archipelago (Indonesia and Philippines) may be conceived as composed of three components. The western islands are Asian and were joined to Asia from at times of low sea stand and so have an Asiatic fauna. The eastern islands originated on the Indo-Australian Plate near Australia and New Guinea and so have an Australasian fauna. And the intermediate islands, which presumably arose from the seafloor as the tectonic plates collided, were settled by immigrants from both major landmasses. One of these immigrants apparently was the ancestral lineage of *V. komodoensis.* Volcanic eruptions on both Komodo and Flores started during the Early Cretaceous, at least 130 mya (Auffenberg 1980). Fossil wood shows that large trees grew there between this time; and when volcanism started ~130 mya, then stopped, and then began again ~50 mya. Forest grew between 130 and 150 mya—volcanism recommenced in the Eocene (~50 mya). Auffenberg concluded that both Komodo and Flores existed since Cretaceous times. Presumably, ancestral oras arrived in the Late Miocene or later, as the oldest Australian monitors are Miocene, but probably before the Pleistocene.

Living oras eat fairly large mammals such as deer, boar, and goats. These animals were introduced by people around 5000 years ago, and oras have been there much longer. So what did oras eat before they were provided with convenient artiodactyls? Diamond (1987) suggested they ate the native pygmy stegodonts of these islands. These small elephants were also found on Timor, also home to also large Pleistocene monitors, perhaps oras, and large mammals were certainly available on Java. But pygmy elephants were not the only potential prey of Pleistocene oras. Giant tortoises, thought to be *Geochelone atlas,* lived on several Indonesian islands. Fossil *G. atlas* shells from the Siwalik Hills of northwestern India and adjacent Pakistan are thought to reach about 1.7 m in length and about 1.5 m wide (Brown 1931; Fenton and Fenton 1958). Although none so large have yet been reported from Indonesia, the Indonesian shells were still reasonably big. A fossil shell fragment from Flores, together with a hind leg, of one of these tortoises was found associated with six ora teeth (Sondaar et al. 1994). Even in the absence of introduced artiodactyls, oras had at least two large animals as potential prey.

Varanus sivalensis

Most people who think about monitors at all—including a few who really should know better—think that there were only two giant monitors, oras and *Megalania.* But there were others, even one from continental Asia, *Varanus sivalensis,* from the Pliocene rocks of the Siwalik Hills. Even by fossil varanid standards, it is poorly known, with only two vertebrae (Fig. 2.16) and part of a humerus.

The humerus is 35 mm across its distal articulation, close to that of a large *V. komodoensis,* which may be 38 mm across the distal articulation (Estes 1983). *V. sivalensis* lived in a fauna of continental mammals (and other animals), unlike the other giant monitors that lived on isolated landmasses, Indonesian islands in the case of the ora, and Australia for the rest.

Monitor Zoogeography and Evolution

This volume has a separate chapter on monitor zoogeography and evolution, but a few comments on that subject will not be out of place here. Throughout most of their history, varanoids seem to have had a range of sizes comparable to those seen in modern monitors. One might think the ecological factors that permitted monitors to become giants were associated with isolation on islands or island continents, if one considered only the two better-known giant forms, *V. priscus* and the ora. But *Varanus sivalensis* shows that monitors could also become giants (presumably) in the face of competition with advanced mammalian predators. Given the maybe 100-million-year history of platynotans, giant monitors are a rather recent development, at best less than 25 million years old. Platynotans include the largest land-dwelling lizards—and probably the largest aquatic ones (mosasaurs) as well. *Varanus* are basically carnivorous lizards, but even the oldest known species (*V. rusingensis*) already had blunt teeth, suggesting that the potential to prey on insects and mollusks, and even maybe fruit, developed early in this lineage. Tantalizingly, there is no fossil record for the unusual and problematic *Lanthanotus borneensis.*

Annotated List of Fossil Terrestrial Varanoids

The following list of fossil terrestrial varanoids is listed alphabetically for convenience of use. The format of the entries in this list is as follows: taxonomic name, author, and date; synonyms (if any); type material, referred material (if any); formation; age; locality. This may be followed by comments. Material from localities or strata other than type is listed as "additional material." Specimens not assigned to species are listed. Co., county; Fm., formation; Memb., member.

Aiolosaurus oriens Gao and Norell, 2000; incomplete skull, mandible, incomplete postcranial skeleton; Djadokhta Fm.; Santonian-Campanian, Ukhaa Tolgod, Mongolia.

Cherminotus longifrons Borsuk-Białynicka, 1984; incomplete skull, two referred incomplete skulls (Alifanov 1993) and fragments; Baruungoyot Svita; Santonian-Campanian; Khermeen Tsav, E Mongolia.

Colpodontosaurus cracens Estes, 1964; dentary, referred maxillary and dentary fragments; Lance Fm.; Maastrichtian; Niobrara Co., Wyoming, USA.

Ekshmer bissektensis Nessov, 1981; maxilla; Bissekty Fm.; Coniacian(?); Dzhara-Kuduk, Uzbekistan.

This taxon seems generally to have been ignored by later workers.

Eosaniwa koehni Haubold, 1979; skull and partial skeleton; Mittelkohle of Geiseltal Braunkohle; Middle Eocene; Geiseltal, Halle, Germany.

Estesia mongoliensis Norell et al., 1992; incomplete skull with mandibles (one incomplete), referred skull with mandible and 12 vertebrae; Baruungoyot Svita; Santonian-Campanian; Khulsan, S Mongolia.

Eurheloderma gallicum Hoffstetter, 1957; maxilla, referred maxillae, pterygoid, parietals, dentaries, vertebrae; Phosphorites de Quercy; ?; Quercy, France.

cf. *Eurheloderma* sp.; parietal; ?; Late Paleocene; Wyoming; Pregill et al. (1986).

Gobiderma pulchrum Borsuk-Białynicka, 1984; skull and mandible, two referred incomplete skulls with mandibles, one with incomplete vertebrae and pectoral girdle fragments; Baruungoyot Svita; Santonian-Campanian; Khermeen Tsav, E Mongolia.
Additional material:
1—incomplete skull; Baruungoyot Fm.; Santonian-Campanian; Khulsan, S Mongolia.

Heloderma texana Stevens, 1977; nearly complete skull, referred cranial bones, including frontal and maxilla, vertebrae; Delaho Fm.; Early Miocene; Big Bend National Park, Texas, USA.

Heloderma suspectum:
1—osteoderm fragments, skin; cave deposits; Late Pleistocene; Clark Co., Nevada, USA; Estes (1983).
2—?; cave deposits; Late Pleistocene; Hidalgo Co., New Mexico, Harris (1993).

Heloderma? sp.; trunk vertebra, vertebral fragments, one complete and one fragmentary femur; ?; Early Miocene; Thomas Farm, Gilchrist Co., Florida, USA; Estes (1983).

Occurrence of indeterminate Helodermatidae:
vertebra; Landénian; Early Eocene; Dormaal (="Orsmael"), Belgium: Augé (1995).

Iberovaranus catalaunicus Hoffstetter, 1969; vertebra; ?; Middle Miocene; Valles-Penedes Basin, Catalonia, Spain.
Additional material:
1—trunk and caudal vertebrae; ?; Middle Miocene; Lisbon, Portugal; Telles Antunes and Rage (1974).

Lowesaurus matthewi (Gilmore, 1928); incomplete maxilla, referred maxilla; Cedar Creek Memb., White River Fm.; Middle Oligocene; Logan Co., Colorado.

Additional material:

1—incomplete skull, dentary, isolated osteoderms; Brule Memb., White River Fm.; Late Oligocene; Morrill Co., Nebraska; Estes (1983).

2—frontal, two trunk vertebrae; Mitchell Pass Memb., Gering Fm.; Early Miocene; Morrill Co., Nebraska; Pregill et al. (1986).

Occurrences of indeterminate Necrosauridae:

1—incomplete right maxilla; Cedar Mountain Fm.; Albian; Emery Co., Utah; Cifelli and Nydam (1995).

2—left maxilla; Djadokhta Fm.; Campanian; Bayan Mandahu, Nei Mongol, China; Gao and Hou (1996).

Necrosaurus cayluxi (Filhol, 1873) = *Paloeosaurus cayluxi* Filhol, 1877 = ?*Odontomophis atavus* de Rochebrune, 1884? = *Palaeovaranus filholi* de Stefano, 1903; dentary; Phosphorites de Quercy; Late Eocene; Quercy, France.

 Although the holotype is a dentary, much material—most of it undescribed (Estes 1983)—is known for this species. Further details of the material, localities, age, and systematic history are given by Estes (1983).

Necrosaurus cf. *N. cayluxi* (Rage 1988); one trunk and seven caudal vertebrae and osteoderm; Phosphorites de Quercy; Late Eocene; Quercy, France.

Necrosaurus eucarinatus Kuhn, 1940a = *Melanosauroides giganteus* Kuhn, 1940a = "aff. *Ophisauriscus* (*Melanosauroides*) *eucarinatus*" (Kuhn 1940a) = cf. *Glyptosaurus hillsi* (Kuhn, 1940a) = *Necrosaurus giganteus* Hoffstetter, 1943; hind limb with associated osteoderms, referred skeleton (disarticulated), vertebrae and scales; ?; Middle Eocene; Geiseltal, Germany.

Occurrences of *Necrosaurus* sp.:

1—?; ?; Late Paleocene; Cernay-les-Reims, France; Hoffstetter (1943).

2—vertebra; ?; Late Paleocene; Wahlbeck, Germany; Kuhn (1940b). This specimen was originally referred to *Saniwa* and is now lost. Estes (1983) gives reasons for assigning it to *Necrosaurus*.

3—?; ?; Late Paleocene; Tsagan-Hushu, Mongolia; Alifanov (1993).

4—?; ?; Early Eocene; Cuis, France; Hoffstetter (1943).

5—caudals, osteoderms; Sparnacian; Early Eocene; Dormaal, Belgium; Hecht and Hoffstetter (1962) and Godinot et al. (1978).

6—vertebrae, osteoderms(?); ?; Middle (or Late) Eocene; Egerkingen, Switzerland; Hoffstetter (1943, 1962) and Augé (1990).

7—incomplete trunk vertebra and incomplete caudal vertebra; Lower Headon beds; Late Eocene; Isle of Wight, UK; Rage and Ford (1980).

8—frontal, parietal, postorbital–postfrontal, palpebral, maxilla, premaxilla, palatine, dentary, vertebrae, osteoderms; Tongrian; Early Oligocene; Hoogbutsel and Hoeleden, Belgium; Hecht and Hoffstetter (1962).

This material differs from that of *N. cayluxi* in that the osteoderms are not fused to the cranial bones, the vertebrae are relatively shorter, and the overall size is about two-thirds as large (Hecht and Hoffstetter 1962).

Palaeosaniwa canadensis Gilmore, 1928 = *Megasaurus robustus* Gilmore, 1928; trunk vertebra, referred vertebra; Oldman Fm.; Campanian; Alberta, Canada.
Additional material:
1—isolated teeth, trunk, and caudal vertebrae; Lance Fm.; Maastrichtian; Niobrara Co., Wyoming, USA; Estes (1983).
2—vertebra; Lance Fm.; Maastrichtian; Richland Co., Montana, USA; Estes (1983).
3—incomplete skull, vertebrae; Hell Creek Fm.; Maastrichtian; Garfield Co., Montana, USA; Estes (1983).

Palaeosaniwa cf. *P. canadensis;* dentary; Fort Union Fm.; Middle Paleocene; Carbon Co., Wyoming, USA; Sullivan (1982).
This specimen has neither been figured nor described; thus, the assignment is tentative.

Parasaniwa wyomingensis Gilmore, 1928 = *Parasaniwa obtusa* Gilmore, 1928; incomplete dentary, referred maxilla, dentaries (including type of *P. obtusa*), parietal, frontal; Lance Fm.; Maastrichtian; Niobrara Co., Wyoming, USA.
Additional material:
1—?; Hell Creek Fm.; Maastrichtian; McCone Co., Montana, USA; Estes (1983).
Estes notes that there is possibly other material.

Occurrences of *Parasaniwa* sp.:
1—?; ?; Early Eocene; Craig Co., Wyoming, USA; Estes (1983).
2—?; Wasatch Fm.; Early Eocene; Sweetwater Co., Wyoming, USA; Estes (1983).
3—?; ?; Early Eocene; Colorado, USA; McKenna (1960).
4—maxillary fragment; Bridger Fm.; Middle Eocene; Sublette Co., Wyoming, USA; Estes (1983).
Estes (1983) comments that Jacques Gauthier thought this material was not *Parasaniwa*.

Primaderma nessovi Nydam, 2000; incomplete right maxilla, parietal, dentary, cervical, dorsal; Cedar Mountain Fm.; Albian-Cenomanian; Emery Co., Utah, USA.

Paravaranus angustifrons Borsuk-Białynicka, 1984; incomplete skull and mandible; Baruungoyot Fm.; Santonian-Campanian; Khulsan, S Mongolia.

Parviderma inexacta Borsuk-Białynicka, 1984; incomplete skull with mandible; Baruungoyot Svita; Santonian-Campanian; Khermeen Tsav, E Mongolia.

Provaranosaurus acutus Gilmore, 1942; maxilla, referred dentary and maxilla; Fort Union Fm.; Late Paleocene; Park Co., Wyoming, USA.

Occurrences of cf. *Provaranosaurus* sp.:

1—maxillae, fragmentary dentaries; Tongue River Fm.; Middle Paleocene; Carter Co., Montana (USA); Estes (1983).
The teeth here are less high crowned, slender, and needlelike than
in the type, and the intramandibular septum lacks a free ventral
border. Estes (1983) was undecided whether or not this represented individual variation or a second, undescribed species.

2—dentaries; Regina Memb., San Jose Fm.; Early Eocene; Carbon Co.,
Wyoming, USA; Sullivan (1982) and Kues (1993).
Because this material lacks teeth, reference to *Provaranosaurus* is
tenuous (Estes 1983).

Paraderma bogerti Estes, 1964; maxilla, referred premaxilla, maxillary
fragments, dentary (and incomplete dentary), and parietal; Lance
Fm.; Maastrichtian; Niobrara Co., Wyoming.

Proplatynotia longirostrata Borsuk-Białynicka, 1984; Incomplete skull
with mandible; Baruungoyot Fm.; Santonian-Campanian; Khulsan, S Mongolia.

Saniwa agilis (Marsh 1872); dentary, referred vertebrae, distal femur
and distal phalanx; Bridger Fm.; Middle Eocene; Uinta Co., Wyoming, USA.

Saniwa brooksi Brattstrom, 1955; three nearly complete and one fragmentary vertebrae; Poway Conglomerate; Late Eocene; Mission
Valley, San Diego, California, USA.

Additional material:

1—maxilla, 3 trunk vertebrae, 10 caudals; Sespe Fm.; Late Eocene;
Ventura Co., California, USA; Brattstrom (1958).

2—?; Friars Fm.; Late Eocene; San Diego Co., California, USA; Estes
(1983) and Schatzinger (1975).

3—?; Mission Valley Fm.; Late Eocene; San Diego Co., California,
USA; Estes (1983) and Schatzinger (1975).
The material found at the latter two localities is only vaguely
described in the literature. It consists of dentary fragments, vertebrae, and doubtfully referred teeth.

Saniwa crassa Marsh, 1872 = *Saniwa major* Leidy, 1873, in part;
vertebra, referred vertebrae, humeral shaft; Bridger Fm.; Middle
Eocene; Uinta Co., Wyoming, USA.

Additional material:

1—vertebrae (holotype of *Saniwa major*—lost, Estes, 1983); Bridger
Fm.; Middle Eocene; Wyoming, USA; Estes (1983).

2—dentary, vertebrae, appendicular and other(?) fragments; Bridger
Fm.; Middle Eocene; Wyoming, USA; Estes (1983).

3—vertebrae, fragments; Uinta Fm.; Late Eocene; Uinta Co., Wyoming, USA; Estes (1983).

Saniwa ensidens Leidy, 1870 = *Thinosaurus leptodus* Marsh, 1872;
skeleton (disarticulated); Bridger Fm.; Middle Eocene; Sweetwater
Co., Wyoming, USA.

Additional material:

1—vertebrae and limb elements; Wasatch Fm.; Early Eocene; Clark's Fork Basin, Wyoming, USA; Gilmore (1928).

2—fragmentary skeleton and vertebrae, limb and girdle elements (registered under seven catalog numbers); Bridger Fm.; Middle Eocene; Uinta Co., Wyoming, USA; Gilmore (1928) and Estes (1983).

Saniwa feisti Stritzke, 1983; skeleton, referred second skeleton; Lutetian; Middle Eocene; Messel, Darmstadt, Germany.

Saniwa grandis Marsh, 1872; incomplete trunk vertebra, referred caudal, pelvis; Bridger Fm.; Middle Eocene; Uinta Co., Wyoming, USA.

Additional material:

1—caudal; Bridger Fm.; Middle Eocene; Uinta Co., Wyoming, USA; Estes (1983).

2—?; Bridger Fm.; Middle Eocene; Sublette Co., Wyoming, USA; Estes (1983).

3—vertebrae, femur; Wind River Fm.; Middle Eocene; Fremont Co., Wyoming, USA; Estes (1983).

Saniwa orsmaelensis Dollo, 1923; seven trunk vertebra (lectotype; Estes 1983), referred maxilla, femur, vertebrae, ilia; Landénian; Early Eocene; Dormaal (="Orsmael"), Belgium.

Additional specimen:

1—four trunk vertebrae; Landénian; Early Eocene; Erquelinnes, Belgium; Hecht and Hoffstetter (1962).

Saniwa paucidens Marsh, 1872; incomplete skeleton (lost—Estes 1983); Bridger Fm.; Middle Eocene; Uinta Co., Wyoming, USA.

Additional specimen:

1—22 vertebrae, fragments; ?; Late Eocene; Uinta Co., Wyoming, USA; Estes (1983).

Occurrences of *Saniwa* sp.:

1—vertebra; Baruungoyot Fm.; Santonian-Campanian; Khermeen Tsav, E Mongolia; Alifanov (1993).

2—vertebra; ?; Early Eocene; Cuis, France; Estes (1983) and Augé (1990).

3—vertebra; ?; Early Eocene; Monthelon, France; Hecht and Hoffstetter (1962).
Similar to those of *S. grandis,* but with a low and more oblique cotyle.

4—vertebra, fragments; San Jose Fm.; Early Eocene; Sandoval Co., New Mexico, USA; Gilmore (1928).

5—?; Regina Memb., San Jose Fm.; Early Eocene; Sandoval Co., New Mexico, USA; Kues (1993).

6—?; Tapicitos Memb., San Jose Fm.; Early Eocene; Rio Arriba Co., New Mexico, USA; Kues (1993).
Kues (1993) also lists cf. *Saniwa* sp. from localities 5 and 6.

7—vertebrae; ?; Early Eocene; Fremont Co., Wyoming, USA; White (1952).

8—vertebrae; ?; Early Eocene; Shoshone Co., Wyoming, USA; White (1952).

9—vertebra; Bumban Memb., Naran Bulak Fm.; Middle Eocene; Tsagan-Hushu, Mongolia; Alifanov (1993); Averianov and Danilov (1997).

10—vertebra; Bumban Memb., Naran Bulak Fm.; Middle Eocene; Tsagan-Hushu, Mongolia; Alifanov (1993, table 1); Averianov and Danilov (1997).
This site is given twice because two forms are apparently represented.

11—vertebrae; Huerfano Fm.; Middle Eocene; Gardner Buttes, Colorado, USA; Gilmore (1928).

12—dentary, vertebrae; Bridger Fm.; Middle Eocene; Sublette Co., Wyoming, USA; Hecht (1959).

13—?; Chadron Fm.; Late Eocene: southwestern North Dakota; Smith and Gauthier (2002).

14—vertebra; Kusto Fm.; Early Oligocene; Kiin-Kerish, Kazakhstan; Zerova and Chkhikvadze (1986); Averianov and Danilov (1997).

15—vertebrae; White River Fm.; Early Oligocene; Natrona Co., Wyoming, USA; Estes (1983).

16—caudal vertebra; White River Fm.; Middle Oligocene; Sioux Co., Nebraska, USA; Gilmore (1928).

Occurrence of cf. *Saniwa* sp.:

1—two vertebrae; Fort Union Fm.; Middle Paleocene; Carbon Co., Wyoming, USA; Sullivan (1982).

2—three vertebrae; ?; Early Eocene; Andarak 2, Fergana Valley, Kirghizia; Averianov and Danilov (1997).

Saniwides mongoliensis Borsuk-Białynicka, 1984; skull and incomplete cervicals; Baruungoyot Fm.; Santonian-Campanian; Khulsan, S Mongolia.

Telmasaurus grangeri Gilmore, 1943; incomplete skull and vertebrae, referred specimen with vertebrae, forelimb material, pelvis, hind limbs, pedes; Djadokhta Fm.; Santonian-Campanian; Bain Dzak, S Mongolia.
Additional material:

1—incomplete skull, with two incomplete cervicals; Baruungoyot Fm.; Santonian-Campanian; Khulsan, S Mongolia.

Occurrences of indeterminate Varanidae:

1—front of skull with mandible, vertebra; Djadokhta Fm.; Campanian; Bayan Mandahu, Nei Mongol, China; Gao and Hou (1996).
Given as "?Varanidae."

2—vertebrae; Eureka Sound Fm.; Eocene; Ellesmere Island, Canada; Estes and Hutchison (1980).
This represents the northernmost occurrence of varanids.

3—vertebrae; Kusto Fm.; Early Oligocene; Kiin-Kerish, Kazakhstan; Zerova and Chkhikvadze (1986).

4—?; Upper Site Local Fauna; Early? Miocene; Godthelp Hill, Riversleigh, Queensland, Australia; Archer et al. (1991).

5—?; cave deposit; Pliocene; Rackham's Roost, Riversleigh, Queensland, Australia; Archer et al. (1991).

6—?; Waite Fm.; Pliocene; Alcoota, Northern Territory, Australia; Hecht (1975).

7—vertebrae?; Katipiri Fm.; Pleistocene; Kalamurina, South Australia, Australia; Tedford and Wells (1990).

8—?; unnamed pond deposits; Late Pleistocene; Cuddie Springs, New South Wales, Australia; Dodson et al. (1993).

In addition to *V. priscus*.

Occurrences of indeterminate Varanoidea:

1—incomplete trunk vertebra; ?; middle Turonian; Sainte-Maure, Indre-et-Loire, France; Rage (1989).

2—incomplete cervical vertebra; Zhirkindek Fm.; Turonian-Coniacian; Tyul'kili, Kazakhstan; Kordikova et al. (2001).

3—incomplete maxilla, vertebrae; Djadokhta Fm.; Santonian-Campanian; Ukhaa Tolgod, Mongolia; Gao and Hou (1996).

4—vertebra; Djadokhta Fm.; Santonian-Campanian; Bayan Mandahu, China; Gao and Norell (2000).

5—incomplete dentary; fissure fills; Late Eocene or Early Oligocene; Dielsdorf, Kt. Zurich, Switzerland; Hünermann (1978).

Varanus bengalensis; ?; cave deposit; Pleistocene; ? Cave, Karnool, central India; Prasad and Yadagiri (1986).

Varanus cf. *V. bengalensis;* jaws, vertebrae, and limb fragments; cave deposit; Pleistocene, Billa Surga Caves, Madras, S India; Lydekker (1886).

Varanus bolkayi de Fejervary, 1935; vertebrae; Trinil beds; Late Pleistocene; Trinil, Java, Indonesia.

Varanus darevskii Levshakova 1986; incomplete skull (including snout, partial skull roof, and partial palate); Magianskaia Svita; Early Pliocene; Magian Hollow, Tadjikistan.
Levshakova attributes this species to the subgenus *Psammosaurus*.

Varanus dirus De Vis, 1889; tooth; Darling Downs soil horizon; (probably Late) Pleistocene; Darling Downs, Queensland, E Australia.

Varanus "dirus"; maxilla; Chinchilla Sand; Early Pliocene; Chinchilla, Queensland, E Australia; De Vis (1900).

Varanus emeritus, De Vis, 1889; distal humerus, referred proximal tibia; Darling Downs soil horizon; (probably Late) Pleistocene; Darling Downs, Queensland, E Australia.

Varanus cf. *V. giganteus;* vertebrae; cave deposit?; Pleistocene, Wellington Valley, New South Wales, E Australia; Lydekker (1888).

Varanus "gouldii"; ?; cave deposit; Late Pleistocene; Victoria Fossil Cave, South Australia, Australia; Smith (1976).

Varanus cf. *V. "gouldii";* ?; cave deposit; Pleistocene; Henschke's Fossil Cave, South Australia, Australia; Pledge (1990).

Varanus hofmanni Roger, 1898; vertebrae, referred ?; Dinotheriensande; Late Miocene; Statzling, Germany.

Occurrences of *Varanus* cf. *V. hofmanni*:
1—vertebrae; Burdigalian; Middle Miocene; Artenay, France; all references from Hoffstetter (1969).
2—vertebrae; Helvetian; Late Miocene; La Grive-St. Alban, France.
3—vertebrae; Vindobonian; Late Miocene; Vieux-Collonges, France.
4—vertebra; Vallesian; Late Miocene; St. Miguel del Taudell, Catalonia, Spain.
  Items 1, 2, and 3 were considered as probably *V. hofmanni* by
  Hoffstetter (1969). They were suggested to be *V. hofmanni* from
  the shared possession of depressions on the dorsal face of laminae
  (of neural arch) anteriorly that are well developed, with an angulate
  bottom.

Varanus hooijeri Brongersma, 1958; parietal, referred maxilla and
prootic; cave deposit; (presumably Early) Holocene; Michael Cave,
Flores, Indonesia.
Additional material:
1—maxilla and dentary; cave deposit; (presumably Early) Holocene;
Toge Cave, Flores, Indonesia.

Varanus komodoensis:
1—two vertebrae; gravel deposits; Pleistocene; Atambua, Timor, Indonesia; Hooijer (1972).
2—teeth; Ola Bula Fm., Memb. A; Pleistocene; Tangi Talo, Flores; van
den Bergh et al. (2001).
3—tooth; Ola Bula Fm., Memb. B; Pleistocene; Mata Menge, Flores;
van den Bergh et al. (2001).

Varanus lungui Zerova and Chkhikvadze, 1986; trunk vertebra; middle
Sarmatian; Miocene; Bujor, Moldava.

Varanus marathonensis Weithofer, 1888 = *Varanus atticus* Nopcsa,
1908 = *Varanus deserticolus* Bolkay, 1913; vertebra, referred cranial bones; ?; Pliocene; Pikermi, Greece.
Additional material:
1—vertebrae; ?; Pliocene; Csarnota, Hungary; de Fejervary (1918).
2—dentary, vertebrae (holotype of *V. deserticolus*); ?; Pliocene; Bere-
mend, Hungary Bolkay (1913) and de Fejervary (1918).
3—vertebrae; ?; Pliocene; Calta, Turkey; Rage and Sen (1976).
4—dentary; ?; Late Pleistocene; Grotte Arene Candide, Italy; Morelli
(1891).
  Estes (1983) expressed doubts about the identification of this
  material.

Varanus priscus (Owen, 1859) = *Megalania prisca* Owen, 1859 = *Notiosaurus dentatus* Owen, 1884b = *Varanus warburtonensis* Zietz,
1889; cranial elements, teeth, cervical, dorsals, caudals, fragmentary shoulder girdle elements, humerus, ulna, ilia, pubes, ischium,
femora, tibia, fibulae, metatarsals (details given by Hecht 1975);
eastern Darling Downs soil horizons, Katipiri Fm., Wyandotte

Fm.; Early Pliocene? to Late Pleistocene; central and E Australia (Queensland, E New South Wales, E South Australia, Victoria). Most fossils of *V. priscus* have been found in unnamed soil horizons or alluvial sediments; only formally or informally named deposits are given above. Murray and Megirian (1992) reported, but did not describe or figure, giant varanid material from the Camfield Beds and Waite Fm. (Miocene) of the Northern Territory, Australia.

Varanus pronini Zerova and Chkhikvadze, 1986, trunk vertebra; ?; (Middle) Miocene; NW shore of Aral Sea, Kazakhstan.

Varanus rusingensis Clos, 1995, incomplete skeleton; Grit Memb., Hiwegi Fm.; Early Miocene; Rusinga Island, Kenya.

Varanus cf. *V. salvator;* two mandibles, vertebra, metatarsal; cave deposits; probably subrecent; Tarung-tung and Merasuen caves, Quezon Municipality, Palawan, Philippines; Reis and Garong (2001); Reis (personal communication, 2002).
The material is from two sites and thus represents at least two individuals.

Varanus semjonovi Zerova and Chkhikvadze, 1986; trunk vertebra; Meotian; Miocene; Odessa district, Ukraine.

Varanus sivalensis Falconer, 1868; incomplete humerus, two referred vertebrae; ?; Early Pliocene; Siwalik Hills, India.

Varanus tyrasiensis Lungu et al., 1983; two trunk vertebrae, with referred caudal; upper Middle Sarmatian; Miocene; Varnitza, Moldava.

Varanus varius; ?; cave deposit; Late Pleistocene; Victoria Fossil Cave, South Australia, Australia; Smith (1976).

Occurrences of *Varanus* sp.:
1—trunk vertebrae; ?; Middle Eocene; Tsagan-Hushu, Mongolia; Alifanov (1993).
2—vertebrae; ?; Middle Eocene; Tsagan-Hushu, Mongolia; Alifanov (1993, table 1).
3—?; ?; Early Oligocene; ?, Mongolia; Alifanov (1993).
    A large form (Alifanov 1993).
4—vertebrae; Etadunna Fm.; Late Oligocene to Early Miocene; Lake Ngapakaldi, South Australia, Australia; Estes (1984).
5—?; Camfield beds; Miocene; Bullock Creek, Northern Territory, Australia; Murray and Megirian (1992).
6—jaw fragments and vertebrae; "zeolitized tuffaceous sediments"; Early Miocene; Songhor, Kenya; Clos (1995).
    A small taxon or individual with blunt teeth. Clos (1995) allowed the possibility that these were immature *V. rusingensis.*
7—vertebrae; Helvetian; Late Miocene; La Grive-St. Alban, France; Estes (1983).
    This isn't cf. *hofmanni,* but a form with more elongate vertebrae

(Hoffstetter 1969). It was originally reported as *Necrosaurus* sp. by de Fejervary (1935).

8—?; cave deposit; Late Miocene–Early Pliocene; Corra Lynn Cave, South Australia, Australia; Pledge (1992).

9—vertebrae, humeri; ?; Pliocene; Layna, Spain; Sanz (1977).
Sanz regarded these as probably related to *V. marathonensis*.

10—?; ?; Early Pliocene; Tadjikistan; Zerova and Chkhikvadze (1986).

11—?; Allingham Fm.; Early Pliocene; upper Burdekin River, Queensland, Australia; Archer and Wade (1976).

12—dentary, teeth, vertebrae; Chinchilla Sand; Early Pliocene; Chinchilla, Queensland, E Australia; unpublished material in the Queensland Museum collections.

13—?; ?; Middle Pliocene; Moldava; Zerova and Chkhikvadze (1986).

14—?; ?; Late Pliocene; Turkmenia; Zerova and Chkhikvadze (1984).

15—vertebra; ?; Late Pliocene; Kotlovina, Odessa District, Ukraine; Zerova and Chkhikvadze (1986).

16—?; unconsolidated (and unnamed) riverine sediments; Late Pliocene–Pleistocene; upper Gregory River, Queensland, Australia; Rich et al. (1991).

17—sacrum; cave breccia; Pleistocene; Tea Tree Cave, Queensland, Australia; Molnar (1978).
This specimen was collected for the Queensland Museum but when last sighted was at the University of Adelaide in South Australia.

18—?; unnamed fluviatile sediments; Late? Pleistocene; Terrace Site, Riversleigh, Queensland, Australia; Archer et al. (1991).

19—?; ?; Pleistocene; Seton Rockshelter, Kangaroo Island, Australia; Meredith (1991).

20—?; cave deposit; Pleistocene, Cathedral Cave, Wellington Valley, New South Wales, E Australia; Dawson and Augee (1997).
Remains found at two levels, thus very likely representing at least two individuals.

21—vertebrae; soil horizon; Pleistocene; Darling Downs, Queensland, Australia; Wilkinson (1995).
Related to *V. gouldii* "group."

22—?; rock-shelter deposits; Late Pleistocene; Baturong Massif, Sabah, Borneo; Bellwood (1985).

23—cranial elements, vertebrae; cave deposits; Late Pleistocene or Holocene; Ayamaru Lakes region, Irian Jaya; Aplin et al. (1999). "Most of the remains are from quite large monitors" (Aplin et al. 1999).

24—vertebra; ?; Holocene; Harappa, India; Prashard (1936).

25—?; cave deposits; Holocene; Gua Lawa, Java; Bellwood (1985).

26—?; Holocene; Gua Sireh, Sarawek, Borneo; Datan (1993).
Cited as "monitor lizards and/or snakes" (Datan 1993).

27—?; cave deposits; ?; Niah Cave, Sarawak, Borneo; Bellwood (1985).

28—maxilla, parietal, dentary, postdentary bones, axis, dorsal, sacral, scapulocoracoid, humerus, pelvis, and other pieces; cave deposit; Holocene; Pindai, New Caledonia; Balouet (1991).
This, and the following material, may represent a new species.

24—vertebra; cave deposit; Holocene; Gilles, New Caledonia; Balouet (1991).

Acknowledgments

In various ways, Tracy L. Ford, David Gillette, Gao Keqin, Mike Lee, Jim I. Mead, Eric Pianka, Kelley Reis, and Joanne E. Wilkinson assisted with the composition of this chapter and with furthering my understanding of monitor lizards.

References

Alifanov, V. A. 1993. Some peculiarities of the Cretaceous and Palaeogene lizard faunas of the Mongolian People's Republic. In *Monument Grube Messel—Perspectives and relationships, pt. 2*, ed. F. Schrenk and K. Ernst. *Kaupia* 3:9–13.

———. 2000. The fossil record of Cretaceous lizards from Mongolia. In *The age of dinosaurs in Russia and Mongolia*, ed. M. J. Benton, M. A. Shishkin, D. M. Unwin, and E. N. Kurochkin, 368–389. Cambridge: Cambridge University Press.

Aplin, K. P., J. M. Pasveer, and W. E. Boles. 1999. Late Quarternary vertebrates from Bird's Head Peninsula, Irian Jaya, Indonesia, including descriptions of two previously unknown marsupial species. *Rec. W. Austr. Mus. Suppl.* 57:351–387.

Arambourg, C. 1952. Les vertébrés fossiles des gisements de phosphates (Maroc-Algerie-Tunisie). *Notes Mém. Serv. Géol. Maroc* 92:1–372.

Archer, M., and M. Wade. 1976. Results of the Ray E. Lemley Expeditions, part 1: The Allingham Formation and a new Pliocene vertebrate fauna from northern Queensland. *Mem. Queensl. Mus.* 17:379–397.

Archer, M., S. J. Hand, and H. Godthelp. 1991. *Riversleigh*. Sydney: Reed Books.

Ast, J. C. 2001. Mitochondrial DNA evidence and evolution in Varanoidea (Squamata). *Cladistics* 17:211–226.

Auffenberg, W. 1980. The herpetofauna of Komodo, with notes on adjacent areas. *Bull. Florida State Mus. Biol. Sci.* 25:39–156.

———. 1981. *The behavioral ecology of the Komodo monitor.* Gainesville: University Presses of Florida.

Augé, M. 1990. La faune de Lézards et d'Amphisbaenes de l'Eocène inférieur de Condé-en-Brie (France). *Bull. Mus. Nat. Hist. Nat. Paris* 4, 12, C, 2:111–141.

———. 1995. Un helodermatidé (Reptilia, Lacertilia) dans l'Eocène inférieur de Dormaal, Belgique. *Bull. Inst. R. Sci. Nat. Belgique Sci. Terre* 65:277–281.

Averianov, A. O. 2001. The first find of a dolichosaur (Squamata, Dolichosauridae) in central Asia. *Paleontol. Zhurnal* 35:525–527

Averianov, A. O., and I. G. Danilov. 1997. A varanid lizard (Squamata: Varanidae) from the early Eocene of Kirghizia. *Russ. J. Herpetol.* 4:143–147.

Balouet, J. C. 1991. The fossil vertebrate record of New Caledonia. In *Vertebrate palaeontology of Australasia,* ed. P.-V. Rich, J. M. Monaghan, R. F. Baird, T. H. Rich, E. M. Thompson, and C. Williams, 1383–1409. Melbourne: Pioneer Design Studio and Monash University Publications Committee.

Bartels, W. S. 1983. A transitional Paleocene-Eocene reptile fauna from the Bighorn basin, Wyoming. *Herpetologica* 39:359–374.

Bell, B. A., P. A. Murry, and L. W. Osten. 1982. *Coniasaurus* Owen, 1850 from North America. *J. Paleontol.* 56:520–524.

Bellwood, P. 1985. *Prehistory of the Indo-Malaysian archipelago.* London: Academic Press.

Bennett, D. 1998. *Monitor lizards: Natural history, biology and husbandry.* Frankfurt am Main: Edition Chimaira.

Bolkay, S. 1913. Additions to the fossil herpetology of Hungary from the Pannonian and Praeglacial periode. *Mitt. Jahrbüch. Königl. Ungar. Reichsansalt* 21:217–230.

Borsuk-Białynicka, M. 1984. Anguimorphans and related lizards from the Late Cretaceous of the Gobi Desert, Mongolia. Results of the Polish Mongolian Palenotological Expeditions Part 10. *Palaeontol. Polonica* 46:5–105.

Brattstrom, B. H. 1955. New snakes and lizards from the Eocene of California. *J. Paleontol.* 29:145–149.

———. 1958. New records of Cenozoic amphibians and reptiles from California. *Bull. South. Calif. Acad. Sci.* 57:5–12.

Brongersma, L. 1958. On an extinct species of the genus *Varanus* (Reptilia, Sauria) from the island of Flores. *Zool. Mededelingen Rijksmus. Nat. Hist. Leiden* 36:113–125.

Brown, B. 1931. The largest known land tortoise. *Nat. Hist.* 31:183–187.

Caldwell, M. W. 1999a. Description and phylogenetic relationships of a new species of *Coniasaurus* Owen, 1850 (Squamata). *J. Vertebr. Paleontol.* 19:438–455.

———. 1999b. Squamate phylogeny and the relationships of snakes and mosasauroids. *Zool. J. Linnean Soc.* 125:115–147.

Caldwell, M. W., R. L. Carroll, and H. Kaiser. 1995. The pectoral girdle and forelimb of *Carsosaurus marchesetti* (Aigialosauridae), with a preliminary phylogenetic analysis of mosasaurids and varanoids. *J. Vertebr. Paleontol.* 15:516–553.

Caldwell, M. W., and M. S. Y. Lee. 1997. A snake with legs from the marine Cretaceous of the Middle East. *Nature* 386:705–709.

Caldwell, M. W., and J. A. Cooper. 1999. Redescription, palaeobiogeography and palaeoecology of *Coniasaurus crassidens* Owen, 1850 (Squamata) from the Lower Chalk (Cretaceous; Cenomanian) of SE England. *Zool. J. Linnean Soc.* 127:423–452.

Carroll, R. L., and M. Debraga. 1992. Aigialosaurs: Mid-Cretaceous varanoid lizards. *J. Vertebr. Paleontol.* 12:66–86.

Cifelli, R. L., and R. L. Nydam. 1995. Primitive helodermatid-like platynotan from the Early Cretaceous of Utah. *Herpetologia* 51:286–291.

Clos, L. M. 1995. A new species of *Varanus* (Reptilia: Sauria) from the Miocene of Kenya. *J. Vertebr. Paleontol.* 15:254–267.

Dal Sasso, C., and G. Pinna. 1997. *Aphanizocnemus libanensis* n. gen. n. sp., a new dolichosaur (Reptilia, Varanoidea) from the Upper Cretaceous of Lebanon. *Paleontol. Lombarda* 7:1–31.

Datan, I. 1993. Archaeological excavations at Gua Sireh (Serian) and Luvbang Angin (Gunung Mulu National Park), Sarawak, Malaysia. *Sarawak Mus. J.* 45:1–192.

Dawson, L., and M. L. Augee. 1997. The Late Quaternary sediments and fossil vertebrate fauna from Cathedral Cave, Wellington Caves, New South Wales. *Proc. Linnean Soc. New South Wales* 117:51–78.

de Fejervary, G. J. 1918. Contributions to a monography on fossil Varanidae and on Megalanidae. *Ann. Mus. Nat. Hungarici*, pars zool., 16:341–467.

———. 1935. Further contributions to a monograph of the Megalanidae and fossil Varanidae—With notes on recent varanians. *Ann. Mus. Nat. Hungarici*, pars zool., 29:1–130.

de Rochebrune, A. 1884. Faune ophiologique des phosphorites du Quercy. *Mem. Soc. Sci. Nat. Sâone Loire* 5:149–164.

de Stefano, G. 1903. I sauri del Quercy appartenenti alla collezione Rossignol. *Atti Soc. Ital. Sci. Nat. Mus. Civico* 42:414–415.

De Vis, C. W. 1885. On bones and teeth of a large extinct lizard. *Proc. R. Soc. Queensl.* 2:25–32.

———. 1889. On *Megalania* and its allies. *Proc. R. Soc. Queensl.* 6:93–99.

———. 1900. A further trace of an extinct lizard. *Ann. Queensl. Mus.* 5:6.

Diamond, J. M. 1987. Did Komodo dragons evolve to eat pygmy elephants? *Nature* 326:832.

Dodson, J., R. Fullagar, J. Furby, R. Jones, and I. Prosser. 1993. Human and megafauna in a Late Pleistocene environment from Cuddie Springs, northwestern New South Wales. *Archaeol. Oceania* 28:94–99.

Dollo, L. 1923. *Saniwa orsmaelensis,* varanidé nouveau de Landénien supérieur d'Orsmael (Brabant). *Bull. Soc. Belge Géol. Paléontol. Hydrol.* 33:76–82.

Dunn, E. R. 1927. Results of the Douglas Burden Expedition to the Island of Komodo. I.—Notes on *Varanus komodoensis. Am. Mus. Novitates* 286:1–10.

Estes, R. 1964. Fossil vertebrates from the Late Cretaceous Lance Formation, eastern Wyoming. *Univ. Calif. Publ. Geol. Sci.* 49:1–180.

———. 1983. Sauria Terrestria, Amphisbaenia. In *Handbuch der Paläoherpetologie,* ed. P. Wellnhofer, 10A:179–189. Stuttgart: Gustav Fischer Verlag.

———. 1984. Fish, amphibians and reptiles from the Etadunna Formation, Miocene of South Australia. *Austr. Zool.* 21:335–343.

Estes, R., and J. H. Hutchison. 1980. Eocene lower vertebrates from Ellesmere Island, Canadian Artic Archipelago. *Palaeogeogr. Palaeoclimatol. Palaeoecol.* 30:325–347.

Evans, S. E. 1994. A new anguimorph lizard from the Jurassic and Lower Cretaceous of England. *Palaeontology* 37:33–49.

———. 1996. *Parviraptor* (Squamata: Anguimorpha) and other lizards from the Morrison Formation at Fruita, Colorado. *Mus. Northern Arizona Bull.* 60:243–248.

Evans, S. E., G. V. R. Prasad, and B. K. Manhas. 2002. Fossil lizards from the Jurassic Kota Formation of India. *J. Vertebr. Paleontol.* 22:299–312.

Falconer, H. 1868. *Palaeontological memoirs and notes of the late Hugh Falconer, A.M., M.D.* London: R. Hardwicke.

Fenton, C. L., and M. A. Fenton. 1958. *The fossil book.* Garden City, N.Y.: Doubleday.

Filhol, H. 1873. Sur les vertébrés fossiles trouvés dans les dépots de phosphate de chaux de Quercy. *Bulletin, société philomathique de Paris* 10:85–89.

———. 1877. Recherches sur les phosphorites du Quercy. Pt. II. *Ann. Sci. Géol.* 8:1–338.

Fraser, N. C. 1997. Genesis of snakes in exodus from the sea. *Nature* 386:651–652.

Fuller, S., P. Baverstock, and D. King. 1998. Biogeographic origins of goannas (Varanidae): A molecular perspective. *Mol. Phylogenet. Evol.* 9:294–307.

Furby, J. H., R. Fullagar, J. R. Dodson, and I. Prosser. 1993. The Cuddie Springs bone bed revisited. In *Sahul in review*, ed. M. A. Smith, M. Spriggs, and B. Fankhauser, 204–210. Canberra: Research School in Pacific Studies, Australian National University.

Gao, K., and L. Hou. 1996. Systematics and taxonomic diversity of squamates from the Upper Cretaceous Djadochta Formation, Bayan Mandahu, Gobi Desert, People's Republic of China. *Can. J. Earth Sci.* 33:578–598.

Gao, K., and M. A. Norell. 1998. Taxonomic revision of *Carusia* (Reptilia: Squamata) from the Late Cretaceous of the Gobi Desert and phylogenetic relationships of anguimorphan lizards. *Am. Mus. Novitates* 3230:1–51.

———. 2000. Taxonomic composition and systematics of Late Cretaceous lizard assemblages from Ukhaa Tolgod and adjacent localities, Mongolian Gobi Desert. *Bull. Am. Mus. Nat. Hist.* 249:1–118.

Gilmore, C. W. 1928. Fossil lizards of North America. *Mem. Natl. Acad. Sci.* 22:1–201.

———. 1942. Paleocene faunas of the Polecat Bench Formation, Park County, Wyoming. Part II. Lizards. *Proc. Am. Phil. Soc.* 85:159–167.

———. 1943. Fossil lizards of Mongolia. *Bull. Am. Mus. Nat. Hist.* 81:361–384.

Godinot, M., F. de Broin, E. Buffetaut, J.-C. Rage, and D. Russell. 1978. Dormaal: Une des plus anciennes faunes éocenès d'Europe. *Compt. Rendus Acad. Sci. Paris* 287:1273–1276.

Harris, A. H. 1993. Quaternary vertebrates of New Mexico. *New Mexico Mus. Nat. Hist. Sci. Bull.* 2:179–197.

Haubold, H. 1979. Zur Kenntnis der Sauria (Lacertilia) aus dem Eozän des Geiseltals. In *Eozäne Wirbeltiere des Geiseltales,* ed. H. Matthes and B. Thales, 107–112. Wissenschaftliche Beitäge, Martin-Luther Universität.

Hecht, M. 1959. Amphibians and reptiles. *Am. Mus. Nat. Hist. Bull.* 117:130–146.

———. 1975. The morphology and relationships of the largest known terrestrial lizard, *Megalania prisca* Owen, from the Pleistocene of Australia. *Proc. R. Soc. Vic.* 87:239–250.

Hecht, M., and R. Hoffstetter. 1962. Note preliminaire sur les amphibiens et les squamates du Landenien superieur et du Tongrien de Belgique. *Inst. R. Sci. Nat. Belgique Bull.* 38(39):1–30.

Hoffstetter, R. 1943. Varanidae et Necrosauridae fossiles. *Bull. Mus. Nat. Hist. Nat. Paris* (2nd ser.) 15:134–141.

———. 1957. Un Saurien hélodermité (*Eurheloderma gallicum* nov. gen. et sp.) dans la faune fossile des phosphorites du Quercy. *Bull Soc. Geol. France* 7:775–786.

———. 1969. Présence de Varanidae (Reptilia, Sauria) dans le Miocène de Catalogne. Considerations sur l'histoire de la famille. *Bull. Mus. Nat. Hist. Nat.* 40:1051–1064.

Hooijer, D. 1972. Varanus (Reptilia, Sauria) from the Pleistocene of Timor. *Zool. Mededelingen* 47:445–448.

Hünermann, K. A. 1978. Ein varanoider Lacertilier (Reptilia, Squamata) aus einer alttertiären Spaltenfüllung von Dielsdorf (Kt. Zürich). *Ecol. Geol. Helvetiae* 71:769–774.

Kordikova, E. G., P. D. Polly, V. A. Alifanov, Z. Rocek, G. F. Gunnell and

A. O. Averianov. 2001. Small vertebrates from the Late Cretaceous and Early Tertiary of the northeastern Aral Sea region, Kazakhstan. *J. Paleontol.* 75:390–400.

Krumbiegel, G., L. Rüffle, and H. Haubold. 1983. *Das Eozäne Geiseltal.* Wittenberg Lutherstadt: A Ziemsen.

Kues, B. S. 1993. Bibliographic catalogue of New Mexico vertebrate fossils. *New Mexico Mus. Nat. Hist. Sci. Bull.* 2:199–279.

Kuhn, O. 1940a. Die Placosauriden und Anguiden aus dem mittleren Eozän des Geiseltales. *Nova Acta Leopoldina,* n.f., 8:461–486.

———. 1940b. Crocodilier und Squamatenreste aus dem oberen Palaocän von Wahlbeck. *Zentral. Mineral. Geol. Paläontol.* B 1940:21–25.

———. 1958. Ein neuer Lacertilier aus dem fränkischen Lithographie-schiefer. *Neues Jahrbuch für Geologie und Paläontologie, Monatshefte* 1958:380–382.

Lee, M. S. Y. 1996. Possible affinities between *Varanus giganteus* and *Megalania prisca. Mem. Queensl. Mus.* 39:232.

———. 1997. The phylogeny of varanoid lizards and the affinities of snakes. *Phil. Trans. R. Soc. Lond.* B 352:53–91.

Lee, M. S. Y., and M. W. Caldwell. 1997. Anatomy and relationships of *Pachyrhachis problematicus,* a primitive snake with hindlimbs. *Phil. Trans. R. Soc. Lond.* B 353:1521–1552.

———. 2000. *Adriosaurus* and the affinities of mosasaurs, dolichosaurs, and snakes. *J. Paleontol.* 74:915–937.

Leidy, J. 1870. [Untitled.] *Proc. Acad. Nat. Sci. Phil.* 1872: 277.

———. 1873. Contributions to the extinct vertebrate fauna of the western territories. *Report of the United States Geological Survey of the Territories* 1:14–358.

Levshakova, I. U. 1986. Novii varan iz Nizhnego Pliotsena Tadzhikistana. *Trudi Zoologicheskogo Instituta AN SSSR* 157: 101–106.

Lingham-Soliar, T. 1998. Unusual death of a Cretaceous giant. *Lethaia* 31:308–310.

———. 1991. Predation in mosasaurs—A functional approach. In *Natural structures: Principles, strategies, and models in architecture and nature: Proceedings of the Second International Symposium of the Sonderforschungsbereich* 230, pt. 1, 168–177. Stuttgart: Tübingen.

Lingham-Soliar, T., and D. Nolf. 1989. The mosasaur *Prognathodon* (Reptilia, Mosasauridae) from the Upper Cretaceous of Belgium. *Bull. Inst. R. Sci. Nat. Belgique Sci. Terre* 59:137–190.

Lungu, A. N., G. A. Zerova, and V. M. Chkhikvadze. 1983. Pervie svedeniia o miotsenovom varane severnogo prichernomoriia. *Sovshcheniia Akademii Nauk Gruzinskoi SSR* 110:417–420.

Lydekker, R. 1886. Fauna of the Karnul caves. *Palaeontol. Indica* 10(4): 1–58.

———. 1888. *Catalogue of the fossil Reptilia and Amphibia in the British Museum (Natural History), Cromwell Road, S.W. Pt. 1. The Orders Ornithosauria, Crocodilia, Dinosauria, Squamata, Rhychocephalia, and Proterosauria.* London: The Trustees.

Marsh, O. C. 1872. Preliminary description of new Tertiary reptiles: Parts I and II. *Am. J. Sci.* 4:298–309.

McDowell, S., and C. Bogert. 1954. The systematic position of *Lanthanotus* and the affinities of the anguimorphan lizards. *Bull. Am. Mus. Nat. Hist.* 105:1–142.

McKenna, M. 1960. Fossil mammalia from the Early Wasatchian Four Mile Fauna, Eocene of northwest Colorado. *Univ. Calif. Publ. Geol.* 37:1–130.

McNab, B. K., and W. Auffenberg 1976. The effect of large body size on the temperature regulation of the Komodo dragon, *Varanus komodoensis*. *Comp. Biochem. Physiol.* 55A:345–350.

Meredith, C. 1991. Vertebrate fossil faunas from islands in Australasia and the southwest Pacific. In *Vertebrate palaeontology of Australasia,* ed. P.-V. Rich, J. M. Monaghan, R. F. Baird, T. H. Rich, E. M. Thompson, and C. Williams, 1345–1382. Melbourne: Pioneer Design Studio and Monash University Publications Committee.

Miller, G. H., J. W. Magee, B. J. Johnson, M. L. Fogel, N. A. Spooner, M. T. McCulloch, and L. K. Ayliffe. 1999. Pleistocene extinction of *Genyornis newtoni*: Human impact on Australian megafauna. *Science* 283:205–208.

Molnar, R. E. 1978. Age of the Chillagoe crocodile. *Search* 9:156–158.

———. 1990. New cranial elements of a giant varanid from Queensland. *Mem. Queensl. Mus.* 29:437–444.

———. 1991. But what about the mosasaurs? *Aust. Nat. Hist.* 23(10):755.

Morelli, N. 1891. Resti organici rinvenuti nella caverna delle Arene Candide. *Atti Soc. Ligure Sci. Nat. Geogr.* 2:171–175.

Morwood, M. J., F. Aziz, P. O'Sullivan, Nasruddin, D. R. Hobbs, and A. Raza. 1999. Archaeological and palaeontological research in central Flores, east Indonesia: Results of fieldwork 1997–98. *Antiquity* 763:273–286.

Murray, P. F., and D. Megirian. 1992. Continuity and contrast in middle and late Miocene vertebrate communities from the Northern Territory. *Beagle* 9:195–217.

Nessov, L. A. 1981. Nakhodka cheloosti nazemnoi iashcheritsi v Verkhnem Melu Uzbekistana. *Vestnik Leningradskogo Univ.* 2:105–107.

———. 1997. *Nemorskie Pozvonochnie Melovogo Perioda Severnoi Evrazii* [Cretaceous nonmarine vertebrates of northern Eurasia]. St. Petersburg: Institute of the Earth's Crust.

Nessov, L. A., V. I. Zhegallo, and A. O. Averianov. 1998. A new locality of Late Cretaceous snakes, mammals and other vertebrates in Africa (western Libya). *Ann. Paleontol.* 84:265–274.

Nopcsa, F. 1908. Zur Kenntnis der Fossilen Eidechsen. *Beit. Paläontol. Geol. Österreichische-Ungarns Orients* 21:33–62.

———. 1925. Ergebnisse der Forschungsreisen Prof. E. Stromers in den Wüsten Ägyptens. II. Wirbeltier-Reste der Baharije-Stufe (unterstes Cenoman). 5. *Die Symoliophis-Reste. Abhandlungen der Bayerischen Akademie der Wissenschaften, Mathematisch-naturwissenshaftliche Abteilung,* 30 Bd, 4 Abh., 1–27.

Norell, M. A., M. C. McKenna, and M. J. Novacek. 1992. *Estesia mongoliensis,* a new fossil varanoid from the Late Cretaceous Barun Goyot Formation of Mongolia. *Am. Mus. Novitates* 3045:1–24.

Norell, M. A., and K. Gao. 1997. Braincase and phylogenetic relationships of *Estesia mongoliensis* from the Upper Cretaceous, Gobi desert and the recognition of a new clade of lizards. *Am. Mus. Novitates* 3211:1–25.

Nydam, R. L. 2000. A new taxon of helodermatid-like lizard from the Albian-Cenomanian of Utah. *J. Vertebr. Paleontol.* 20:285–294.

Owen, R. 1854. On some fossil reptilian and mammalian remains from the Purbecks. *Q. J. Geol. Soc. Lond.* 10: 420–422.

———. 1859. Description of some remains of a gigantic land-lizard (*Megalania prisca*, Owen) from Australia. *Phil. Trans. R. Soc. Lond.* 149:43–48.

———. 1884a. *A history of British fossil reptiles.* 4 vols. London: Cassell.

————. 1884b. Evidence of a large extinct lizard (*Notiosaurus dentatus*) from Pleistocene deposits, New South Wales, Australia. *Phil. Trans. R. Soc. Lon.* 175: 249–251.

Pledge, N. S. 1990. The upper fossil fauna of the Henschke Fossil Cave, Naracoorte, South Australia. *Mem. Queensl. Mus.* 28:247–262.

————. 1992. The Curramulka local fauna: A new Late Tertiary fossil assemblage from Yorke Peninsula, South Australia. *Beagle* 9:115–142.

Prasad, K. N., and P. Yadagiri. 1986. Pleistocene cave fauna, Karnool District, Andhra Pradesh. *Rec. Geol. Surv. India* 115:71–77.

Prashard, B. 1936. Animal remains from Harappa. *Mem. Archaeol. Survey India* 51:1–62.

Pregill, G. K., J. A. Gauthier, and H. W. Greene. 1986. The evolution of helodermatid squamates, with description of a new taxon and an overview of Varanoidea. *Trans. San Diego County Soc. Natl. Hist.* 21:167–202.

Rage, J.-C. 1983. *Palaeophis colossaeus* nov. sp. (le plus grand Serpent connu?) de l'Eocene du Mali et la probleme du genre chez les Palaeopheinae. *Compt. Rendus Acad. Sci. Paris* (Ser. 2) 296:1741–1744.

————. 1988. Le gisement du Bretou (Phosphorites du Quercy, Tarn-et-Garonne, France) et sa faune de vertébrés de l'Eocene supérieur. I. Amphibiens et reptiles. *Palaeontographica A* 205:3–27.

————. 1989. Le plus ancien lézard varanoïde de France. *Bull. Soc. Études Sci. Anjou* 13:19–26.

————. 1993. Squamates from the Cainozoic of the western part of Europe: A review. *Rev. Paléobiol.* 7:199–216.

Rage, J.-C., and S. Sen. 1976. Les amphibiens et les reptiles du Pliocène supérieur de Çalta (Turquie). *Geol. Mediter.* 3:127–134.

Rage, J.-C., and R. L. E. Ford. 1980. Amphibians and squamates from the Upper Eocene of the Isle of Wight. *Tertiary Res.* 3:47–60.

Rage, J.-C., and A. Richter. 1994. A snake from the Lower Cretaceous (Barremian) of Spain: The oldest known snake. *Neues Jahrbuch Geol. Palaeont. Monatshefte* 1994:561–565.

Reis, K. R., and A. M. Garong. 2001. Late Quaternary terrestrial vertebrates from Palawan Island, Philippines. *Palaeogeogr. Palaeoclimatol. Palaeoecol.* 171:409–412.

Rich, T. H., M. Archer, S. Hand, H. Godthelp, J. Muirhead, N. S. Pledge, E. L. Lundelius, T. F. Flannery, L. S. V. Rich, M. O. Woodburne, J. A. Case, M. J. Whitelaw, R. H. Tedford, A. Kemp, W. D. Turnbull, and P. V. Rich. 1991. Australian Mesozoic and Tertiary terrestrial mammal localities. In *Vertebrate palaeontology of Australasia,* eds. P. V.-Rich, J. M. Monaghan, R. F. Baird, T. H. Rich, E. M. Thompson, and C. Williams, 1005–1058. Melbourne: Pioneer Design Studio and Monash University Publications Committee.

Roger, O. 1898. Wirbelthierreste aus dem Dinotheriensande, II. Thiel. *Bericht Nat. Vereins Schwaben Neuburg Augsburg* 34: 53–70.

Rothschild, B. M., and L. D. Martin. 1993. *Paleopathology.* Boca Raton, Fla.: CRC Press.

Russell, D. A. 1967. Systematics and morphology of American mosasaurs. *Peabody Mus. Nat. Hist. Bull.* 23:1–241.

Sanz, J. 1977. Presencia de *Varanus* (Sauria, Reptilia) en el Plioceno de Layna (Soria). *Trabajos N/Q* 8:113–125.

Scanlon, J. D., M. S. Y. Lee, M. W. Caldwell, and R. Shine. 1999. The palaeoecology of the primitive snake *Pachyrhachis*. *Hist. Biol.* 13: 127–152.

Schatzinger, R. 1975. Late Eocene (Uintan) lizards from the San Diego area, southern California. Presented at the Paleontological Society, Pacific Coast Section, Annual Meeting, Los Angeles.

Smith, K. T., and J. Gauthier. 2002. Squamate diversity across the Eocene/Oligocene boundary: New data and the applicability of phylogeny (abstract). *J. Vertebr. Paleontol.* 22 (suppl. to 3):108A.

Smith, M. J. 1976. Small fossil vertebrates from Victoria Cave, Naracoorte, South Australia. IV. Reptiles. *Trans. R. Soc. South Aust.* 100:39–51.

Sondaar, P. Y., G. D. van der Bergh, B. Mubroto, F. Aziz, J. de Vos, and U. L. Batu. 1994. Middle Pleistocene faunal turnover and colonization of Flores (Indonesia) by *Homo erectus. Compt. Rendus Acad. Sci. Paris* (Ser. 2) 319:1255–1262.

Stevens, M. S. 1977. Further study of the Castolon local fauna (Arikareean: Early Miocene) Big Bend National Park, Texas. *Pearce-Sellards Series 28.* Texas Memorial Museum.

Stritzke, R. 1983. *Saniwa feisti* n. sp., ein Varanide aus dem Mittel-Eocene von Messel bei Darmstadt. *Senck. Leth.* 64:497–508.

Sullivan, R. 1982. Fossil lizards from Swain Quarry, "Fort Union Formation," Middle Paleocene (Torrejonian), Carbon County, Wyoming. *J. Paleontol.* 56:996–1010.

Tchernov, E., O. Rieppel, H. Zaher, M. J. Polcyn, and L. L. Jacobs. 2000. A fossil snake with limbs. *Science* 287:2010–2012.

Tedford, R. H., and R. T. Wells. 1990. Pleistocene deposits and fossil vertebrates from the "Dead Heart of Australia." *Mem. Queensland Mus.* 28:263–284.

Telles Antunes, M., and J.-C. Rage. 1974. Notes sur la géologie et la paléontologie du Miocène de Lisbonne. XIV—Quelques Squamata (Reptilia). 19:47–60. *Boletim: Sociedad Geologica de Portugal.*

van den Bergh, G., J. de Vos, and P. Y. Sondaar. 2001. The Late Quaternary palaeogeography of mammal evolution in the Indonesian Archipelago. *Palaeogeogr. Palaeoclimatol. Palaeoecol.* 171:385–408.

Watanabe, N., and D. Kadar. 1985. *Quaternary geology of the hominid fossil bearing formations in Java.* Special Publication 4. Bandung: Geological Research and Development Centre.

Weithofer, K. 1888. Beiträge zur Kenntnis der Fauna von Pikermi. *Beit. Paläontol. Osterreichische-Ungarns Orients* 6:225–292.

White, T. 1952. Preliminary analysis of the vertebrate fossil fauna of the Boysen Reservoir area. *Proc. U.S. Natl. Mus.* 102:185–207.

Wilkinson, J. E. 1995. Fossil record of a varanid from the Darling Downs, southeastern Queensland. *Mem. Queensland Mus.* 38:92.

Yadagiri, P. 1986. Lower Jurassic lower vertebrates from Kota Formation, Pranhita-Godavari Valley, India. *J. Palaeontol. Soc. India* 31:89–96.

Zerova, G. A., and V. M. Chkhikvadze. 1984. Obzor Kainozoiskikh iashcherits i zmei SSSR. *Izvestiia Akademia Nauk GSSR* 10:319–326.

———. 1986. Neogene varanids of the USSR. In *Studies in herpetology,* ed. Z. Rocek, 689–694. Prague: Charles University.

Zietz, A. 1889. Notes upon some fossil reptilian remains from the Warburton River, near Lake Eyre. *Trans. Proc. Rep. R. Soc. South Australia* 23:208–210.

3. Biogeography and Phylogeny of Varanoids

RALPH E. MOLNAR AND ERIC R. PIANKA

The lizard clade Varanoidea includes three extant families, each with but a single living genus: North American Gila monsters and beaded lizards (family Helodermatidae, genus *Heloderma*), the earless monitor of Borneo (family Varanidae, genus *Lanthanotus*), and monitor lizards (family Varanidae, genus *Varanus*), as well as a host of extinct forms including mosasaurs and probably snakes (see below). Exact affinities within this large clade have yet to be fully resolved, although many have been hypothesized (McDowell and Bogert 1954; Estes et al. 1988; Norell et al. 1992; Clos 1995; Lee 1997; Norell and Gao 1997; Gao and Norell 1998, 2000; Fuller et al. 1998; Pepin 1999; Ast 2001). It is generally agreed that the living varanoids fall into two clades (Pregill et al. 1986; Norell and Gao 1997; Lee 1997; Lee and Caldwell 2000), the Monstersauria, which includes *Heloderma,* and the Varanidae, which includes *Varanus* and *Lanthanotus.*

Varanid ancestors are probably unknown. The earliest fossils attributed to varanoids are the very fragmentary remains of *Paikasaisaurus* from India (Yadagiri 1986). Although the jaw pieces are reminiscent of those of varanoids, the resemblance seems to be superficial (Evans et al. 2002). *Proaigialosaurus,* from the Late Jurassic of Germany (Kuhn 1958), seems related to the marine varanoids, but the type and only material, a skull, is lost. The most convincing potential ancestor, or sister group, is *Parviraptor* from the Middle and late Jurassic of western Europe and North America (Evans 1994, 1996). Although lacking the derived features of varanoids, *Parviraptor* was a predatory lizard, with similar trophic adaptations to the later terrestrial varanoids.

The oldest varanoid fossils are all from Laurasian deposits. Oddly, the oldest, from the Early Cretaceous of Europe and Asia (~95 my BP), all pertain to marine forms. These are thought to be related to the origins of mosasaurs and snakes (Lee 1997; Lee and Caldwell 2000). But shortly after the appearance of these marine varanoids, evidence of land-dwelling forms is seen in the fossil record. Fossil terrestrial varanoids appear in the Late Cretaceous (~70–80 my BP) and include varanid relatives such as *Saniwides* and *Telmasaurus* (Mongolia), the monstersaur *Estesia mongoliensis* (Norell et al. 1992), and the plesiomorphic *Palaeosaniwa* (North America). These include the oldest well-preserved specimens, those from Mongolia with skulls and jaws in varying states of completeness (Borsuk-Białynicka 1984). Fragmentary fossils from central Asia (Nessov 1981; Kordikova et al. 2001) and Europe (Rage 1989) indicate that land-dwelling varanoids did exist earlier, about 90 my BP. More advanced varanoids are seen in the Cenozoic, with the Eocene to Oligocene *Saniwa* (North America and Europe) and *Necrosaurus* (Europe).

Estes (1983) suggested that varanoids arose during the Cretaceous in eastern Asia, then split into two groups, one of which crossed over into North America via Beringia, whereas the other lineage dispersed throughout Southeast Asia as well as west to colonize south-central Asia, Europe, and Africa. As far as can be seen from a less complete fossil record, monstersaurs did more or less the same, originating in Asia and presumably dispersing east into North America and west into Europe. Varanoids ultimately became extinct in North America during or after the Late Oligocene. Monstersaurs show the complementary pattern, becoming extinct in Eurasia at about the same time—again, so far as we can tell from the fossil record. Fossil varanoid vertebrae from Ellesmere Island in the Canadian arctic (Estes and Hutchinson 1980) show how far north this group reached during a peak period of tropical climate in the Eocene. In the Southern Hemisphere, even today monitors extend south to the Cape of Good Hope in Africa, to Sri Lanka in southern Asia, and to the Great Australian Bight (Southern Ocean) in Australia. Varanids, at their widest distribution, lived on all continents, except South America and Antarctica. Fossil helodermatids are known from Wyoming, Nebraska, New Mexico, Colorado, and France (Estes 1983), and hence presumably once lived in the regions between.

Two competing hypotheses for the origin of varanid lizards are as follows: (1) they arose in Laurasia and subsequently dispersed to Africa and Australia; or (2) they arose in Gondwana and subsequently spread to other regions. The first hypothesis is most strongly supported by the fossil record and phylogenetic evidence, although the second cannot be completely ruled out because varanids are currently most diverse in Australia (Estes 1983; King et al. 1999; Pepin 1999). Yet if varanids did originate in the Australian region, they must have dispersed from that region to the other end of Gondwana, across the Sea of Tethys to Laurasia, and then across Laurasia to what is now Mongolia by the mid–Late Cretaceous. Since the oldest varanoid fossils, of the aquatic forms, are found in Asia, Europe, and North America (along the shores of the Sea of Tethys) and—so far—are conspicuously absent from the Gondwanan continents, this implies that terrestrial varanids would have had to have migrated all the way across Gondwana to what is now Australia, and then all the way back again. This is surely not impossible, but then again, it is not as simple as the hypothesis that they arose in Laurasia. In their dispersal, they would have crossed what is now South America, before it became an island continent in the Tertiary, and so raise the question of why, with their successes elsewhere, monitors are not found in South America. From current evidence, the hypothesis of Laurasian origin seems the more parsimonious.

Most students of *Varanus* believe that this genus arose in Eurasia, where descendants of basal ancestral stocks still occur today. However, the oldest fossil nominate species of *Varanus, V. rusingensis,* had reached Kenya, Africa, by the Early Miocene, 17.8 my BP or more (Clos 1995), suggesting that the genus could have arisen in Africa. There is even older fossil material attributed to *Varanus* sp. (~19 million years old) from Kenya (Clos 1995), but the oldest material attributed to *Varanus* is about 45 million years old, from central Asia (Alifanov 1993). Other such material has been found in later deposits, approximately 30 million years old, also in central Asia (Alifanov 1993) and around 24 million years old in Australia (Estes 1984). The problem with this material not attributed to species is that it consists of vertebrae, and most of the characteristics used in phylogenetic analyses of varanids are from the skull and jaws. Thus some doubt that this material truly represents *Varanus* is possible. *V. rusingensis* includes vertebrae, portions of the skull, and limb elements, thus providing a more secure basis for identification and phylogenetic analysis. If the vertebral material, however, does not represent *Varanus,* it indicates that some other varanoid taxa arrived in Australia and then died out entirely, being completely replaced by *Varanus* itself, a scenario for which there is presently no evidence. However, further support for an African origin of the genus *Varanus* comes from several phylogenetic studies that show that several African monitor species form a basal clade that is the sister group to all other living varanids (Fuller et al. 1998; Ast 2001).

The age of the African material seems to conflict with that of the oldest Australian material, but perhaps *Varanus* originated much earlier in Africa than indicated by the Kenyan material. After all, the oldest terrestrial varanoid fossils are European (Rage 1989), and even during the Late Cretaceous, it should not have been impossible for these lizards to cross into Africa. The current phylogenetic hypothesis for varanoid relationships can be nicely overlaid on a map of the Earth as an area cladogram (Fig. 3.1).

An unexplained anomaly of varanid paleozoogeography is the near-absence of varanid fossils from China. Modern monitors (*V. bengalensis* and *V. salvator*) occur along the south coast of China and fossils of a Late Cretaceous varanoid have been in found in Nei Mongol (Gao and Hou 1996), the Chinese province bordering Mongolia. But in spite of a good vertebrate fossil record and a land that would seem to

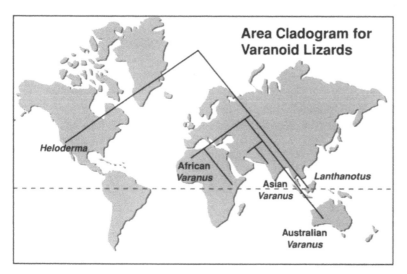

Figure 3.1. Varanoid phylogeny corresponds to geography.

have been amenable to monitors, there are no other Chinese fossil varanids or varanoids (Estes 1983).

Australian fossils are relatively recent and all are true varanids. Stirton et al. (1961) and Estes (1984) described fossil *Varanus* from the Etadunna Formation of South Australia, now considered to have been laid down at the Oligocene–Miocene boundary, about 25 my BP. *Varanus* are usually considered to be a Late Tertiary invader of Australia because the earliest Australian fossils are from the Oligocene–Miocene boundary of South Australia, the beginning of the Late Tertiary (or Neogene). Archer and Wade (1976) mentioned an Early Pliocene *Varanus* from the Allingham Formation of north Queensland. De Vis (1889) described a Pleistocene *Varanus dirus* from Kings Creek, east Darling Downs, Queensland, that Hecht (1975) suggests should be referred to *Megalania* (see following). A Pleistocene fossil varanid from Australia, *Megalania prisca,* sometimes placed in the genus *Varanus* (Molnar, this volume), was the largest land-dwelling lizard known, reaching a total length of about 6 m and an estimated weight of 600 kg (Hecht 1975; Rich 1985). (The mosasaurs, marine varanoids, were larger and reached a length of 15 m; Lingham-Soliar 1992).

Early Pliocene material referred to *Megalania* indicates an equally large lizard, and material from the Miocene of South Australia also indicates an unusually large monitor (Murray and Megirian 1992; Murray, personal communication to R.E.M.). Furthermore, De Vis (1900) described a Late Pliocene *Varanus "dirus"* from Chinchilla, west Darling Downs, Queensland, that may represent another taxon of giant monitor, and his *Varanus emeritus* (De Vis 1889) is also thought to have been a large form, reaching 3 or 4 m in length (Dunn 1927). Unfortunately for visions of an Australia overrun with massive monitors, these latter two taxa are represented by fragmentary material, a maxilla and two halves of limb bones, respectively, so the possibility that they represent juvenile specimens of *Megalania* remains viable.

Fossils of several living species of *Varanus*—*V. bengalensis, V. giganteus, V.* "*gouldii,*" *V. komodoensis, V. salvator,* and *V. varius*—have been reported, with only a single exception (*V. bengalensis*), from Australia and the adjacent Malayan Archipelago. Unfortunately, none of this material has been subjected to a rigorous phylogenetic analysis; indeed, most of it is probably too fragmentary for such analysis. Thus the material may well belong to the referred modern species, but this should not be taken for granted. Levshakova (1986) believed the Early Pliocene *Varanus darevskii* from Tadjikistan was closely related, and perhaps ancestral, to *V. griseus.* Again, this is not based on phylogenetic analysis but similarity—although on zoogeographical grounds, it would not be surprising if this were correct. Thus, with reference to the history and origin of the living species of monitors, the fossil record is suggestive but requires further work. Also, at least one monitor, *Varanus hooijeri,* has seemingly become extinct recently (Brongersma 1958). Fossils of this Indonesian form were found in post-Pleistocene cave deposits on Flores, but it has not been seen alive. Today, about 50 species of *Varanus* are recognized worldwide: all occur in Africa, Asia, Southeast Asia, and Australia (the New World is sadly impoverished, although the sister group to monitors, Helodermatids, still survives in North America).

Two major adaptive radiations occurred in Australia: one evolved small body size (subgenus *Odatria* = pygmy monitors), whereas the other lineage (subgenus *Varanus*) remained large and several members evolved gigantism. Small body size also evolved in an arboreal Asian clade. Two members of this Asian clade, *V. indicus* and *V. prasinus,* appear to have dispersed to northern Australia in the recent past. An Indonesian insular monitor has also evolved gigantism (*Varanus komodoensis*).

However, another giant monitor, *Varanus sivalensis,* lived in continental Asia in what are now the Siwalik Hills of Pakistan (Falconer 1868). It was at least the size of the modern ora (*V. komodoensis*) and demonstrates that gigantism in monitors was not limited to forms on isolated landmasses, such as the ora and *Megalania.* The range of body size in varanids has remained more or less constant over their evolution, although small forms, comparable to the pygmy monitors (*Odatria*), are not seen in fossils, possibly because of their small size. *Varanus rusingensis* is estimated to have been at least 2 m long, and even the Late Cretaceous varanids from Mongolia seem to have been around half a meter to 2 m long. Living monitors, excepting the ora, also fall pretty much into the range. Giant varanids seem to have been a relatively recent development, appearing perhaps 5 to 10 my BP in a history 45 to 50 million years long.

Compared with most other lizards, varanoids are relatively large, which may confer a degree of inertial homeothermy upon them, allowing them to maintain more even and possibly higher body temperatures than smaller lizards. However, their large size also restricts their ability to dump excess body heat during the hottest times of the year.

Modern monitors eat a variety of foods, from soft-bodied vertebrates to insects to shelled creatures, such as crabs and snails. Gray's monitor even goes so far as to feed on fruit. To some extent, these differences in diet are reflected in different tooth forms, from compressed, serrate, recurved teeth to robust, blunt crowns. The (presumably) genetic potential for such differences in tooth form goes far back in varanoid history: the mosasaurs show similar differences in their teeth, and the mosasaur lineage diverged from that of monitors at least 100 my BP. The oldest known dentitions of *Varanus,* those of *V. rusingensis,* show a variety of tooth forms, including blunt teeth, suggesting to Clos (1995) that this lizard may have preyed on shellfish. Swollen, blunt teeth are found again in *Varanus hooijeri,* a post-Pleistocene extinct species from Indonesia. This all suggests that monitors and varanoids throughout their known history had the ability to live on a wide variety of prey, and even—perhaps—plant material. A much more speculative notion is evoked by the claim of Norell et al. (1992) that monstersaurs had venom grooves on their teeth at their first appearance (*Estesia*) in the Late Cretaceous. These are, however, not the oldest grooved teeth among varanoids, for the marine *Coniasaurus crassidens* also had some grooved teeth (Caldwell and Cooper 1999) approximately 10 million years earlier. Modern oras infect their prey with virulent bacteria (Auffenberg 1981), which act basically in a similar fashion to the injection of venom. Thus we might speculate that even early varanoids had either toxic saliva, or saliva that carried infective bacteria, that aided in subduing their prey.

The success of monitors is probably not due to their prey-catching features alone. Although this is a large topic deserving a more extensive treatment than possible here, a few more comments will not be amiss. Bartholomew and Tucker (1964) found that monitors had a greater metabolic rate than other lizards. During activity, the metabolic rate of a varanid could be increased until it matched or surpassed the basal rates of mammals. Monitors could sustain higher levels of activity because their blood did not lose its capacity to transport oxygen during activity as quickly as that of other lizards (Bennett 1973). Thus monitors could be more active, and tire less quickly, than other lizards of the same size. Four out of eight of the distinguishing features of modern monitors involve the vertebral column. One of these, increasing the length of the cervical vertebrae, presumably increases the length of the neck, and so also increases the volume in the trachea available for air for the "gular pump." Gular pumping forces air into the lungs, rather than sucking it by expanding the rib cage, the method used by other lizards (Owerkowicz et al. 1999). This is done while running to overcome the trade-off between breathing and running found in other lizards, and so provides greater stamina. Thus a longer neck suggests the ability to force more air into their lungs, and hence the possibility of greater activity levels and stamina. Furthermore, the hearts of monitors are different from those of other lizards. In those of monitors, the blood flow is more like that of mammals (Millard and Johansen 1974), in that the systemic circulation is supplied under greater pressure than the pulmonary circulation. So monitors can more effectively supply oxygen to their lungs during exercise, and more effectively carry it to their muscles, than other lizards.

Another very important key innovation of varanids is their long, snakelike, hydrostatic, protrusible, forked tongue, which they exploit, in concert with their acute vomeronasal system, to sample heavy, non-airborne scent trails (Schwenk 1995, 2000). Indeed, evidence suggests that they used their forked tongues as edge detectors (Schwenk 1994, 2000). When a monitor locates a scent trail, it invariably follows it in the correct direction rather than backtracking.

Also, varanids are more effective at seizing fast-moving prey animals than related (anguimorph) lizards (Losos and Greene 1988). All these factors indicate enhanced prey-catching ability in monitors and may, taken together, account for the success of these lizards. Although the future of all monitors is far from clear, their past success suggests that they will most likely be doing well 50 million years from now.

References

Alifanov, V. A. 1993. Some peculiarities of the Cretaceous and Palaeogene lizard faunas of the Mongolian People's Republic. In *Monument Grube Messel—Perspectives and relationships, pt. 2*, ed. F. Schrenk and K. Ernst. *Kaupia* 3:9–13.

Archer, M., and M. Wade. 1976. Results of the Ray E. Lemley expeditions, Part 1. The Allingham Formation and a new Pliocene vertebrate fauna from northern Queensland. *Mem. Queensl. Mus.* 17:379–397.

Ast, J. C. 2001. Mitochondrial DNA evidence and evolution in Varanoidea (Squamata). *Cladistics* 17:211–226.

Auffenberg, W. 1981. *The behavioral ecology of the Komodo monitor.* Gainesville: University Presses of Florida.

Bartholomew, G. A., and V. A. Tucker. 1964. Size, body temperature, thermal conductance, oxygen consumption, and heart rate in Australian varanid lizards. *Physiol. Zool.* 37:341–354.

Bennett, A. F. 1973. Blood physiology and oxygen transport during activity in two lizards, *Varanus gouldii* and *Sauromalus hispidus. Comp. Biochem. Physiol.* 46A:673–690.

Borsuk-Białynicka, M. 1984. Anguimorphans and related lizards from the Late Cretaceous of the Gobi Desert, Mongolia. *Palaeontol. Polonica* 46:5–105.

Brongersma, L. 1958. On an extinct species of the genus *Varanus* (Reptilia, Sauria) from the island of Flores. *Zool. Mededelingen Rijksmus. Nat. Hist. Leiden* 36:113–125.

Caldwell, M. W., and J. A. Cooper. 1999. Redescription, palaeobiogeography and palaeoecology of *Coniasaurus crassidens* Owen, 1850 (Squamata) from the Lower Chalk (Cretaceous; Cenomanian) of SE England. *Zool. J. Linnean Soc.* 127:423–452.

Clos, L. M. 1995. A new species of *Varanus* (Reptilia: Sauria) from the Miocene of Kenya. *J. Vertebr. Paleontol.* 15:254–267.

De Vis, C. W. 1889. On *Megalania* and its allies. *Proc. R. Soc. Queensland* 6:93–99.

———. 1900. A further trace of an extinct lizard. *Ann. Queensland Mus.* 5:6.

Dunn, E. R. 1927. Results of the Douglas Burden expedition to the island of Komodo. I.—Notes on *Varanus komodoensis. Am. Mus. Novitates* 286:1–10.

Estes, R. 1983. The fossil record and early distribution of lizards. In

Advances in herpetology and evolutionary biology, ed. A. G. J. Rhodin and K. Miyata, 365–398. Cambridge, Mass.: Museum of Comparative Zoology, Harvard University.

———. 1984. Fish, amphibians and reptiles from the Etadunna Formation, Miocene of South Australia. *Aust. Zool.* 21:335–343.

Estes, R. K., and J. H. Hutchinson. 1980. Eocene lower vertebrates from Ellesmere Island, Canadian Arctic Archipelago. *Palaeogeogr. Palaeoclimatol. Palaeoecol.* 30:325–347.

Estes, R., K. de Queiroz, and J. Gauthier. 1988. Phylogenetic relationships within Squamata. In *Phylogenetic relationships of the lizard families,* ed. R. Estes and G. Pregill, 119–281. Stanford, Calif.: Stanford University Press.

Evans, S. E. 1994. A new anguimorph lizard from the Jurassic and Lower Cretaceous of England. *Palaeontology* 37:33–49.

———. 1996. *Parviraptor* (Squamata: Anguimorpha) and other lizards from the Morrison Formation at Fruita, Colorado. *Mus. Northern Arizona Bull.* 60:243–248.

Evans, S. E., G. V. R. Prasad, and B. K. Manhas. 2002. Fossil lizards from the Jurassic Kota Formation of India. *J. Vertebr. Paneongol.* 22:299–312.

Falconer, H. 1868. *Palaeontological memoirs and notes of the late Hugh Falconer, A.M., M.D.* Vol. 1, *Fauna antiqua sivalensis.* London: R. Hardwicke.

Fuller, S., P. Baverstock, and D. King. 1998. Biogeographic origins of goannas (Varanidae): A molecular perspective. *Mol. Phylogenet. Evol.* 9:294–307.

Gao, K., and L. Hou. 1996. Systematics and taxonomic diversity of squamates from the Upper Cretaceous Djadochta Formation, Bayan Mandahu, Gobi Desert, People's Republic of China. *Can. J. Earth Sci.* 33:578–598.

Gao K., and M. Norell. 1998. Taxonomic revision of *Carusia* (Reptilia: Squamata) from the last Cretaceous of the Gobi Desert and phylogenetic relationships of anguimorphan lizards. *Am. Mus. Novitates* 3230:1–51.

———. 2000. Taxonomic composition and systematics of Late Cretaceous lizard assemblages from Ukhaa Tolgod and adjacent localities, Mongolian Gobi Desert. *Bull. Am. Mus. Nat. Hist.* 249:1–118.

Hecht, M. 1975. The morphology and relationships of the largest known terrestrial lizard, *Megalania prisca* Owen, from the Pleistocene of Australia. *Proc. R. Soc. Vic.* 87:239–250.

King, D., S. Fuller, and P. Baverstock. 1999. The biogeographic origins of varanid lizards. *Mertensiella* 11:43–49.

Kordikova, E. G., P. D. Polly, V. A. Alifanov, Z. Rocek, G. F. Gunnell, and A. O. Averianov. 2001. Small vertebrates from the Late Cretaceous and Early Tertiary of the northeastern Aral Sea region, Kazakhstan. *J. Paleontol.* 75:390–400.

Kuhn, O. 1958. Ein neuer Lacertilier aus dem fränkischen Lithographieschiefer. *Neues Jahrbuch Geol. Palaontol. Monasthefte* 1958:380–382.

Lee, M. S. Y. 1997. The phylogeny of varanoid lizards and the affinities of snakes. *Phil. Trans. R. Soc. Lond. B* 352:53–91.

Lee, M. S. Y., and M. W. Caldwell. 2000. *Adriosaurus* and the affinities of mosasaurs, dolichosaurs, and snakes. *J. Paleontol.* 74:915–937.

Levshakova, I. U. 1986. Novii varan iz Nizhnego Pliotsena Tadzhikistana. *Trudi Zool. Inst. AN SSSR* 157:101–106.

Lingham-Soliar, T. 1992. The tylosaurine mosasaurs (Mosasauridae, Reptilia) from the Upper Cretaceous of Europe and Africa. *Bull. Inst. R. Sci. Nat. Belg.* 62:171–194.

Losos, J. B., and H. W. Greene. 1988. Ecological and evolutionary implications of diet in monitor lizards. *Biol. J. Linnean Soc.* 35:379–407.

McDowell, S., and C. Bogert. 1954. The systematic position of *Lanthanotus* and the affinities of the anguimorphan lizards. *Bull. Am. Mus. Nat. Hist.* 5:1–142.

Millard, R. W., and K. Johansen. 1974. Ventricular outflow dynamics in the lizard, *Varanus niloticus*: Responses to hypoxia, hypercarbia and diving. *J. Exp. Biol.* 60:871–880.

Murray, P. F., and D. Megirian. 1992. Continuity and contrast in middle and late Miocene vertebrate communities from the Northern Territory. *Beagle* 9:195–217.

Nessov, L. A. 1981. Nakhodka cheloosti nazemnoi yashcheritsi v Verkhnem Melu Uzbekistana. *Vestnik Leningradskogo Univ.* 2:105–107.

Norell, M. A., M. C. McKenna, and M. J. Novacek. 1992. *Estesia mongoliensis*, a new fossil varanoid from the Late Cretaceous Barun Goyot Formation of Mongolia. *Am. Mus. Novitates* 3045:1–24.

Norell, M. A., and K. Gao. 1997. Braincase and phylogenetic relationships of *Estesia mongoliensis* from the Late Cretaceous of the Gobi Desert and the recognition of a new clade of lizards. *Am. Mus. Novitates* 3211:1–25.

Owerkowicz, T., C. G. Farmer, J. W. Hicks, and E. L. Brainerd. 1999. Contribution of gular pumping to lung ventilation in monitor lizards. *Science* 284:1661–1663.

Pepin, D. J. 1999. The origin of monitor lizards based on a review of the fossil evidence. *Mertensiella* 11:11–42.

Pregill, G., J. Gauthier, and H. Greene. 1986. The evolution of helodermatid squamates, with description of a new taxon and an overview of Varanoidea. *Trans. San Diego Soc. Nat. Hist.* 21:167–202.

Rage, J.-C. 1989. Le plus ancien lézard varanoïde de France. *Bull. Soc. Études Sci. Anjou* 13:19–26.

Rich, T. H. 1985. *Megalania prisca* (Owen, 1859), the giant goanna. In *Kadimakara: Extinct vertebrates of Australia*, ed. P. V. Rich and G. F. van Tets, 152–155. Lilydale, Australia: Pioneer Design Studio.

Schwenk, K. 1994. Why snakes have forked tongues. *Science* 263:1573–1577.

———. 1995. Of tongues and noses: Chemoreception in lizards and snakes. *Trends Ecol. Evol.* 10:7–12.

———. 2000. Feeding in lepidosaurs. In *Feeding: Form, function and evolution in tetrapod vertebrates*, ed. K. Schwenk, 175–291. San Diego: Academic Press.

Stirton, R. A., R. H. Tedford, and A. H. Miller. 1961. Cenozoic stratigraphy and vertebrate paleontology of the Tirari Desert, South Australia. *Geogr. Rev.* 62:40–70.

Yadagiri, P. 1986. Lower Jurassic lower vertebrates from Kota Formation, Pranhita-Godavari Valley, India. *J. Palaeontol. Soc. India* 31:89–96.

4. Tempo and Timing of the Australian *Varanus* Radiation

W. Bryan Jennings and Eric R. Pianka

Australia is well known for its spectacularly rich lizard communities, many of which may contain more than 40 locally sympatric species (Pianka 1969, 1986; Morton and James 1988; James 1994; James and Shine 2000; Jennings, personal observation), some as many as 54 species (Pianka 1996). Although most previous studies have focused on the community ecological aspects of this phenomenon, phylogenetic approaches to this problem are now catching up (Melville et al. 2001; Jennings et al. 2003). Key questions pertaining to the evolution of Australian lizard diversity are as follows: What was the tempo of speciation? When did these radiations occur? And what caused these radiations?

Before these questions can be meaningfully addressed, an attempt should be made to ascertain whether or not these radiations actually occurred in situ—that is, within Australia. Phylogenetic hypotheses showing that at least the majority of Australian species within each putative radiation are monophyletic would constitute such support. Not surprisingly, phylogenetic evidence bolstering this idea is accumulating rapidly, as evidenced by studies on Australian agamids (Macey et al. 2000; Melville et al. 2001), varanids (Ast 2001), and pygopodids (Kluge 1976, 1987; Jennings et al. 2003).

At least three families of lizards in Australia contain clades that have independently undergone radiations, raising the possibility that these radiations were driven by similar environmental factors and therefore occurred at about the same time. But exactly when? Australia has been isolated as a continental entity ever since its separation from

Antarctica some 42 mya (White 1994). Aridification of central Australia began in the Early Miocene about 23 my BP (White 1994), as forests gave way first to grasslands and then to deserts (Bowler 1976; Bowler et al. 1976; Galloway and Kemp 1981; Vickers-Rich and Rich 1993). Presumably, these desert lizard groups diversified during this time interval. If these radiations occurred simultaneously, then they probably arose sometime relatively recently—the Early to mid-Miocene period (~15–23 mya)—because agamid and varanid lizards apparently didn't arrive in Australia until Australia drifted sufficiently northward to allow for faunal exchange between Australia and Southeast Asia (Frakes et al. 1987; Heatwole 1987; Main 1987; Macey et al. 2000; Evans et al. 2002; but see Hutchinson and Donnellan 1993). Indeed, recent phylogenetic studies suggest that the Australian agamid (Melville et al. 2001) and pygopodid (Jennings et al. 2003) radiations occurred in the past 15 to 23 mya, respectively. Although both studies relied on molecular clock estimates to date speciation events, which may be a questionable practice (Hillis et al. 1996), both estimates are similar despite the use of independent taxa, data, and tree-calibration methods. Although little is known about the tempo and mode of Australian lizard speciation, preliminary insights offered by pygopodid lizards suggest that speciation rates may not have been constant with time, but instead peaked relatively early in the group's history before gradually leveling off over the past 10 my BP (Jennings et al. 2003). Moreover, the vast majority of extant pygopodid species originated only in the last 23 my BP (Jennings et al. 2003).

The lizard family Varanidae currently contains about 50 species, which are distributed throughout Africa, Asia, and Australia. Fossil evidence indicates that varanoid lizards have existed as far back as 90 mya (Molnar 2004), although the oldest known representatives of the genus *Varanus* are Early Miocene (23 mya) *Varanus* sp. from South Australia (Estes 1984; Hutchinson and Donnellan 1993) and the Early Miocene (17 mya) *V. rusingensis* from Kenya (Clos 1995). The Australian record is of particular interest here because it firmly establishes the presence of *Varanus* in Australia at least as far back as Early Miocene times. Therefore, the Australian *Varanus* radiation is at least 23 million years old.

Although some progress toward understanding the evolution of Australian lizard diversity has been made, phylogenetic-based studies of other Australian lizard groups are needed to further corroborate preliminary findings. Here, we examine evolution of diversity in Australian varanid lizards by synthesizing a combination of paleontological and neontological data to estimate the tempo and timing of speciation events in the group.

Analysis of Tempo of Speciation Events

Generation of Ultrametric Tree for Varanidae

Although the topology of Ast's tree seems entirely reasonable to us, we nevertheless reanalyzed her molecular data, consisting of 2800 nucleotide characters for 48 species of varanids and 6 outgroup taxa, to generate an ultrametric (molecular clock) tree. To do this, we first used PAUP 4.0 0d64 (Swofford 2000) to conduct a parsimony analysis. Tree searches were conducted heuristically with TBR branch swapping in a random stepwise addition of taxa repeated 10 times. This analysis yielded a tree with a topology consistent to the one found in Ast (2001). We then pruned all subspecific taxa (i.e., *V. bengalensis nebulosus,* Solomon Islands *V. indicus,* Papua New Guinea *V. indicus, V. panoptes horni, V. salvator bivittatus, V. salvator cumingi, V. salvator togianus, V. scalaris timorensis*) from the tree so that each extant species was represented by one sample, which resulted in 40 ingroup taxa. We then used PAUP to score this tree by using maximum likelihood criterion and a GTR + γ + I substitution model with all parameters estimated via maximum likelihood and enforcing the molecular clock. A new tree resulted that was consistent with Ast's (2001) topology but now had maximum likelihood branch lengths scaled to time. We labeled all nodes in the ultrametric tree with letters for ease of referencing in text (some of these correspond to subgenera of Ziegler and Böhme 1997).

Test of the Molecular Clock Assumption

Use of an ultrametric tree carries with it the assumption that the DNA sequences evolved in a clocklike manner. We evaluated this assumption by performing likelihood ratio tests (Goldman 1993). We used PAUP 4.0 0d64 (Swofford 2000) to individually score our tree; we used the GTR + γ + I model with and without the molecular clock enforced. We assessed the significance of the test statistic, which is the quantity equal to twice the difference between the two log-likelihood scores, by referring to the X^2 table in Rohlf and Sokol (1981) and using $N - 2$ (N = number of species) degrees of freedom (Huelsenbeck and Rannala 1997).

Test of Rapid Speciation

We subjected our ultrametric tree of the Varanidae to the rapid cladogenesis test implemented in the computer program End-Epi (Rambaut and Harvey 1994). In this analysis, the program calculates a "cladogenesis statistic" for each internal branch, which represents the probability that a particular lineage occurring at time t will further diversify into k tips under a constant-rates birth–death model (Rambaut and Harvey 1994). Any clade with a probability value less than 0.05 is considered to have undergone a higher than expected rate of speciation.

Species Accumulation Profiles

In a second analysis of speciation rates, we qualitatively examined so-called lineage-through-time profiles (Barraclough and Nee 2001). These plots show the number of lineages (on a log scale) as a function of time (Rambaut and Harvey 1994; Schluter 2000; Barraclough and Nee 2001). If speciation rates are constant through time, then the plot is predicted to be linear, an expectation based on a pure-birth stochastic model (Barraclough and Nee 2001). We used End-Epi to generate such profiles for the Varanidae and also for the large subclade involving only the "Australian" *Varanus* clade. To facilitate future comparisons with other major Australian lizard clades, we define this clade as that containing species of the subgenera *Odatria*, *Varanus* (*V. giganteus*, *V. panoptes*, *V. gouldii*, *V. rosenbergi*, *V. mertensi*, *V. spenceri*, *V. varius*, and *V. komodoensis*), and *Papusaurus* (*V. salvadorii*). All are Australian, except for *V. komodoensis* and *V. salvadorii*, which occur in Indonesia and New Guinea, respectively. Although these plots provide a subjective viewpoint of speciation rates, they may still be useful in the study of Australian lizard diversity, particularly if profiles from multiple independent radiations are found to converge on a common pattern (Jennings et al. 2003).

Results

The ultrametric tree we generated from Ast's (2001) DNA sequence data is shown in Figure 4.1. This tree contains eight fewer taxa (all subspecies) than Ast's original tree. If the DNA sequences evolved in a strict clocklike manner, then the branch lengths in this tree reflect relative time. However, statistical evaluation of the clock assumption indicated that the clock version of the GTR + γ + I substitution model fits these data significantly worse than its nonclock equivalent (=169, X^2_{38} = 53 [5% significance], $P < 0.001$). Although this result certainly warrants some caution in any interpretations made from this ultrametric tree, how severely a given data set must deviate from the clock in order to invalidate conclusions is not yet known.

The entire *Varanus* clade, in addition to several nested subclades, seems to have undergone rapid cladogenesis, as indicated by statistically significant P values at nodes A, B, C, D, and B'. One clade whose species have collectively been subsumed within subgenus *Odatria* may also have undergone rapid speciation, but its P value was not quite statistically significant (node E; Fig. 4.1; see also below).

The species accumulation profile for Varanidae is mostly linear, suggesting that speciation rates may have been relatively constant over much of the group's history. The portion of the profile that begins to plateau toward the present can be explained by several processes, which include a general slowdown in diversification rate toward the present, an increase in extinction rate, or taxonomic sampling artifacts. Interestingly, the profile's linearity is punctuated at several time intervals by sharp upturns, which may be the signature of either an elevated speciation rate or a diminished extinction rate relative to speciation rate. The five most obvious upturns in the profile are highlighted in Figure 4.2 to reveal the cladogenetic events producing the pattern. The first upturn is explained by the splitting of node B into Asian and Australian lineages, with African lineage at node L' splitting to form *V. griseus* and the blunt-toothed monitor lineage (*V. exanthematicus* and *V. niloticus*). The second upturn is due to the Australian lineage at node C giving rise to three new lineages. These three new lineages eventually formed their own clades: Australian pygmy monitors of the subgenus *Odatria* (node E), a clade of large-bodied *Varanus* species in-

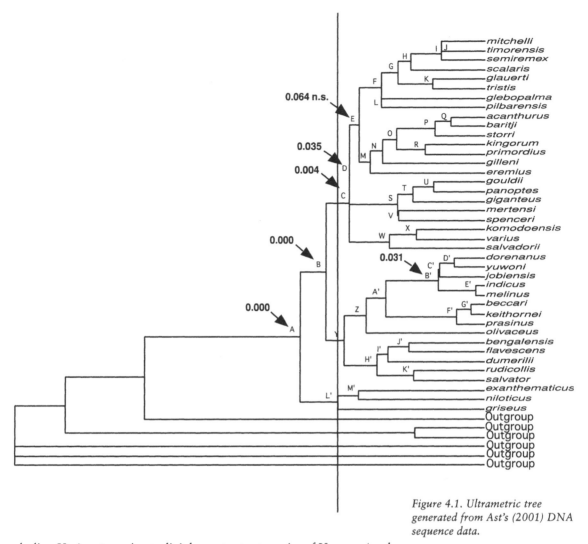

Figure 4.1. Ultrametric tree generated from Ast's (2001) DNA sequence data.

cluding *V. giganteus,* Australia's largest extant species of *Varanus* (node S), and another large-bodied *Varanus* clade including *V. komodoensis* the largest known extant species of *Varanus* (node W). Speciation events at nodes F, L, and N within the odatrian clade (node E) are responsible for the third upturn. Splitting of nodes W, A', and I' account for the fourth upturn, and the fifth upturn is explained by splitting of nodes G, O, S, and V (Fig. 4.2).

On a finer scale, the lineage-through-time plot for the Australian *Varanus* clade (node C) shows five possible increases in speciation rates. The initial burst may have occurred when the most recent common ancestor (MRCA) to clade C rapidly split into lineages, leading to clades E, S, and W. The second apparent upturn can be explained by the MRCA to clades F, L, and N each splitting into new lineages. The third increase in speciation rate occurred when MRCA of G, O, S, and V speciated into new lineages. The fourth increase is explained by the speciation events at node X, which gave rise to the *V. komodoensis* and *V. varius* lineages, and node T, which split into the *V. giganteus* and lineage leading to *V. gouldii/V. panoptes*. The fifth increase is attributable to speciation events at nodes K, P, and U, which gave rise to the *V. glauerti, V. tristis, V. acanthurus/V. baritji, V. storri, V. gouldii,* and *V. panoptes* lineages (Fig. 4.3).

Discussion

The DNA sequence data from which our ultrametric tree was derived apparently did not evolve in a strictly clocklike manner. Nevertheless, interpretations made from such an imperfect clock tree may remain valid because the effects of violating this assumption are not yet known. Also, a recent simulation study by Huelsenbeck et al. (2002), which dealt with clock rooting of phylogenetic trees, suggests that violations of clock assumption may not adversely affect results. Our analysis of speciation rates in the Varanidae suggests that the family, as well as several of its nested subclades, may have diversified rapidly in evolutionary time. Interestingly, these particular clades do not appear to be randomly dispersed in the varanid tree but instead mostly form a continuum between the root of *Varanus* up to the clade containing *Odatria* and its sister clade. Of particular interest to us is the statistical evidence supporting the notion that the Australian *Varanus* clade underwent rapid cladogenesis. Indeed, some of the more derived clades, such as Australian *Varanus,* could be driving the statistical significance of nodes located in more basal positions in the tree. Although test of *Odatria* was marginally nonsignificant, it would almost certainly have been significant if one of its suspected polytypic species, *V. scalaris,* had been further split into new taxa (L. Smith, personal communication; Smith estimates "*scalaris*" could consist of as many as 10 as yet undescribed new species).

Our species accumulation profiles for Varanidae as well as its Australian *Varanus* clade were both roughly linear, suggesting that throughout much of varanid history, the group has exhibited a relatively constant rate of speciation. However, two additional aspects of these plots may reveal other important phases of *Varanus* history. First, several intervals on each curve are punctuated by apparent upward spikes in speciation rates, which may be indicative of past episodic bursts in speciation. This observation therefore suggests that there may have been times when speciation rates were anomalously high. The second important finding to glean from the curve concerns the tendency for the line to plateau toward the present. Such a tapering off of speciation rates is consistent with the signature of a slowdown in diversification rate (Barraclough and Nee 2001). In fact, our curves show remarkable resemblance to the curve generated for pygopodid lizards (Jennings et al. 2003) and one for *Dendroica* warblers (Lovette and Bermingham 1999).

Although a recent slowdown in *Varanus* speciation rates may have occurred, the plateau in our plots could also be due to taxonomic sampling artifacts (underestimated species richness) or an increase in extinction rate relative to speciation rate (Barraclough and Nee 2001). The significance of the aforementioned upturns should also be tem-

Figure 4.2. (opposite page) The five most obvious upturns in speciation rate, highlighted to reveal cladogenetic events producing the pattern.

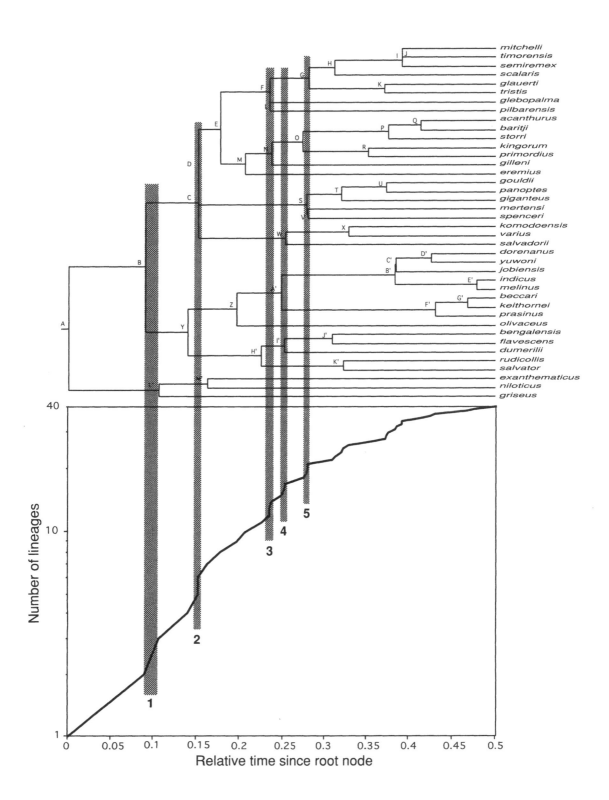

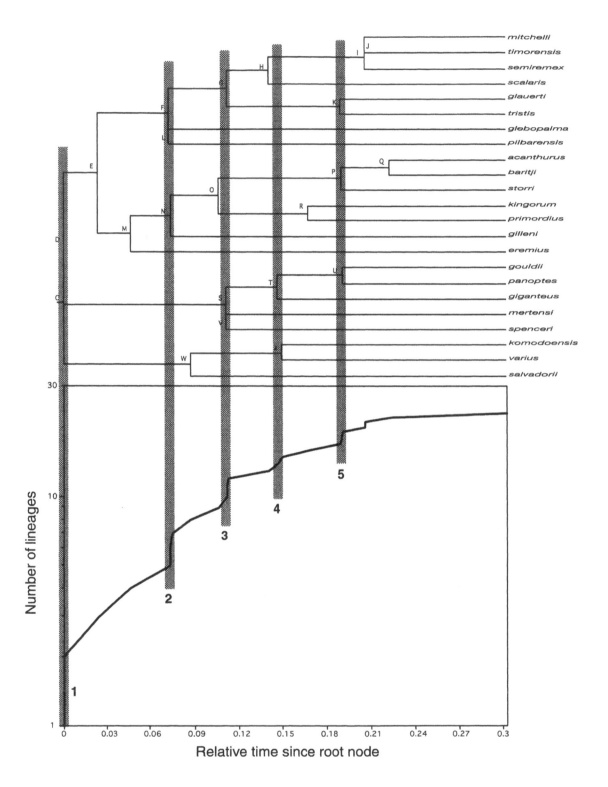

pered by the fact that the upturns may only represent illusory increases in speciation rate due to a reduction in extinction rate alone (Barraclough and Nee 2001). Although we already highlighted these upturns in our plots, to go beyond that and to attempt to further interpret these would be inappropriate speculation. However, these descriptive plots may ultimately play an important role in establishing robust macroevolutionary patterns pertaining to the origination of Australia's lizard diversity, particularly if profiles from other groups are found to agree with each other (Jennings et al. 2003). Indeed, the study of such curves may be the neontological equivalent to paleontological stratigraphy whereby lineage-through-time plots are used to estimate the tempo of diversification for various groups.

How old is the Australian *Varanus* radiation? Before we can estimate the age of this clade, we must note that organisms presently found in Australia are either descendants from Gondwanan relatives, which were present at least as far back as 45 mya, when Australia first became a continental entity (White 1994), or are descendants from Asiatic lineages, which dispersed to Australia no earlier than Early Miocene times (23 mya; Frakes et al. 1987; Heatwole 1987; Main 1987; but see Hutchinson and Donnellan 1993). Because fossil (Molnar 2003) and phylogenetic data (Ast 2001) indicate that *Varanus* arose in Asia and subsequently dispersed to Africa and Australia, it appears as though *Varanus* probably didn't exist in Australia before Miocene times. Moreover, discovery of the oldest Australian *Varanus* fossil, dated to Early Miocene times (Estes 1984), shows that *Varanus* have roamed Australia at this early date as well, suggesting the time of 23 mya as the starting point for this lineage's history in Australia. If this timing assessment is correct, then the tempo and timing of the Australian *Varanus* radiation closely mirrors that estimated for another Australian lizard radiation—the pygopodid lizards (Jennings et al. 2003).

Indeed, over the past 23 mya, both groups seem to have undergone repeated episodic bursts of speciation leading to the vast majority of their extant species richness. The time frame for diversification of Australian *Varanus* is also consistent with the theory that aridification may have played a crucial role in the diversification of Australian lizards. When these results are compared with similar studies of other clades of Australian vertebrates, commonalities may emerge that will aid in reconstructing the probable history of this island continent's animal life.

Acknowledgment

We are grateful to Dr. Jennifer Ast for generously providing us with her aligned molecular data set for *Varanus* lizards.

References

Ast, J. C. 2001. Mitochondrial DNA evidence and evolution in Varanoidea (Squamata). *Cladistics* 17:211–226.

Barraclough, T. G., and S. Nee. 2001. Phylogenetics and speciation. *Trends Ecol. Evol.* 16:391–399.

Figure 4.3. (opposite) Lineage-through time plot for the Australian Varanus *clade (node C) showing five possible increases in rates of speciation.*

Bowler, J. M. 1976. Aridity in Australia: Age, origins and expression in Aeolian land-forms and sediments. *Earth-Sci. Rev.* 12:279–310.

Bowler, J. M., G. S. Hope, J. N. Jennings, G. Singh, and D. Walker. 1976. Late Quaternary climates of Australia and New Guinea. *Quat. Res. N. Y.* 6:359–394.

Clos, L. M. 1995. A new species of *Varanus* (Reptilia: Sauria) from the Miocene of Kenya. *J. Vertebr. Paleontol.* 15:254–267.

Estes, R. 1984. Fish, amphibians and reptiles from the Etadunna Formation, Miocene of South Australia. *Aust. Zoologist* 21:335–343.

Evans, S. E., G. V. R. Prasad, and B. K. Manhas. 2002. Fossil lizards from the Jurassic of Kota Formation of India. *J. Paleontol.* 22:299–312.

Frakes, L. A., B. McGowran, and J. M. Bowler. 1987. Evolution of Australian environments. In *Fauna of Australia*, ed. G. R. Dyne and D. W. Walton, 1A:1–16. Canberra: Australian Government Publishing Service.

Galloway, R. W., and E. M. Kemp. 1981. Late Cainozoic environments in Australia. In *Ecological biogeography in Australia*, ed. A. Keast, 51–80. The Hague: Junk.

Goldman, N. 1993. Statistical tests of models of DNA substitution. *J. Mol. Evol.* 36:182–198.

Heatwole, H. 1987. Major components and distributions of the terrestrial fauna. In *Fauna of Australia*, ed. G. R. Dyne and D. W. Walton, 1A:101–135. Canberra: Australian Government Publishing Service.

Hillis, D. M., B. K. Mable, and C. Moritz. 1996. Applications of molecular systematics and the future of the field. In *Molecular systematics*, 2nd ed., ed. D. M. Hillis, C. Moritz, and B. K. Mable, 515–543. Sunderland: Sinauer.

Huelsenbeck, J. P., and B. Rannala. 1997. Phylogenetic methods come of age: Testing hypotheses in an evolutionary context. *Science* 276: 227–232.

Huelsenbeck, J. P., J. P. Bollback, and A. M. Levine. 2002. Inferring the root of the phylogenetic tree. *Syst. Biol.* 51:32–43.

Hutchinson, M. N., and S. C. Donnellan. 1993. Biogeography and phylogeny of the Squamata. In *Fauna of Australia*, ed. C. J. Glasby, G. J. B. Ross, and P. L. Beasley, 2A:210–220. Canberra: Australian Government Publishing Service.

James, C. D. 1994. Spatial and temporal variation in structure of a diverse lizard assemblage in arid Australia. In *Lizard ecology: Historical and experimental perspectives*, ed. L. J. Vitt and E. R. Pianka, 287–317. Princeton, N.J.: Princeton University Press.

James, C. D., and R. Shine. 2000. Why are there so many coexisting species of lizards in Australian deserts? *Oecologia* 125:127–141.

Jennings, W. B., E. R. Pianka, and S. Donnellan. 2003. Systematics of the lizard family Pygopodidae with implications for the diversification of Australian temperate biotas. *Syst. Biol.* 52:757–780.

Kluge, A. G. 1976. Phylogenetic relationships in the lizard family Pygopodidae: An evaluation of theory, methods and data. *Misc. Publ. Mus. Zool. Univ. Michigan* 152:1–72.

———. 1987. Cladistic relationships in the Gekkonoidea (Squamata, Sauria). *Misc. Publ. Mus. Zool. Univ. Michigan* 173:1–54.

Lovette, I. J., and E. Bermingham. 1999. Explosive ancient speciation in the New World *Dendroica* warblers. *Proc. R. Soc. Lond. B Biol. Sci.* 266:1629–1636.

Macey, J. R., J. A. Schulte II, A. Larson, N. B. Ananjeva, Y. Wang,

R. Pethiyagoda, N. Rastegar-Pouyani, and T. J. Papenfuss. 2000. Evaluating trans-Tethys migration: An example using acrodont lizard phylogenetics. *Syst. Biol.* 49:233–256.

Main, A. R. 1987. Evolution and radiation of the terrestrial fauna. In *Fauna of Australia: General articles,* ed. G. R. Dyne and D. W. Walton, 1A:136–155. Canberra: Australian Government Publishing Service.

Melville, J., J. A. Schulte, and A. Larson. 2001. A molecular phylogenetic study of ecological diversification in the Australian lizard genus *Ctenophorus. J. Exp. Zool. Mol. Dev. Evol.* 291:339–353.

Molnar, R. E. 2004. The long and honorable history of monitors and their kin. In *Varanoid lizards of the world,* ed. E. R. Pianka and D. R. King. Bloomington: Indiana University Press.

Morton, S. R., and C. D. James. 1988. The diversity and abundance of lizards in arid Australia: A new hypothesis. *Am. Natur.* 132:237–256.

Pianka, E. R. 1969. Habitat specificity, speciation, and species density in Australian desert lizards. *Ecology* 50:498–502.

———. 1986. *Ecology and natural history of desert lizards: Analyses of the ecological niche and community structure.* Princeton, N.J.: Princeton University Press.

———. 1996. Long-term changes in lizard assemblages in the Great Victoria Desert: Dynamic habitat mosaics in response to wildfires. In *Long-term studies of vertebrate communities,* ed. M. L. Cody and J. A. Smallwood, 191–215. New York: Academic Press.

Rambaut, A., and P. H. Harvey. 1994. *End-Epi—Phylogenetic process analysis for the Apple Macintosh.* Computer Applications in Biological Sciences.

Rohlf, F. J., and R. R. Sokal. 1981. *Statistical tables.* 2nd ed. New York: W. H. Freeman.

Schluter, D. 2000. *The ecology of adaptive radiation.* Oxford: Oxford University Press.

Swofford, D. L. 2000. *PAUP*. Phylogenetic analysis using parsimony (*and other methods).* Ver. 4. Sunderland: Sinauer.

Vickers-Rich, P., and T. H. Rich. 1993. *Wildlife of Gondwana.* Reed.

White, M. E. 1994. *After the greening, the browning of Australia.* Kenthurst: Kangaroo Press.

Ziegler, T., and W. Böhme. 1997. Genitalstrukturen und Paarungsbiologie bei squamaten Reptilien, speziell den Platynota, mit Bermerkungen zur Systematik. *Mertensiella* 8:1–207.

Part II

INTRODUCTION TO PART II:
Species Accounts

Fifty-six detailed species accounts follow. Most are written by people with personal field experience with a particular species, although a few accounts had to be based on literature reviews. We tried to standardize species accounts, but because little or nothing is known about some subject areas for many species, accounts vary considerably in level of detail. Species accounts are separated into three sections, one for 6 African species, another for 23 Asian and Southeast Asian species (a few of these species also reach tropical northern Australia), and finally a section for 24 endemic Australian species, one of which (*Varanus panoptes*) has reached southern New Guinea. Subspecies are recognized for several species, especially for the widespread Asian water monitor *V. salvator,* for which the nominate form is given its own species account and its various subspecies another.

5. African Varanid Species

Six *Varanus* species belong to the African clade, which is the sister group to all other *Varanus*. Although Ast (2001) included only three of these six, *V. griseus, V. exanthematicus,* and *V. niloticus,* in her phylogenetic study, *V. albigularis* and *V. yemenensis* are clearly closely related to *V. exanthematicus,* and *V. ornatus* is closely related to *V. niloticus* (*V. ornatus* was formerly considered a subspecies of *V. niloticus* but has recently been elevated to full species status).

Figure 5.1. Juvenile Varanus albigularis *eating a grasshopper (photo by John Phillips).*

5.1. *Varanus albigularis*

JOHN ("ANDY") PHILLIPS

Nomenclature

1802 *Tupinambis albigularius*—Daudin, *Hist. Nat. Rept.*, 3 72, Taf. 32.—Type locality not known (Fig. 5.1).
1988 *Varanus albigularis albigularis*—Böhme 1988.

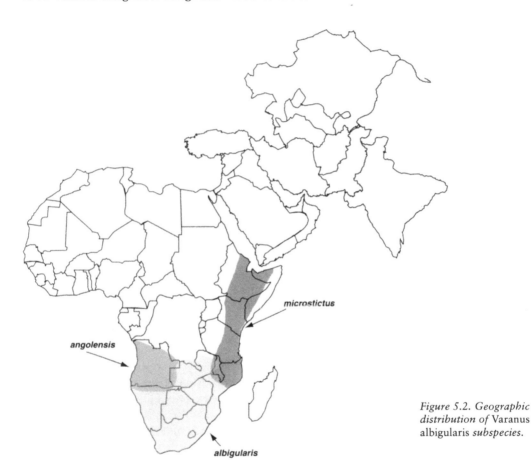

Figure 5.2. Geographic distribution of Varanus albigularis *subspecies.*

Geographic Distribution

Varanus albigularis is widely distributed throughout southwestern, south-central, and eastern Africa. There are three suggested subspecies (see Böhme 1988; and Bennett 1998) (Fig. 5.2).

Fossil Record

Varanus rusingensis is considered to be a possible ancestor of *V. albigularis* (Clos 1995). Fossils date to the early Miocene of Kenya, Africa, about 18 mya.

Diagnostic Characteristics

- Distinctive dark temporal stripe originates at the orbit and continues to the shoulder, where it commonly branches.
- Head color is solid gray or brown above, with throat much lighter, especially below the temporal stripe.
- Snout is blunt and bulbous with distinct thickening superior to the nares, especially in adults.
- Body pattern with dark rosettes with cream-colored center; in adults, the rosettes are much less distinct and can give the illusion of bands around the ribcage.
- Sharp, recurved claws.

Size

Size is variable depending on geographic location, with larger specimens in the range of *V. a. albigularis* and *V. a. microstictus* (Bennett 1998). Sexual dimorphism is apparent only after onset of reproduction, with males being larger. Before reproduction, the sexes do not differ in mass/snout–vent length (SVL) ratio. For *V. a. albigularis*, reproductive status was noted in individuals with SVL >500 mm (Phillips 1995). Body length is generally 45% of total length. Body mass of reproductive class males ranged from 5 to 8 kg, whereas females ranged from 4.5 to 6.5 kg in *V. a. albigularis* (Phillips 1995). Captive animals can become extremely obese, with individuals weighing over 20 kg (Branch 1991). Hatchlings are about 120 mm SVL and weigh 20 to 25 g (Phillips and Packard 1994; Horn and Visser 1989; Visser 1981).

Habitat and Natural History

Terrestrial, Arboreal, Aquatic

V. albigularis is mainly a terrestrial species that nevertheless spends a considerable amount of time in trees, especially during the reproductive season and while escaping predators. *V. albigularis* will actively hunt for prey in trees as well as on the ground, and will also use trees as refuges from midday heat or as overnight retreats (Branch 1988).

Time of Activity

During the summer, *V. albigularis* is active throughout the daylight hours, except at midday, when ambient temperatures are extreme. During winter months, the animals are far less active, although they remain alert. The lizards generally remain in their overnight refuges during the coldest months but often have their heads protruding from the refuge entrance. Winter inactivity appears to be the result of lack of available prey, as experimentally increasing food supply resulted in a 30-fold increase in activity (Phillips 1995).

Foraging Behavior and Diet

V. albigularis is an active predator. The species will eat anything it can subdue. When invertebrate prey are abundant during the wet season, snails, millipedes, grasshoppers, beetles, and crickets form the bulk of the diet. When insect flushes are great, especially grasshoppers and crickets, lizards need to move only short distances to acquire sufficient food (Branch 1988; Phillips 1995).

Snakes are a favored food item (Phillips and Alberts 1992). In Namibia, cobras (*Naja nigricollis, N. haje*), adders (*Bitis arietans, B. caudalis*), and sand and grass snakes (*Psammophis* spp.) were commonly subdued. Only rock pythons appeared to be avoided, perhaps as a result of their extreme size. Even hatchlings would attack snakes (Phillips and Alberts 1992).

V. albigularis also preys on the eggs and chicks of both ground- and tree-nesting birds. The largest bird consumed by this lizard in Namibia was a barn owl. Farmers also report chickens being eaten by *V. albigularis* (Branch 1991).

In Namibia, *V. albigularis* did not appear to consume any small mammals, even though the lizards often shared underground refuges with ground squirrels (*Xerus inauris*). Ground squirrels were the only rodent species with a daily activity pattern that was coincident with that of the lizards (Phillips 1995).

V. albigularis appears to discriminate between prey items using visual and chemical cues (Phillips and Alberts 1992; Kaufman et al. 1994, 1996). A result of their selectivity is acquisition of a high caloric reward at a low energetic cost (Kaufman et al. 1996).

Reproduction

On the basis of circulating hormones, both sexes appear to become sexually mature at a body size of 500 mm SVL. In southwestern Africa, mating occurs in early spring (August and September), about 3 months before the onset of significant rainfall. In Namibia, onset of mating activity was 1 month earlier in eastern populations than western populations and was aligned to the earlier rainfall in the east (Phillips and Millar 1998).

Males will mate with more than one female. Male–male interactions only occur in the presence of estrus females. At all other times of the year, males and females are solitary and do not interact. Males often move 1 to 4 km per day to locate receptive females. Once located, mating occurs over a 1- to 2-day period. When receptive, females are almost always located in trees, which suggests that a pheromone attracts the males (Phillips and Millar 1998).

Oviposition occurs about 35 days after mating, about 2 months before the onset of significant rainfall. Clutches are large (up to 50 eggs), with larger females producing larger clutches. Numerous clutches of *V. albigularis* have been incubated with various results (Staedeli 1962; Horn and Visser 1989; Visser 1981). In controlled experiments, hatchlings pip after about 135 days of incubation, which coincides with the middle of the rainy season (Phillips and Packard 1994). The sex ratio of hatchlings is 1:1.

Growth rates of individuals in Namibian populations suggest that sexual maturity is reached at about 4 to 5 years (Phillips 1995).

Movement

Home range of this species is large, with males averaging 18.3 km^2 and females 6.1 km^2 in southwestern Africa. The entire home range was used during the rainy season when invertebrate prey was abundant. Individual lizards often traveled up to 5 km per day in search of prey. During the dry season, individual lizards utilized less than 10% of their wet season home range. In Namibia, individual lizards did not appear to have a preference among grassland, mixed grassland–scrub, scrub, and woodland habitats. Lizards utilized refuges in all habitat types. Specific refuges, both terrestrial and arboreal, were utilized on multiple occasions (Phillips 1995; Bayless 1997).

During the rainy season, the sex ratio of lizards encountered along transects was 1:1; however, during the mating season, because only males exhibited substantial movement, the apparent (and misleading) sex ratio was 6:1 (Phillips 1995).

Physiology

Both sexes exhibit a definite annual cycle of circulating testosterone. Levels of testosterone peak during the period of observed mating in both sexes (males 100 ng/mL, females 40 ng/mL). Peak levels of estrogen in females also coincided with the mating period (1000 pg/mL). After mating, testosterone levels in females dropped to baseline, whereas testosterone levels in males returned to baseline within 2 months of mating (Phillips and Millar 1998).

Although *V. albigularis* is an active predator, it exhibits the digestive physiology of a sit-and-wait hunter (Secor and Phillips 1997). This may be the result of the long period of gut quiescence during the dry season, as well as the ability of the species to consume large meals (>10% of total body mass).

Mean active temperature was 32.3°C (Bowker 1984).

Body mass of individual lizards was greatest immediately after the rainy season. Once prey populations decreased, lizards lost 4% to 5% of their body mass per month. By the end of the dry season, males were about 60% of rainy season peak body mass, whereas females were about 50% (Phillips 1995). Because reproduction occurred during the dry season, weight loss in females also included production of clutch mass (generally representing 25% to 40% of the female's dry season body mass) (Phillips and Packard 1994). Both sexes recover their dry season loss in body mass within 2 months of the onset of the rainy season. Skeletal growth in reproductive individuals appears to occur only after fat levels (as reflected in body mass) have been reestablished. Body fat is stored in the tail and abdomen. Individuals that have lost a substantial portion of the tail are generally in poorer condition than individuals with intact tails (Phillips, unpublished data).

5.2. *Varanus exanthematicus*
DANIEL BENNETT

Figure 5.3. Varanus exanthematicus *(photo by Daniel Bennett).*

Nomenclature

First described as *Lacerta exanthematica* by Bosc in 1792 (*Act. Soc. Hist. Nat. Paris* 1: 25) from a specimen collected on the Senegal River and corrected to *Varanus exanthematicus* by Merrem in 1820, a name it shared with *Varanus albigularis* until 1989 (Böhme et al. 1989) (Fig. 5.3). Mertens (1942) placed the species in the subgenus *Empagusia* along with *Varanus flavescens*; subsequently Ziegler and Böhme (1997) placed it alongside *V. albigularis*, *V. niloticus*, *V. ornatus,* and *V. yemenesis* in the subgenus *Polydaedalus*. The status of *Varanus exanthematicus ocellatus* from Sudan (Heyden in Ruppell, *Atlas Reise N. Afr. Rep.* 1:25) remains uncertain, with some authors attesting a close relationship with *V. albigularis* (Anderson 1898; Müller 1905), others with *V. exanthematicus* (Schmidt 1919; Mertens 1942), and recent authors preferring to leave it unclassified (Böhme 1997). Specimens I have examined (British Museum of Natural History [BMNH] 1924.5.21.5, 1933.11.17.48, 1969.11.4.1, and 60 trimmed and tanned skins imported to Egypt from Sudan) are very similar to typical *V. exanthematicus* from other parts of Africa. In the absence of further information, I consider *V. ocellatus* a synonym of *V. exanthematicus.*

Geographic Distribution

V. exanthematicus occurs in grasslands and woodlands of sub-Saharan Africa north of the equator. It has been recorded from Benin, Burkina Faso, Cameroon, Central African Republic, Chad, Congo, The

Figure 5.4. Geographic distribution of Varanus exanthematicus *(based on map in Bennett 1998 with permission of the publisher).*

Gambia, Ghana, Guinea, Guinea-Bissau, Ivory Coast, Liberia, Mali, Mauritania, Niger, Nigeria, Senegal, Sierra Leone, Sudan, Togo, and Zaire (Mertens 1942; Luxmoore et al. 1988; Röder and Horn 1994). The limits southward, particularly with relation to the distribution of *V. albigularis,* remain unclear (Fig. 5.4).

Fossil Record

No fossils of *V. exanthematicus* are known. The only fossil varanid reported from Africa is the 18-million-year-old *Varanus rusingensis,* from Lake Victoria, which shows close affinity to the present-day *Polydaedalus* group (Clos 1995).

Diagnostic Characteristics

• Distinguished from *V. flavescens* by the slit-shaped nares and from *V. albigularis* by the flattened scales on the neck.

• Nares almost equidistant between eye and tip of snout.

• Large scales all over the body.

Description

V. exanthematicus is a stockily built monitor lizard with a short tail and a large head. Nares are slitlike and are situated slightly closer to the eye than the tip of the snout. Tail is short (90%–120% of SVL), with a low double keel on the median third. Body color is basically brown, sometimes gray or dull orange, with varying numbers of lighter, dark-edged spots over the dorsum that sometimes form bands or a reticulum. The tail is uniform or shows inconspicuous banding. Newly hatched individuals often have a noticeable yellow tinge to the snout that disappears within weeks.

Supraoculars are not enlarged; other scales on the top of the head are larger and irregularly shaped, increasing in size and becoming more regularly oval shaped and flattened on the neck. Each nuchal scale is surrounded by clusters of small scales. Scales on the dorsum have a pit on their smooth surface and one or two dark patches, most prominent on the hind legs and at the base of the tail. Ventral scales are smooth, rectangular, and regular. There are 81 to 103 scales around the body and 58 to 73 scales from the gular fold to the insertion of the hind limbs.

Size

Hatchlings measure about 70 mm SVL and weigh 6 to 7 g. Animals over 50 cm SVL (100 cm total length) are rare in nature, although captive specimens are known to reach about 75 cm (150 cm total length). Thirty-one adults I have caught in Ghana had a mean SVL of 320 ± 61.9 mm (mean ± SE; range, 237–470 mm) and a mean mass of 709 ± 379 g (range, 300–1698 g). Cansdale (1951) remarked that after many years in Ghana, he had never seen one larger than 70 cm total length. Thus, I consider Yeboah's (1993) statement that the mean length of 16 adults caught at Cape Coast was 130 cm to be in error.

Seventy-six juvenile animals caught in March and April 1996 had a mean SVL of 90 ± 11.4 mm (range, 70–118 mm) and a mean mass of 15 ± 6.2 g (range, 6–36 g). Juveniles in the same area caught in August and September 1994 had a mean SVL of 159 ± 26.7 mm (range, 103–243 mm, n = 206) and a mean mass of 87 ± 46.1 g (range, 17–287 g, n = 181). Assuming similar conditions in these years suggests average growth rates of about 13 mm and 12 g per month over the first 6 months of life (Bennett 2000c). Increases in length are slower than for neonate *V. bengalensis* of a similar size (Auffenberg 1994).

Habitat and Natural History

The vast majority of reports on *Varanus exanthematicus* in the literature actually refer to *Varanus albigularis*. The primary sources are Cissé (1971, 1972, 1976), who reported on the diet, activity, and reproductive cycles of animals from Senegal and my studies, based mainly in the coastal plain of Ghana (Bennett 2000a, 2000b, 2000c). The species occurs in Sahel, Sudan, and coastal grasslands and woodlands, but appears to be absent from the rain forest belt. *V. exan-*

thematicus is broadly sympatric with the larger *V. niloticus,* but it is not restricted to areas with surface water and can exist in many places that lack sufficient moisture to sustain *V. niloticus.*

At Fissel-Diaganiao in Senegal, Cissé (1971) found *V. exanthematicus* in woodland areas with seasonal pools. Where large trees are common, the lizards sheltered in tree hollows; in sparse woodland, they used burrows. In Ghana, the species appears to be most common on sandy soils, especially in areas with mosaics of cultivated and uncultivated land. In these areas, many juvenile animals live in fields, whereas adults occupy grasslands. They are also common on rocky hillsides favored by *Python sebae.* It is not well known from riverine forests and other closed-canopy woodlands, although it is recorded from savanna–forest transitional zones along the Togo border (Bennett 2000c).

Schmidt (1919) considered the species to be rare in grasslands around Garamba (Belgium Congo); Angel (1933) thought it was rare around Goudam, Sudan. At the end of the 19th century, it was considered common in Guinea Bissau (Manaças 1955). Dunger (1967) stated it was common in northeastern Nigeria. In a mark-recapture study, Bennett (2000b) calculated densities of 357 juveniles (biomass 30,900 g) per square kilometer in a sandy farm and grassland mosaic near Accra, Ghana, and suggested they were the offspring of about 24 reproductively active females. In good habitat, an experienced searcher finds a mean of 0.54 ± 0.47 juveniles (range, 0–1.81 animals) or 0.43 ± 0.28 adults (range, 0–0.83 animals) per hour (Bennett 2000b and unpublished data).

Predators of *V. exanthematicus* include rock pythons *Python sebae* (Lenz 1995), garter snakes *Elapsoidea guentheri* (Bennett and Basuglo 1998), birds of prey (personal observation), other monitor lizards (Cissé 1972), and people (Luxmore et al. 1988; Bennett 2000c).

Schmidt (1919) noted a posture adopted by threatened lizards in which the animal grasps its rear right leg in its jaws and feigns death. I have witnessed this behavior only once and presume it serves to make the animal difficult for a snake to swallow. Specimens confined with pythons sometimes grasp the snake behind the head and remain locked in this position until the potential predator has suffocated.

The relationship between juvenile *V. exanthematicus* and the cricket *Brachytrupes membranaceus* in parts of the coastal plain of Ghana appears to be important. Lizards hatching in farmland invariably inhabit the crickets' burrows as soon as they leave their nest and rarely use other shelters until they reach a weight of over 200 g. When the lizards are a few weeks old, *Brachytrupes* becomes their most important prey species, and remains so apparently until the lizards are too large to enter the burrows and leave farmland to seek shelters in uncultivated areas.

Terrestrial, Arboreal, Aquatic

V. exanthematicus shelters in burrows and on the branches of trees and bushes. In Senegal, Cissé found them mainly in tree hollows or in burrows dug by other animals. In Ghana, juveniles are commonly found in burrows of the large orthopteran *Brachytrupes*. Mean depth of 82 occupied burrows was 26 ± 9.3 cm (range, 12–57 cm). Adults often shelter in abandoned termitaria or the burrows of rodents, squirrels, and other mammals. Mean depth of 12 burrows occupied by adults was 59 ± 28.6 cm (range, 17–100 cm). Although they regularly use trees in many areas, they are not particularly agile climbers. Like all monitors, *V. exanthematicus* is a strong swimmer.

Time of Activity (Daily and Seasonal)

Cissé (1971) reported strongly seasonal behavior in *V. exanthematicus,* with most feeding occurring during the wet season and activity greatly reduced between late December and April. In Ghana, juveniles probably feed throughout the first year of life, whereas adult activity is greatly reduced toward the end of the dry season (February–March), when animals may remain under the same bush for more than a month at a time.

In Senegal, Cissé (1971) suggested that emergence from burrows occurred around 0900 and animals returned around 1730, suspending activity between 1400 and 1500 hours. Yeboah (1993) reported emergence from burrows between 0630 and 1000 hours at air temperatures of 27 to 30°C. My observations suggest time of emergence depends on climatic conditions, but that monitors are least likely to be found in burrows from 1100 to 1400 hours.

Thermoregulation

Little information is available about the thermoregulatory behavior of *V. exanthematicus*. Suitable temperatures for activity occur throughout the year in all parts of its range. Cloacal temperatures taken from 106 animals in Ghana (mainly juveniles) ranged between 26.2 and 39.1°C (mean, 31.5 ± 2.72°C) (Fig. 5.5). Cloacal temperatures were almost always higher than ambient temperatures, but there was no apparent relationship between them ($r^2 = 0.09$). Thirty-six burrows examined at all times of the day had a temperature of 27.2 ± 1.5 °C (range, 25.0–31.0°C), similar to lowest cloacal temperatures encountered (unpublished data). The low r^2 value suggests that *V. exanthematicus* is an active thermoregulator.

Foraging Behavior

V. exanthematicus finds most of its food by foraging on or below the surface. The animals root through debris with their snout, dig for prey with their feet, and regularly explore burrows, often dragging prey to the surface to swallow it. In Ghana, almost all important prey of juveniles are taken from burrows. Larger animals often excavate burrows in search of prey, and find most food here and in leaf litter. Cissé (1972) suggested most food items at his sites in Senegal were taken from the base of plants, among *Ipomea* flowers, and beneath ruminant dung.

Prey eaten by young *V. exanthematicus* are considerably larger than those consumed by juvenile *V. bengalensis* (Auffenberg and Ipe 1983; Auffenberg 1994) or juvenile *V. niloticus* (Bennett 2002). Individual prey up to 28% of body weight have been recorded from 2-month-old animals that consume *Brachytrupes* crickets and *Pandinus*

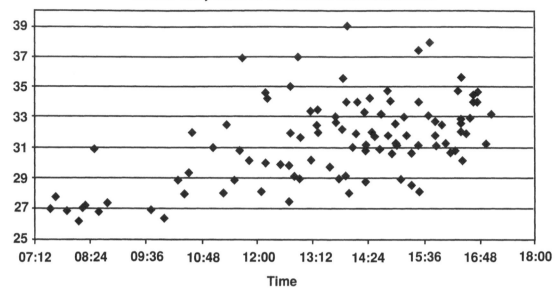

Cloacal Temperatures of *Varanus exanthematicus*

Time

Figure 5.5. Body temperatures plotted against time of day.

scorpions. On a good day, year-old lizards consume about 18% of their body weight, typically consisting of two or three prey items.

Along with *V. niloticus, V. ornatus,* and the Filipino *V. olivaceus, V. exanthematicus* shows adaptations that facilitate the crushing of hard prey, principally snails (Lönnberg 1903; Mertens 1942; Auffenberg 1988). However, hard-shelled prey are not common in diets of studied animals (see below), and the extent to which molluscivory is important to the species is unknown.

Diet

V. exanthematicus feeds almost exclusively on arthropods and mollusks. It is a voracious feeder. Cissé (1972) reported on stomach contents of 28 animals caught around Fissel-Diaganiao, Senegal, over 12 months and found that animals fasted during the driest part of the year, between mid-November and February. The most common prey was the large millipede *Iulus,* followed by beetles (13 different species), lepidopteran larvae (mainly *Chenilles* and sphingids), orthopterans, lizard eggs (of its own species and *Agama*), snails (Helicarionines), and buthid scorpions. Millipedes were replaced by orthopterans in the diet from the end of September until suspension of feeding in December. Bennett (unpublished data) investigated the diet of 71 hatchling *V. exanthematicus* (1–8 weeks old, mean SVL 90 ± 10.7 mm, range, 73–118 mm, mean mass 15.5 ± 5.7 g, range, 8–36 g) in farmland at Katomensa in the Greater Accra Region of Ghana using dissections, stomach flushing, and fecal collection. Results are summarized in Table 5.1. Most important prey by weight were small crickets, followed by locusts, *Brachytrupes* crickets, slugs, centipedes, coleopterans, pupae,

TABLE 5.1.
Diet of juvenile *Varanus exanthematicus* in Ghana determined by stomach flushings, fecal samples, and dissections.[a]

Stomach Contents	Feces and Flushings (% of total)	Occurrence (%) (n = 57)	Dissections (% of total)	Occurrence (%) (n =14)
Crickets	41.8	59.6	55	46.7
Brachytrupes	25.4	10.5	5	13.3
Locusts	28.7	19.3	2.5	6.7
Centipedes	1.0	1.8	5	13.3
Beetles	2.1	7.0	2.5	6.7
Wasps	0.7	1.8	0	0
Pupae	0.3	1.8	2.5	6.7
Scorpions	0	0	2.5	6.7
Slugs	0	0	25	26.7

[a] Data are based on weights of intact prey items.

scorpions, and wasps. Maximum predator–prey ratio (the ratio of the mass of the empty lizard to original mass of intact prey items in the gut) was 0.24. The smallest animals fed mainly on crickets; the smallest individual to have eaten *Brachytrupes* was 92 mm SVL (16 g). Scorpions were represented by a single pedipalp from *Pandinus imperator* eaten by a 105-mm-SVL (23 g) individual that was apparently unable to swallow the entire prey. Juvenile animals in the same population at the age of approximately 6 to 7 months fed mainly on *Brachytrupes* and other orthopterans, *Pandinus* scorpions, amphibians, snails, and beetles (Bennett 2000a).

Overall predator–prey ratios (ratio of intact prey mass to empty predator mass) for some individuals were higher than 0.3. These data suggest an ontogenetic shift in diet over the first few months of life from smaller orthopterans and soft mollusks to giant crickets, scorpions, and snails. Juvenile lizards apparently maintain this diet until they outgrow available burrows in farmland and move into uncultivated areas. The diet of adult *V. exanthematicus* in Ghana between March and September consists largely of *Iulus* millipedes, *Pandinus* scorpions, beetles, orthopterans (but only rarely *Brachytrupes*), and snails (Bennett 2000c).

Reproduction

Like other *Polydaedalus* monitors, *V. exanthematicus* produces large clutches of eggs that hatch relatively quickly. Cissé (1976) examined 162 males and 121 females caught over 12 months. Testes weight was lowest from December to June and greatest from August to October, with spermatogenesis occurring from April to October. Ovaries began to increase in size from April, with explosive growth in September followed by a rapid decrease from October to December. Seventeen females examined in September and October all contained oviductal eggs or had recently ovulated. The relative weight of mature ovaries was much higher than for *V. niloticus* (*exanthematicus* mean of 7.4% of body weight vs. *niloticus* mean of 0.6% at the height of the breeding season). The smallest female containing eggs found by Cissé was 500 g; the smallest I have found in Ghana with enlarged follicles measured 274 mm SVL and weighed 350 g.

In Ghana, mating and egg laying occur mainly in November and December. Eggs hatch early in the rainy season (March–April), although in unusually dry years, eggs may be only half developed by the end of April. A female from Niger (BMNH 99.8.23.6) also contained shelled eggs in November. In Senegal, hatchlings emerge in July (Cissé 1976), and neonate animals were collected from Sudan in June (BMNH 1924.5.21.5).

Cissé records a maximum clutch size of 41 eggs. Switak (1998) reported 27 eggs laid by a female near Accra in November. Four nests I found in Ghana contained an average of 18 eggs (range, 6–29 eggs), all showing 100% hatch rate. They were shallow (20–25 cm) burrows 30 to 50 cm long, dug in sunny locations around farmland, and refilled (once with a plug of vegetation at the distal end of the shaft). Neonates appear through a hole they dig out of the roof of the nest chamber. In captivity, up to four clutches of eggs can be produced in a year.

Movement

Where food is abundant or conditions unduly dry, animals' movements are restricted to a small area. Spool and line tracking of adults during the usually dry period of March and April 2001 revealed no daily movement further than 6 m over periods of up to 18 days (unpublished data). Juveniles remained in the same 1.5-ha plots for up to 2 months despite being regularly excavated and recaught (Bennett 2000a), suggesting an abundance of both food and refugia. Observations of adult males crossing the Black Volta River at the height of the rainy season suggest some adults must have large activity areas. In Ghana, the normally secretive animals are regularly observed by people during the hot months of November and December, presumably corresponding to an increase in activity during the breeding season.

Physiology

V. exanthematicus is easily available in Europe and North America, and there is a substantial body of literature on various aspects of its anatomy and physiology. However, some of these reports actually refer to *V. albigularis,* and in many cases the identity of the study animals is unclear. They include studies of the skull (Mertens 1942), brain and nervous system (Hoogland 1981; Bangma and ten Donkelaar 1984; ten Donkelaar et al. 1985, 1987; ten Donkelaar and de Boer-van Huizen 1981, 1984a, 1984b, 1988; Bangma et al. 1984; Barbas 1982; Barbas and Lohman 1984, 1986, 1988; Wolters et al. 1986a, 1986b; Barbas and Wouterlood 1988; Luebke et al. 1992; Sakai et al. 1995), respiratory and circulatory systems (Wood et al. 1978, 1981; Glass et al. 1979; Burggren and Johansen 1982; Heisler et al. 1983; Maina et al. 1989; Hopkins et al. 1995; Hicks et al. 2000), hemipeneal morphology (Böhme 1988, 1991b), physiological aspects of starvation (Adjovi 1981; Dupe-Godet and Adjovi 1981a, 1981b, 1983; Dupe-Godet 1984; Godet et al. 1983, 1984), shoulders (Jenkins and Goslow 1983), forearm (Landsmeer 1983), and skull musculature and kinetics (Smith 1982; Smith and Hylander 1985).

Fat Body Cycles

Few data on fat bodies in wild *V. exanthematicus* are available. Juvenile *V. exanthematicus* in Ghana do not accumulate fat deposits as described by Auffenberg (1994) for *V. bengalensis*. Cissé (1971, 1973) hints that fat reserves are greatest at the end of the rainy season, just before estivation, but does not provide details.

Parasites

Most records of parasites in this species come from captive individuals, and many reports probably refer to *V. albigularis*. Cumming (1999) reported *Amblyomma sparsum* infesting *V. exanthematicus,* and species of *Aponomma* infest many wild individuals in Ghana (87% of animals more than 1 year old, 11% of juveniles).

5.3. *Varanus griseus*

MICHAEL STANNER

Figure 5.6. Varanus g. griseus
(photo by Adel Ibrahim).

Nomenclature and Systematics

Varanus griseus was first described by Daudin (1803) as *Tupinambis griseus* (Fig. 5.6). The type locality is Egypt. Boulenger (1885) renamed the species to its current status. *Varanus griseus* is the only species in the subgenus *Psammosaurus* (Mertens 1942). Mertens (1954) distinguished three subspecies: *V. g. griseus* (Daudin 1803): type locality—Egypt; *V. g. caspius* (Eichwald 1831): type locality—Halbinsel Dardsha on the eastern coast of the Caspian Sea; and *V. g. koniecznyi* (Mertens 1954): type locality—Korangee near Karachi, Pakistan. On the basis of morphological features, Mertens (1942) concluded that the closest relative of *Varanus griseus* is the south Asian *Varanus bengalensis* of the *Indovaranus* subgenus. More recent studies (King et al. 1991, 1999; Ast 2001) of morphological features, chromosomal groups, rbDNA, and mtDNA sequencing conclude that the closest relatives of *V. griseus* are *V. niloticus* and *V. exanthematicus* (subgenus *Polydaedalus*), with which it forms a distinct basal Afro-Asian clade. Ziegler and Böhme (1997) removed *V. exanthematicus* from *Empagusia* to *Polydaedalus* (*Empagusia* now includes *V. bengalensis*, *V. dumerilii*, *V. flavescens*, and *V. rudicollis*).

Geographic Distribution and Habitat

Varanus griseus is the northernmost *Varanus* species, and its distribution area is one of the largest among monitor lizards, extending from

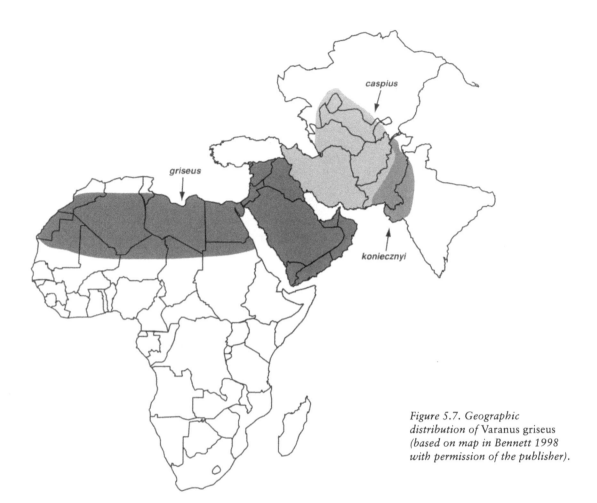

Figure 5.7. Geographic distribution of Varanus griseus *(based on map in Bennett 1998 with permission of the publisher).*

the Rio De Oro eastward throughout the Sahara Desert (except for coastal areas north of the Atlas Mountains in northern Morocco, Algeria, and Tunisia), to Egypt and northern Sudan, the Arabian Peninsula, central Asian deserts, Pakistan, and northwestern India (Fig. 5.7). The northern distribution area reaches the Caspian and Aral Seas (latitude 46°N; Makayev 1982). *Varanus griseus* is mainly a desert species, but it also inhabits semiarid as well as nonarid areas. Typical habitats are sand dunes, clay steppes, riverbeds, and savanna plains up to altitudes of 1300 m (Anderson 1963).

   Varanus griseus is distributed over an array of climates: dry subtropical climate (Sahara Desert), mild Mediterranean climate (coastal plain of Israel), dry tropical climate (northwestern India) and dry terrestrial climate (Kyzylkum Desert). Besides climate, the main factor limiting its distribution seems to be sandy or light soils in which it can burrow. Moreover, various traction marks in the sand were found to play an important role in intraspecific communication and courtship (Tsellarius and Menshikov 1994; Tsellarius and Tsellarius 1996), as well as in orientation and navigation (Vernet 1977; Stanner 1983) and in tracking of prey (Tsellarius et al. 1997a). Home ranges of individual desert monitors may include stretches of regs and hard soils, but the presence of sandy or light soils seems to be essential. In Israel, *Varanus griseus* avoids seashore areas affected by saltwater spray, and its distribution is limited to more inland areas.

The nominal subspecies is mainly a Saharo-Arabic lizard. Its distribution area extends across northern Africa, the Arabian Peninsula, west Asia: Israel, Syria, Lebanon, southeastern Turkey, Jordan, Iraq eastward to the Zagros Mountains (Mertens 1942; Leviton and Anderson 1970). Martens and Kock (1992) claim that its occurrence in Lebanon is questionable and not supported by more recent surveys. The eastern subspecies *V. g. koniecznyi* is found in Pakistan and northwestern India (Auffenberg et al. 1990). The distribution area of the northern subspecies *V. g. caspius* extends from the eastern shores of the Caspian Sea and southern Kazakhstan to the central areas of Asian Russia, throughout the plateau of Iran east of the Zagros Mountains, western Baluchistan and Afghanistan, and western Pakistan (Auffenberg et al. 1990; Anderson 1999). In eastern and southern Baluchistan, *V. g. caspius* and *V. g. koniecznyi* probably intergrade (Minton 1966).

Fossil Record

No fossils of *Varanus griseus* are known (Kuhn 1963), but Levshakova (1986) described a skull in comparatively good condition of a new *Varanus* species (*Varanus darevskii*) from the lower Pliocene found near the Sor village in Tajikistan. Levshakova (1986) postulates that *V. darevskii* is a possible ancestor of *V. griseus* and places it in the subgenus *Psammosaurus*. *Varanus darevskii* is somewhat smaller than *V. griseus,* its skull is higher, the snout is shorter and more pointed, and its teeth are bigger. Furthermore, Levshakova (1986) argues that the Pliocene *Varanus marathonensis* from Pikermi in Greece should be placed in the same subgenus. Hence, the formerly monotypic subgenus *Psammosaurus* should include three species, the Pliocene *V. darevskii* and *V. marathonensis* and the recent *V. griseus*.

Diagnostic Characteristics

Varanus griseus is a medium-sized varanid, usually up to 1 m in total length. The general color is light brown or yellow to dark gray, with or without darker transverse bands on the back and tail, with or without yellow patches on the back. Nostrils are diagonal slits situated closer to the eyes than to the tip of the snout. Juveniles are vividly orange colored with distinct transverse black bands on the back and tail.

Size

Varanus griseus conforms to a certain extent with the Bergmann Law: the northern subspecies *V. g. caspius* is the largest, and the southern (and easternmost) subspecies *V. g. koniecznyi* is the smallest (Table 5.2). In the *Varanus* literature, morphometrics usually refer to length (SVL or total length) but tend to ignore weight. Weight of free-

TABLE 5.2.
Average length and weight of adult free-ranging desert monitors, *Varanus griseus*.[a]

Subspecies	TOL (cm)				Weight (g)				TL/SVL
	Mean ± SD	Range	n	Reference	Mean ± SD	Range	n	Reference	
griseus	83 ± 8	62–102	21	(1)	647± 252	239–1261	21	(1)	1.31–1.46 (4)
caspius	105 ± 10	83–130	106	(2)	1482 ± 582	610–3320	99	(2)	1.48 (3); 1.23–1.67 (4)
koniecznyi	62 ± 4	41–84	94	(3)	277 ± 118	62–580	94	(3)	1.17–1.27(4); 1.09–1.28 (3)

[a] TOL, total length; TL, tail length; SVL, snout–vent length; SD, standard deviation; n, sample size.
References are as follows: (1) Stanner (1983); (2) Tsellarius (personal communication); (3) calculated from SVL data in Auffenberg et al. (1990) on the basis of an assumption of an average 1.215 TL/SVL ratio; (4) Mertens (1954).

ranging desert monitors usually fluctuates in accordance with reproductive state, environmental conditions, and other factors, and adult monitors can almost double their weight in the course of several months (Stanner 1999). Mertens (1942c) mentions that *V. griseus* may reach 150 or even 155 cm (probably *V. g. caspius*). Khalaf (1959) states that desert monitors in Iraq reach 120 cm (probably *V. g. griseus*). The largest monitor I measured during my study of *V. griseus* in Israel was an old captive male from the former Tel Aviv Zoo (total length, 109.8 cm; weight, 2918 g). Desert monitors encountered in the field are usually much smaller (Table 5.2).

Body sizes of desert monitors of the same subspecies may vary in areas with different environmental conditions. Mosauer (1934) reports that in Tunisia, specimens from Gafsa (dry and harsh desert) are bigger than those from Tozeur (wetter and more productive). In Algeria, specimens larger than 80 cm are rare (Mammeir in Bennett 1998). On average, males in the coastal plain of Israel are 15% larger than females (90 vs. 78 cm) and 72% heavier (873 vs. 508 g) (Stanner 1983). In the western Kyzylkum Desert, males of *V. g. caspius* are 7% larger than females (106 vs. 99 cm) and 34% heavier (1570 vs. 1174 g) (Tsellarius, personal communication). Males have larger heads and are more robust in general appearance. Females, on the other hand, look more gentle. These differences enable an experienced observer to distinguish between the sexes from a distance (Stanner, unpublished observations). In contrast, Smirina and Tsellarius (1996) argue that males do not necessarily grow faster than females, and that at the population level, males are larger than females because their survival in the field is higher and they are therefore older on average. Auffenberg et al. (1990) found that males of *V. g. koniecznyi* are only slightly, but significantly, larger than females. Tsellarius et al. (1991) report that in the Kyzykum Desert, *V. g. caspius* grow quickly and reach adult size within 2 years. Smirina and Tsellarius (1996) report that up to 3 years old, the age of *V. g. caspius* can be determined accurately by measuring body length, and less accurately up to 4 years old. From age 4 years onward, size classes are indistinguishable. Perry et al. (1993) report that in captivity, neonates and juvenile desert monitors grow slowly. I have noticed the same phenomenon (personal observation), which might be caused by the uniform and dull diet of captive monitors compared with the wide variety of prey animals in the field (see "Foraging Behavior and Diet") and/or by the lack of physical exercise.

Description

The main morphological features of *Varanus griseus* are the diagonal slit-shaped nostrils situated closer to the eyes than to the tip of the snout, and the tail that is rounded in cross section. The latter statement is correct only for the nominal subspecies. The tail of *V. g. caspius* is laterally compressed (especially distally) with a distinct keel on the back. The tip of the tail is rounded. In the eastern subspecies *V. g. koniecznyi,* the shape of the tail is more rounded (Auffenberg et al. 1990) with a weak lateral compression in the distal half (Mertens 1954). However, in ill-nourished or emaciated monitors, the tail may have a somewhat triangular cross section due to depletion of fat bodies inside the tail. Relative tail length varies among the three subspecies (Table 5.2), but this trait is not a useful diagnostic characteristic because relative tail length decreases with age, especially in old individuals (Mertens 1942). In many cases, the tip of the tail is cut off and lost —for example, 12% of *V. g. koniecznyi* studied by Auffenberg et al. (1990) in Pakistan had broken tails, as did 40% of the males (but only 9% of the females) of *V. g. caspius* studied by Tsellarius (personal communication) in the western Kyzylkum Desert in Uzbekistan.

Color and Pattern

The general color is gray, light brown, light yellow, or light orange. The ventrum is whitish. The back is covered with transverse dark bands that bifurcate at the flanks. At the tail, the dark bands are almost circumferential and are discontinuous at the ventral side of the animal. The dark bands are very vivid and distinct in juveniles, but they tend to fade away or even disappear in adults. These bands—their number, color, and location (or absence)—are one of the main diagnostic characteristics for the three subspecies of *Varanus griseus* (Table 5.3). Between the dark bands, if present, yellowish or white patches are arranged in nonorderly transverse lines on the back, tail, and legs. Light-colored individuals may lack such patches and may have instead dark grainlike dots dispersed all over the dorsal aspects of the body, tail, and legs, as well as on the ventral side of the head and throat. The tail of some individuals is covered with narrow yellowish transverse bands between the dark bands. Very often, the tail (especially the distal part) is almost uniformly black or brown. In some individuals, the

TABLE 5.3.
Color patterns of *Varanus griseus* of the transverse bands
on the back and tail.[a]

	Subspecies		
	griseus	*caspius*	*koniecznyi*
Bands on the back	6 (5–8); black, gray, or absent, usually narrow	6 (5–8); black, dark brown always present	4 (3–5); black or gray, broad
Bands on the tail	19–28, banded to the tip[b]	13–19, only proximal two-thirds is banded; distal parts yellow[c]	7–12 or 8–15; distal parts not banded

[a] Average (range) number of bands. Data based on Mertens (1942, 1954) and Auffenberg et al. (1990).
[b] In adults, the distal parts of the tail may be uniformly brown or black.
[c] In juveniles, the distal parts of the tail may be banded.

distal part of the tail has a somewhat reddish appearance. At the temporal region and on the lateral side of the neck, there are two to three longitudinal dark bands. The middle band on each side of the head extends backward and upward and merges with the opposite band to form a common dorsal transverse band at the proximal part of the neck.

Several (up to six) blurred dark lines encircle the snout laterally and dorsally between the eyes and the tip of the snout. Juveniles, especially neonates, are more vividly colored. The general color is yellow or orange, with very distinct intense black bands. As the monitor grows older, its general color becomes blurred, and the dark bands and lines tend to fade away and disappear. In the northern subspecies *V. g. caspius,* dark transverse bands are retained throughout a monitor's life, and in most cases bands bifurcate in a variety of patterns. Tsellarius and Cherlin (1991) use this trait to identify monitors individually. Mertens (1954) considers retention of transverse bands as a juvenile/primitive characteristic indicating that *V. g. caspius* is the most primitive subspecies. Hence, he concludes that *Varanus griseus* has radiated from central Asia westward to the Middle East and North Africa, and eastward to Pakistan and northwestern India.

Corkill (1928) reports that in early summer, bellies of *V. griseus* males in Iraq have a pinkish tinge. I noticed the same phenomenon in one male in Israel, but the pinkish tinge was very indistinct and was restricted to the ventral side of the head and throat. Auffenberg et al. (1990) also recorded it in *V. g. koniecznyi* in Pakistan, but they could not confirm seasonal color changes in either sex.

Activity Patterns

Varanus griseus displays strictly seasonal patterns of activity, being active from March–April to October–November and hibernating underground during the rest of the year. Peak activity is in May–June (Stanner and Mendelssohn 1991). Curiously, throughout most, if not all, of its large distribution area and across several climatic zones, the activity pattern is very uniform, with very little variation. However, within the activity season, time allocated to various types of activity by various population subgroups (juveniles, adults, males, reproductive females, nonreproductive females) may vary between different geographical areas, or even within the same geographical area where monitors are subjected to different environmental conditions. For instance, activity patterns of males held in outdoor enclosures in the southern coastal plain of Israel were not identical to males in the field. In the field, males terminated activity by the end of July, but in outdoor enclosures, they were active until October–November (Stanner 1999). *Varanus griseus* seems to be an obligatory hibernator, and if it is deprived (in captivity) of hibernation, it is most likely doomed to die, although death may be prolonged up to 1.5 years (Stanner 1999).

Vernet (1977) reports that in the Sahara, *V. griseus* undergoes a partial estivation. Activity level of *V. griseus* in both the southern coastal plain of Israel (Stanner and Mendelssohn 1991) and the western Kyzylkum Desert in Uzbekistan (Smirina and Tsellarius 1996) is greatly reduced after June, but in neither place were *V. griseus* found to estivate.

Corkill (1928) reports that *V. griseus* in the Shamiyah Desert in Iraq are active throughout the year, and that they are more abundant during the winter, but his reports are based on tales of local tribesmen ("For their truth I cannot vouch . . .") rather than on direct observations.

Intersexual Differences

In the southern coastal plain of Israel, males start activity somewhat earlier than females and cease activity much earlier than females. Females remain active until October–November, but males terminate activity by the end of July (Stanner and Mendelssohn 1991). Such intersexual differences may arise because postoviposition females must continue to feed to restore fat bodies to sustain themselves through hibernation. However, because males were in good nutritional condition, they could enter hibernation any time. This phenomenon may have adaptive value in terms of survivorship in the field because after copulation, well-nourished males need not continue activity and expose themselves to predators. In Algeria, Vernet (1977) did not find intersexual differences, maybe because productivity in the Sahara is lower than in the coastal plain of Israel and males must also continue to feed to restore fat bodies for hibernation. A somewhat different conclusion might be drawn from Vernet et al. (1988b) who report that the largest individuals (males?) estivate and then enter hibernation without reemergence. In the western Kyzylkum Desert in Uzbekistan, males of the Caspian monitor do not terminate activity as early as they do in the southern coastal plain of Israel, but from mid-August onward, males and nonreproductive females spend most of the time inside burrows (Tsellarius and Tsellarius 1996) and/or may migrate and disappear temporarily (Tsellarius et al. 1991). Smirina and Tsellarius (1996) report that the Caspian monitor enters hibernation over a prolonged period, from mid-August to early October, but do not report on inter-sexual differences.

Molting Cycle, Feeding Period, Growth, and Metabolite Balance

Like most lizards, *V. griseus* molts gradually, and like many other desert reptiles, it molts annually. In captivity, the process of molting may take several months, but in the field, it is synchronized over a shorter period, and most monitors start and finish molting between mid-June and early August (Stanner 1999). Unlike snakes, after molting is completed, most adult *V. griseus* stop feeding (found in captivity by Stanner 1999). Whether juveniles or neonates do this is unknown. A neonate at the Tel Aviv University (TAU) Zoo was feeding quite intensely in October, 1 month after it had completed its molting cycle. Reproductive females also feed intensely until October–November to restore fat bodies depleted during yolking up their eggs.

Shammakov (1981) assumes that *V. g. caspius* molts three times a year. His assumption is somewhat speculative and is based on molting frequencies at the population level rather than on direct observations of individual monitors. Igolkina (1975) reports that in the Leningrad Zoo, *V. griseus* molts three times a year.

Adult monitors held in outdoor enclosures at the TAU Zoo did not feed before late April to mid-May or after September–October, and peak feeding activity was observed in mid-May to late June. Conspecifics in the field had already started feeding in March (Stanner and Mendelssohn 1991) and continued to feed quite intensively until late October. In the western Kyzylkum Desert in Uzbekistan, *V. g. caspius* start to feed in late April to early May (Tsellarius and Tsellarius 1996) and males and nonreproductive females stop feeding from mid-August to mid-September; reproductive females continue to feed intensely until early October (Tsellarius and Menshikov 1995). Auffenberg et al. (1990) recovered food remains from stomachs of *V. g. koniecznyi* in Pakistan during April–November, whereas all individuals caught in December–March had empty stomachs and intestines.

Tsellarius et al. (1991) report that in the field *V. g. caspius* grow mainly in spring and early summer. Stanner (1999) found that in captivity growth (length increase) takes place from April–May to August–September or even October. During hibernation, little or no growth takes place. Adults gain weight from April–May to July–August or August–September (even in October in juveniles). During late summer, males and nonreproductive females usually lose weight at a rate higher than during hibernation (Stanner 1999).

Metabolites, electrolyte levels, and water turnover rates of *V. griseus* fluctuate with various phases of the annual cycle as well as with extent of activity (Haggag et al. 1965, 1966a, 1966b; Lemire and Vernet 1985; Vernet et al. 1988b). Khalil and Abdel-Messeih (1962) and Auffenberg et al. (1990) have studied seasonal changes of liver and abdominal fat bodies of desert monitors.

Daily Activity Patterns

Like most other varanids, *V. griseus* is strictly diurnal. In the southern coastal plain of Israel, the latest activity was recorded at 1740 on May 28, approximately 2 hours before sunset. Activity closest to sunset was at 1640 on October 14, approximately 1 hour before sunset (Stanner and Mendelssohn 1991). In Turkmenistan, Shammakov (1981) recorded active desert monitors as late as 1900 or even 2000 hours.

Varanus griseus in the southern coastal plain of Israel (Stanner and Mendelssohn 1991) display a unimodal activity pattern in April, May, and October and a bimodal pattern in June–September. Similar or close patterns were found also in the western Kyzylkum Desert in Uzbekistan (Tsellarius et al. 1991), Turkmenistan (Shammakov 1981), and north Sinai, Egypt (Ibrahim 2002). These patterns are probably true only at the population level. Individual monitors do not necessarily follow these rules, but they are rather active or inactive during many time intervals throughout the day (Stanner 1983; Stanner and Mendelssohn 1991). I have seen active individuals running over scorching stretches of sand during the hottest midday hours of July and August. On the other hand, monitors often stayed in burrows when environmental conditions seemed optimal. Vernet et al. (1988a) assume that contrary to common opinion, in the Sahara Desert, climatic factors play only a secondary role in determining daily activity patterns and extent of activity of *V. griseus*. They report that monitors were seen active on cloudy days or even during sandstorms. Conversely, other monitors remained inside burrows for several days during periods of calm and warm weather despite the fact that these conditions are favorable for hunting activity. According to Vernet et al. (1988a), the major factor influencing extent of daily activity is the type of biotope, that is, the density of vegetation. In areas of sparse vegetation (and low prey density) in the barren sand dunes of the Guir Hamada, monitors were active 2 to 8 hours per day and moved 0.2 to 2.5 km per day. In the more lush and wet areas of the Saoura Wadi, where vegetation and prey were more dense, monitors were less active (1–6 hours per day; 0.1–2 km per day). Yet the durations of these activities were less than the 9 to 12 hours per day reported by Sokolov et al. (1975) and Shammakov (1981) for *V. g. caspius* studied during the same time of the year (May) in Turkmenistan.

Thermoregulation

Varanus griseus is heliothermic and a precise thermoregulator. Its preferred body temperature (pbt) is 35 to 38°C. It is somewhat less thermophilic and less heat tolerant than the large vegetarian agamid *Uromastyx acanthinurus,* with whom it shares the same distribution area in the Sahara Desert. In the field, climatic conditions are not always favorable for *V. griseus*; hence, it often cannot maintain its body temperature within the preferred 35 to 38°C range (Grenot and Vernet 1977). Vernet et al. (1988b) report seasonal differences in mean pbts of free-ranging *V. griseus* in spring, 33.2°C; summer, 36.1°C; and autumn, 35.1°C. Below 20°C, the monitors become inactive, and hibernation temperature is between 16 to 18°C. At about 41°C, *V. griseus* starts to hyperventilate at about 120 breaths per minute, and cardiac arrhythmia occurs at a body temperature of 42°C. This arrhythmia is reversible if the temperature of individuals is returned to the pbt but becomes irreversible at 43°C. The lethal temperature is 44 to 47°C. Ibrahim (2000) reports that in north Sinai *V. griseus* hibernate at body temperatures of 15–30.5° and relates it to a behavioral response to the ambient temperatures or to prey availability. The pbt of *V. g. caspius* studied telemetrically in southeastern Turkmenistan by Sokolov et al. (1975) is 32.1 to 38.8°C, and by Tsellarius and Tsellarius (1997b), 37.2 ± 0.1°C (range, 33.8–41°C). Ibrahim (2000) reports on seasonal differences in mean dial body temperatures of free-ranging *V. griseus* in north Sinai. The body temperatures of free-ranging active *V. griseus* fluctuate between 33.7 and 41.9°, and the body temperatures before the morning emergence ranged between 27.7 and 34.5°. Moreover, he reports on intersexual differences in body temperatures (higher in males), as well as between resident and introduced monitors of both sexes (higher in introduced monitors), and attributes the higher body temperatures to higher activity levels of males and introduced monitors. According to Tsellarius and Tsellarius (1997b), in the Kyzylkum Desert in Uzbekistan, pbt is 36.9 ± 2.8°C (range, 33–41.5°C). They found that body temperatures of free-ranging *V. g. caspius* could be estimated from environmental temperatures as an average of the sand temperature and the air temperature.

Desert monitors invariably begin their daily activity with a period of basking at burrow entrances (Stanner and Mendelssohn 1991; Tsellarius and Tsellarius 1997b). In Israel, emergence time in May–October was 0615 to 0850 with no significant monthly differences. No correlation could be found between emergence time and sunrise. In all observations, monitors emerged at least 1 hour after sunrise when radiation was already intense. Ibrahim (2002) reports that desert monitors in north Sinai emerge 2–4 hours after sunrise, and in June they emerge earlier than in other months. In the southern coastal plain of Israel, basking period in May–October lasted 18 to 47 minutes. Time spent basking in July and August was significantly shorter than in May, June, September, and October. In Algeria (Vernet 1977), *Varanus griseus* has much the same emergence time as in Israel except in October, when Algerian monitors emerged later (0900–1000). In April, emergence time of Algerian monitors was 1000 to 1030 (Vernet 1977). In southeastern Turkmenistan, *V. g. caspius* basks (May–June) for 25 to 77 minutes (Sokolov et al. 1975). Like all heliothermic reptiles, the heating rate during the morning basking is higher than the cooling rate during the night when monitors stay inside burrows. Heating and cooling rates of *V. g. caspius* in May were 0.33°C and 0.062°C per minute, respectively, and in September, 0.13°C and 0.038° per minute, respectively (Sokolov et al. 1975).

Reproduction

Sexual Maturity and Frequency of Reproduction

V. griseus in Tunisia may reach sexual maturity as early as 2 years, but most individuals require 4 to 5 years (Thilenius 1897). Tsellarius and Tsellarius (1996) report that males of *V. g. caspius* in the Kyzylkum Desert become sexually active—that is, start to court females—at the age of 3 to 4 years, and that females are courted by males by the age of 2 to 3 years, but do not lay eggs until they are 4 to 5 years old (Tsellarius and Menshikov 1994). Hence, many females are being courted but do not reproduce, and in any given year, only a small portion of adults actually reproduce. Like many other desert reptiles, in the Kyzylkum Desert, female *V. g. caspius* do not reproduce every year but they do not have a regular bi- or triannual reproductive cycle but rather irregular cycles. Tsellarius and Tsellarius (1996) assume that intraspecific factors such as population density and long-term pairing, rather than climatic factors, play the major role in determining frequency of reproduction. Auffenberg et al. (1990), on the other hand, report that all females of *V. g. koniecznyi* have synchronous ovarian cycles and that all of them lay one clutch per season. Vernet (1977) reaches the same conclusion for *V. g. griseus* in Algeria.

Gonadal Cycles

Annual patterns of gonadal cycles of *V. g. griseus* in Algeria (Vernet 1977) and *V. g. koniecznyi* in Pakistan (Auffenberg et al. 1990) are similar but do not occur at the same times. In Algeria, testes reach maximum size in mid-May and minimum size in early August; in Pakistan, maximum is reached in March, minimum in May–June. From 10 to 20 follicles are found in each ovary (more in the left ovary) (Vernet 1977). In Algeria, follicles reach maximum size and ovulate in late April–May and are smallest in August. In Pakistan, the maximum is reached in August, the minimum in June–July; ovulation occurs in August–September. In both Algeria and Pakistan, follicles increase in size in autumn and remain in this stage until January (Pakistan) or February (Algeria) of the following year. Curiously, in Pakistan, gonadal cycles of males and females are asynchronous. The mechanism inhibiting ovarian activity during the early part of the year is unknown (Auffenberg et al. 1990). More data on the gonadal cycle can be found in Yadgarov (1968), Shammakov (1981), Stanner (1983), and Dujsebayeva (1995).

Courtship and Mating

Tsellarius and Tsellarius (1996) reveal interesting details on courtship and mating of *V. g. caspius* in the Kyzylkum Desert in Uzbekistan. Courtship is characterized by a prolonged pursuit of the female by a courting male. Tsellarius and Tsellarius (1996) recognize two types of pursuits: (1) tracking and (2) escorting. In the first case, the male follows a female's spoor and does not have eye contact with her. In the second case, the male sees the female and pursues her. The mating period (both tracking and escorting) lasts for 15 to 20 days during the first two-thirds of June. All ritual combats between males occur during this period or shortly before it.

Tracking may last many hours or even days. Tracking males are remarkably persistent; if they lose the track, they begin to search for it in loops and circles. In several instances, males stopped tracking at nightfall and resumed the next morning. In one case, immediately after the morning emergence from the night shelter (a burrow), the male walked immediately to the female spoor located 560 m from its burrow, to the spot where tracking had been interrupted the day before. In spite of this, in most cases, tracking is futile. In only 5 of 26 tracking instances did the male succeed in catching up with the female. Sometimes, males abandoned tracking for no obvious reason, but usually tracking was abandoned because the male lost tracks of the female at nightfall.

Males escort both females overtaken by tracking and those that are accidentally encountered. In several incidents, males abandoned the escort of their own accord. In other cases, males lost sight of the females when they suddenly spurted into a complex of rodent burrows (*Rhombomys opimus*). None of the escorts was ever observed to culminate in copulation; hence Tsellarius and Tsellarius (1996) assume that it might have occurred underground inside a burrow, or did not occur at all. The latter assumption is reinforced because during the study period, none of the females that escaped into burrows during the escort ever laid eggs. Escorts were observed only in areas of dense monitor population. In low-density areas, only tracking was observed; however, in both areas, females reproduced and laid eggs.

In 48 of the 49 cases of pursuits (both tracking and escorting) recorded during the study, the partners belonged to the same "settlement" (areas of a fixed-monitor population—see below, "Spacing Patterns"). Because of this, Tsellarius and Tsellarius (1996) conclude that the partners are well acquainted with each other. Moreover, they concluded that long escorts occur only when the courting male is rejected. When a female is receptive to a courting male, courtship is usually short. Tsellarius and Tsellarius (1996) speculate that in high-density populations, females have a steady sexual partner or a limited number of partners. In low-density populations, stable pairs are less common or lacking and females mate with almost any male. Vernet (1977) reports that during the mating season, males in the Sahara cover 6 to 7 km per day in search of females, and they often share the same burrow with the courted female. In my study of *V. griseus* in the coastal plain of Israel, I was unable to measure these distances because of the terrain and the density of vegetation, but in one case (on June 11, 1981) I dug up a pair of monitors from the same burrow (Stanner 1983). The female was inflated and seemed to be in an advanced stage of pregnancy. This female laid eggs between June 21 and July 27, 1981 (see below).

Oviposition and Nesting Burrows

In both Israel and Uzbekistan, females lay eggs between late June and early July (Perry et al. 1993; Tsellarius and Cherlin 1994). In Algeria, eggs are laid between June 15 and July 10 (Vernet 1977), and in Pakistan eggs are laid in September–October (Auffenberg et al. 1990).

Tsellarius and Menshikov (1995) reveal interesting details on nesting sites of *V. g. caspius* in the Kyzylkum Desert and protection of these sites by females. They report that nesting burrows are composed of two or rarely three, relatively straight (rarely winding) shafts with one common opening. One shaft is used for the female's rest, and the second one, 80 to 330 cm long and 83 to 114 cm deep, leads to the nest chamber. The shaft leading to the nest chamber is tightly packed with sand by the female. The nest chamber is unfilled with sand, and eggs are laid on the floor of the chamber in one layer.

Excavation of Nesting Burrows

About 10 to 12 days before egg laying, the female becomes less mobile (males and nonreproductive females do not reduce activity), and 2 to 7 days before egg laying, the female starts to dig trial nesting burrows 20 to 30 cm deep. The distance between trial burrows varied from 5 to 10 m apart to 200 to 300 m apart. One or 2 days before egg laying, the female chooses one of the trial burrows, although not necessarily the last dug, as her final nesting burrow. First she digs the shaft leading to the nest chamber, then the second shaft, which may take another 3 to 5 days. During the whole period of the excavation, except on the egg-laying day, the female spends nights in another burrow (the base burrow) located up to 35 m away from nesting burrow. During excavation, females become emaciated. The caudal neural spines protrude under the skin, which is noticeable from a distance. Stanner (1983) also reports this, but curiously, even though the female did not feed during this period, she did not lose much weight. Nutrients were presumably mobilized from caudal fat bodies to the eggs. Auffenberg et al. (1990) reach a similar conclusion and add that caudal fat bodies (rather than abdominal fat bodies) are the main lipid source for vitellogenesis in *V. g. koniecznyi* as well as in other varanids.

Protection of Nest Sites

Females start to hunt and feed immediately or shortly after oviposition. During 15 to 32 days after oviposition, the female spends nights in the base burrow and checks the nest site at least once a day. This frequency declines as time passes. A female protects her nesting burrow only from conspecifics. Curiously, she runs away or pays no attention to other species (e.g., humans, jackals, horses, lizards, rodents). During 2 to 72 days after oviposition, Tsellarius and Menshikov (1995) recorded 18 cases in which conspecifics of both sexes approached a nesting burrow. In over 80% of cases, the intruding monitor examined or even tried to dig a passage to the nest chamber, but in the rest passed the nesting burrow without paying attention to it. In 83% of the cases, the female displayed aggression toward the intruder and chased it away. No aggressive interactions occurred the other times. In four cases, an intruder approached a nesting burrow when the female was absent. Three of the intruders examined the burrow or tried to dig it partially and then moved away, and the fourth ignored the nesting burrow completely. Interestingly, Tsellarius and Menshikov (1995) state that the behavior of intruders that tried to dig the nesting burrow did not resemble the regular foraging behavior of *V. g. caspius*, speculating that the different reaction of the female to intruding monitors (aggression vs. nonaggression) depends on the behavior of intruders as well as on previous contacts with them.

Clutch Size, Incubation Period, Hatching, and Neonate Behavior

Comparatively few studies document oviposition, clutch size, incubation period, and related parameters in *V. griseus* (Tables 5.4 and 5.5). These studies conclude that in late June to early July, 10 to 20 eggs are laid. Eggs are laid in areas of slightly damp sand in which relative humidity in the spring and early summer may reach 100% (Tsellarius and Cherlin 1994). The temperature in the nest chamber is 30 to 32°C in July, 25 to 27°C in October, and 17 to 20°C in early May of the following year (Tsellarius and Menshikov 1995). Tsellarius and Cherlin (1994) observed development of two clutches (Tables 5.4 and 5.5). They conclude the following:

• Average incubation time is 110 days.

• Eggs usually hatch during the first half of October.

• Hatching is not simultaneous—up to 10 days may elapse between hatching of the first and last eggs.

Hatching itself is a long process. Several hours or even days may elapse between the time that a hatchling pokes its head through the eggshell and the entire evacuation of the egg. Hatchlings do not immediately break their way to the surface but stay in the nest chamber until the spring of the following year; hence, the youngest "neonates" encountered in the field are actually 5- to 6-month-old juveniles that have already passed one hibernation period underground (Tsellarius and Cherlin 1994). Curiously, despite climatic differences, the scenario is probably the same in Israel (H. Mendelssohn, personal communication). Contrary to *V. g. griseus* in Israel and *V. g. caspius* in Uzbekistan, Auffenberg et al. (1990) conclude that the eggs of *V. g. koniecznyi* in Pakistan and northwestern India hatch between July and September after a 10-month incubation period, but their conclusion is speculative and based on circumstantial evidence rather than on direct observations. Vernet (1977) speculates that neonates of *V. g. griseus* in Algeria hatch in August (after a 2-month incubation period—my conclusion), and Gauthier (1967) suggests that they hatch in the autumn.

One female that reproduced in captivity at the TAU Zoo lost 65% of her postoviposition body weight (Perry et al. 1993). Stanner and Mendelssohn (1991) report that two radio-equipped females in the field in mid- to late July lost at least 39% and 47% of their (preoviposition) body weight after oviposition and were emaciated. This large relative clutch mass is typical of all varanids and represents a large maternal reproductive effort (Auffenberg 1981). However, by the end

TABLE 5.4.
Varanus griseus oviposition, clutch size, and incubation period.[a]

Date of Oviposition	No. of Eggs	Date First Egg Hatched	Date Last Egg Hatched	Incubation Time (days)	Site	Temp. (°C)	RH(%)	Ref.[b]
25 Jun 1991	18	2 Oct 1991	10 Oct 1991	99–107	I, N	30–32	100	2
2 Jul 1992	19	11–15 Oct 1992	?	110–115?	N			2
End Jun 1956	10[c]							1
End Jun 1988	12[c]							1
24 Jun 1989	18	23 Oct 1989	?	121	I	30–32		1

[a] Conditions are as follows: I, incubator; N, in the field; RH, relative humidity.
[b] References: 1, Perry et al. (1993); 2, Tsellarius and Cherlin (1994).
[c] Eggs did not develop.

of September, the loss is regained and the emaciated appearance disappears (Stanner and Mendelssohn 1991). The postoviposition female in captivity described above regained weight within a month after oviposition (Perry et al. 1993).

Spacing Patterns

Population Density

In its desert ecosystem, *V. griseus* is a top diurnal predator (Vernet and Grenot 1972), and as such, its population density is low (Stamps 1977). Because of this, and because of the elusive and secretive nature of *V. griseus,* it is hard to measure its population density with accuracy. Consequently, population densities, including those cited in Table 5.6, are very variable (1–18.8 monitors per square kilometer) and should be taken as rough estimates. In some instances, differences between population densities can be attributed to environmental factors such as habitat productivity. For example, in comparatively wet areas of the Saoura Wadi in south Algeria, the density is six monitors per square kilometer, and in the harsh Guir Hamada, two monitors per square kilometer (Vernet et al. 1988a). In other instances, the causes are less clear. The Saoura Wadi has an average annual rainfall of 32 mm, but in my study area in the coastal plain of Israel, with 535 mm average rainfall, the population density is only four monitors per square kilometer. Auffenberg (1989) reports exceptionally high densities of *V. g. koniecznyi* in various habitats in Pakistan (minimum estimates of adults per square kilometer): thornbrush—16; deciduous forest—28; wetland borderlands—32. Tsellarius et al. (1995) and Tsellarius (personal communication) outline a complex spatial distribution pattern of *V. g.*

TABLE 5.5.
Morphometrics of eggs and neonates of *Varanus griseus*.[a]

No. of Eggs	Clutch Weight (g)	Egg Weight (g), Mean ± SD	Length (mm) × Diameter of Egg	Weight of Mother (g)	Clutch/ Mother Weight (%)	Neonate SVL (mm), Mean ± SD	Neonate TL (mm), Mean ± SD	Weight of Neonates (g), Mean ± SD	Ref.[b]
18	463	26 ± 1	52 × 32[c]	707[d]	65[d]	109 ± 2	144 ± 4	19 ± 1	2
18						105 ± 3			3
5–12		~15	(30–47) × (25–30)		15–40[e]				1

[a] SVL, snout–vent length; TL, tail length; SD, standard deviation; n, sample size.
[b] References: 1, Vernet (1977); 2, Perry et al. (1993); 3, Tsellarius and Cherlin (1994).
[c] Average values in mm.
[d] Postoviposition weight.
[e] Preoviposition weight.

caspius in the Kyzylkum Desert in Uzbekistan. They conclude that the spatial distribution of the monitors is uneven, with areas of high density and low density. In high-density areas ("settlements"—see below) density varies spatially over time from 1.3 to 3.5 monitors per square kilometer at the periphery to 5.5 to 18.8 monitors per square kilometer in "core areas" of settlements (calculated from raw data in Tsellarius et al. 1995).

Sex Ratio

Tsellarius et al. (1995) and Auffenberg et al. (1990) report on unequal sex ratios favoring males in *V. g. caspius* and *V. g. koniecnyi* (Table 5.6). In neither case do the authors attribute the unequal sex ratio to artefacts such as differences in activity levels or patterns. Stanner and Mendelssohn (1987) also report unequal sex ratio favoring females (1:1.6) in *V. g. griseus,* but contrary to Tsellarius et al. (1995) and Auffenberg et al. (1990), they do conclude that this is an artefact caused by intersexual differences in activity patterns. On the coastal plain of Israel, the activity season of females averages 1.7 times longer than that of males, and hence, females are subject to capture during a longer period, a factor that may alter a 1:1 ratio to 1:1.6.

Movements and Home Ranges

Varanus griseus is an active forager that may cover 5 to 6 km per day during its daily forays, and during the mating season, males may cover 6 to 7 km per day (Vernet 1977). Distances between recapture points during 3-to 9-month intervals in the same year and/or in two consecutive years were 1 to 9 km (Vernet 1977). Despite its agility and swiftness, Vernet (1977) concludes that *V. griseus* lives in definite home ranges (Table 5.6), although he does not exclude the possibility that the monitors may change their home ranges in the course of several years (no individual was ever recaptured after more than 2 years). In the Sahara, annual fluctuations in rainfall often change the spatial distribution of the vegetation and consequently prey animals; hence, *V. griseus* may adjust its movement patterns and home range accordingly (Vernet 1977).

As expected, habitat productivity affects movement patterns and home range area—for example, in harsh Guir Hamada, the average home range area is 2.4 times larger than in the comparatively wet Saoura Wadi (Table 5.6). In Guir Hamada, monitors usually move in straight lines with comparatively few turns, but in Saoura Wadi, turns and loops are more common. In the comparatively productive coastal plain of Israel with its high annual rainfall (535 mm/y), the home range area of *V. griseus* is smaller (Table 5.6). Curiously, despite of higher aridity, in north Sinai, home ranges of *V. griseus* studied by Ibrahim (2002) were smaller than in the coastal plain of Israel (Table 5.6). Ibrahim (2002) relates it to a rich food supply in his study site and/or to the smaller body size of the monitors studied by him. Average home range area of males in the coastal plain of Israel is 3.1 times larger than that of females (98 vs. 32 ha) (Stanner and Mendelssohn 1987). Stability of home ranges of *V. griseus* in the coastal plain of Israel is still questionable. Home range areas of four of the nine monitors radio tracked by Stanner and Mendelssohn (1987) started to level off after 30 to 70 locations, but the remaining five showed no tendency to reach a final plateau. In many cases, the addition of a single location could considerably increase the home range area by up to 46%. Distances between recapture points of females in the coastal plain of Israel during 1- to 2-year intervals were 20 to 310 m, which are considerably shorter than the longest distances covered by females within the same activity season. Therefore, I assume that females maintain a fairly constant home range for several years. In contrast, no male was ever observed or recaptured in two consecutive years, leaving the home range permanency of males an open question. In the coastal plain of Israel, monitors do not necessarily explore their entire home range area every day; most explored a comparatively small area for several days before moving to another area and so on. In the Kyzylkum Desert, *V. g. caspius* explores 13% to 61% of its home range area per day (Tsellarius et al. 1991).

Vernet (1977) concludes that *V. griseus* is nonterritorial. I have never observed encounters between desert monitors in the field, but indirect evidence, such as extensive overlap of home ranges among and within sexes, suggests nonterritoriality (Stanner and Mendelssohn 1987). This conclusion is reinforced because neither fixed basking sites nor fixed foraging trails were found, and the interchange of burrows was high (Stanner 1983), all indicating nonterritoriality (Stamps 1977). Furthermore, in captivity, desert monitors exhibit considerable mutual tolerance (see below, "Behavior"). Tsellarius and Menshikov (1994) observed both fights and ritual combats of *V. g. caspius* in the field (see

TABLE 5.6.
Density, sex ratio, and home ranges of *Varanus griseus*.

Country	Study Area	Density (monitors/ km²)	Sex Ratio(M:F)	Home Range Area (ha), mean(range)	Ref.[a]
Algeria	Saura Wadi	6		100–150	1
Algeria	Guir Hamada	2		200–400	1
Israel	Coastal plain	4	1:1.6	61 (14–160)	2
Turkmenistan	Southern Karakum Desert	Up to 5			3
Turkmenistan	Valley of Murgab River	1–1.5, 10–12			3
Uzbekistan	Saihan Valley	10.5			3
Turkmenistan	Karakala	5			3
Turkmenistan	Suyunaksak	9–12			4
Uzbekistan	Kyzylkum Desert	2–18.8	1.3–5:1	190 (64–1950)	5, 8
Pakistan	General		2.23:1		6
Algeria	Sahara Desert			200–500	7
Egypt	North Sinai			16.2 (7.5–22.8)	10
Uzbekistan	Kyzylkum Desert			66 (50–115)	9

[a] References: 1, Vernet (1977); 2, Stanner and Mendelssohn (1987); 3, Makayev (1982); 4, Zarhidze (in Makayev 1982); 5, Tsellarius et al. (1995), Tsellarius (personal communication); 6, Auffenberg et al. (1990); 7, Saint Girons and Saint Girons (1959); 8, Tsellarius and Menshikov (1994); 9, Tsellarius et al. (1991); 10, Ibrahim (2002).

below, "Behavior"), but neither were related to territoriality. The winner was not always the owner of the home range, the loser was never chased away, and defeat did not affect movement patterns in that the loser did not avoid the area where it had been defeated.

Tsellarius and Menshikov (1994) and Tsellarius et al. (1995) outline a complex spatial distribution pattern of *V. g. caspius* in the western Kyzylkum Desert in Uzbekistan. They coin the term "settlement" for areas with high monitor density. The area of the settlements averaged 5.53 km², about 2 km in cross section). Distances between centers of neighboring settlements did not exceed 5 km. The composition of the monitor population in each settlement was comparatively stable. During the 3 years of field study, 76% of the monitors were old residents (1–3 years or more), and only 24% were newcomers that had not been recorded in previous years. Within each settlement, monitors live in overlapping home ranges averaging 190 ha. Tsellarius and Menshikov (1994) conclude that members of the settlement are acquainted with each other. "Nomads" may enter settlements, but they do not live there regularly.

Home ranges of nomads (mean, 1606 ha) are elongated (7–10 km) and include one core area (mean, 39 ha) at each end of the range. All nomads were large males (8–10 years or older). They often disappear temporarily, possibly migrating to remote areas of the desert, far beyond the study area's boundaries. Home ranges of settlers (mean, 190 ha) are more compactly shaped and include a core area averaging 37 ha. An average home range of a male settler (257 ha) is 2.1 times larger than that of a female settler (124 ha). Settlers may visit other settlements for short periods. The average number of monitors per settlement is 9 ± 2.2. Tsellarius et al. (1995) assume that in the course of several years, environmental changes over time, such as annual fluctuations in prey populations (mainly rodents), may affect the spatial distribution of the monitors. In less favorable conditions, monitors become more mobile and cover longer distances during their daily forays. Consequently, settlements tend to disintegrate or undergo a reshuffle of their population composition.

Behavior

Solitary Mode of Life and Forced Socialization

Like other varanids, *V. griseus* is a solitary species. Except for a short period during the mating season in which the sexes may share the same burrow, *V. griseus* leads a completely solitary mode of life. In spite of that, Tsellarius and Menshikov (1994) state that *V. g. caspius* are tolerant of each other in the field, and on two occasions, Ibrahim (2002) found two and even three monitors in the same burrow. Desert monitors kept in outdoor enclosures at the TAU Zoo also tolerated each other. Notwithstanding their mutual tolerance, antagonistic interactions such as tail slaps, bites, and threatening postures did occur occasionally. Such behaviors seemed to be caused by factors such as an abrupt movement by one of the monitors in the enclosure (Stanner and Mendelssohn 1987). Vernet (1977) reports that in captivity *V. griseus* are less tolerant of each other, and in overcrowded conditions, aggressive and violent interactions occur frequently, often leaving the neck and limbs wounded and bleeding. Group rearing of desert monitors in outdoor enclosures of the TAU Zoo had negative effects on several monitors, which underwent a behavior change diagnosed by Stanner (1999) as psychotic depression. Symptoms included the following:

• Depressed individuals spent almost all their time inside wooden boxes and did not emerge.

• They did not feed, and they became emaciated and flaccid (reduced muscle tonus).

• They did not molt. Depressed individuals that were not treated eventually died within several months.

If depressed monitors were treated and separated from the rest of the group and reared alone, they fed, gained weight, molted, and resumed a natural activity cycle.

Interactions with Humans

Varanus griseus is one of the most secretive and elusive varanids. In the field, it usually spots an observer first and evades it. Consequently, even in areas of dense monitor population where spoor is abundant, active monitors are only rarely encountered. In my comparatively few encounters with active desert monitors, they either darted into a thicket or shelter, sprinted and escaped, or assumed various threatening postures (see below) before running away.

In captivity, *V. griseus* maintains its usual anthropophobic behavior. In the TAU Zoo, whenever humans approached the outdoor enclosure, they darted into a hiding box (provided for that purpose) or assumed a threatening posture. In other instances, monitors seemed to be indifferent to the approach and did not change posture. If reared in small enclosures or cages with no hiding places, they either assume a threatening posture or lie motionless. The latter may be interpreted as a habituation to humans, but in my opinion, this motionlessness is actually a psychomotor retardation, one of the subsidiary symptoms of psychotic depression probably caused by stressful captive conditions (Stanner 1999). In spite of this, *V. griseus* seems to be able to habituate itself somewhat to the proximity of humans. One juvenile reared at the TAU Zoo became accustomed to my presence and approached me whenever I approached to feed it. However, one day, this behavior abruptly reverted to the usual anthropophobia. Igolkina (1975) observed the same phenomenon in captive Caspian Desert monitors reared in the Leningrad Zoo. Andres (1904) describes a desert monitor raised in his home that became accustomed to his family and waited at the dining table for chunks of meat to be thrown to it. The monitor apparently distinguished between family members and strangers. It let his "foster parents" approach (but never touch), but it avoided strangers and neighbors.

In the field, *V. griseus* seems to be able to distinguish between size classes of potential enemies. Once, while driving in the northwestern Negev in southern Israel, a desert monitor crossed the road about 50 m ahead of me. I stopped the car, and we stared at each other motionlessly for more than 5 minutes. The instant I stepped out of my car, it darted away in its usual manner and escaped. Tail whips from a large and warm-bodied desert monitor are unpleasant, and its bites can be painful.

Sopiev et al. (1987) describes a case of envenomation by *V. griseus*. Symptoms included general weakness, dizziness, lack of appetite, muscle pains, pain during swallowing, eye pain, breathing difficulties (but no oedema), and increased pulse rate. Symptoms persisted for several hours, and then the patient's general condition improved. The following day, the patient experienced only general weakness, and 4 days later, the wounds were completely healed. Similar symptoms that included pain in the facial bones and difficulty walking (due to muscle soreness) are described by Ballard and Antonio (2001). Sixteen hours after the bite, almost all symptoms had dissipated.

Indirect Communication by Spoor

Varanus griseus creates four main types of spoor:

- While running fast, the monitor raises its body and leaves only footprints. Occasionally the tail may slightly touch the sand, leaving a short, straight traction mark.

- While walking or running slowly, the monitor leaves a continuous sinuous traction mark of the tail with footprints on both sides.

- After defecation, *V. griseus* wipes its cloacal region in the sand, leaving a broad traction mark. During this act, the monitor propels itself forward with its forelimbs only. The hind legs are raised and do not touch the sand.

- Tsellarius and Menshikov (1994) coined the term "drag" for the fourth type of traction mark. The monitor presses its tail and/or its cloaca and hind part of the belly into the sand and leaves a distinct traction mark of up to 10 m long, using all four limbs to propel itself forward.

All four types are probably used for intraspecific communication and/or for other purposes such as hygiene (the third type) or orientation (the first two types). Vernet (1977) observed monitors moving back and forth across barren dunes of the Guir Hamada (Algeria), frequently backtracking their own spoor. The last type is used exclusively for intraspecific communication (Tsellarius and Menshikov 1994). Monitors examine spoor of conspecifics both visually and by olfaction with tongue flicks, but the latter plays the major role. Tsellarius and Menshikov (1994) observed that courting males always follow fresh (several hours old) tracks of females in the right direction, but old (1 or more days old) tracks are sometimes followed in the wrong direction even if the tracks are visually distinct. By examining the spoor of conspecifics, *V. griseus* can distinguish age and sex, and it can distinguish between different individuals inhabiting an area. For instance, adult monitors always ignore spoor of juveniles younger than 2 years, and courting males ignore the spoor of other males. Response of adults to "drags" (the fourth type listed above) left by members of the same settlement differs from their response to drags of strangers. Tsellarius and Menshikov (1994) believe that the drags of *V. griseus* are comparable to markings of mammalian carnivores, and therefore might be considered as an expression of excitation. Nevertheless, drag marking is not an automatic stimulus–response reaction, but rather a complex behavior dependent on many factors including the immediate emotional state of the monitor, its temperament, and previous experience and encounters. The intensity of drag markings fluctuates monthly, and it varies between population subgroups (e.g., more in males, less in females, more in nonreproductive females and in large males, intensity declines with age), among settlers, newcomers, visitors, and nomads. Drags are also performed by juveniles and even neonates. Drags are performed in various circumstances such as encounters with the spoor of conspecifics, but never during direct encounters between conspecifics.

Displays

Tsellarius and Tsellarius (1997a) analyze nine displays:

- *Confident gait.* The head, body, and tail are raised and aligned parallel to the ground. This posture is exhibited during regular daily activities as well as during intraspecific encounters and reflects "a state of a psychological comfort."

- *Threatening gait.* This posture was established only from tracks and not by direct observation. The monitor moves slowly toward its adversary, dragging its feet in the sand, leaving distinct traction marks of the feet.

- *Sitting dog posture.* This posture is displayed when the monitor encounters a "disturbing object" (e.g., a human) located far away from it. This display reflects a low degree of fear. The monitor also uses this posture during intraspecific encounters, where it functions as a display of dominance.

- *Stooping.* The monitor lowers its head while simultaneously inflating the throat and pressing its tail into the sand during encounters with potential threats of an uncertain nature (e.g., a backpack placed on the sand, a shirt hung on a bush, a careless movement of an observer in a hideout) or during encounters with strange conspecifics. This display reflects a low degree of "fear, unease and lack of self confidence." The stooping display can also be unmistakably established from spoor. The tail leaves a distinct straight furrow in the sand.

- *Zatir.* Zatir is synonymous with drag (see above the fourth kind of traction mark).

- *Showing of the back.* The body is oriented laterally to the threat and flattened dorsoventrally. The throat might be slightly inflated and the back slightly arched. This posture reflects an "alarmed" state displayed, for example, as a response to active self-defense of prey to a "nonaggressive human" (who poses no apparent threat) in the vicinity of the monitor, and during encounters with conspecifics. These encounters with other monitors excite a low degree of "apprehension" that may, but usually does not, cause one of the monitors to flee. Generally, however, both monitors are inclined to continue with their current activities.

- *Arching of the back.* This posture displays "readiness for active self defense." The monitor orients itself laterally to the adversary, tail raised ready for a whip, and head turned toward the adversary. This display is often combined with elements of "passive intimidation." The monitor tries to look larger than it is by inflating its body and throat and raising the body on straightened front legs, a posture displayed in cases of grave danger with no option for escape. It is a common display during close encounters with humans, but never during encounters with conspecifics. If the human adversary does not "manifest hostility," the monitor slowly retreats. If he does, it inflicts a tail whip and escapes, or if escape is impossible, it lunges toward the adversary. In this case, the lateral body orientation is disrupted, and the "passive-intimidation" elements are attenuated. Only rarely does the monitor attack.

- *Gape.* In cases of an unexpected or a very grave danger, arching of the back is combined with a gape, and the monitor lunges toward the adversary. This display was observed only during encounters with humans and is often followed by an attack.

- *Lurking.* The monitor presses itself tightly to the ground with all the ventral parts in contact with the substratum. Lurking is a display of appeasement performed in various social interactions. The same posture is also assumed in nonsocial contexts, such as during basking, but in the latter case, the eyes are usually closed (eyes are always open during social interactions).

Typical Behaviors During Encounters with Conspecifics

Tsellarius and Tsellarius (1997a) analyze the following four behaviors:

- *Flight.* Usually, one of the monitors escapes to a distance of 2 to 3 m (rarely 10–15 m), after which it resumes its previous activity. Flight can be interpreted as a ritualized activity that terminates any previous interaction. The escaper is only rarely pursued, and if it is, only for a short distance, and the pursuer never catches up with the escaper.

- *Mutual sniffing.* Mutual sniffing occurs almost invariably during intraspecific encounters, except shortly before or after hibernation or in cases of an immediate attack (e.g., while the female protects the nesting site, when "the individual distance is violated"). Tongue flicks are directed to the snout and sacrum. Frequently, monitors try to avoid being licked at the sacrum, resulting in the monitors "waltzing" in a circle. The waltz is often accompanied by stooping and showing of the back.

- *Fight.* Fight follows mutual sniffing and waltzing (combined with back showing and stooping). One monitor whips its opponent with its tail, which invariably results in the latter fleeing. Tsellarius and Tsellarius (1997a) did not observe monitors biting each other during fights, although wounds and scars at the sacrum and shoulders indicated that bites did occur.

- *Ritual combat.* Bipedal ritual combat has been observed in many *Varanus* species (Horn et al. 1994), but Tsellarius and Tsellarius (1997a) are the first to document it in *V. griseus*. In *V. griseus,* ritual combats are invariably preceded by mutual sniffing. Ritual combat can be terminated at any stage of the combat, although in most cases, it stops during the first stage of "neck crossing." Ritual combats were observed directly only in males, but indirect evidence established from tracks indicated that they probably also occur among females. Combatants in all observed cases were members of the same settlement. The causality and background of ritual combats are still unclear, but Tsellarius and Tsellarius (1997a) do not believe it to be a ritualization of a fight, but rather a ceremony directly related to the social hierarchy within the monitor population in the settlement. All-out ritual combats are rare, probably because they depend on an unlikely set of circumstances, namely interrelations between combatants have to include elements of uncertainty, the social status of the combatants should be more or less equal but not yet established. In any given settlement, members of the settlement know each other so that in most cases, dominant/subordinate interrelations are rather obvious. Disputes are rare and can be resolved by mutual sniffing, and ritual combat is unnecessary. Extreme antagonistic encounters between settlers and intruders do not end with a ritual combat but with a fight. In summary, contrary to a fight, the ritual combat is basically a social ceremony reflecting at the most a low level of aggression.

Foraging Behavior and Diet

Like other varanids, *V. griseus* is a carnivorous lizard. Vegetative material was found in both stomach contents and fecal pellets of *V. griseus* (Vernet and Grenot 1972; Stanner and Mendelssohn 1986, 1987), but it was attributed to secondary predation. *Varanus griseus* is an active forager that may cover 5 to 6 km during its daily forays in the Algerian Sahara (Vernet 1977) or 10 km per day or more in the Kyzylkum Desert in Uzbekistan (Tsellarius and Cherlin 1991). Once the prey is spotted, the monitor either immediately rushes to it and attacks it, or stalks to it to a distance of 3 to 5 m and then sprints and attacks it. During stalking, the monitor lowers its body, often using shrubs and ground mounds as cover (Tsellarius and Tsellarius 1997a). Prey is killed by a strong bite powerful enough to squeeze the thorax of rodents and disrupt breathing (Tsellarius et al. 1997). The monitor holds on to the prey until it ceases to struggle, and then it relaxes the grip and starts to swallow. If the prey flutters, the monitor renews the grip. Prey is also killed by violently shaking it from side to side or by pushing it to the ground, shrub, etc. This behavior is probably innate. Monitors feeding on dead laboratory mice at the TAU Zoo almost invariably demonstrate this behavior (Stanner, unpublished observation). However, Tsellarius et al. (1997) conclude that hunting behavior of *V. griseus* is only partially innate and assume that experience and learning play a major role in structuring the hunting pattern of an individual monitor. They report that *V. griseus* examines the spoor of lizards (but not of mammals and amphibians) by tongue flicks and tracks them to distances of up to 15 m. If the tracking leads the monitor to a lizard's burrow, it usually starts to dig it out.

Varanus griseus probably also scavenges in the wild. A claw of a large domestic cat and spines of adult hedgehogs (*Hemiechinus auritus*) found by Stanner and Mendelssohn (1986, 1987) in fecal pellets of *V. griseus* in the coastal plain of Israel were attributed to scavenging on carcasses of animals run over by cars or discarded at a nearby garbage dump. Conversely, Tsellarius et al. (1997) conclude that *V. griseus* rarely if ever eats carrion, and when encountering carcasses, it examines them by tongue flicks but does not eat them.

Varanus griseus hunts on the ground surface or extricates prey from beneath debris. It also digs prey from burrows. Hence it regularly feeds on both diurnal and nocturnal species. Anderson (1963) reports that in Iran, *V. griseus* systematically examines rodent burrows, and Tsellarius et al. (1997) report that differential hunting behavior targets different rodent species inhabiting central Asian deserts. For example, underground colonies of *Rhambomys opimus* are examined systematically, but burrows of *Spermophilopsis leptodactylus* are seldom examined, and monitors prefer to hunt them on the ground surface. *Varanus griseus* is an opportunistic predator consuming almost any prey of an appropriate size, including unlikely prey such as hedgehogs, which are swallowed with the spines, up to 2-year-old tortoises (*Testudo graeca*) swallowed whole (Stanner and Mendelssohn 1986, 1987) and large snakes (see below).

In contrast with the wide variety of prey animals consumed in the field, captive *V. griseus* are often fussy. At the TAU Zoo, desert monitors accepted laboratory mice and chick, but were reluctant to eat laboratory rats and chunks of chicken meat. *Varanus griseus* can eat big meals; for example, one juvenile ate a large *Lacerta sicula*, 53% of its body weight (Stanner 1983).

There are reports that *V. griseus* drinks water (e.g., Minton 1966; Igolkina 1975; Shammakov 1981; Chandra 1987). I have never observed drinking in captivity. Furthermore, in many areas, drinking water is not readily available for desert monitors during the activity season; hence, the water content of their prey has to suffice. *Varanus griseus* has efficient water metabolism enabling it to endure long periods without water (Khalil and Abdel-Messeih 1959a). Moreover, during periods of food scarcity, these lizards can produce metabolic water at rates of 0.26 to 0.32 mL/100 g/d (Vernet et al. 1988a). On the other hand, during periods of replete food, *V. griseus* can store up to 15% of its body weight in excess water in various body tissues (Khalil and Abdel-Messeih 1959b).

In general, dietary analyses of *V. griseus* in the field reflect the relative abundance of the prey species in the research area (Vernet 1977; Stanner 1983), although certain prey species may not conform with this statement. For example, Vernet (1977) notes the total absence of *Eremias pasteuri* from the diet of *V. griseus* in the Algerian Sahara. Dietary analyses also indicate seasonal and annual fluctuations in relative abundance of prey species (Stanner 1983; Stanner and Mendelssohn 1986, 1987; Auffenberg et al. 1990; Tsellarius et al. 1997). Stanner and Mendelssohn (1986, 1987) report remnants of chameleons in 9% of the fecal pellets collected in 1981 (vs. 0% in 1980), whereas vipers were six times more common in 1980 than in 1981. Yadgarov (1968) reports that in the spring *V. g. caspius* in the Surhandarja basin feed mainly on tortoise hatchlings (size, 4–8 cm) but only 7% of fecal pellets collected by Stanner in 1981 contained remnants of tortoises, and in any given month, tortoise occurrence did not exceed 6.3% (Stanner 1983).

Tsellarius et al. (1997) report that seasonal and annual fluctuations in relative abundance of prey species affect hunting methods and movement patterns of monitors, including individual differences in hunting behavior among monitors inhabiting their research area. Stanner and

Mendelssohn (1986, 1987) conclude that vertebrates constitute the main bulk of food of *V. griseus* in the coastal plain of Israel, although remnants of invertebrates (mainly insects) were found in more than 90% of fecal pellets. Only about 10% of these were of large arthropods, the remainder being small- to medium-sized species attributed to secondary predation of insectivorous lizards. A similar conclusion is reached by Tsellarius et al. (1997), who report that more than 95% of the prey weight of *V. g. caspius* in central Asian deserts is composed of rodents, snakes, and agamid lizards. Large arthropods are hunted intensively only when vertebrate prey is scarce (Tsellarius et al. 1997). On the other hand, Auffenberg et al. (1990) examined stomach contents of *V. g. koniecznyi* from Pakistan and concluded that insects (and not vertebrates) are the main source of food of this subspecies. Generally, birds do not constitute an important component of the diet, although eggs and/or chicks of ground-nesting species are frequently eaten; 19% of fecal pellets examined in 1981 by Stanner and Mendelssohn (1986, 1987) contained eggshells and/or feathers, beaks, and scales of chicks of the Chukar partridge (*Alectoris chukar*), eaten either shortly before or shortly after hatching. Tsellarius et al. (1997) reached the same conclusion concerning *V. g. caspius* preying on eggs and chicks of pheasants (*Phasianus* sp.). Repeated observations of footprints of *V. g. caspius* at the entrance of nesting holes of rollers (*Coracias garrulus*) in the Atrek River Valley in Turkmenistan are also reported. Vernet (1977) reports that birds form 11% of stomach contents of *V. griseus* in the comparatively wet areas of the Saoura Wadi in south Algeria (vs. 3% in the harsh Guir Hamada) but could identify with certainty only one species (*Sylvia cantillans*). Desert monitors have been observed to climb trees and ravage nests. Tsellarius et al. (1997) observed a large male ravaging the nest of a *Pica pica* built on a Turanga tree about 8 m high.

Varanus griseus is a good swimmer and diver (Tsellarius et al. 1991) and may sporadically eat fish. Fish bones and scales were found in fecal pellets of *V. griseus* inhabiting the Karabelent sand dunes near the Karakum Channel in Turkmenistan, but generally, this species does not feed on fish even if it is abundant and easily caught (Tsellarius et al. 1997). Snakes constitute an important component of their diet, making up 19.4% of the occurrences in fecal pellets collected in 1981 by Stanner and Mendelssohn (1986, 1987). Large snakes are eaten regularly—for example, 1.5-m *Coluber jugularis* (Stanner and Mendelssohn 1986, 1987) and 80- to 140-cm *Vipera lebetina* (estimated from SVL data in Tsellarius et al. 1997). Curiously, Vernet (1977) reports that in the Algerian Sahara, snakes play a minor role in the diet of *V. griseus* (<2%). Snakes eaten include large venomous snakes of both the Viperidae and Elapidae families (Rjumin 1968), but Mertens (1942) does not believe that monitor lizards are immune to snake venom. Rjumin (1968), however, disagrees, and reports that *V. griseus* is 30 to 200 times more immune than humans to the venom of *Vipera lebetina* and 400 to 4000 times less vulnerable to the venom of *Naja oxiana*.

Varanus griseus may occasionally be cannibalistic. Vernet (1977) reports on cannibalism in captive conditions. Makarov (1985) palpated remains of a 72-cm *V. griseus* from the stomach of a 108-cm male. Makarov (1985) speculates that it may have been eaten while the two monitors were seizing (and beginning to swallow) the same prey animal, or during feeding, when monitors are aroused and aggressive and often devour even inedible objects. Fecal pellets of equal number and sex analyzed by Tsellarius et al. (1997) revealed that cannibalism on 2- to 3-year-old conspecifics occurs in frequencies of up to 5.5%. Cannibalism was recorded mainly in periods or areas with low rodent abundance, but Tsellarius et al. (1997) do not relate cannibalism directly to food shortage and/or nonselective predation caused by starvation, but rather see it as "a social phenomenon" caused by re-shuffling of spacing patterns and social structure of monitor populations.

Conservation

Varanus griseus is distributed over a large area. Nevertheless, as a result of habitat destruction, it has become rare or even extinct in many areas. In Israel until a century ago, *V. griseus* could be found south of the Yarkon River that constituted its northernmost distribution border in the coastal plain of Israel (Mendelssohn, personal communication). Because of the expansion of the city of Tel Aviv and satellite townships, its distribution area was pushed southward, and now it has become rare or even extinct north of the town of Rishon Lezion. South of Rishon Lezion, there are still several isolated populations, many of which are probably nonviable (Stanner, unpublished observation). Unlike larger animals (e.g., gazelles, foxes) that usually flee and evacuate development areas at the first appearance of tractors and bulldozers, desert monitors will enter underground burrows in which they are doomed to die and be crushed by the heavy machinery (Stanner, unpublished observation). In southern Israel, in the northwestern Negev and Arava Valley, expansion of agricultural areas in the last 20 to 30 years has decimated *Varanus griseus* habitats (Mendelssohn 1983). Makayev (1982) argues that conversion of natural *Varanus* habitats to farmland does not necessarily mean the total disappearance of *V. griseus* from these areas. Desert monitors can survive at edges of cultivated fields, and if not persecuted by humans, they can live there for many years, but it is uncertain whether they can form viable populations. Conversion of natural habitats to pastureland has little effect on desert monitors (Makayev 1982). Nevertheless, Makayev (1982) reports that in central Asia, it has disappeared almost completely from Ferganskaya Valley, as well as from the Golodanya and Dalverzinskaya steppes, because of expansion of cotton fields. Among the decimating factors, Makayev (1982) includes road kills and deliberate killings. After the establishment of the USSR Red Data Book in 1974 and adoption of the law of "Conservation and Use of Animals in the USSR," which came into effect in 1981, conservation of *V. griseus* in the former USSR was at least backed legally (Makayev 1982).

In the last 20 to 30 years, the number of wild desert monitors has fallen drastically as a result of extensive hunting targeting the international skin trade (Anonymous 1983). Minimum estimates of whole skins of *V. griseus* reported in the trade were as follows: 1975, 10,793; 1976, 73,997; 1977, 9008; 1978, 2400; 1979, ?; and 1980, 61 (Anonymous 1983).

V. griseus has been included in the Convention on International Trade in Endangered Species (CITES) appendix 1 since 1975. In the CITES 2000 list, it is included in appendix 1 together with *V. bengalensis, V. komodoensis,* and *V. flavescens. Varanus griseus* is not included in the IUCN Red List of Threatened Species of 2000, but *V. komodoensis* and *V. olivaceus* are included under the status "vulnerable." In 1994, it was listed in the Red List as vulnerable, and in 1996 it was not evaluated. *Varanus griseus* has been included in lists of the USA Fish and Wildlife Service since 1976. In the 2001 list, *V. griseus* is classified as "endangered." In the European Community (EC) Regulation 2724/2000 (amending Regulation 338/97), *V. griseus* is included in annex A (referring to all species listed in CITES appendix 1, several CITES appendix 2 and 3 species, and several non-CITES species), which strictly regulates wildlife transportation into and out of the EC, within the EC, as well as its utilization, keeping, and maintenance inside the EC. In the "Arab Man and Biosphere Program" (ArabMab), *V. griseus* is given the status "endangered." According to the Oregon Administrative Rule 635-056-0060 (certified to be effective on December 22, 2000), all monitor lizards, except *V. griseus,* are noncontrolled species (i.e., may be imported, possessed, sold, purchased, and exchanged in the state of Oregon without a permit).

On the other hand, Auffenberg et al. (1990) conclude that in India and Pakistan, *V. g. koniecznyi* are not much hunted for their skins. Tanners and manufacturers of leather products in Pakistan do not like skins of *V. griseus,* which they claim to be thin and easily torn (Auffenberg 1989). Auffenberg et al. (1990) have never seen leather products made of the skins of *V. griseus,* as they have of other varanids, and among the thousands of *Varanus* skins (most of them of *V. bengalensis* and *V. flavescens*) confiscated by the authorities, *V. g. koniecznyi* skins were never included. Since *V. g. koniecznyi* is abundant in India and Pakistan, these findings are striking and may be related to the fact that by comparison with *V. bengalensis* and *V. flavescens, V. g. koniecznyi* is scattered sparsely in its remote desert habitats and is thus harder to find and hunt (Auffenberg et al. 1990). Moreover, although habitats of *V. bengalensis* and *V. flavescens* are destroyed extensively by development projects as a result of aridity and remoteness, habitats of *V. g. koniecznyi* remain largely intact. Continued soil salination (and consequent abandonment for agricultural purposes) of various irrigation projects further ensures the future of *V. g. koniecznyi* and attenuates damage caused by urbanization and highway construction. Hence, Auffenberg et al. (1990) conclude that of the three *Varanus* species found in arid habitats in southern Asia, *V. g. koniecznyi* is the least threatened. In spite of that, according to the Indian law, *V. griseus* is listed (together with *V. salvator*) in part 1 of schedule 2 of the Wildlife Protection Act, which prohibits wildlife hunting without a special permit (other *Varanus* species are not mentioned in the Act).

Acknowledgments

I thank Mark K. Bayless for his help with hard-to-find reprints and his instantaneous response to all my requests; Alexej Tsellarius for providing body-size data of *V. g. caspius,* for sending me reprints of his *Varanus* articles and spectacular photographs of this subspecies, and for his comments on the article in manuscript; Daniel Bennett for sending me copies of his *Varanus* books and his compiled collection of English-language translations of *Varanus* articles; Iryna Medvedeva for translating Russian-language *Varanus* articles; Dennis King and Jennifer Ast for E-mailing me information on the phylogeny of varanids; Yehudah L. Werner for allowing me to use his library and reprint collection and for his comments on the chapter in manuscript; and Daniel J. Greenhill for reviewing the chapter. Their kind help is gratefully acknowledged.

5.4. *Varanus niloticus*

Sigrid Lenz

Figure 5.8. Varanus niloticus *(photo by Jeff Lemm).*

Nomenclature

The species was first described by Linnaeus in 1766 under the name *Lacerta monitor* (Fig. 5.8).

Since 1997, two traditional subspecies of the Nile monitor have been regarded as full species (Böhme and Ziegler 1997a), *V. niloticus* and *V. ornatus* (Daudin 1803).

Geographic Distribution

The distribution area covers the whole of sub-Saharan Africa, from 18°N longitude to the most southern part of the continent. Additionally, it extends along the River Nile up to Egypt (Fig. 5.9). The species does not inhabit the great deserts of Somalia, Namibia, Botswana, and South Africa. In the West and Central African rain forest (Congo, Gabon, Cameron, Guinea, Equatorial Guinea, Togo, Zaire), *V. niloticus* is replaced by *V. ornatus* (see map after Buffrénil 1993 and Böhme and Ziegler 1997a). An overview of different ecological regions of Africa inhabited by *V. niloticus* is given by Bayless (1997).

Fossil Record

Stromer (1910) mentions a jaw fragment of *Varanus niloticus* in Cretaceous sediments in Togo and referred it to the living form. Estes (1993) is skeptical about the origin of this "fossil."

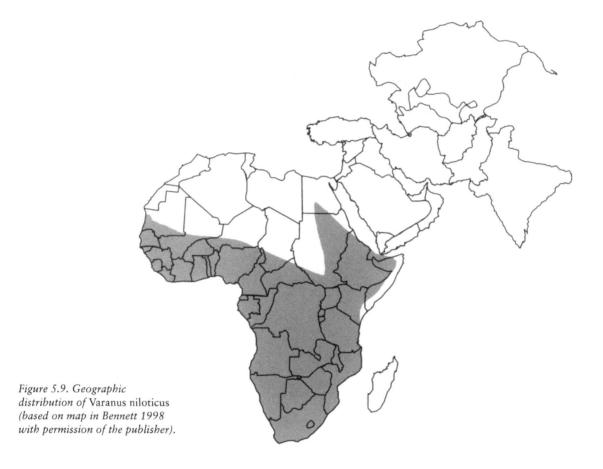

Figure 5.9. Geographic distribution of Varanus niloticus *(based on map in Bennett 1998 with permission of the publisher).*

Diagnostic Characteristics

- Large lizard up to approximately 80 cm SVL and about 200 cm total length.
- Tail laterally compressed with a low dorsal crest.
- A total of 136 to 183 scales around the midbody.
- Basic dorsal color of adults gray brown to olive-brown with light yellow ocelli and bands on head, back, limbs, and tail. Belly and throat are paler, with black bars.

V. niloticus is characterized by six to nine crossbands or rows of yellow ocelli on the back between fore- and hind limbs, whereas *V. ornatus* normally has only five bands.

Description

A widespread, terrestrial, arboreal, and aquatic monitor lizard. No sexual dimorphism in color is evident. Hatchlings and juveniles are patterned with black and bright yellow.

Size

West African specimens reach a total length of about 1.7 m (Cissé and Karns 1978; Lenz 1995), whereas monitor lizards in the eastern and southern parts of its range grow larger. A maximum length of 2.14 m was measured in Tschad and of 2.4 m in South Africa (Haacke and Groves 1995; Branch 1991; Buffrénil et al. 1994; Böhme and Ziegler 1997b) for males. Large lizards are also reported from South Africa with a total length of 1.7 m. Average SVL of 60 adult lizards caught in Gambia was 60 cm (52.3 cm for females and 64.4 cm for males).

Tail length averaged 1.5 times SVL. There is no sexual dimorphism in the proportions of the body, tail, or limbs. Juveniles have proportionally larger heads than adults. The head is shorter and broader than high, and the nares lie closer to the eye than to the tip of the snout. Comparative studies in the Tchad region demonstrate that males reach their asymptotic size (207 cm) at a later age (9 years) than females (154.8 cm reached at age of 6 years) (Buffrénil et al. 1994).

Habitat and Natural History

The presence of Nile monitors is tied to a few necessary habitat structures, which can be variably shaped in different habitats and vegetation zones. As described by Bayless (1997), the different habitats of *V. niloticus* in Africa consist of forest–savanna mosaic, woodland, dry savanna, bushland and thickets, evergreen thickets, scrub, swamps, mangroves, lakes, and rivers. Exposed and poorly covered areas form an essential part of the habitat, serving as basking localities. In natural areas, this condition is offered, for example, by embankments, termite mounds, and sandy banks, and also by trunks and tops of trees. In areas altered by humans, roofs, streets, and walls fulfill this demand. The (at least) temporary availability of water is also essential. Because of its wide food range, *Varanus niloticus* finds prey in nearly all vegetation and climatic zones, even in localities with a reduced fauna (e.g., urban areas, mangroves).

Habitat structures preferred by juveniles differ in several ways. In Gambia, hatchlings were mainly found close to permanent water (Lenz 1995). They are also more arboreal. Night resting sites were often found under vegetation cover, under loose bark of trees, or in shelters of human origin. The search for food takes place in areas rich in insects or small vertebrates, such as dense riverine forests, as well as human-influenced areas (e.g., rice fields or hotel gardens).

In all natural habitats, Nile monitors use burrows as shelter, which they dig themselves or enlarge, if these already exist. Burrows are mainly located in sandy ground on slightly inclined surface profiles and have an oval entrance. Tunnels are sometimes branched, up to 6 m long, and open into an enlarged chamber. If refuges of anthropogenic or other origin are available, these are preferred over self-dug burrows.

Terrestrial, Arboreal, Aquatic

The high ecological plasticity of the Nile monitor is characterized by the relative unpretentiousness and adaptability in habitat choice; a wide and flexible range of prey; and excellent climbing, diving, and swimming ability. As a generalized reptile species, it inhabits almost all given habitats, sometimes in high population densities.

Nile monitors are superb swimmers. In the wild, they have been observed to remain underwater for more than an hour (Ingleton 1929). Juveniles spend most of the time in trees and are better at climbing than adults. However, even large Nile monitors will sometimes readily climb high trees to bask, to search for food, and also to sleep.

Time of Activity (Daily and Seasonal)

Varanus niloticus is active over most of the year. Cowles (1930) reported a resting period during the cooler winter months, whereas Cissé (1971) mentions estivation in habitats without permanent water during the hot dry season.

Investigations in Gambia (Lenz 1995) demonstrate that seasonal activity reaches a maximum during the rains, with a nearly constantly high observation rate from July to November. Male and female monitors show different activity peaks, caused by the reproductive cycle. Males were observed mostly in September and October, during the mating period, whereas females are mainly active in November, during the time of egg laying. Time spent per day foraging is directly positively correlated with seasonally varying prey availability. A highly reduced prey quantity leads to an increase in the daily time searching for prey (~55.7% of the daily activity period during the hot dry season compared with ~29.8% during the rains). In the rainy season, enough time remains for reproductive activities and extended basking and resting periods of the monitor lizards. During the cold dry season, besides the long-lasting search for food (~33.9%), extended basking periods (~54.2%) are necessary to maintain body temperature because of the low ambient temperature. During hotter months, thermoregulatory activities occupy less time (~27.1%) because of constantly high ambient temperatures.

In general, Nile monitors show a marked circadian rhythm with a peak of activity in early afternoon (Cloudsley-Thompson 1966).

Thermoregulation

Hirth and Latif (1979) recorded an average body temperature of 28.8°C (range, 20–37.2°C) with activity generally when body temperature is above 26°C. Field observations indicate a temperature optimum between 29 and 30°C (measured in the shade) for *Varanus niloticus*. Highest activity rates occurred when humidity is between 75% and 85% (Lenz 1995).

Foraging Behavior

Correlated with seasonal changes in food availability, Nile monitors show varying hunting strategies. During the rains, active foraging in the habitat (which means on or under the ground, in trees, and in the water) is the dominant strategy, whereas during the dry season, ambushing in attractive localities—for example, near water holes—is often observed.

Some specific hunting strategies have been observed for *Varanus niloticus*. Cooperation between two monitors is described by Pitman (1931) and Horn (1999, after a television film). One lizard provokes a female Nile crocodile (*Crocodylus niloticus*), guarding its nesting place, to chase it into the water. Meanwhile, the second monitor digs up the crocodile's nest and feeds on eggs and hatchlings, later joined by its accomplice. Horn also mentions similar teamwork with two monitors robbing birds' nests.

Diet

Nile monitors are carnivorous, eating nearly all animal species that they can overpower and catch or which they find as carrion. Diets vary with habitats and seasons, and with age.

During the dry season, approximately 50% of the feeding requirements are covered by carrion because of the highly limited availability of invertebrates. During the rains, utilization of carrion is distinctly reduced. Arthropods, amphibians, birds, and small mammals or their juveniles occur in high densities and characterize the food composition of Nile monitors. With the transition to the cold, dry season, food resources become scarce, so carrion again becomes a higher percentage in the diet (Lenz 1995).

The diet includes insects, arachnids, mollusks, fish, amphibians, reptiles, birds, and mammals (Cissé 1972; Branch 1991). Cannibalism is also known. *V. niloticus* has a tolerance to poison, especially from snakes (Auerbach 1987).

Ontogenetic changes in dentition (Mertens 1942) are directly correlated to the highly insectivorous (80%) diet of hatchlings. In subadult specimens, prey include a high percentage of arthropods (~75%) as well as amphibians, annelids, mollusks, and occasionally carrion, which represent a transition to the adult diet.

Reproduction

Both sexes reach sexual maturity at a total length of about 90 cm in West Africa and with about 120 cm in southern and central Africa (Cissé 1971; Branch 1991; Buffrénil 1993) at an age of 3 to 4 years.

The reproductive cycle can be correlated with climate, with mating taking place at the end of the rainy season, for example, September to November in West Africa (Cissé 1971), or March to May in South Africa (Cowles 1930; Branch 1991). Hagen et al. (1995) described and illustrated behavioral patterns of courtship and copulation of *V. niloticus* in Kenya, during which two males fight each other, but without the clinch phase, which suggests a shortened ritual combat in *V. niloticus*. Horn et al. (1994) explained this as secondary loss. Clutch size varies between 5 and 60 eggs depending on the size of a female. Eggs are deposited in termite mounds (Branch 1991; Cowles 1930) or burrows (Cissé 1971). Enright (1992, cited in Horn and Visser 1997) observed in captivity a clutch size of 14 eggs and an incubation period of 120 to 126 days at 30°C. Average clutch size is approximately 20 eggs.

In the wild, juveniles hatch after an incubation period of 6 to 9 months at the beginning of the rains with a total length of 16.5 to 30 cm (Enright 1989; Cowles 1930; Cissé 1971).

Movement

In Abuko Nature Reserve (Gambia, West Africa), Nile monitors use home ranges up to 50,000 m², the size depending on the season and the sex and age of the lizard. Hatchlings inhabit an area of about 30 m²; subadults occupy an average area of approximately 5000 m². Adult females use a mean area of about 15,000 m², whereas adult males claim an average of 50,000 m² (Lenz 1995). Home ranges of individuals overlap extensively, and territorial behavior is seldom observed. All individual home ranges in and around the study area show varying portions of savanna and forest habitats, and all of them include at least one permanent water hole. Radiotelemetry studies have not been carried out for this species until recently.

Physiology

Millard and Johansen (1974) studied the ventricular outflow dynamics as responses to hypoxia, hypercarbia, and diving and found high pressures in the systemic arteries coupled with low pulmonary pressures in a ratio of 4.0 to 6.0. This ratio approaches values in homeothermic vertebrates. Diving increased pulmonary blood pressure while the systemic blood pressure fell markedly.

Fat Bodies, Testicular Cycles

Cissé (1976) described the development of gonads and testes during the annual cycle of *V. niloticus* in Senegal via histological techniques. A resting period during the dry season is followed by an active period during the rains.

Parasites

Two species of ticks, *Aponomma flavomaculatum* and *A. arcanum,* are known as ectoparasites of *Varanus niloticus* in West Africa (Aeschlimann 1967). *Aponomma flavomaculatum* of different stages and sex use specific niches on the body of the monitor lizard in forest and gallery forest habitats, whereas *A. arcanum* dominates in savanna areas. Additionally, Nile monitors are heavily infested with tsetse flies (*Glossina palpalis gambiense*). Two species of endoparasitic cestodes were isolated from fecal pellets; these unidentified nematodes were found in a southern African specimen (Branch 1991). Additionally, a probable infestation with *Sambuca lohrmanni* (Pentastomidae) has been described, causing irritations and discolorations of the mucous membrane of the mouth (Lenz 1995).

5.5. *Varanus ornatus*

WOLFGANG BÖHME AND THOMAS ZIEGLER

Figure 5.10. Varanus ornatus *(photo by Jeff Lemm)*.

Nomenclature

Described as *Tupinambis ornatus* by Daudin (1803), the forest-inhabiting Nile monitor lizard was later considered to be a subspecies of *Varanus niloticus* (Linnaeus 1766) (Mertens 1942, 1963) (Fig. 5.10). The type locality is Malimbe, Cameroon. The holotype is stored in the Muséum National d'Histoire Naturelle in Paris, France, cataloged as MHNP 7471 (1570) (Brygoo 1987). Böhme and Ziegler (1997a) identified numerous areas where it occurs in sympatry with *V. niloticus* (e.g., Liberia, Ghana, Nigeria, Gabon, southwestern Democratic Republic of Congo), without any hint of a possible intergradation, so that both have to be regarded as two distinct, although closely related, species. The sympatric occurrence in the Lagos area, Nigeria, was challenged by Angelici and Luiselli (1999) and needs further clarification.

Geographic Distribution

V. ornatus is widely distributed in the African tropical rain forest areas, i.e., the Congo Basin and the Upper Guinean forested areas (Fig. 5.11). It has been recorded from Sierra Leone, Liberia, Guinea,

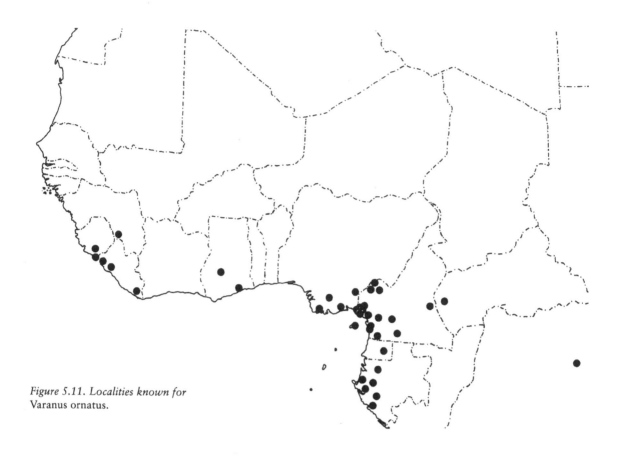

Figure 5.11. Localities known for
Varanus ornatus.

Ghana, Nigeria, Cameroon, Equatorial Guinea (including Bioko Island), Gabon, the Central African Republic, and the Democratic Republic of Congo (Kivu Province). It is expected to occur also in the Ivory Coast. Localities in Nigeria, Cameroon, and Gabon are concentrated in coastal areas. Only two central localities in the Congo Basin exist, and the Kivu record marks its eastern margin. This disjunct pattern does not seem to be the result of simple collecting gaps only but may rather reflect the dynamic vegetational history of the African forests during the Pleistocene (Böhme and Ziegler 1997a).

Fossil Record

None.

Diagnostic Characteristics

- Stoutly built, big-headed species, surpassing 200 cm total length, and thus among the biggest monitor lizards of the world.
- Very stout skull, with extremely blunt, crushing teeth in adults.
- Only (three to) five oblique rows of light ocelli on dorsum.
- Whitish tongue.

- Mean midbody scale count (S value) 162.
- Up to 12 paryphasmata on hemipenis, the one to two proximal ones uninterrupted medially on the asulcal side (vs. interrupted in *V. niloticus*).

Description

V. ornatus is a large, stoutly built monitor lizard with a relatively big head in adult males. It grows to more than 200 cm total length. The tail is relatively longer than in its savanna sibling *V. niloticus* (head–body/tail length ratio 1:1.71 vs. 1:1.6 in the latter; Mertens 1942). The dorsal ground color is blackish with a bright yellow pattern of supratemporal and temporal lines as well as vertical bars on the sides of head, some faint cross lines on the head passing into chevron-shaped nuchal lines, and mostly five (rarely less) distinct crossbands that consist of black-centered light ocelli. Between these bands, there is often a fine yellow stippling. Underparts are yellow with a wider-meshed dark reticulum. Tail has 12 broad yellow crossbands. Tongue is whitish or pink. Dentition changes ontogenetically from normal, sharp, and recurved teeth in juveniles to very blunt, crushing teeth in adults.

Size

The largest reliably measured specimen of *V. ornatus* was a male from Ikoyi, Nigeria (Dunger 1967). Its total length was 190 cm, but its tail tip was lacking. Two considerably bigger specimens from southern Cameroon are known from their skulls only, with condylobasal lengths of 13.6 cm (Lönnberg 1903) and 14.2 cm (Böhme and Ziegler 1997a, 1997b). The total length of these two specimens, as calculated from their skull measurements, is greater than 250 cm (Böhme and Ziegler 1997b).

Habitat and Natural History

V. ornatus is mostly confined to the west and central African lowland rain forest. However, it also penetrates secondary forests, clearings, human settlements (Schmitz et al. 2000), and, next to deltaic swamps and mangrove, even savanna-like habitats at the edge of the forest (Angelici and Luiselli 1999). In mountainous areas (e.g., Cameroon), it occurs up to at least 1200 m (Herrmann et al. 1999). Its habitats are typically closely associated with watercourses.

Terrestrial, Arboreal, Aquatic

Although basically terrestrial, *V. ornatus* is also an excellent swimmer and a good climber.

Time of Activity (Daily and Seasonal)

V. ornatus is diurnal. In spite of its savanicolous sibling *V. niloticus*, which has an annual estivation period during the dry season in most areas (and to which the great majority of ecological literature data on Nile monitors applies; see Lenz 1995; Bennett 1998), *V. ornatus* is active throughout the year (Buffrénil 1993), although seasonal activity peaks during the wet months (Angelici and Luiselli 1999).

Foraging Behavior and Diet

V. ornatus is an active forager. The only study on its food habits is that of Angelici and Luiselli (1999) in Southeast Nigeria. These authors found that "crabs" (not specified taxonomically) formed 56% of all prey items. Vertebrate prey (in total just 10%) included juvenile conspecifics and hatchlings of the dwarf crocodile (*Osteolaemus tetraspis*), in both cases 1.7% of the dietary spectrum. In juveniles, crabs made up nearly 78% of the diet. Diet composition was significantly different between males and females, but not between juveniles and both males and females. The trophic niche breadth of juveniles was less wide than in adults, but nearly equal in males and females (Angelici and Luiselli 1999). The lack of a significant dietary difference between adults and juveniles in Southeast Nigeria does not reflect the ontogenetic change in dentition which is typical for Nile monitors.

Reproduction

In Southeast Nigeria, Angelici and Luiselli (1999) found six gravid females between late March and mid-April and hatchlings in early May, and they deduced a pronounced reproductive seasonality from this observation. However, newly hatched juveniles were found in October in Southeast Guinea and in southern Cameroon (Böhme, unpublished data) so that at least a bimodality of the reproductive phase exists.

Movement

Data on home range for this species, in spite of its sibling *V. niloticus,* are still completely lacking.

Nothing is known about thermoregulation, physiology, fat bodies, testicular cycles, or parasites of *Varanus ornatus*.

5.6. *Varanus yemenensis*

WOLFGANG BÖHME

Nomenclature

Described by Böhme et al. (1988) in the still valid combination *Varanus yemenensis* (Fig. 5.12). The first discovery took place 2 years before by means of a TV movie that had been produced in (former Northern) Yemen to show the nature, landscape, and biodiversity of this country. By chance, a monitor lizard was filmed and recognized on TV as an undescribed species, of which some voucher specimens were subsequently brought to Europe some months later (Böhme et al. 1987). The holotype is deposited in the Zoologisches Forschungsinstitut und Museum Alexander Koenig (ZFMK) in Bonn, Germany, as ZFMK 46500, together with one paratype (a subadult). Additional paratypes are in the collections of the natural history museums of Geneva, London, and Dresden.

Geographic Distribution

V. yemenensis is restricted to the southwest coast of the Arabian Peninsula from Ta'izz and Al Khobar (Yemen) in the south to Wadi Maraba, Asir Mountains (Saudi Arabia), in the north. It inhabits foothills of the southwest Arabian mountain ranges (montane Tihama). The vertical distribution ranges from 200 to 300 m (As Soknah, Yemen, type locality) to 1700 to 1800 m above sea level (Manakha, Yemen) (Böhme et al. 1987, 1989; Böhme 1999) (Fig. 5.13).

Figure 5.12. Varanus yemenensis *(photo by Wolfgang Böhme).*

Figure 5.13. Geographic distribution of Varanus yemenensis.

Fossil Record

No fossils of *V. yemenensis* are known.

Diagnostic Characteristics

- Large, stoutly built, up to 110 cm in total length.
- Nostril oblique, in front of eye.
- Nasal, temporal, and parietal region swollen.
- Dark brownish, with more or less distinct crossbands.
- Tail also with dark crossbands fading in the last third, laterally compressed with a median double crest.
- Conspicuous yellow band across snout (may be blackish in specimens from the highlands near Ta'izz).

Moreover, some characteristics of lung and hemipenis morphology are diagnostic for this species (Becker 1991; Böhme 1991a, 1991b; Ziegler and Böhme 1997).

Description

Adult *V. yemenensis* are dark brownish with indistinct dark cross-bands particularly on the sacrum, which are more distinct on the tail; last third of tail fading into yellowish. The juvenile is more conspicuously banded with somewhat irregularly arranged crossbars. Both adult and juvenile specimens have a conspicuous yellow band across the snout in front of the eyes. The slitlike nostrils are obliquely situated in front of the eyes. Forelegs are more strongly developed than the hind legs; they also bear stronger claws than the latter; and they are yellowish, whereas the hind legs have the same ground color as the body, but with distinct yellow spotting. The tail strongly compressed laterally, with a double median crest. Nuchal scales are only slightly larger than those on the dorsum.

Size

Males grow larger than females. The holotype has a head–body length of 47 cm, and head–body lengths of the two adult male paratypes are 43 and 37.4 cm long, respectively. Total length cannot be given for two of these males because their tails are mutilated and incomplete, respectively. The largest specimen recorded so far is an adult male from the Asir Mountains, Saudi Arabia (Gasperetti in Böhme et al. 1989; Schätti and Gasperetti 1994), with a head–body length of 59 cm and a total length of 115 cm. Another male, caught with a head–body length of 45.8 and a total length of 99.9 cm, grew in captivity during 20 months to 101.8 cm (head–body, 47.5 cm). The two largest females known have 39.5 and 39 cm head–body length and 86.8 and 81.4 cm total length, respectively (Böhme et al. 1989; Schätti and Gasperetti 1994). The only known hatchling of *V. yemenensis* has a head–body length of 14.5 cm and a total length of 32 cm.

Habitat and Natural History

Varanus yemenensis inhabits the margin of the montane Tihama in the southwest of the Arabian Peninsula, from 300 up to 1700 to 1800 m above sea level. It is not present in the coastal lowland. Low-elevation habitats (e.g., type locality As Soknah, Yemen, 300 m above sea level) often consist of light thornbush savanna and degraded dry forest, with a warm climate throughout the year. Annual precipitation is about 300 mm, concentrated in the late summer (August, September) that corresponds to the rainy season in the opposite African coast. Substrate is sand and gravel (Böhme et al. 1987). Habitats in the foothill range of the montane Tihama are characterized by basaltic rocks partially covered by dense vegetation (e.g., *Euphorbia cactus, E. inarticulata, E. triaculeata*). Cultivated fields (maize, millet) surrounded by dense shrubs and groups of *Acacia* trees are also inhabited (Schätti and Fortina 1987; Böhme et al. 1989).

Terrestrial, Arboreal, Aquatic

V. yemenensis is a terrestrial lizard that occasionally also climbs trees. It shelters in self-dug burrows (excavated using its strong forelegs), rock crevices, or hollow trees, and will even pass through shallow water when fleeing.

Time of Activity (Daily and Seasonal)

The scant observations available indicate that *V. yemenensis* estivates during the dry season. Despite intensive searches, no individual could be found during 3 months from January to March in an area where they turned out to be rather common in October of the same year (Böhme et al. 1987). Despite the claim of some locals, they are strictly diurnal and have their active phases in the morning and in the afternoon, with a peak after 4 P.M. (Böhme et al. 1989).

Foraging Behavior and Diet

The animals actively forage beneath stones as well as dead leaves, branches, and other vegetation, particularly along places with shallow water. The bulk of prey is formed by insects, in particular beetles, but also by mollusks such as the enid snail *Passamella (Euryptyxis) candida,* which has been recovered from *V. yemenensis* excrement.

Captive specimens readily feed on mice, chicken and other birds, fish, crickets, and large cockroaches as well as dog food.

Parasites

Two species of ixodid ticks have been collected from *V. yemenensis*: *Amblyomma sparsum* and *Aponomma flavomaculatum* (Böhme et al. 1989).

Nothing is known about thermoregulation, physiology, movement, or reproduction in *V. yemenensis*.

References

Adjovi, Y. 1981. Evolution of glucose induced hyperglycemia during the annual cycle of a Sahelian lizard (*Varanus exanthematicus*). *Comp. Biochem. Physiol. A Comp. Physiol.* 69:529–536.

Aeschlimann, A. 1967. Biologie et ecologie des tiques (Ixodoidea) de Cote d'Ivoire. *Acta Trop.* 24:282–405.

Anderson, J. 1898. *Zoology of Egypt.* Vol. 1, *Reptiles and batrachia.* London: Bernard Quaritch.

Anderson, S. C. 1963. Amphibians and reptiles from Iran. *Proc. Calif. Acad. Sci.* 31:417–498.

———. 1999. *The lizards of Iran.* Society for the Study of Amphibians and Reptiles.

Andres, A. 1904. Intelligentz eines Wustenwarans (*Varanus griseus*). *Bull. Aqua. Terrar. Kunde* 15:269.

Angel, M. F. 1933. Sur quelques reptiles et batrachiens du nord du Soudan Francais. *Bull. Mus. Hist. Nat. Paris,* ser. 2, 5:68–69.

Angelici, F. M., and L. Luiselli. 1999. Aspects of the ecology of *Varanus niloticus* (Reptilia, Varanidae) in southeastern Nigeria, and their contribution to the knowledge of the evolutionary history of *V. niloticus* species complex. *Rev. Ecol. (Terre Vie)* 54:29–42.

Anonymous. 1983. International trade in the skins of monitor and tegu lizards 1975–1980. *Traffic Bull.* 4:71–79.

Ast, J. C. 2001. Mitochondrial DNA evidence and evolution in Varanoidea (Squamata). *Cladistics* 17:211–226.

Auerbach, R. D. 1987. *The amphibians and reptiles of Botswana.* Mokwepa Consultants.

Auffenberg, W. 1981. *The behavioral ecology of the Komodo monitor.* Gainesville: University Presses of Florida.

———. 1988. *Gray's monitor lizard.* Gainesville: University Press of Florida.

———. 1989. Utilization of monitor lizards in Pakistan. *Traffic Bull.* 11: 8–12.

———. 1994. *The Bengal monitor.* Gainesville: University Presses of Florida.

Auffenberg, W., and I. M. Ipe. 1983. The food and feeding of juvenile Bengal monitors. *J. Bombay Nat. Hist. Soc.* 80:119–124.

Auffenberg, W., H. Rehman, F. Iffat, and Z. Perveen. 1990. Notes on the biology of *Varanus griseus koniecznyi. J. Bombay Nat. Hist. Soc.* 87:26–36.

Bangma, G. C., and H. J. ten Donkelaar. 1984. Cerebellar efferents in the lizard *Varanus exanthematicus.* 1. Corticonuclear projections. *J. Comp. Neurol.* 228:447–459.

Bangma, G. C., H. J. ten Donkelaar, P. J. W. Dederen, and R. de Boer-van Huizen. 1984. Cerebellar efferents in the lizard *Varanus exanthematicus.* 2. Projections of the cerebellar nuclei. *J. Comp. Neurol.* 230:218–230.

Barbas, H. A. 1982. The motor nuclei and primary projections of the facial nerve in the monitor lizard *Varanus exanthematicus. J. Comp. Neurol.* 207:105–113.

Barbas, H. A., and A. H. M. Lohman. 1984. The motor nuclei and primary projections of the IXth, Xth, XIth, and XIIth cranial nerves in the monitor lizard, *Varanus exanthematicus. J. Comp. Neurol.* 226:565–579.

———. 1986. The motor complex and primary projections of the trigeminal nerve in the monitor lizard, *Varanus exanthematicus. J. Comp. Neurol.* 254:314–392.

———. 1988. Primary projections and efferent cells of the 8th cranial nerve in the monitor lizard, *Varanus exanthematicus. J. Comp. Neurol.* 277:234–249.

Barbas, H. A., and F. G. Wouterlood. 1988. Synaptic connections between primary trigeminal afferents and accessory abducens motor neurons in the monitor lizard, *Varanus exanthematicus. J. Comp. Neurol.* 267:387–397.

Bayless, M. K. 1997. The distribution of African lizards (Sauria: Varanidae). *Afr. J. Ecol.* 35:374–377.

Becker, H. O. 1991. The lung morphology of *Varanus yemenensis* Böhme, Joger and Schätti, 1989, and its bearing on the systematics of the Afro-Asian monitor radiation. *Mertensiella* 2:29–37.

Ballard, V., and F. B. Antonio. 2001. *Varanus grisens* (desert monitor): Toxicity. *Herpetol. Rev.* 32(4):261.

Bennett, D. 1998. *Monitor lizards: Natural history, biology and husbandry.* 2nd ed. Frankfurt am Main: Edition Chimaira.

———. 2000a. Preliminary data on the diet of juvenile *Varanus exanthe-*

maticus (Sauria: Varanidae) in the coastal plain of Ghana. *Herpetol. J.* 10:75–76.

———. 2000b. The density and abundance of juvenile Bosc's monitor lizards (*Varanus exanthematicus*) in the coastal plain of Ghana. *Amphibia-Reptilia* 21:301–306.

———. 2000c. Observations on Bosc's monitor lizard (*Varanus exanthematicus*) in the wild. *Bull. Chicago Herpetol. Soc.* 35:177–180.

———. 2002. The diet of juvenile *Varanus niloticus* (Sauria: Varanidae) on the Black Volta River in Ghana. *J. Herpetol.* 36:116–117.

Bennett, D., and B. Basuglo. 1998. Lacertilia: *Varanus exanthematicus* (Bosc's monitor): Predation. *Herpetol. Rev.* 29:240–241.

Böhme, W. 1988. Zur Genitalmorphologie der Sauria: Funktionelle und stammesgeschichtliche Aspekte. *Bonn Zool. Monogr.* 27:1–176.

———. 1991a. Artbildung bei Waranen. *Mitt. Zool. Mus. Berl.* 67:81–83.

———. 1991b. New findings on the hemipenial morphology of monitor lizards and their systematic implications. *Mertensiella* 2:42–49.

———. 1997. Robert Mertens' Systematik und Klassifikation der Warane: Aktualisierung seiner 1942 er Monographie und eine revidierte Checkliste. In *Die Familie der Warane (Varanidae)*, ed. R. Mertens, i–xxii. Frankurt am Main: Edition Chimaira.

———. 1999. New records of SW Arabian monitor lizards, with notes on the juvenile dress of *Varanus yemenensis. Mertensiella* 11:267–276.

Böhme, W., J. P. Fritz, and F. Schütte. 1987. Neuentdeckung einer Großechse (Sauria: *Varanus*) aus der Arabischen Republik Jemen. *Herpetofauna* 9(46):13–20.

Böhme, W., U. Joger, and B. Schätti. 1988. A new monitor lizard (Reptilia: Varanidae) from Yemen, with notes on ecology, phylogeny and zoogeography. *Fauna Saudi Arabia* 10:433–448.

Böhme, W., and U. Joger. 1991. Zur systematischen Stellung eines neu entdeckten Großwarans aus Südwestarabien: *Varanus yemenensis. Verh. Dt. Zool. Ges.* 1991:440.

Böhme, W., and T. Ziegler. 1997a. A taxonomic review of the *Varanus* (*Polydaedalus*) *niloticus* (Linnaeus 1766) species complex. *Herpetol. J.* 7:155–162.

———. 1997b. Großwarane im Museum Koenig, mit Bemerkungen zu Afrikas größter Echse. *Tier Mus.* 5:65–74.

Boulenger, G. A. 1885. *Catalogue of the lizards in the British Museum.* Vol. 2. London: Taylor and Francis.

Bowker, R. G. 1984. Precision of thermoregulation of some African lizards. *Physiol. Zool.* 57:401–412.

Branch, W. R. 1988. *Field guide to the snakes and other reptiles of Southern Africa.* Cape Town: Struik.

———. 1991. The Regenia registers of "Gogga" Brown (1869–1909), "Memoranda on the species of monitor or varan." *Mertensiella* 2:57–110.

Brygoo, E.-R. 1987. Les types de Varanidés (Reptiles, Sauriens) du Muséum national d'Histoire maturelle. Catalogue critique. *Bull. Mus. Nat. Hist. Nat. Paris,* 4th ser., 9, A, 2, suppl., 21–38.

Buffrénil, V. de. 1993. *Les varans africains:* Varanus niloticus *et* Varanus exanthematicus. *Données de synthèse sur leur biologie et leur exploitation.* Gland, Switzerland: Programmes des Nations Unies pour l'environnement (PNUE), CITES Sécretariat.

Buffrénil, V. de, C. Chabanet, and J. Chabanet. 1994. Données preliminaires sur la taille, la croissance et la longevite du varan du Nil

(*Varanus niloticus*) dans la region du lac Tchad. *Can. J. Zool.* 72:262–273.

Burggren, W., and K. Johansen. 1982. Ventricular haemodynamics in the monitor lizard *Varanus exanthematicus*: Pulmonary and systemic pressure separation. *J. Exp. Biol.* 96:343–354.

Cansdale, G. S. 1951. Some Gold Coast lizards. *Nigerian Field* 16:21–24.

Chandra, H. 1987. Some observations on the predatory behaviour of desert monitor *Varanus griseus* Daudin (Lacertilia: Varanidae). *Plant Prot. Bull.* 39:22–23.

Cissé, M. 1971. La diapause chex les varanides du Sénégal. In *Notes Africaines,* 131:57–67. Dakar: Institut Français d'Afrique Noire.

———. 1972. L'Alimentaire des Varanides au Sénégal. *Bull. Inst. Fond. Afr. Noire,* ser. A, *Sci. Nat.* 34:503–515.

———. 1973. Evolution de la graisse de reserve et cycle génital chez *Varanus niloticus. Bull. Inst. Fond. Afr. Noire,* ser. A, *Sci. Nat.* 35:169–179.

———. 1976. Le cycle génital des Varans du Sénégal. *Bull. Inst. Fond. Afr. Noire,* ser. A, *Sci. Nat.* 38:188–205.

Cissé, M., and D. R. Karns. 1978. Les Sauriens du Sénégal. *Extrait Bull. Inst. Fond. Afr. Noire* T40, 1.

Clos, L. M. 1995. A new species of *Varanus* from the Miocene of Kenya. *J. Vertebr. Paleontol.* 15:254–262.

Cloudsley-Thompson, J. L. 1966. Water relations and diurnal rhythm of activity in the young Nile monitor. *Br. J. Herpetol.* 3:296–300.

Corkill, N. L. 1928. Notes on the desert monitor (*Varanus griseus*) and the spiny tailed lizard (*Uromastix microlepis*). *J. Bombay Nat. Hist. Soc.* 32:608–610.

Cowles, R. B. 1930. The life history of *Varanus niloticus* (Linnaeus) as observed in Natal, South Africa. *J. Entomol. Zool.* 22:1–31.

Cumming, G. S. 1999. Host distributions do not limit the range of most African ticks. *Bull. Entomol. Res.* 89:303–327.

Daudin, F. M. 1803. *Histoire Naturelle, Générale et Particulière des Reptiles.* Vol. 8. Paris: F. Dufart.

Dujsebayeva (Matveyeva), T. 1995. *Varanus griseus* (desert monitor lizard) reproduction. *Herpetol. Rev.* 26:100.

Dunger, G. T. 1967. The lizards and snakes of Nigeria. Part 3. The monitors and a plated lizard. *Nigerian Field* 32:170–178.

Dupe-Godet, M. 1984. Characterization and measurement of plasma somatostatin-like immunoreactivity in a Sahelian lizard (*Varanus exanthematicus*) during starvation. *Comp. Biochem. Physiol. A Comp. Physiol.* 78:53–58.

Dupe-Godet, M., and Y. Adjovi. 1981a. Seasonal variations of immunoreactive insulin contents in pancreatic extracts of a Sahelian lizard (*Varanus exanthematicus*). *Comp. Biochem. Physiol. A Comp. Physiol.* 69:717–729.

———. 1981b. Seasonal variations of immunoreactive glucagon contents in pancreatic extracts of a Sahelian lizard (*Varanus exanthematicus*). *Comp. Biochem. Physiol. A Comp. Physiol.* 69:31–42.

———. 1983. Somatostatin-like immunoreactivity in pancreatic extracts of a Sahelian lizard (*Varanus exanthematicus*) during starvation. *Comp. Biochem. Physiol. A Comp. Physiol.* 75:347–352.

Eichwald, C. E. I. 1831. *Zool. Spec.* 3:190. Vilnae: Josephi Zawadzki.

Enright, B. 1989. Notes on breeding the Nile monitor (*Varanus niloticus*) in captivity. *Ontario Herpetol. Soc. Newsl.* 21:8–9.

Estes, R. 1993. Sauria Terrestria, Amphisbaenia. In *Handbuch der Paläo-herpetologie,* ed. P. Wellnhofer, 10A:179–189. Stuttgart: Gustav Fischer Verlag.

Gauthier, R. 1967. Écologie et éthologie des Reptiles du Sahara Nord-occidental (région de Beni-Abbes). Museé Royal de l'Afrique centrale, Tervuren, Belgique. *Ann. Sci. Zool.* 155:24–26.

Glass, M. L., S. C. Wood, R. W. Hoyt, and K. Johansen. 1979. Chemical control of breathing in the lizard, *Varanus exanthematicus. Comp. Biochem. Physiol. A Comp. Physiol.* 62:999–1003.

Godet, R., X. Mattei, and M. Dupe-Godet. 1983. Ultrastructure of the exocrine pancreas in a Sahelian reptile (*Varanus exanthematicus*) during starvation. *J. Morphol.* 176:131–139.

———. 1984. Alterations of endocrine pancreas B cells in a Sahelian reptile (*Varanus exanthematicus*) during starvation. *J. Morphol.* 180:173–180.

Grenot, C., and R. Vernet. 1977. Rythme d'activité et regulation thermique chez deux lézards sahariens *Uromastix acanthinurus* Bell et *Varanus griseus* Daud. *Bull. Soc. Ecophysiol.* 2:54–57.

Hagen, H., H. G. Horn, and W. Hagen. 1995. Zu Balz- und Kopulations-verhalten des Nilwarans (*Varanus n. niloticus*)—Eine fotografische Dokumentation. *Herpetofauna* 17(98):29–33.

Haggag, G., A. R. Khamis, and F. Khalil. 1965. Hibernation in reptiles I. Changes in blood electrolytes. *Comp. Biochem. Physiol.* 16:457–465.

———. 1966a. Hibernation in reptiles II. Changes in blood glucose, hae-moglobin, red blood cells count, protein and non-protein nitrogen. *Comp. Biol. Physiol.* 17:335–339.

———. 1966b. Hibernation in reptiles III. Tissue analysis for glycogen and high energy phosphate compound. *Comp. Biol. Physiol.* 17:341–347.

Heisler, N., P. Neumann, and G. M. Maloiy. 1983. The mechanism of intracardiac shunting in the lizard *Varanus exanthematicus. J. Exp. Biol.* 105:15–31.

Herrmann, H.-W., P. A. Herrmann, and W. Böhme. 1999. Die ALSCO-Kamerun-Expedition I: Amphibien und Reptilien vom Mt. Nlonako. *Z. Köln. Zoo.* 42:181–197.

Hicks, J. W., T. Wang, and A. F. Bennett. 2000. Patterns of cardiovascular and ventilatory response to elevated metabolic states in the lizard *Varanus exanthematicus. J. Exp. Biol.* 203:2437–2445.

Hirth, H. F., and E. M. A. Latif. 1979. Deep body temperatures of the Nile monitor (*Varanus niloticus*) taken by radiotelemetry. *J. Herpetol.* 13:367–368.

Hoogland, P. V. 1981. Spinothalamic projections in a lizard, *Varanus exanthematicus*: An HRP study. *J. Comp. Neurol.* 198:7–12.

Hopkins, S. R, J. W. Hicks, T. K. Cooper, and F. L. Powell. 1995. Ventila-tion and pulmonary gas exchange during exercise in the savannah monitor lizard (*Varanus exanthematicus*). *J. Exp. Biol.* 198:1783–1789.

Horn, H.-G. 1999. Evolutionary efficiency and success in monitors: A survey on behaviour and behavioural strategies and some comments. *Mertensiella* 11:167–180.

Horn, H. G., and G. J. Visser. 1989. Review of reproduction of monitor lizards in captivity. *Int. Zoo Yrbk.* 28:140–150.

———. 1997. Review of reproduction of monitor lizards *Varanus* spp. in captivity. *Int. Zoo Yrbk.* 35:227–246.

Horn, H. G., M. Gaulke, and W. Böhme. 1994. New data on ritualized combats in monitor lizards (Sauria, Varanidae), with remarks on their function and phylogenetic implication. *Zool. Garten N.F.* 64:265–280.

Ibrahim, A. A. 2000. A radiotelemetric study of the body temperature of *Varanus griseus* (Sauria: Varanidae) in Zaranik Protected Area, north Sinai, Egypt. *Egy. J. Biol.* 2:57–66.

———. 2002. Activity area, movement patterns, and habitat use of the desert monitor, *Varanus griseus*, in the Zaranik Protected Area, north Sinai, Egypt. *Afr. J. Herpetol.* 51:35–45.

Igolkina, V. A. 1975. *Varanus griseus* in the Leningrad Zoo (in Russian). *Prioroda Mosk.* 9:95–96.

Ingleton, E. C. 1929. A seven foot monitor from South Africa. *Field* 53(213):486.

Jenkins, F. A., Jr., and G. E. Goslow. 1983. The functional anatomy of the shoulder of the savannah monitor lizard (*Varanus exanthematicus*). *J. Morphol.* 175:195–216.

Kaufman, J. D., G. M. Burghardt, and J. A. Phillips. 1994. Density-dependent foraging strategy of a large carnivorous lizard, the savanna monitor (*Varanus albigularis*). *J. Comp. Physiol.* 108:381–384.

———. 1996. Sensory cues and foraging decisions in a large carnivorous lizard, *Varanus albigularis*. *Anim. Behav.* 52:727–736.

Khalaf, K. T. 1959. *Reptiles of Iraq.* Baghdad: Ar-Raqbitta Press.

Khalil, F., and G. Abdel-Messeih. 1959a. Effects of starvation on contents of water, nitrogen and lipids of tissues of *Varanus griseus* Daud. *Z. Vergl. Physiol.* 42:410–414.

———. 1959b. The storage of extra water by various tissues of *Varanus griseus* Daud. *Z. Vergl. Physiol.* 42:415–421.

———. 1962. Tissue constituents of reptiles in relation to their mode of life—II. Lipid contents. *Comp. Biol. Physiol.* 6:171–174.

King, D. R., M. King, and P. Baverstock. 1991. A new phylogeny of the Varanidae. *Mertensiella* 2:211–219.

King, D., S. Fuller, and P. Baverstock. 1999. The biogeographic origins of varanid lizards. *Mertensiella* 11:43–49.

Kuhn, O. 1963. Sauria Supplementum I. In *Fossilium Catalogus. I. Animalia, Part 104*, ed. F. Westphal, 45–50. The Hague: Junk.

Landsmeer, J. M. F. 1983. The mechanism of forearm rotation in *Varanus exanthematicus*. *J. Morphol.* 175:119–130.

Lemire, M., and R. Vernet. 1985. L'élimination nasale d'electrolytes chez le varan gris, *Varanus griseus* Daud (Sauria, Varanidae). *Bull. Soc. Herpetol. France* 36:20–27.

Lenz, S. 1995. Zur Biologie und Ökologie des Nilwarans, *Varanus niloticus* (Linnaeus 1766) in Gambia, Westafrika. *Mertensiella* 5:1–265.

Leviton, A. E., and S. C. Anderson. 1970. The amphibians and reptiles of Afghanistan: A checklist and key to the herpetofauna. *Proc. Calif. Acad. Sci.*, 4th ser., 38:163–206.

Levshakova, I. J. 1986. A new *Varanus* from the lower Pliocene of Tajikistan (in Russian). *Tr. Zool. Inst. Akad. Nauk. SSSR* 157:102–116.

Lönnberg, E. 1903. On the adaptations to a molluscivorous diet in *Varanus niloticus*. *Ark. Zool.* 1:65–83.

Losos, J., and H. W. Greene. 1988. Ecological and evolutionary implications of diet in monitor lizards. *Biol. J. Linnean Soc. Lond.* 35:379–407.

Luebke, J. I., J. M. Weider, R. W. McCarley, and R. W. Greene. 1992. Distribution of NADPH, diaphorase positive somata in the brainstem of the monitor lizard *Varanus exanthematicus*. *Neurosci. Lett.* 148:129–132.

Luxmoore, R., B. Groombridge, and S. Broad, eds. 1988. Significant trade in wildlife: A review of selected species. In *Reptiles and invertebrates,* appendix 2, vol. 2, 182–207. Cambridge: IUCN.

Maina, J. N., G. M. O. Maloiy, C. N. Warui, E. K. Njogu, and E. D. Kokwaro. 1989. Scanning electron microscope study of the morphology of the reptilian lung: The savanna monitor lizard *Varanus exanthematicus* and the pancake tortoise *Malacochersus tornieri*. *Anat. Rec.* 224:514–522.

Makarov, A. N. 1985. A case of cannibalism in a desert monitor lizard (*Varanus griseus*) (in Russian). In *Voprosy Gerpetologii: Abstracts of the Sixth All-Union Herpetological Conference,* 130. Leningrad: Nauka.

Makayev, V. M. 1982. Present conditions and problems of the conservation of the desert monitor (*Varanus griseus*) (in Russian). In *Scientific basis of conservation and rational use of wildlife,* 36–42. Moscow: All-Union Research Institute of Nature Protection and Conservation. Ministry of Agriculture, USSR.

Manaças. 1955. Saurios e ofidos de Guiné Portuguesa. *Anais Junta Invest. Unltramar (Estudios Zool.)* 10:190–193.

Martens, H., and D. Kock. 1992. The desert monitor, *Varanus griseus* (Daudin 1803), in Syria. *Senck. Biol.* 72:7–11.

Mendelssohn, H. 1983. Herpetological nature protection. *Israel Land Nature* 9:21–27.

Merrem, B. 1820. Tentamen systematis Amphibiorum: 60.

Mertens, R. 1942. Die Familie der Warane (Varanidae). *Abh. Senck. Naturf. Ges.* 462, 466 467:1–391.

———. 1954. Über die Rassen des Wustenwarans (*Varanus griseus*). *Senck. Biol.* 35:353–357.

———. 1963. Helodermatidae, Varanidae, Lanthanotidae. *Tierreich* 79:1–26.

Millard, R. W., and K. Johansen. 1974. Ventricular outflow dynamics in the lizard *Varanus niloticus*: Reponses to hyperoxia, hypercarbia and diving. *J. Exp. Biol.* 60:871–880.

Minton, S. A. 1966. A contribution to the herpetology of west Pakistan. *Bull. Am. Mus. Nat. Hist.* 134:27–184.

Mosauer, W. 1934. The reptiles and amphibians of Tunisia. *Publ. Univ. Calif. Los Angeles Biol. Sci.* 1:49–64.

Müller, L. 1905. Der westafrikanische Steppenwaran (*Varanus exanthematicus*). *Blätter Aquarien Terrarienfreunde (-kunde)* 16:266–268.

Perry, G., R. Habani, and H. Mendelssohn. 1993. The first captive reproduction of the desert monitor *Varanus griseus griseus* at the research zoo of the Tel Aviv University. *Int. Zoo Yrbk.* 32:188–190.

Phillips, J. A. 1995. Movement patterns and density of *Varanus albigularis*. *J. Herpetol.* 29:407–416.

Phillips, J. A., and A. C. Alberts. 1992. Naïve ophiophagus lizards recognize and avoid venomous snakes using chemical cues. *J. Chem. Ecol.* 18:1775–1783.

Phillips, J. A., and G. C. Packard. 1994. Influence of temperature and moisture on eggs and embryos of the white-throated savanna monitor *Varanus albigularis*: Implications for conservation. *Biol. Conserv.* 69:131–136.

Phillips, J. A., and R. P. Millar. 1998. Reproductive biology of the white-throated savanna monitor, *Varanus albigularis*. *J. Herpetol.* 32:366–377.

Pitman, C. R. S. 1931. *A game warden among his charges*. London: Nisbet.

Rjumin, A. V. 1968. The ecology of the desert monitor (*Varanus griseus*) in southern Turkmenistan (in Russian). In *Herpetology of Middle Asia*, 27–31. Tashkent: Academy of Science, Uzbek, USSR.

Röder, A., and H.-G. Horn. 1994. Uber zwei Nachzuchten des Steppenwarans (*Varanus exanthematicus*). *Salamandra* 30:97–108.

Saint Girons, H., and M. C. Saint Girons. 1959. Espace vital, domaine et territoire chez les vertebres terrestres (reptiles et mammiferes). *Mammalia* 23:448–476.

Sakai, K., N. Karasawa, and K. Yamada. 1995. Immunocytochemical localization of APUD cells in the hypothalamus of the grass parakeet (*Melopsittacus undulatus*) and the lizard (*Varanus exanthematicus*). *Biogenic Amines* 11:7–17.

Schätti, B. 1989. Amphibien und Reptilien aus der Arabischen Republik Jemen und Djibouti. *Rev. Suisse Zool.* 96:905–937.

Schätti, B., and R. Fortina. 1987. Herpetologische Beobachtungen in der Arabischen Republik Jemen. *Jemen Rep.* 18:28–31.

Schätti, B., and J. Gasperetti. 1994. A contribution to the herpetofauna of southwest Arabia. *Fauna Saudi Arabia* 14:348–423.

Schmidt, K. P. 1919. Contributions to the herpetology of the Belgium Congo based on the collection of the American Museum Expedition 1909–1915. *Bull. Am. Mus. Nat. Hist.* 39:385–624.

Schmitz, A., O. Euskirchen, and W. Böhme. 2000. Zur Herpetofauna einer montanen Regenwaldregion in SW-Kamerun (Mit. Kupe und Bakossi-Bergland). III. Einige bemerkenswerte Vertreter der Familien Lacertidae, Scincidae, Varanidae, Elapidae und Viperidae. *Herpetofauna* 22(124):16–27.

Secor, S. M., and J. A. Phillips. 1997. Specific dynamic action of a large carnivorous lizard, *Varanus albigularis*. *Comp. Biol. Physiol.* 117A: 515–522.

Shammakov, S. 1981. Reptiles of the plains of Turkmenistan. (in Russian). 144–150. Ashkhabad: Academy of Science, Turkmen USSR.

Shammakov, S. 1981. *Varanus griseus caspius* (Eichwald 1831) (in Russian). In *The reptiles of Turkmenistan*. Acad. Nauk. Turkmen. SSR. Ashkabad, 144–150.

Smirina, E. M., and A. Yu. Tsellarius. 1996. Aging, longevity, and growth of the desert monitor (*Varanus griseus* Daud). *Russ. J. Herpetol.* 3:130–142.

Smith, K. K. 1982. An electromyographic study of the function of the jaw adducting muscles in *Varanus exanthematicus* (Varanidae). *J. Morphol.* 173:137–158.

Smith, K. K., and W. L. Hylander. 1985. Strain gauge measurement of mesokinetic movement in the lizard *Varanus exanthematicus*. *J. Exp. Biol.* 114:53–70.

Sokolov, V. E., V. P. Sukhov, and Yu. M. Chernyshov. 1975. A radio telemetric study of diurnal fluctuations of body temperature in the desert monitor (*Varanus griseus*) (in Russian). *Zool Zh. Moscow* 54:1347–1356.

Sopiev, O., V. M. Makeyev, S. V. Krudrjstev, and A. N. Makarov. 1987. A case of intoxication from a bite of *Varanus griseus*. *Izv. Akad. Nauk. Turm. SSR. Ser. Biol. Nauk.* 87:78.

Staedeli, J. H. 1962. Our very own monitors. *Zoonooz* 35:10–15.

Stamps, J. A. 1977. Social patterns and spacing patterns in lizards. In *Biology of the Reptilia*, vol. 7, *Ecology and behavior,* ed. D. W. Tinkle and W. C. Gans, 265–334. New York: Academic Press.

Stanner, M. 1983. The etho-ecology of the desert monitor (*Varanus griseus*) in the sand dunes south of Holon, Israel (in Hebrew). M.Sc. thesis, Tel Aviv University.

———. 1999. Effects of overcrowding on the annual activity cycle and the husbandry of *Varanus griseus*. *Mertensiella* 11:51–62.

Stanner, M., and H. Mendelssohn. 1986. The diet of *Varanus griseus* in the southern coastal plain of Israel (Reptilia, Sauria). *Isr. J. Zool.* 34:67–75.

———. 1987. Sex ratio, population density and home range of the desert monitor (*Varanus griseus*) in the southern coastal plain of Israel. *Amphibia-Reptilia* 8:153–164.

———. 1991. Activity patterns of the desert monitor (*Varanus griseus*) in the southern coastal plain of Israel. *Mertensiella* 2:253–262.

Stromer, E. V. 1910. Reptilien und Fischreste aus dem marinen Alttertiär von Südtogo. *Z. Deutsch. Geol. Ges.* 62:478–507.

Switak, K. H. 1998. Living in peril—Africa's savannah and white-throated monitors. *Reptiles* Feb.:76–89.

ten Donkelaar, H. J., and R. de Boer-van Huizen. 1981. Ascending projections of the brain stem reticular formation in a nonmammalian vertebrate (the lizard *Varanus exanthematicus*), with notes on the afferent connections of the forebrain. *J. Comp. Neurol.* 200:501–528.

———. 1984a. Ascending and descending axon collaterals efferent from the brainstem reticular formation: A retrograde fluorescent tracer study in the lizard, *Varanus exanthematicus*. *Brain Res.* 34:184–188.

———. 1984b. The fasciculus longitudinalis medialis in the lizard *Varanus exanthematicus*. 1. Interstitiospinal, reticulospinal and vestibulospinal components. *Anat. Embryol.* 169:177–184.

———. 1988. Brain stem afferants to the anterior dorsal ventricular ridge in a lizard (*Varanus exanthematicus*). *Anat. Embryol.* 177:465–475.

ten Donkelaar, H. J., G. C. Bangma, and R. de Boer-van Huizen. 1985. The fasciculus longitudinalis medialis in the lizard *Varanus exanthematicus* 2. Vestibular and internuclear components. *Anat. Embryol.* 172:205–215.

ten Donkelaar, H. J., G. C. Bangma, H. A. Barbas, R. de Boer-van Huizen, and J. G. Wolters. 1987. The brain stem in a lizard, *Varanus exanthematicus*. *Adv. Anat. Embryol. Cell Biol.* 107:i, xiii, 1–168.

Thilenius, G. 1897. Herpetologische Notizen aus Sud Tunis. *Zool. Yrbk. Syst.* 10:219–236.

Tsellarius, A. Yu. 1994. Behavior and mode of life of desert monitors in sand deserts (in Russian). *Priroda* 5:26–35.

Tsellarius, A. Yu., and V. A. Cherlin. 1991. Individual identification and new method of marking of *Varanus griseus* (Reptilia, Varanidae) in field conditions (in Russian). *Herpetol. Res. Leningrad* 1:104–118.

———. 1994. Duration of egg incubation in *Varanus griseus* (Reptilia, Sauria) and emergence of hatchlings on surface in the sand deserts of middle Asia (in Russian). *Selevinia* 2:43–46.

Tsellarius, A. Y., V. A. Cherlin, and Y. A. Menshikov. 1991. Preliminary report on the study of biology of *Varanus griseus* (Reptilia, Varanidae) in middle Asia (in Russian). *Herpetol. Res. Leningrad* 1:61–103.

Tsellarius, A. Yu., and Yu. G. Menshikov. 1994. Indirect communication and its role in the formation of social structure in *Varanus griseus* (Sauria). *Russ. J. Herpetol.* 1:121–132.

―――. 1995. Construction of nest burrows and clutch protection by females of *Varanus griseus* (in Russian). *Zool. Zh.* 74:119–129.

Tsellarius, A. Yu., and E. Yu. Tsellarius. 1996. Courtship and mating in *Varanus griseus* of western Kyzylkum. *Russ. J. Herpetol.* 3:122–129.

―――. 1997a. Behavior of *Varanus griseus* during encounters with conspecifics. *Asiatic Herpetol. Res.* 7:108–130.

―――. 1997b. Thermal conditions of *Varanus griseus* (Reptilia, Sauria) activity (in Russian). *Zool. Zh.* 76:206–211.

Tsellarius, A. Yu., Yu. G. Menshikov, and E. Yu. Tsellarius. 1995. Spacing pattern and reproduction in *Varanus griseus* of western Kyzylkum. *Russ. J. Herpetol.* 2:153–165.

Tsellarius, A. Yu., E. Yu. Tsellarius, and Y. G. Menshikov. 1997. Notes on the diet and foraging of *Varanus griseus*. *Russ. J. Herpetol.* 4:170–181.

Vernet, R. 1977. Recherches sur l'ecologie de *Varanus griseus* Daudin (Reptilia, Sauria, Varanidae) dans les ecosystems sableaux du Sahara nord-occidental (Algerie). Ph.D. thesis, l'Université Pierre et Marie Curie Paris IV.

Vernet, R., and C. Grenot. 1972. Étude du milieu et structure trophique du peuplement reptilien dans le Grand Erg Occidental (Sahara Algerien). *Comp. Rendus Soc. Biogeogr.* 433:97–104.

Vernet, R., M. Lemire, and C. Grenot. 1988a. Field studies on activity and water balance of a desert monitor *Varanus griseus* (Reptilia, Varanidae). *J. Arid Environ.* 15:81–90.

Vernet, R., M. Lemire, C. J. Grenot, and J. M. Francaz. 1988b. Ecophysiological comparisons between two large Saharan lizards, *Uromastix acanthinurus* (Agamidae) and *Varanus griseus* (Varanidae). *J. Arid Environ.* 14:187–200.

Visser, G. J. 1981. Breeding the white-throated monitor *Varanus exanthematicus albigularis* at the Rotterdam Zoo. *Int. Zoo Yrbk.* 21:87–91.

Wolters, J. G., R. de Boer-van Huizen, H. J. ten Donkelaar, and L. Leenen. 1986a. Collateralization of descending pathways from the brainstem to the spinal cord in a lizard, *Varanus exanthematicus*. *J. Comp. Neurol.* 251:317–333.

Wolters, J. G., H. J. ten Donkelaar, and A. A. J. Verhofstad. 1986b. Distribution of some peptides (substance P, (leu) enkephalin, (met) enkephalin) in the brain stem and spinal cord of a lizard, *Varanus exanthematicus*. *Neuroscience* 18:917–946.

Wood, S. C., K. Johansen, M. L. Glass, and G. M. O. Maloiy. 1978. Aerobic metabolism of the lizard *Varanus exanthematicus*: Effects of activity, temperature, and size. *J. Comp. Physiol.* 127:331–336.

Wood, S. C., K. Johansen, M. L. Glass, and R. W. Hoyt. 1981. Acid–base regulation during heating and cooling in the lizard, *Varanus exanthematicus*. *J. Appl. Physiol.* 50:779–783.

Yadgarov, T. Y. 1968. Information on the ecology of the desert monitor (*Varanus griseus*) from the river basin of the Surhandarja, Col (in Russian). In *Herpetology of Middle Asia*, 24–28. Tashkent: Academy of Science Uzbek, USSR.

Yeboah, S. 1993. Aspects of biology of two sympatric species of monitor lizards *Varanus niloticus* and *Varanus exanthematicus* (Reptilia, Sauria) in Ghana. *Afr. J. Ecol.* 32:331–333.

Ziegler, T., and W. Böhme. 1997. Genitalstrukturen und Paarungsbiologie bei squamaten Reptilien, speziell der Platynota, mit Bemerkungen zur Systematik. *Mertensiella* 8:3–207.

6. Asian Varanid Species

Twenty-three varanid species belong in the Asian clade of *Varanus*, 14 of which were included in Ast's (2001) phylogenetic study. Two of those not included, *V. kordensis* and *V. spinulosus*, have recently been elevated to species status. Eight other new species from remote Indonesian and Philippine Islands in Southeast Asia have only recently been described (*boehmei, caerulivirens, cerambonensis, finschi, mabitang, macraei, melinus,* and *yuwonoi*). Still other new species doubtlessly remain to be described from the *indicus* and *prasinus* species complexes.

Figure 6.1. Varanus bengalensis *(photo by Walter Erdelen).*

6.1. *Varanus bengalensis*

ERIC R. PIANKA

Nomenclature

Varanus bengalensis (Fig. 6.1) was described by Daudin in 1802. The type locality is Bengal, India. Gray (1831) described the currently recognized subspecies *Varanus b. nebulosus* in 1831 (type locality Java). Several supraocular scales are enlarged in *V. b. nebulosus*. Mertens relegated *V. b. nebulosus* to subspecific status. *Varanus b. bengalensis* has more smaller scales than does *V. b. nebulosus*. Auffenberg (1994) reported continuous clinal variation in scale counts. Hence, one cannot easily draw a line on a map that separates the two subspecies. However, Böhme and Ziegler (1997c) reported the two types to sympatric in the Phuket Island region of Thailand, suggesting that they deserve full species status at this locality.

Figure 6.2. Geographic range of Varanus bengalensis.

Geographic Distribution

V. bengalensis occurs from southeastern Iran through most of Pakistan, India, Sri Lanka, and Bangladesh eastward throughout Southeast Asia, including Nepal, Bhutan, southern China, Burma, Thailand, Laos, Cambodia, Vietnam, Malaysia, Sumatra, and Java. They do not reach the lesser Sunda island chain or Borneo. The subspecies *Varanus b. bengalensis* is western, and *V. b. nebulosus* is found in the eastern parts of its geographic range (Fig. 6.2).

Fossil Record

Fossils tentatively identified as *V. bengalensis* have been found from Pleistocene or Recent deposits in the Billa Surga (Lydekker 1886) and at nearby Karnool Caves (Prasad and Yadagiri 1986) near Madras in India.

Diagnostic Characteristics

- Moderately large, up to about 160 cm in total length (TL).
- Black, dark gray, or brown, with variable amounts of lighter pattern.
- Nares slit shaped, closer to tip of snout than to eye.
- Nuchal scales smooth, larger than middorsals and about the size of posterior head scales.
- Belly scales smooth in 75 to 120 transverse rows (from gular fold to insertion of hind limb at its anterior edge).
- Tail laterally compressed with a double row of keeled scales dorsally.

Description

Adult *V. b. bengalensis* are black, dark gray, or brown, with variable amounts of lighter pattern over the back. Adult *V. b. nebulosus* are somewhat lighter, ranging from light gray to dirty yellow with variable amounts of speckling. Some juveniles are darker than adults, and most are crisper.

Juvenile *V. b. bengalensis* have 11 light yellow bands, whereas juvenile *V. b. nebulosus* are darker, with broader dark crossbanding extending farther ventrally well onto the belly (Auffenberg 1994).

In juveniles, the dorsal surface of the neck has several prominent dark Vs. Head and color patterns of juveniles vary considerably across the geographic range, with paler animals in drier areas and darker ones in wetter, more forested, regions. Much of the juvenile coloration is lost as these monitors mature.

Size

V. bengalensis is a moderately large monitor, but its size varies throughout its range. In Bangladesh, they reach only about 100 cm in TL, but Sri Lankan lizards are larger, at about 140 cm TL. The largest animals are found in Malayasia (160 cm) and Burma (240 cm).

Average size of adult males is 58 cm snout–vent length (SVL; 150 cm TL), weight 2.7 kg, whereas that of females is only 46 cm SVL (120 cm TL), weight 1.5 kg (Auffenberg 1979). Much larger specimens have been recorded.

Habitat and Natural History

The prime habitat of *V. bengalensis* is scrub woodland along river floodplains.

Terrestrial, Arboreal, Aquatic

V. bengalensis is primarily terrestrial, but they can climb well when necessary. Juveniles spend more time in trees than adults.

Time of Activity (Daily and Seasonal)

V. bengalensis is diurnal. In warm tropical areas, these monitors are active throughout the entire year, but in temperate regions, activity ceases during cooler months. Inactivity periods are longest in the coldest areas.

Thermoregulation

In the field, active body temperatures averaged 32.6°C (range, 22.1–40.2°C, n = 102) (Auffenberg 1994). The difference between body temperature and ambient temperature is greater during colder months (7.3°C) than during warmer months (5.6°C), showing that these monitors are active thermoregulators.

Foraging Behavior

V. bengalensis are active foragers, moving about incessantly, constantly flicking their tongues, and poking their snouts into every nook and cranny. They flip over dried cow patties to find prey.

Diet

V. bengalensis are generalists but strict carnivores. They feed largely on relatively small invertebrate prey (Losos and Greene 1988), including beetles, bugs, cicadas, earwigs, grubs, hymenoptera, lepidopterans, orthopterans, termites, earthworms, spiders, scorpions, centipedes, crustaceans, and snails (Auffenberg 1984, 1994). When available, vertebrate prey are eaten, including fish, frogs, lizards, snakes, birds, and small mammals. Reptile and bird eggs are also often eaten. Adult males are sometimes cannibalistic, killing and eating other smaller *V. bengalensis*. Occasionally, they will even eat carrion, but they do not appear to be attracted to it, like some other monitors are. However, one *V. bengalensis* walked 3 km into the wind to carrion (Subba Rao and Rao 1984).

Reproduction

Sexual maturity is reached in 3 to 4 years in captivity, but may require 5 years or more in natural environments. Males reach larger sizes than females. In India, mating occurs during the monsoon season in June–July, and eggs are laid in August. Oviposition sites are chosen with care. Eggs are laid inside earthen embankments, rotten tree trunks, and termite nests. They incubate over the winter and hatch in early March. Neonates hatch with little yolk reserve and must begin feeding immediately. Juveniles stay close to termitaria and use them as retreats for several days after hatching before dispersing. Average clutch size is 20.2 (n = 23) and relative clutch mass 21% (Auffenberg 1994). Eggs and hatchlings are relatively small compared with a female's size.

Neonate survivorship is low—about half perish by the end of their second year. Horn and Visser (1989) report three clutches in captivity of 20, 29, and 8 with incubation periods of 235 to 254 days at 29°C and 172 to 173 days at 30 to 34°C.

Movement

Males are active on more days (mean, 252 days; n = 16) than females (mean, 182 days; n = 9). Males are more active than females, spending about 4.5 hours per day moving, as compared with 2.85 hours for females (Auffenberg 1979). Average distance moved per day is similar in males and females (191 vs. 210 m, respectively), but males move over greater maximal distances (529 vs. 342 m) (Auffenberg 1994). Juveniles are much more sedentary (mean, 113 m). Males have larger home ranges than females. Home range sizes vary geographically and with habitat. Males move the greatest distances during the last half of June, probably in search of females for mating. Females carrying shelled eggs move extensively during the first week of July, probably searching for the most suitable nesting site.

Physiology

Metabolic rates at a body temperature of 37°C are twice as high as those of other reptiles and one-third the mean value for mammals (Bartholomew and Tucker 1964). Earll (1982) studied heating, cooling, and oxygen consumption rates of *Varanus bengalensis*. Rates of many such biochemical reactions double when temperature increases by 10°C and fall to half the original rate when temperature is decreased by 10°C (this is known as the Q_{10} effect). A Q_{10} value of 2 indicates such doubling. Q_{10} values for *V. bengalensis* were estimated to be 1.91 (20–30°C) and 1.74 (30–40°C) (Earll 1982).

Fat Bodies, Testicular Cycles

In both males and females, fat bodies are small during summer when mating and oviposition occur and enlarge during October–November. Fat bodies diminish in size as winter progresses, becoming small by spring. Testes are largest in July and August and regress during other times of the year (Auffenberg 1994).

Parasites

Auffenberg (1994) lists dozens of endoparasites of *V. bengalensis*, which include trematodes, cestodes, nematodes, filaria, and sporozoan protozoans. Various parasites live in the blood and lymph, lungs, body cavities, and intestines. In most cases, prey are intermediate hosts. Four species of ectoparasitic ticks infest a high percentage (21% to 90%) of these monitors. Tick infestation varies seasonally, peaking in October in males and from August to November in females.

6.2. *Varanus caerulivirens*

THOMAS ZIEGLER, WOLFGANG BÖHME,
AND KAI M. PHILIPP

Figure 6.3. Varanus caerulivirens
(photo by Kai Philipp).

Nomenclature

The turquoise monitor lizard, *Varanus caerulivirens* (Fig. 6.3), was described by Ziegler et al. in 1999a. The male holotype is deposited in the Zoologisches Forschungsinstitut und Museum Alexander Koenig (ZFMK 68874); its type locality is Halmahera, Moluccas, Indonesia.

Varanus caerulivirens belongs within the *Varanus indicus* group (Ziegler et al. 1999a), and according to its genital morphology characteristics, it is assignable to the subgenus *Euprepiosaurus* Fitzinger (see Böhme 1988; Ziegler and Böhme 1997). In genital morphology, because of the reduced and therefore derived condition of unilaterally developed hemipenial and hemiclitorial paryphasmata, *V. caerulivirens* is linked with the following currently known representatives of the *V. indicus* group: *V. cerambonensis, V. indicus* sensu stricto, *V. juxtindicus,* and *V. melinus* (Philipp et al. 1999a; Ziegler et al. 1999b; Böhme et al. 2002). *V. caerulivirens* is a monotypic species.

Geographic Distribution

At present, *Varanus caerulivirens* is only known from Halmahera, Moluccan Islands, Indonesia, where it might occur sympatrically with

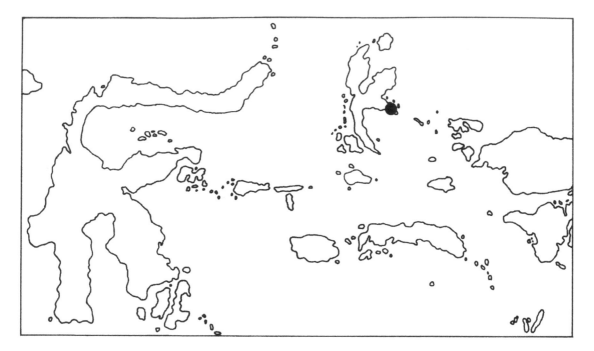

Figure 6.4. Locality of collection of Varanus caerulivirens.

its relatives *V. indicus* sensu stricto and *V. yuwonoi.* In recent times, *V. caerulivirens* was repeatedly exported by the Indonesian pet trade (Ziegler et al. 1998). Reinvestigation of long-time preserved material from museum collections revealed further specimens of *V. caerulivirens* previously assigned to *V. indicus.* Among them is the only specimen with precise locality data for Halmahera: a juvenile collected in the year 1895 at Patani on the eastern coast of Halmahera (Ziegler et al. 1999a) (Fig. 6.4).

Fossil Record

No fossils of *Varanus caerulivirens* are known.

Diagnostic Characteristics

Within the subgenus *Euprepiosaurus* (see Ziegler and Böhme 1997 for diagnosis), *Varanus caerulivirens* can be distinguished from the remaining members of the *V. indicus* group by having a light blue to turquoise tinge to the head, neck, body, dorsum, limb, and tail coloration; a high midbody scale count (170–185); a pink-colored, light tongue that may have dark pigment only on its tips and/or at its bifurcation point; and differentiated paryphasmata only on one side of the sperm groove of the hemipenis and hemiclitoris, respectively (Ziegler et al. 1999a).

Description

Varanus caerulivirens is a medium-sized monitor lizard of slender habitus. Ground color of the upper side of head, neck, dorsum, limbs, and tail is dark in life (grayish-brown to black) with lighter (beige to turquoise or sky blue) patterning; the ground color of the underside of the respective parts is light (yellowish-beige to turquoise) with gray-blackish patterning. The upper side of the head is blackish with turquoise spots. A dark temporal band consisting of grayish-turquoise scales on both sides is usually framed by yellowish to weakly turquoise scales. Tympanum may have a light anterior margin. The tongue is pink-colored and light; it may have dark pigment only at the tips and/ or their bifurcation point. Neck and dorsum are usually grayish turquoise with distinct dark ocelli surrounding several turquoise scales each in their centers. These ocelli tend to become larger posteriorly and may enclose a dark center. The upper side of limbs as well as the upper neck region usually have marked, black ocelli surrounding grayish to turquoise scales. Toward the fore- and hind feet, the turquoise coloration becomes more intense. In darker specimens, the ocellation of the upper side of the anterior neck and of the limbs may be less distinct, but usually at least indicated. The dorsum generally has up to 20 transverse rows of ocelli between the limbs, which become more and more inconspicuous posteriorly. Depending on the angle of incidence of the light, several dark transverse bands are discernible on the back. The tail usually has broad grayish-turquoise and blackish crossbands on the last two-thirds of the tail. The throat and the lower part of the neck are yellowish, more or less dark marbled or with a few dark transverse bands. The chest has grayish-turquoise to beige marbling, similar to the lower parts of limbs. The belly is beige-colored, with more or less developed black marginal crossbands. The underside of the tail is beige-colored, except for weakly contrasting dark crossbands in the last two-thirds of the tail.

In spite of more than 100 years of alcohol preservation, the turquoise to sky-blue color tinge in the juvenile paratype (see Ziegler et al. 1999a), and the characteristic pattern, are rather well preserved. The dorsal ocellation is particularly distinct and is otherwise hardly different from the color pattern of the adults. However, the predominantly dark belly is remarkable.

Size

The largest known male was a live pet trade specimen, which was imported to Germany. This large male had a TL of 104 cm, with a SVL of 40 cm. The largest known female is one of the paratypes, the folded oviducts of which indicated that eggs had already been laid. This female had a TL of 98.5 cm and a SVL of 37.5 cm (Ziegler et al. 1999a). The only known juvenile has a TL of 42.3 cm and a SVL of 16.9 cm (Ziegler et al. 1999a).

Habitat and Natural History

Currently, no field data on the habitat and life habits of the species are available. Lemm (1998, p. 72) reported a very rare second blue-tailed mangrove-type monitor besides *V. yuwonoi* from Halmahera, which he said was "extremely flighty and aggressive." In captivity, *Varanus caerulivirens* are excellent climbers and are also very well adapted to semiaquatic habitats (Philipp et al. 1999b). First results of stomach content analyses (which include whip scorpions, crustaceans, grasshoppers, and anurans) of preserved museum specimens suggest that the species most probably inhabits tropical lowland forests (Phillip et al., in preparation). As already stated, a coexistence of *V. caeruli-virens*, *V. indicus* sensu stricto, and *V. yuwonoi* on Halmahera raises the interesting question of niche segregation among members of the *V. indicus* group (see Philipp 1999a).

6.3. *Varanus cerambonensis*

KAI M. PHILIPP, THOMAS ZIEGLER,
AND WOLFGANG BÖHME

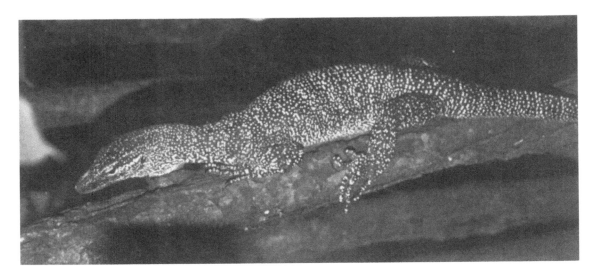

Figure 6.5. Varanus cerambonensis *(photo by Kai M. Philipp).*

Nomenclature

The banded Pacific monitor lizard, *V. cerambonensis* (Fig. 6.5), was described by Philipp et al. in 1999a. The holotype originates from Laimu (3°19'S; 129°44'E), south coast of Ceram (now known as Seram), Moluccas, Indonesia, and is stored in the Zoologisches Forschungsinstitut und Museum A. Koenig, Bonn, Germany as ZFMK (MZB) 70617.

The specific name refers to the islands Ceram and Ambon, from where the type series originates.

V. cerambonensis belongs to the *Varanus indicus* group (sensu Böhme et al. 1994) and is a member of the subgenus *Euprepiosaurus* Fitzinger. According to its derived ornamentation of the outer genitals, *V. cerambonensis* groups with *V. caerulivirens*, *V. indicus*, *V. juxtindicus*, and *V. melinus* (Böhme and Ziegler 1997a; Böhme et al. 2002; Philipp et al. 1999a; Ziegler and Böhme 1999; Ziegler et al. 1999a).

No subspecies of *V. cerambonensis* are recorded.

Geographic Distribution

V. cerambonensis is known from some neighboring central Moluccan Islands: Ambon, Banda, Buru, Obi, and Seram. Philipp et al. (1999a) mentioned a museum voucher from New Guinea, but detailed specific locality data are lacking.

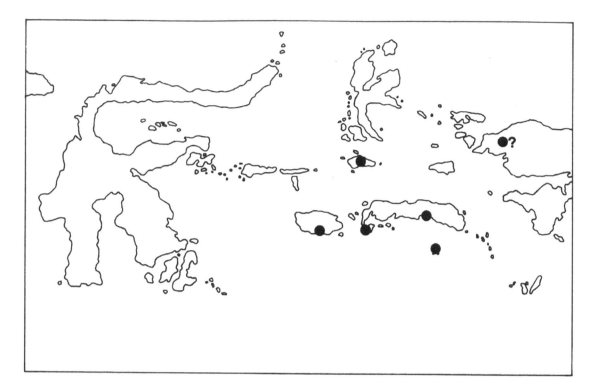

Figure 6.6. Localities of collection of Varanus cerambonensis.

Sprackland (1999a) includes a picture of a pet trade specimen of *V. indicus* from western Irian Jaya that obviously resembles *V. cerambonensis*.

On Ambon and probably on New Guinea, *V. cerambonensis* occurs sympatrically with *V. indicus*. Sympatry with *V. indicus* in New Guinea is based on a single specimen without detailed locality data. More collections from New Guinea, especially its western parts, are needed to confirm the distribution of *V. cerambonensis* in New Guinea (Fig. 6.6).

Fossil Record

No fossils of *V. cerambonensis* are known.

Diagnostic Characteristics

- Moderately large monitor of slender habitus.
- Tail with yellow transverse bands, laterally compressed with an apical keel.
- Well-defined yellow temporal band.
- Crossbanded dorsal pattern of adults.
- Outer genitals with unilaterally developed paryphasmata rows, stretching only to the outer apical lobe.

Description

V. cerambonensis is a moderately large monitor lizard of slender habitus. The upper parts of head, neck, dorsum limbs, and tail are deep brown to black and intensively speckled with small yellow spots. On the back, these spots are arranged in oblique areas of changing intensity, thus creating the characteristic crossbanded dorsal pattern. On the underside, this monitor is dirty white. The temples bear a characteristic yellow band, about two scales wide, running from the eye to the upper margin of the tympanum. The tongue is pink; only the tips and an ill-defined area behind the bifurcation point are dark pigmented. The tail is an indistinct deep brown/black and yellowish banded, lacking any blue pigmentation.

Juvenile coloration is remarkably different from that of adults. The dorsal pattern consists of black ocelli containing a yellow center. These yellow spots occupy 5 to 12 scales each. The yellow temporal band is somewhat thicker and even more contrasting than in adults. The entire tongue is pink. Thus, dorsal pattern and tongue coloration change during ontogeny.

The tail of *V. cerambonensis* is laterally compressed and bears a double keel along the dorsal apex, as seen in most other species of the *V. indicus* group.

V. cerambonensis has medium sized scales: 129 to 150 scales around its midbody and 126 to 163 dorsal scales from the back of the head to the hind legs.

Size

Only a few individuals of *V. cerambonensis* are known. The largest is a male 98.4 cm in TL, with SVL 40.9 cm and tail length 57.5 cm, with the tip of the tail missing. Tail length averages 1.3 to 1.7 times SVL. The smallest specimen of the type series measures 24.8 cm (9.8 cm SVL). Its slightly open umbilicus indicates that it is a newborn hatchling.

Habitat and Natural History

Currently, scant field data of this species are available. *V. cerambonensis* inhabits lowland rain forest with adjacent gardens on Seram. *V. cerambonensis* further seems to be closely associated with freshwater streams that are also inhabited by *Hydrosaurus amboinensis* (Philipp et al. 1999a). Yuwono (personal communication) stated that *V. cerambonensis* could be found in the vicinity of small human settlements on Ambon.

Diet

First stomach content analyses of preserved museum specimens revealed that *V. cerambonensis* feeds on crabs, centipedes, cockroaches, beetles, skinks, and reptile eggs (Philipp et al., in preparation). In captivity, this species readily accepts chicks (Yuwono, personal communication).

6.4. *Varanus doreanus*

WOLFGANG BÖHME, KAI M. PHILIPP,
AND THOMAS ZIEGLER

Figure 6.7. Varanus doreanus
(photo by Wolfgang Böhme).

Nomenclature

Described from Dore (=Doreh), Berou Peninsula, northwest New Guinea, as *Monitor doreanus* by Meyer, 1874, this nominal species has been considered a synonym of the widespread *Varanus indicus* for 120 years (Boulenger 1885; Mertens 1942, 1963). However, its name had to be revalidated for a large New Guinean monitor lizard, which is easily distinguished from *V. indicus* by a light, yellowish-whitish tongue, a densely marbled throat, and a bluish tail (note the vernacular name "blue-tailed monitor") (Fig. 6.7). Böhme et al. (1994) demonstrated the name *V. "kalabeck,"* which was occasionally in use for these monitors to be inapplicable, and they resurrected A. B. Meyer's name for this species, which had already been discussed under the name *Varanus* sp. by Böhme (1991). Because the holotype, formerly in the Staatliches Museum für Tierkunde, Dresden, was destroyed during World War II, they redefined *V. doreanus* by designating a neotype that is deposited in the Zoologisches Forschungsinstitut und Museum A. Koenig, Bonn, under ZFMK 52922. Moreover, they described a new subspecies (*V. d. finschi*) that was subsequently (Ziegler et al. 1999b) raised to the rank of a full species and thus has its own chapter in this book.

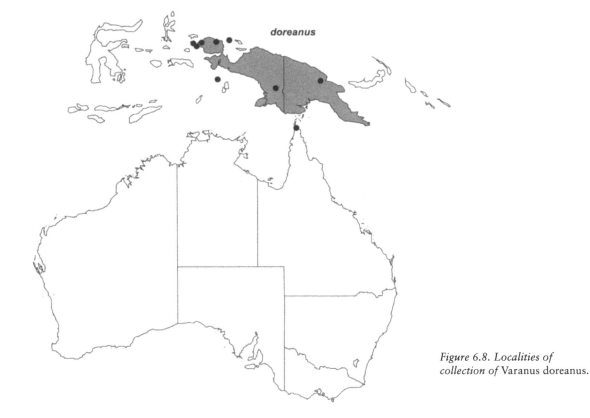

Figure 6.8. Localities of collection of Varanus doreanus.

Geographic Distribution

Varanus doreanus is distributed over large parts of New Guinea and some offshore islands (Salawati, Biak), in broad sympatry with the relatively closely related *V. indicus* (Böhme et al. 1994). It is also documented for Wamar (=Wammer), Aru Islands, and for northern-most Queensland, Australia (Ziegler et al. 1999b, 2001). A record from Halmahera, Moluccas (Yuwono 1998; Ziegler et al. 1999a, 1999b), however, must still be verified (Fig. 6.8).

Fossil Record

None.

Diagnostic Characteristics

• Habitus slender, but with a relatively big and bulky head, particularly in adult males.

• Yellowish-white tongue.

• Dark, densely marbled throat.

• Bluish tail.

• Increased scale counts around midbody (154–180), tail base, and body side.

- On the outer genitals, paryphasmata ornamentation present on both sides of the asulcal, apical notch.

Description

V. doreanus is a slender but nevertheless robust lizard with a relatively big head, strong limbs, and sharp claws. TL is up to 135 cm. Adult *V. doreanus* have a brownish or blackish ground color with a dense speckling of yellow spots that may form larger flecks and that are arranged in irregular crossrows. They are remnants of the juvenile pattern, which exhibits much more conspicuous whitish or yellowish spots surrounded by black rings arranged in distinct crossrows. The strongly compressed tail has the characteristic double keel on its mid-dorsal edge. Its pattern consists basically of dissolved crossbars that tend to leave light interspaces. These interspaces increase in size toward the tail tip and are bluish in the adult and bright blue in the juvenile. The throat is distinctly patterned and marbled with dark vermiculations. This marbling is somewhat less conspicuous but nevertheless present in southeast Papuan and north Australian specimens (Ziegler et al. 1999b).

Size

The maximum size is about 135 cm TL (Philipp 1999a). The largest ZFMK voucher is (46.0 SVL + 79.5 tail) 125.5 cm (Ziegler et al. 1999b).

Habitat and Natural History

According to M. Reimann and N. Stronach (in Böhme et al. 1994), *Varanus doreanus* inhabits monsoon forest and is said to have a closer affinity to dense, primary forest than *V. indicus,* with which it was confused in past decades. Philipp (1999a) found it on the Bird's Head (=Vogelkop) Peninsula in mixed alluvium forest and in mixed hill forest, where *V. doreanus* preferred open areas such as riverbanks and small clearings resulting from single fallen trees. Highest density was observed around a fallen tree in a seasonally dry riverbed, where within only 5 hours, 13 specimens were encountered in an area of about 600 m² (Philipp 1999a). According to Bennett (1998), it appears to be restricted to intact forests.

Terrestrial, Arboreal, Aquatic

V. doreanus is not an arboreal species (contra Sprackland 1997), but exhibits a remarkable infraspecific niche segregation. Adults are largely terrestrial and take refuge on the ground in dense shrubs, driftwood, and the like, whereas juveniles and subadults (TL, 40–75 cm) use the upper strata of the vegetation (10 m and more above the forest floor). This arboreality of the subadult stages is likely to avoid food competition and cannibalism by the adults, similar to the case of *V. komodoensis* (see Auffenberg 1981).

Even in close proximity to rivers, *V. doreanus* will not flee into the water (in spite of *V. indicus'* tendency to do so) but in the opposite direction, away from the water (N. Stronach in Böhme et al. 1994).

Foraging Behavior and Diet

Corresponding to the infraspecific spatial niche segregation, juvenile *V. doreanus* are expected to forage in the upper vegetation strata. Adults forage on the ground (Stronach in Böhme et al. 1994; Bennett 1998; Philipp 1999a) and have been observed to be attracted by carrion and by turtle eggs (Stronach in Böhme et al. 1994). Because of their bulky heads, they are also expected to feed regularly on hard-shelled prey items, such as the geocarcinid land crabs that are abundant in their habitat (Philipp 1999a).

Nothing is known about time of activity, thermoregulation, reproduction, movement, physiology, or parasites.

6.5. *Varanus dumerilii*

DANIEL BENNETT

A B

Figure 6.9. Varanus dumerilii
hatchling (a) and adult (b)
(photos by Daniel Bennett and
Richard Bartlett).

Nomenclature

First described by Schlegel in 1839 as *Monitor dumerilii,* the type locality is Banjermasin, southern Kalimantan (Abb. Amphib: 78). Mertens (1942) assigned it to the subgenus *Tectovaranus* on the basis of its unique skull (Fig. 6.9). Ziegler and Böhme (1997) reclassified it as *Empagusia* along with *V. bengalensis, V. rudicollis,* and *V. flavescens.* The subspecies *V. dumerilii/heteropholis* was considered invalid by Brandenburg (1983) and Sprackland (1993b).

Geographic Distribution

V. dumerilii is known from Thailand, Burma, Malayan Peninsula, Borneo, and Sumatra as well as some smaller islands (Fig. 6.10). De Rooij (1915) examined material from Stabat, Kaluk, and Indragiri in Sumatra and Sinkawang, Baram, Kuching, Pangkalan Ampat, Mt. Dusit, Bongon, and Tandjong and the Rejang, Akar, and Howong Rivers in Borneo. She also records specimens from Malacca and Tenasserim and one from Tennger Mountains in Java. This is the only specimen known from that island. Brandenburg (1983) examined specimens from Sumatra (Taloek, Serdang, Bangka, and Batu) and Borneo (Balikpapan, Banjermasin, and Nanga Raoen). Bennett and Lim (1995) recorded *V. dumerilii* from peninsular Malaysia (Penang Hill below Belercteiro; Malaysia Johor, Kota Tingii; Rawang, Selangor Gallang, Gombak Reserve, Fraser's Hill, and Cameron Highlands), plus Gallang in the Rhiau Archipelago and Lundu, Sarawak. Nutphand (n.d.) claimed they were "frequently" found in the southern forests of Thailand and occa-

Figure 6.10. Geographic distribution of Varanus dumerilii *(based on map in Bennett 1998 with permission of the publisher).*

sionally found in the west around Kanchanaburi province. The species has not been found in the Philippines, but the limit of its distribution in the Sulu Sea is unknown. Similarly, the limits of its distribution to the east and north are uncertain.

Fossil Record

No fossils of this species are known.

Diagnostic Characteristics

- Enlarged scales on the neck, not arranged in longitudinal rows.
- Slit-shaped nares closer to the eye than the tip of the snout.
- Brown body color in adults.

Description

V. dumerilii has a flat, broad skull with blunt, peglike teeth. Prominent preanal pores exist in both sexes. Some specimens have greater or lesser numbers of enlarged scales on the dorsum so that there are between 66 and 102 scales at midbody and 75 to 87 ventrals from the gular fold to the insertion of the hind limb. Krebs (1979) reported that the nostrils of this species are sealed with flaps of skin and made watertight when the animals are underwater. The tail is laterally compressed with a strong double keel, 130% to 160% of SVL.

Changes in coloration and pattern in this species with age were first documented by Horn and Schulz (1977). Although adults are basically brown with creamish transverse dorsal bands, hatchlings are a glossy black with bright orange or yellow crossbands and a vivid orange or yellow head.

Size

The largest *V. dumerilii* found in Thailand by Taylor (1963) was 50-cm SVL with a 75-cm tail. Seven caught in peninsular Malaysia measured an average of 292 mm SVL and weighed 988 g (Bennett and Lim 1995). Measurement of a number of long-term captives reveals none over 130 cm TL. Males grow larger than females. A breeding pair at the Buffalo Zoo measured 130 cm and 2950 g (male) and 99 cm and 2300 g (female) (Radford and Payne 1989). Hatchlings measure 21 cm TL and weigh 11 g (Hauschild 1998).

Habitat and Natural History

V. dumerilii is known from coastal mangroves and inland forests, including degraded and agricultural areas. It is evidently a rare animal in populated areas because it is very poorly known, but it was considered more abundant than *V. rudicollis* by both Lekagul (1969) and Nutphand (n.d.). According to Nutphand, they are less active than other monitor species in Thailand and tend to be sedentary, sleeping in tree hollows or rock crevices.

Virtually nothing is known of the behavior of *V. dumerilii* in the wild. They are known from captivity to be excellent tree climbers and swimmers and may seek refuge in trees or in water when pursued in the wild (Smith 1931). Biebl (1995) and Hauschild (1998) commented on scent-marking behavior in captive males.

Terrestrial, Arboreal, Aquatic

Stomach contents and observations in captivity suggest *V. dumerilii* is active on and above the ground and in water. The lizard is a very good climber and an active digger, and it spends considerable amounts of time submerged in water.

Foraging Behavior

Evidently *V. dumerilii* finds food on and below the ground and probably also in trees and in water.

Diet

There are no published accounts of stomach contents of free-living *V. dumerilii*. According to Raven (1946, in Neill 1958), they eat turtle eggs; according to Barbour (1921), they eat ants [Editors' note: ants seem unlikely items for monitors to eat]; and according to Loveridge (in Pitman 1962), they eat birds. Auffenberg (1988) claimed that they eat insects on the forest floor but gave no details. Stomach contents of four museum specimens have been analyzed. Losos and Greene (1988) examined two; one contained four crabs, a spider, and an insect larva, and the other contained a single crab. A specimen examined by Brandenburg (1983) also contained a crab. Ziegler and Böhme (1996) reported scorpions, whip scorpions, and beetle larvae in another. Krebs (1979) noted that they were particularly adept at eating crabs and suggested they were specialized for this way of life. The scanty data available suggest that *V. dumerilii* may feed largely on crabs in certain habitats (probably mangroves) but that it has a wider dietary niche in other habitats.

Reproduction

The only data about reproduction are from captive animals. Clutches contain up to 23 eggs, and females can lay two or three clutches over a year (Radford and Payne 1989; Frost 1995; Hauschild 1998). Eggs are 30 mm long and weigh about 34 g; they hatch after 203 to 234 days at 29°C. Hatchlings measure 21 cm TL and weigh 11 g (Hauschild 1998).

Nothing is known about time of activity, thermoregulation, movements, physiology, fat body cycles, or parasites.

6.6. *Varanus finschi*

KAI M. PHILIPP, THOMAS ZIEGLER,
AND WOLFGANG BÖHME

Figure 6.11a. Varanus finschi (photo by Danny Gorman with help from Vincent den Breejen).

Nomenclature

Finsch's monitor lizard, *Varanus finschi,* was described by Böhme et al. in 1994 as subspecies of *V. doreanus.* The holotype was collected in 1891 by Otto Finsch in Blanche Bay, New Britain, Bismarck Archipelago, Papua New Guinea, and is stored in the Zoologisches Forschungsinstitut und Museum Alexander Koenig (ZFMK 26347). Ziegler et al. (1999b) elevated *finschi* to full specific rank because of the sympatric occurrence with *doreanus* in New Guinea and Australia. *V. finschi* is a monotypic species. It belongs to the subgenus *Euprepiosaurus* Fitzinger, within which, because of its genital morphology, it is clearly assignable to the *V. indicus* group (Böhme et al. 1994; Ziegler and Böhme 1997; Ziegler et al. 1999b). No subspecies of *V. finschi* are recorded.

Geographic Distribution

At first, *V. finschi* was only known from Blanche Bay, Ralum, and Massawa in New Britain (Böhme et al. 1994). Additional research on further museum vouchers considerably enlarged the range of the spe-

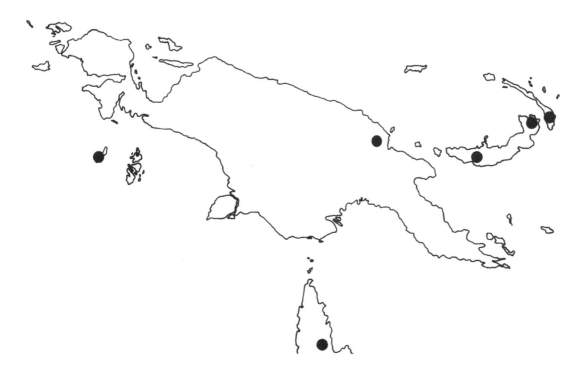

Figure 6.11b. Localities of collection for V. finschi.

cies, which currently includes New Ireland (Bismarck Archipelago), New Guinea, and Queensland, Australia (Fig. 6.11; Ziegler et al. 1999b, 2001). Unfortunately, the vouchers from Queensland have no precise locality data, so the exact distribution of *V. finschi* in northern Australia remains unknown. *V. finschi* from the Kai Islands, Indonesia, are regularly available in the pet trade.

Fossil Record

No fossils of *V. finschi* have been recorded.

Diagnostic Characteristics

• Moderately large.

• Tail laterally compressed, with a double keel along dorsal apex.

• Dorsum dark gray with black, light-centered ocelli.

• Whitish throat.

• Paryphasma rows stretch to both lobes of the outer genitals.

Description

The dorsal pattern of adults consists of black ocelli—often arranged in irregular transverse rows—with a yellowish center on a dark grayish background. These yellow spots often form distinct rings with a black center. The belly and throat are whitish. The dark head is speckled with numerous yellowish spots. The tongue color is pink (Sprackland 1999a).

The pattern of juveniles corresponds largely to that of adults but is more striking. Sometimes the whitish belly shows dark gray transverse bands. Occasionally, throat and neck are speckled with gray. These ventral markings of juveniles fade more and more, and finally disappear when the juveniles mature.

The tail is laterally compressed, more or less banded black and white, with a dorsal double keel.

Compared with other members of the *V. indicus* group, *V. finschi* has quite small scales. Thus, the number of scales around midbody (158–196) and the number of dorsal scales from gular fold to hind legs (171–199) are high.

Size

SVL ranges from 105 to 305 mm, TL from 257 to 820 mm. Tail length averages 1.34 to 1.66 times SVL (Böhme et al. 1994; Ziegler et al. 1999b).

Habitat and Natural History

Hediger (1934) characterized New Britain's Pacific monitors as eurytopic reptiles, inhabiting different types of habitat: mangrove forest, inland forest, fresh-cut clearings, coconut plantations, and rocky beaches. He saw Pacific monitors on the ground, in trees, and sometimes also in the water (both salt water and freshwater). But as *V. indicus* and *V. finschi* occur in broad sympatry on New Britain, the question of which observation belonged to which Pacific monitor species remains open.

Diet

Hediger (1934) analyzed stomach contents of several specimens of New Britain's Pacific monitor lizards. He pointed out that Pacific monitors from New Britain feed on crabs, scorpions, locusts, and beetles. But as stated above, these observations cannot unequivocally be applied to *V. finschi*. Nevertheless, we cite Hediger's detailed observations here, as they represent the only published data referring to Pacific monitors from New Britain.

The first stomach content analyses conducted by us revealed large amounts of skinks, several birds, spiders, and insects (Philipp et al., in preparation).

Varanus gouldii (photo by Jeff Lemm).

Varanus albigularis (photo by John Phillips).

Varanus exanthematicus
(photo by Daniel Bennett).

Varanus g. griseus (photo by Adel Ibrahim).

Varanus niloticus (photo by Jeff Lemm).

Varanus ornatus (photo by Jeff Lemm).

Varanus yemenensis (photo by Wolfgang Böhme).

Varanus bengalensis (photo by Walter Erdelen).

Varanus caerulivirens (photo by Kai M. Philipp).

Varanus cerambonensis
(photo by Kai M. Philipp).

Below: *Varanus doreanus*
(photo by Wolfgang Böhme).

Varanus dumerilii
(photo by Daniel Bennett).

Below: *Varanus finschi*
(photo by Danny Gorman with help
from Vincent den Breejen).

Left: *Varanus flavescens* (photo by Indraneil Das).

Above: *Varanus indicus* (photo by Hal Cogger).

Varanus jobiensis (photo by Louis Porras).

Above: *Varanus komodoensis*
(photo by Dennis King).

Varanus kordensis
(photo by Hans Jacobs).

Varanus mabitang
(photo by Maren Gaulke).

Varanus macraei (photo by Hans Jacobs).

Above: *Varanus melinus*
(photo by Thomas Ziegler).

Left: *Varanus olivaceus*
(photo by Daniel Bennett).

Top right: *Varanus prasinus*
(photo by Jeff Lemm).

Bottom right: *Varanus rudicollis*
(photo by Jeff Lemm).

Above: *Varanus salvadorii* (photo by Jeff Lemm).

Below: *Varanus salvator salvator* (photo by Daniel Bennett).

Above: *Varanus s. cumingi* (photo by Maren Gaulke).

Below: *Varanus s. nuchalis* (photo by Maren Gaulke).

Varanus timorensis (photo by James Murphy).

Varanus yuwonoi (photo by Jeff Lemm).

6.7. *Varanus flavescens*

Gerard Visser

Figure 6.12. Varanus flavescens *adult (a) and brightly marked juvenile (b) (photos by Indraneil Das and Gerard Visser, respectively).*

Nomenclature

Varanus flavescens was first described by T. Hardwicke and J. E. Gray in 1827 as *Monitor flavescens*. Its type locality, India, was restricted to Calcutta, West Bengal, India, by Auffenberg et al. (1989). There is no type specimen known, and the species was described "from a drawing" (Hardwicke and Gray 1827: 226). To date, no subspecies have been described (Fig. 6.12).

Geographic Distribution

In southern Asia, *V. flavescens* are found south of the Himalayas in East Pakistan, Northern India, Nepal, and Bangladesh. The range of the species is restricted to the Indo-Gangetic Plain, mainly along or near the Indus, Ganges, and Brahmaputra Rivers in India, also known from the Brahmani and Mahanadi River tributaries. Rao (n.d.) surveyed the four Indian varanid species between 1994 and 1997: *V. flavescens* was not reported from Haryana, although it is presumed to live there. In the state of Bengal, the species was found to be highly secretive, and it proved to be difficult to locate specimens. It was found in good numbers in the "Ganpur" catchment area. *V. flavescens* was also reported from Orissa and from the state of Uttar Pradesh, Rajasthan.

In Bangladesh, densities in suitable habitats were estimated to be around 7.5 animals per square kilometer (Khan, unpublished data, 1988, in Bennett 1998) (Fig. 6.13).

Fossil Record

No fossils of *Varanus flavescens* are known.

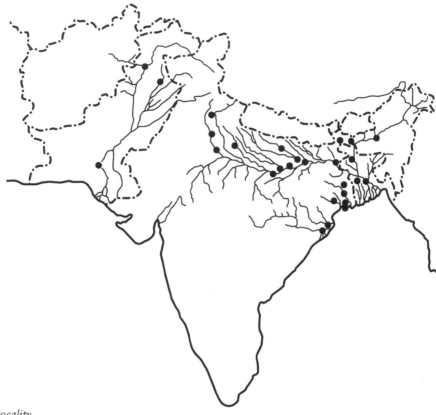

Figure 6.13. Confirmed locality records of V. flavescens. *(After Auffenberg et al. 1989.)*

Diagnostic Characteristics

- Medium size, up to approximately 90 cm.
- Short, broad head.
- Transverse rows of more or less fused yellow spots on the dorsum.
- Large and heavily keeled dorsal scales.
- Relatively short toes.
- Relatively short tail.

Description

This medium-sized, stocky monitor has variable coloration with crossbands of more or less fused, often black-edged yellow spots on the dark to light olive brown or reddish dorsum. A black reticulation can be present over large parts of the body. Ventral parts are yellowish with indistinct crossbars. Correlated with the mating season, the coloration of both sexes becomes more intense during the monsoon season, a feature not known in other monitor species. The temporal stripe is black. Juveniles have bright black and yellow markings.

Compared with most monitors, body scales are large. The laterally compressed tail has a double keel above. The slitlike nostrils are closer to the tip of the snout than to the eye. The skull is short, broad, and relatively high. Toes are short (hence the name "short-toed monitor"), and this is especially evident in the hind feet. Tail length is about 1.3 times the SVL.

Although the size of (almost) a meter is often encountered in the literature, Boulenger's (1885) 920-mm-long specimen still holds the record for size. Individuals (n = 167) measured in India by Auffenberg et al. (1989) averaged almost 700 mm (SVL, 315 mm; tail length, 384 mm). In Bangladesh, adults measure 700 to 800 mm (Sarker 1987).

A captive hatched neonate measured SVL 66 mm, tail length 79 mm, and TL of 145 mm (Visser 1985); 18 neonates measured by Auffenberg et al. (1989) in the field in India averaged SVL 77.6 mm, tail length 85.7 mm, and TL 163.3 mm, with the TL ranging from 143 to 188 mm. Tail length is 1.2 to 1.3 times SVL. There is no significant change in proportionate tail length in relation with age. Adult *Varanus flavescens* weigh between 500 and 1000 g. In contrast to many other varanid species, there is little difference in size between the sexes.

Habitat and Natural History

Contrary to what was widely accepted in the 20th century (i.e., Mertens 1942; Rotter 1963; Neugebauer 1973), *Varanus flavescens* does not inhabit hot steppes or dry grassland. Combining captive (bathing) behavior with the description of its humid habitat in Western Bengal by Sights (1949), it was speculated that this species might live in much wetter habitats than was believed (Visser 1985). This was soon confirmed in the comprehensive study of Auffenberg et al. (1989). The prime habitat of *V. flavescens* is marshland. The animals are also found in small numbers in rice fields and along canals and riverbanks. Usually these populations live in former marshland areas that were converted for agricultural use and are in decline. *Varanus flavescens* is one of the most threatened monitor species in the world.

Terrestrial, Arboral, Aquatic

V. flavescens is mainly terrestrial. Although they may sometimes climb in trees and bushes, their short toes make them inefficient climbers. Between June and October, the monsoon season, they are largely aquatic.

Time of Activity (Daily and Seasonal)

Yellow monitors are diurnal animals. They are least active between November and February, the dry season. They retreat to self-dug burrows, termite mounds, and crevices or cracks in the dry earth. Anecdotal information implying that the animals close off the mouth of their burrows at night, as claimed by Pakistan tribesmen, has not yet been confirmed. In the wet season, they spend much time in water.

Thermoregulation

No field data are available on thermoregulation in *V. flavescens*. In captivity, these lizards bask at temperatures of around 45°C [Editors' note: this is probably an overestimate], only to become active during very short periods in the morning and late afternoon. Lizards rested much of the day in a rock burrow or in the water basin of about 25°C (Visser 1985). In the wild, the species digs burrows to spend cold nights and the cool winter season. In dry periods, it also uses these burrows for estivation (Auffenberg et al. 1989).

Foraging Behavior

Although no field observations are available, it seems to be safe to assume that *V. flavescens* are active foragers, mainly searching for food along the water's edge, approaching their prey both from the land and from the water.

Diet

Stomach contents of 32 wild individuals included frogs, toads, amphibian eggs, turtle eggs, squamate eggs, birds, bird eggs, insects, and rodents. Fifty percent of the food consisted of amphibians: 40% of frogs (*Rana* spp.) and 9% percent of the toad *Bufo stomaticus*. Reptile eggs counted for 15%, rodents 6%, and insects 6% (Auffenberg et al. 1989). Surprisingly, masses of fertilized amphibian eggs were found in the stomachs of a few specimens. This has not been reported in other varanid species. Mollusks were not found in stomachs of these animals, although at least one species of aquatic mollusk is common in these marshland habitats. According to Sarker (1987), crabs, beetles, and earthworms are also taken. Captive animals feed readily on mice and a mix of chicken eggs, minced meat, and day-old chicks.

Reproduction

Male *V. flavescens* become sexually mature at a SVL of approximately 260 mm. Fertile females have SVLs of over 250 mm and weigh about 300 g. Both females and males become sexually mature between their third and fourth year (Visser 1985; Auffenberg et al. 1989). Females produce only one clutch per year. Ovarian tissue volumes are minimal from November to May and increase in June and July as a result of yolk deposition. In September, the tissue volume drops as undeveloped follicles deteriorate, then increases slightly again when new follicles develop during November and December. The first date for shelled eggs in the oviduct in the study of Auffenberg et al. (1989) was August 2, close to Rotterdam Zoo's captive female, which laid eggs on July 21 (Visser 1985).

Testicular volume is lowest from September to February, increasing from March to peak in June and July, dropping in August to the lowest level of the annual cycle. Courtship and mating are therefore presumed to occur in June and July, at the beginning of the wet season. This period was also reported in captivity.

The 4 to 30 eggs (with an average clutch size of 16 eggs) are deposited in burrows, and egg deposition continues into October. Clutch size is probably positively related to female size. Eggs measure about 37 by 21 mm and weigh 10 g. Although Auffenberg et al. (1989)

do not give any data on nest site selection, Das (personal communication in Bennett 1998) reports egg deposition in elevated areas to prevent flooding of the nest.

The very limited difference in size between the sexes suggests a less pronounced male-dominant behavior during courtship. This could also be related to the seasonal change in body coloration. There is a dramatic increase in the pigment of sexually mature animals of both sexes from April to May, and it remains high through June and July, dropping in August and September to the lowest levels in October. Clearly this correlated with development of testes and ovarian tissues and is therefore probably under hormonal control.

Wild *V. flavescens* show very few scars. Although bitten-off tail tips are common in other varanids, very few *V. flavescens* show such an injury. This suggests that courtship behavior in this species does not include extensive biting.

Movement

At present, no field data exist on distances moved per day or home range sizes. This species is most inactive in the cool winter period (November–January).

Physiology

No data are available about physiology.

Fat Bodies, Testicular Cycles

Fat bodies are present in adult *V. flavescens* only and are small or absent in individuals less than 200 mm SVL. Fat deposits may be more than 15% of the total body weight during part of the year. The mean fat weight of *V. flavescens* is greater than in any other species of *Varanus* so far (Auffenberg 1987). The highest values in both sexes are found in winter (December to March), decreasing in April to June, reaching their lowest levels from July to October. A spectacular increase is seen in November, leading to the high winter levels (Auffenberg et al. 1989). Values for males and females are almost identical for all months and therefore are probably not related to yolking of ova, as in some iguanids, but to seasonal food abundance. Fat reserves are being built up in November to enable survival in the inactive winter months.

Parasites

No animals caught by Auffenberg et al. (1989) in India and Pakistan had any ticks, although sympatric *V. bengalensis* were heavily infested by the tick *Aponomma gervaisi*. The animals did not show any tick scars either.

6.8. *Varanus indicus*

GIL DRYDEN AND THOMAS ZIEGLER

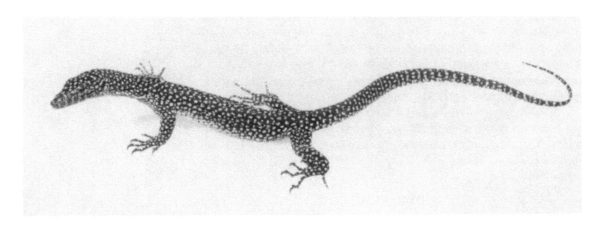

Figure 6.14. Varanus indicus
(photo by Hal Cogger).

Nomenclature

The mangrove or Pacific monitor (Fig. 6.14) was described in 1802 by Daudin. Its taxonomy was reviewed by Mertens (1926, 1942, 1959) and Böhme et al. (1994). Recently, *V. indicus* was redefined and a neotype designated (Philipp et al. 1999a). The juvenile neotype from Ambon, Moluccas, Indonesia, is deposited in the Zoologisches Forschungsinstitut und Museum Alexander Koenig (ZFMK 70650) in Bonn, Germany. The mangrove monitor lizard is now considered to represent a species complex (*V. indicus* group within the subgenus *Euprepiosaurus* Fitzinger; see Böhme 1988; Ziegler and Böhme 1997). Meanwhile, the *V. indicus* group consists of 10 species (see respective species chapters for details), befitting its large, discontinuous distribution: *V. caerulivirens, V. cerambonensis, V. doreanus, V. finschi, V. indicus* sensu stricto, *V. jobiensis, V. juxtindicus, V. melinus, V. spinulosus,* and *V. yuwonoi* (Böhme et al. 1994; Böhme and Ziegler 1997a; Harvey and Barker 1998; Philipp et al. 1999a; Ziegler et al. 1999a, 1999b; Ziegler and Böhme 1999; Böhme et al. 2002).

Geographic Distribution

According to the older literature, scattered populations (some introduced) of *V. indicus* lie roughly 15° either side of the equator in the western Pacific and connecting seas (Fig. 6.15). They are reported to inhabit some but not all Marshall Islands and extend westward some 45° to the Moluccas and Timor. They are on some but not all of the Mariana, Caroline, and Solomon Islands as well as along coastal Arnhem Land, eastern Cape York Peninsula, Torres Straight Islands, and New Guinea (Burt and Burt 1932; Mertens 1959; Uchida 1967; Brandenburg 1983; Luxmoore et al. 1988; Cogger 1992).

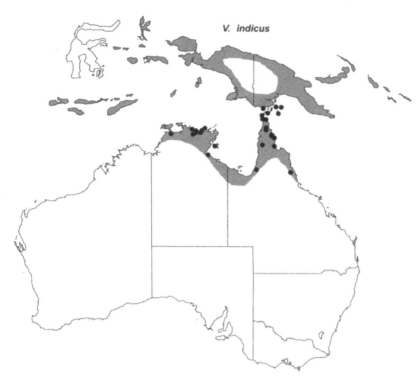

Figure 6.15. Geographic distribution of Varanus indicus *after Bennett (1998), but compare discussion in the text.*

Because of the diverse new descriptions of species within the *V. indicus* group, older citations with respect to the distribution of *V. indicus* sensu lato must be evaluated critically. For example, Ziegler et al. (1999b) recorded *V. doreanus* and *V. finschi* for the first time for Australia, from where only the collective species *V. indicus* was known so far. The existence of "true" *V. indicus* in continental Australia had to remain open at the time, because only records for the Australian Murray Islands, situated between New Guinea and Australia, were available. The first "true" *V. indicus* (sensu stricto) occur in mainland Australia (Maningrida, Arnhem Land, Northern Territory, Ziegler et al. 2001), thus proving the coexistence of three sibling species of the *V. indicus* group in continental Australia.

In their redefinition of *V. indicus,* Philipp et al. (1999a) state the following with respect to the distribution of the species: Sulawesi (some Indonesian herpetologists and animal traders deny such an occurrence), the Moluccan islands Morotai, Ternate, Halmahera, Obi, Buru, Ambon, Haruku, and Seram. Kai Islands and New Guinea, with its offshore islands Salawati, Waigeo, Biak, and Japen, Bismarck Archipelago Islands, New Britain, New Ireland, Duke of York. Solomon Islands Bougainville, Guadalcanal, Ysabel Island, and Hermit Island (Papua New Guinea). Japanese Bonin Islands record represents the northernmost occurrence of the species (and of any other species of the *indicus* group), although an anthropogenic transport cannot be excluded. However, distributional records given for *V. indicus* in Philipp et al. (1999a) were evaluated before the taxonomic recognition of the new species *V. cerambonensis.* Therefore, some of the Moluccan Islands records may actually be based on vouchers of *V. cerambonensis.*

In addition, Böhme et al. (2002) described *V. juxtindicus* from Rennell Island, Solomon Islands. The species seems to be endemic to Rennell Island within the Solomon Archipelago. However, records of "true" *V. indicus* were available only from Guadalcanal; other Solomonese records must still be proven.

Fossil Record

No known fossils exist.

Diagnostic Characteristics

The mangrove monitor lizard is a medium-sized varanid with generalized proportions and lacks distinct sexual dimorphism. It can be distinguished from the related species of the *V. indicus* group (see above) by the combination of the following features (Philipp et al. 1999a; Böhme et al. 2002):

• Dorsal color pattern consisting of irregularly scattered, small whitish to yellowish spots, mostly smaller than an area covered by five scales, on a dark-brownish or blackish background.

• Absence of a light, dark-bordered postocular–supratemporal stripe.

• Light, unpatterned throat.

• Entire tongue dark.

• Low midbody scale counts (106–137).

• Tail shape laterally compressed (not roundish in its first third), with differentiated double keel on its dorsal ridge.

• Hemipenes and hemiclitores, respectively, with unilaterally differentiated paryphasmata rows only.

Mangrove monitors on the southern Marianas and on Angaur (Palau) are dark gray, olive, or black with white to orange flecks dorsolaterally from snout to midtail. The legs are similarly colored. Flecking is generally restricted to individual scales, but aggregations approaching hexagons of light scales frequently appear. The tail is dorsolaterally banded dark with intervening light extending from its venter and forward to the chin. Larger animals and the tail tip are often black.

Size

Mature males on Guam in 1962–1964 were nearly three times the mass of mature females, the maximum being 1900 g (Wikramanayake and Dryden 1988). Maximal lengths reported are 124 cm TL in New Britain (Hediger 1934); 100 cm TL in Australia (Cogger 1981); 58 cm SVL on Guam, where the tail tip often was missing (Wikramanayake and Dryden 1988); and 34 cm SVL in New Guinea (Room 1974).

Habitat and Natural History

V. indicus clearly prefers mangroves (Cogger 1981; Philipp 1999a).

Field observations unequivocally referring to *V. indicus* were made by Philipp (1999a) in Irian Jaya. The species proved to be rather eurytopic, occupying a wide variety of habitats, being most common in forest types influenced by salt water. *V. indicus* proves to be well adapted to (salt) water, often foraging extremely close to water and even fleeing into the water. Böhme et al. (1994) stress that *V. indicus* (sensu stricto) may be also associated with human settlements.

Terrestrial, Arboreal, Aquatic

If the specific name belies its distribution (unknown from India), the vernacular misleadingly suggests a restricted habitat. Fisher (1948) found it in swamps on Yap. It is commonly found on high, dry ground far from water on Guam (McCoid and Hensley 1993) and in the Solomons (McCoy 1980a). Of approximately 180 monitors seen or handled in 1962–1964 on Saipan, Tinian, Guam, and Angaur, only one monitor was seen near water. It was wet, and its tracks suggested that it had just come ashore onto the northwest Guam beach. All other monitors were in open spaces (e.g., roads, abandoned runways), scrub, or forested habitats. Micronesian monitors frequent human habitations (McCoid and Hensley 1991; McCoid and Witteman 1993) and readily take to trees when disturbed or pursued (G. Dryden, personal observation). They swim well and escape into water in Australia (Bustard 1970; Cogger 1992) and Irian Jaya (Philipp 1999a). Many night-time hours spent walking in known monitor habitat on Guam and Saipan revealed no monitors on the surface. They presumably spend the night under cover or in trees there (G. Dryden, personal observation).

Time of Activity (Daily and Seasonal)

V. indicus on Saipan, Angaur (G. Dryden, personal observation), and Guam (Wikramanayake and Dryden 1988) basked and foraged in the morning (especially in hot weather), but some could be seen in the afternoons of cooler days throughout the year. Captive animals slept on elevated tree branches at night (Dryden 1965).

Foraging Behavior

The activity pattern of these generalists in Micronesia suggested that they emerged from refugia in the morning to bask until commencing midmorning foraging. Hunting began earlier (without basking?) on hot days. Foraging consisted of walking (moderate high stance) along, peering, and tongue flicking from side to side, occasionally pausing to scratch the substrate or eat an item. A few animals were trapped in a long, wooden box trap baited at the blind end with carrion, suggesting the importance of olfaction in hunting (G. Dryden, personal observation). Captive animals readily took mice, which they oriented headfirst by tossing the mouse and elevating the head (Dryden 1965).

Diet

As suggested by their unrestricted habitat, this species has an orthodox diet and feeds opportunistically. A wide range of invertebrate and vertebrate prey as well as eggs and carrion comprise its diet. There is little correlation between monitor size and food taken (Dryden 1965; Losos and Greene 1988; McCoid and Hensley 1993). Increased human presence on Guam and increasingly on Saipan (G. Dryden, personal observation) in recent years, as well as an explosion of the brown tree snake *Boiga irregularis* population on Guam, coupled with increased competition from feral hogs and carnivores, have conspired against monitors (McCoid and Witteman 1994). A significant reduction of the shrew population on Guam (probably *Boiga*-induced) has resulted in monitors shifting to smaller invertebrate prey (compared with 1962–1964) as well as to human refuse (McCoid and Witteman 1993).

Reproduction

Apparently only the mangrove monitors on Guam have been investigated for reproductive activity. Males collected there in 1962–1964 achieved anatomic maturity (substantial spermatogenesis) at 320 mm SVL. Gravid monitors or those with maturing follicles or corpora lutea were at least 275 mm SVL. The mean mass of 34 mature males was 1287 ± 542 g and that of 36 mature females was 484 ± 157 g. Ovaries were heaviest during the dry season, when corpora lutea were also present. The sex ratio of 98 animals was 1.3M:1F (Wikramanayake and Dryden 1988). In unsexed, unmarked animals, apparent male–male combat preceded copulation during the 1990 dry season. Interestingly, overnight and prolonged diurnal couplings were suggested (McCoid and Hensley 1991). Disparity between predicted copulatory seasonality (Wikramanayake and Dryden 1988) and observed behavior (McCoid and Hensley 1991) might be due to asynchronous activity of individuals or to nutritionally associated shifts in the monitor population alluded to above (see Diet). Consistent with the former interpretation, Brandenburg (1983) found hatchling mangrove monitors throughout the year in New Guinea. A clutch size of 10 eggs has been suggested (McCoid 1993).

Fat Bodies, Testicular Cycles

During 1962–1964 on Guam, fat bodies of female monitors increased in mass as the wet season progressed, then peaked in the early dry season, to decline in the latter part of the dry season. Males maintained relatively constant fat body mass throughout the year and showed no seasonal cyclicity of testicular activity (Wikramanayake and Dryden 1988).

Parasites

Sixty percent of 20 mangrove monitors from Indonesia bore ticks (*Aponomma trimaculatum*) over most body parts (King and Keirans 1997). Moderate numbers of unidentified nematodes were found in the stomachs of several monitors on Guam, and histologic preparations of testes revealed unidentified worm eggs in the intertubular spaces (G. Dryden, personal observation).

6.9. *Varanus jobiensis*

KAI M. PHILIPP, THOMAS ZIEGLER,
AND WOLFGANG BÖHME

Figure 6.16. Varanus jobiensis
(photo by Louis Porras).

Nomenclature

In 1932, Ernst Ahl described *Varanus indicus jobiensis* (Fig. 6.16) from a single specimen from Jobi Island (=Yapen), Irian Jaya, Indonesia. The type specimen is deposited in the Zoological Museum Berlin, Germany (ZMB 34106). Later on, Mertens (1951) described *V. karlschmidti* from northern Papua New Guinea. Böhme (1991) pointed out that the holotype of *V. indicus jobiensis* is a juvenile of *V. karlschmidti*. Therefore, *V. karlschmidti* is a junior synonym of *V. indicus jobiensis*, which consequently was elevated to full specific rank: *V. jobiensis*.

The species name refers to the type locality, Jobi Island.

V. jobiensis is a member of the *V. indicus* group within the subgenus *Euprepiosaurus* Fitzinger (Böhme et al. 1994; Ziegler and Böhme 1997).

No subspecies of *V. jobiensis* are recorded today, but Horn (1977) and Böhme and Ziegler (1997a) suggest that cryptic forms of *V. jobiensis* may exist.

Geographic Distribution

V. jobiensis is recorded for New Guinea Island and its offshore islands Yapen, Biak, Salawati, and Weigeo. Published data suggest that

Figure 6.17. Localities of collection of Varanus jobiensis.

V. jobiensis is widely distributed across huge parts of lowland New Guinea (Horn 1977; Böhme et al. 1994; Philipp 1999a) (Fig. 6.17).

Fossil Record

No fossils of *V. jobiensis* are recorded.

Diagnostic Characteristics

- Moderately large monitor of slender habitus.
- Head somewhat angular with conspicuous large eyes.
- Tongue pink.
- Throat reddish.
- Dorsum dark olive with numerous light spots.
- Blue banded tail, laterally compressed, with a double keel on the apex.

Description

A slender species, with long neck and distinct pointed head. The dorsum is dark olive to black, with numerous tiny light spots, arranged in more or less distinct broad bands. Tail is blue-turquoise banded. The top of the head is solid slate-gray, often with a blue touch, lacking light speckling. The tongue is pinkish. The ventral side is whitish, except the throat, which shows a remarkable pinkish, reddish, or orange hue, leading to the vernacular name peach-throated monitor lizard. Legs are dark olive, finely speckled with blue to green, with coloring more intense in hind legs (Horn 1977; Bennett 1995).

In juveniles, the coloration is more intensive, and the bands of light markings on the dorsum are in general more widely separated.

Compared with its relatives, *V. jobiensis* bears small scales: 164 to 201 scales around midbody, and 163 to 196 dorsal scales from back of the head to the hind legs (Böhme et al. 1994).

The hemipenes or hemiclitores are strongly asymmetrical, having paryphasmata stretching to both lobes (Ziegler and Böhme 1997).

Size

V. jobiensis grows to about 120 cm. The largest reported male has a TL of 119.5 cm (SVL, 44.5 cm; tail length, 75.0 cm) (Mertens 1951). Horn (1977) reports an almost equal sized female with 45.0 cm SVL and 73.5 cm tail length. Tail length averages 1.48 to 1.83 times SVL (Horn 1977; Böhme et al. 1994).

Captive-bred hatchlings had a SVL of 7.9 to 10.2 cm (Bayless and Dwyer 1997). Engelmann and Horn (2003) report a captive-bred hatchling of 27.2 cm TL (SVL 11.2 cm). Within half a year, this specimen grew to 38.0 cm TL (SVL 15.5 cm) and weighed 55 g. Two years later, its TL was 45.5 cm and it weighed 180g.

Habitat and Natural History

V. jobiensis inhabits mixed alluvium forests and mixed hill forests and prefers areas with dense vegetation. *V. jobiensis* is often encountered on the ground, basking in the sun or searching for food. When alarmed, they seek refuge by climbing trees.

V. jobiensis often can be observed resting on tree trunks displaying their strikingly colored, expanded throats (Philipp 1999a, fig. 5). In the gloomy lowland rain forest, this signal is quite visible and may play an important role in the intra- or interspecific communication (Philipp 1999a).

Engelmann and Horn (2003) describe a remarkable kind of warning posture of a young *V. jobiensis*: the body is raised, the gular fold is extended, and the coiled-up tail is presented toward the intruder.

Time of Activity (Daily and Seasonal)

According to Indonesian reptile traders, *V. jobiensis* is available throughout the year (Yuwono 1998). This suggests that *V. jobiensis* is active all year long.

Diet and Foraging Behavior

Stomach content analyses of seven specimens (Philipp et al., in preparation) revealed that *V. jobiensis* feed on tarantulas, insects, frogs, and reptile eggs. Almost 75% of all prey items were insects. Cockroaches, grasshoppers, rhynchota, beetles, bees, and even butterflies are eaten. None of the stomachs investigated were empty.

In Irian Jaya, *V. jobiensis* have been seen patrolling along almost dried-out brooks, searching for food such as small fish or shrimps that can be found regularly at such locations and are easy to catch (personal observation).

In captivity, this species accepts locusts, chicks, mice, frogs (*Rana temporaria*), and fish (*Carassius carassius*) (Horn 1977). Other specimens in captivity feed on beetles, grasshoppers, crustaceans, eggs, small rodents, fish, frogs, and anoles (*Anolis carolinensis*), but ignored live fiddler crabs, shrimp, goldfish, and superworms (Bayless and Dwyer 1997).

Reproduction

No field data are available. In captivity, mating has been observed after a simulated wet period, and eggs have been laid in August and December (Kirschner et al. 1996; Bayless and Dwyer 1997). Bayless and Dwyer (1997) suggest that females normally produce a maximum of 20 to 30 eggs per year, two to six at a time.

6.10. *Varanus juxtindicus*

WOLFGANG BÖHME, KAI M. PHILIPP, AND THOMAS ZIEGLER

Figure 6.18. The preserved holotype of Varanus juxtindicus *(photo by T. Ziegler).*

Nomenclature

This very recently described species was recovered in the collections of the Copenhagen Zoological Museum, where five specimens (one huge adult male, three subadults, and one juvenile) had been deposited since the return of the Noona Dan Expedition from the Solomon Islands in 1962. The description was published by Böhme et al. (2002). The holotype and three paratypes are deposited in the Zoological Museum of the Copenhagen University (ZMUC E 605, R4223-4224, and E 617), and one paratype is in the Zoologisches Forschungsinstitut und Museum A. Koenig (ZFMK 72865), Bonn. This type series seems to be the only material of this species currently known. A specimen figured by McCoy (1980) as *V. indicus* could also belong to this species (Fig. 6.18).

Geographic Distribution

V. juxtindicus is known with certainty only from Rennell Island, the southernmost of the Solomon Islands, from where the type series

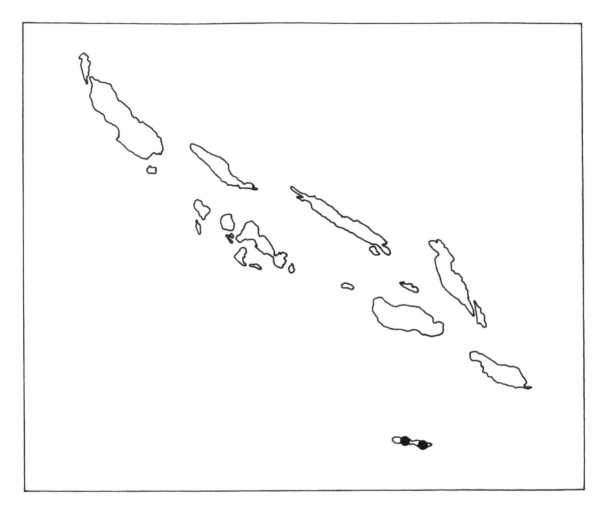

Figure 6.19. Map of the Solomon Islands showing the distribution of Varanus juxtindicus.

originated (Fig. 6.19). It was collected at 2 localities on Rennell: Lake Tegano and Niupani (Böhme et al. 2002). The above-mentioned specimen figured by McCoy (1980), which looks similar to *V. juxtindicus,* is from Malau Olu east of San Cristobal, the easternmost island of the Solomon group. But as long as its identity is not clear, *V. juxtindicus* must be considered an endemic of Rennell Island (from where it may have been also passively transported to other islands).

Fossil Record

No fossils are known.

Diagnostic Characteristics

- Large, very stoutly built, up to 132.5 cm TL.
- Tail roundish in its proximal third, without middorsal double-keeled crest.
- Parietal region extremely swollen in the adult male.

- Blue coloration lacking.
- Tail unbanded.
- Light tongue, with ill-defined pigmentation only in its anteriormost part.
- Hemipenis with paryphasmata on the outer apical lobe only.

Description

The adult male of *V. juxtindicus* has a dark brownish upper side that is densely spotted with yellow; the single spots cover one single dorsal scale, sometimes—particularly on the upper side of head—even less. On the tail, the yellow spots may form indistinct yellow rings, but broad oblique bands and any blue coloration on the tail are lacking. Underparts are a uniform yellow, except a few lighter brown spots on the infralabial scales and the underside of tail, but without any gular or ventral pattern. Subadults have a very indistinct, faded reticulate pattern on the underside of head and body, which is still more distinct in the only known hatchling. All ontogenetic stages, however, have a similar dorsal pattern in common and, most importantly, have tails that are constantly round in cross section in their proximal third, passing into a slight compression in the distal direction. A differentiated mid-dorsal ridge consisting of a double crest is lacking.

Size

The only data available are those from the type series. The holotype, an adult male, has enormous temporal–parietal swellings, so it seems to be full grown. It has a TL of 132.4 cm with 50.4 cm head–body length. The ratio head–body length to TL is similar in the three subadult paratypes. The juvenile, according to its still clearly visible umbilicus, is regarded to be a very young hatchling. It has a head–body length of 14.4 cm and a TL of 36.7 cm.

Habitat and Natural History

Autecological data on *V. juxtindicus* are virtually lacking. Its exclusive distribution area (as far as known), Rennell Island, is a post-Pliocene raised coral atoll and is younger than the majority of the Solomon archipelago. It covers an area of 650 km^2, including Lake Tegano with 155 km^2, and is still predominantly covered with different types of pristine rain forest. Ten plant species, four species and nine subspecies of birds, a sea snake, and a flying fox are known to be endemic to the island (for further information on the fauna of Rennell Island, see Böhme et al. 2002 and references cited therein).

Terrestrial, Arboreal, Aquatic

McCoy (1980) reported that he had seen monitor lizards on Rennell Island and named them *V. indicus,* but they were almost certainly *V. juxtindicus.* He reported that they lived around the shore of

Lake Tegano and that they were mostly aquatic, which agrees with the circumstances of capture of the holotype. According to field notes of the Noona Dan Expedition, it "was shot from a tree trunk within Lake Tegano." This argues for an arboreal/aquatic lifestyle (Bedford and Christian 1996), although the more rounded tail shape of *V. juxtindicus,* as compared with its closest relatives within the *V. indicus* species group, would point to a more terrestrial way of life. However, as pointed out by Böhme et al. (2002), the example of *V. semiremex* versus *V. mitchelli* demonstrates that tail shape and lifestyle might not be as closely correlated in the genus *Varanus* as believed by Bedford and Christian (1996).

No information is available on thermoregulation, movements, reproduction, physiology, or parasites in *Varanus juxtindicus.*

6.11. *Varanus komodoensis*

Claudio Ciofi

Figure 6.20. Varanus komodoensis *(photo by Dennis King).*

Nomenclature

The first description of a "*Varanus* species of an unusual size" was given in 1912 by Peter A. Owens, at that time director of the zoological collection of Bogor, east Java. His account was based on photographic evidence and on the description of a skin of an adult specimen collected on the island of Komodo. Significant genetic diversity has been recorded across the species' range, but no subspecies of *Varanus komodoensis* can be recognized on either genetic or morphological grounds (Fig. 6.20).

Geographic Distribution

V. komodoensis is found on the islands of Komodo, Rinca, Gili Dasami, Gili Motang (part of Komodo National Park since 1980), and Flores, in the Lesser Sunda region, Indonesia. On Flores, its distribution is limited to coastal areas. On the southwest coast, the species occurs from the southern flank of Sano Nggoang to Wae Wuul, a nature reserve located on the west coast, about 10 km south of the town of Labuan Bajo. On the north coast, *V. komodoensis* is found across coastal hills from the village of Pota to the nature reserve of Wolo Tado. Circumstantial evidence suggests a limited distribution on the north-

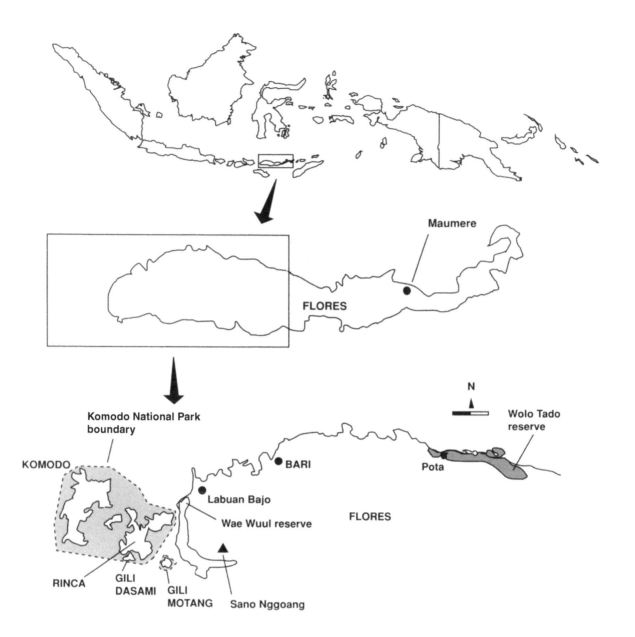

Figure 6.21. Geographic distribution of Varanus komodoensis.

west near the village of Bari, and on the northeast coast, near the town of Maumere (Fig. 6.21).

Fossil Record

Fossil records of varanoid species thought to be possible ancestors of *V. komodoensis* include the large *Megalania prisca* from Pleistocene deposits of Queensland, New South Wales, and South Australia (estimated to be ~7 m in TL and over 600 kg in weight), the smaller *V.*

bolkayi from Java, and *V. hooijeri* from Flores. Fossil records of *V. komodoensis* as old as 900,000 years have been found in central Flores (Morwood et al. 1998).

Diagnostic Characteristics

• Large and bulky lizard, up to 3 m in TL.

• Juveniles are brown with distinct white, yellow, and orange flecks.

• Adults have a uniform earth-brown coloration.

• In adults, SVL is approximately equal to tail length.

Description

Hatchlings are characterized by a brown pattern with large, distinct yellow and orange spots on the dorsal region and on the snout. Light gray coloration with white spots are found from the temporal region to the front legs. Front legs are brown with white flecks disposed in round, horizontal circles. The ventral part of the body is light yellow with large dark spots.

The juvenile pattern is gradually lost with age. Subadult individuals still present lighter coloration on the snout, especially females, but have an earth-brown body. Adults are uniform brown in color. Adult females can have a brown-green coloration on the snout. Males are larger than females. Tail is round in cross section.

Size

Hatchlings average 42 cm in TL and weigh on average 100 g. Adult Komodo monitors can weigh over 70 kg. Maximum SVL and TL recorded in field studies was 154 cm and 302 cm, respectively.

Habitat

V. komodoensis is found from sea level up to about 800 m in altitude, mainly in tropical dry and moist deciduous monsoon forest, savanna, and mangrove forest. In Komodo National Park, the limited annual rainfall and consequent pronounced dry season determine a similar distribution of vegetation ecotypes across islands, strictly linked to topography and elevation. On Komodo and Rinca, savanna and grassland extend inland on the hills rising from the coast to about 500 m of altitude. These same elevations are covered by dry deciduous monsoon forest, which also occurs in narrow strips along a few seasonal water streams between grassland upheavals and in woods over ample valleys. Higher altitudes of north Komodo and southern Rinca have a more humid climate, allowing the formation of moist deciduous monsoon forest and moist semievergreen forest patches. Gili Dasami, off the southern shore of Rinca, is covered mainly by dry deciduous monsoon forest. Gili Motang, on the south east of the archipelago, has extensive patches of savanna on northern and southwestern coastal areas, whereas dry monsoon forest covers the east and central parts of the island.

Main habitat types found across the range of *V. komodoensis* on the western and northern coast of Flores are grassland, savanna, and dry deciduous monsoon forest. Rainfall is more frequent than in Komodo National Park. The Wae Wuul nature reserve, located on the west coast, is a hilly area of less than 3000 ha with peaks of up to 370 m of altitude. It is covered predominantly by savanna, with stretches of dry deciduous monsoon forest occurring along rivers and on lower flanks of hills. The Wolo Tado reserve, on the north coast, covers 4000 ha of savanna and dry deciduous monsoon forest. In the last 20 years, expansion of human settlements and slash-and-burn forest clearance has reduced Komodo monitor habitat in north Flores, near the town of Pota. On the southwest coast of Flores, forest conversion and harvesting was still limited at the time of a survey carried out in 1999.

Terrestrial, Arboreal, Aquatic

Hatchling Komodo monitors are arboreal. Juveniles and subadults are both arboreal and terrestrial. Adult individuals are strictly terrestrial; their large body size hinders climbing trees. *V. komodoensis* has been occasionally observed swimming for short distances.

Time of Activity (Daily and Seasonal)

V. komodoensis is diurnal and active throughout the year. Daily activity is from 0600 hours to 1900 hours. Bimodal activity patterns are predominant for most of the year except between January and March, when activity is concentrated during the warmest hours of the day (Auffenberg 1981). About 70% of morning activity includes basking; movement represented 68% of all activities from 1330 hours to 1730 hours. Movements between 1200 hours and 1400 hours occurred mainly within forested areas. High usage of forested savanna, open forest, and open savanna is recorded mainly between 0800 hours and 1000 hours and after 1600 hours (Sastrawan and Ciofi 2002).

Thermoregulation

Mean ± SE body temperature recorded for adult and juvenile Komodo monitors was 35.4 ± 0.9°C and 35.0 ± 0.6°C, respectively (Wikramanayake et al. 1999). Mean maximum values were similar in juveniles and adult lizards (39.6 ± 1.1°C and 39.2 ± 1.0°C, respectively), whereas mean values of minimum temperature were lower in juveniles (25.9 ± 2.9°C) than in adult individuals (28.1 ± 2.1°C). Heating rates during basking ranged from 0.06 to 0.20°C/min in juveniles to 0.03 to 0.06°C/min in adult specimens (Green et al. 1991). Komodo monitors make use of a number of different microhabitats to regulate body temperature. Differences between body and ambient temperature as high as 7°C and 5°C were recorded in juveniles and adult specimens, respectively.

Foraging Behavior and Diet

Juvenile *V. komodoensis* are active predators, whereas adult specimens ambush their prey. The diet includes insects, small to large vertebrates (lizards, snakes, rodents, monkeys, wild boars, deer, and water buffalo), bird and sea turtle eggs, and carrion. Deer make up about 50% of the diet of adult specimens (Auffenberg 1981). Adult individuals may also prey on young conspecifics. *V. komodoensis* can detect airborne volatile oils released by decomposition of carcasses, which leads lizards on long feeding excursions.

Reproduction

Age of first reproduction has been estimated at about 8 and 9 years for captive females and males, respectively (T. Walsh, personal communication). No record is available for wild specimens. Courtship and mating occur from May to August. Adult males of similar size get involved in ritual combats to have access to females. Using their tail for support, they wrestle in an upright position, grabbing each other with their forelegs as they try to floor the opponent. The male begins courtship by flicking his tongue on the female's snout and then over her body until he reaches her cloaca. He then presses his snout at the base of her tail, starts scratching her body with his claws, and eventually crawls on her back.

Females lay their eggs mainly in September, in burrows located on the slope of hills or in nests of megapode birds. Bird nests consist of heaps of twigs mixed with earth, up to 150 cm in height and 4 to 5 m in diameter. A number of holes are dug into the nest, but usually only one is used to lay eggs. Females have been observed lying on the nest for the first 3 months of incubation, probably guarding the eggs from predators. Eggs hatch in March–April. In captivity, incubation period averages 220 days (Walsh et al., in press). Parental care has not been observed in Komodo monitors. Average clutch size is 18 eggs. A maximum of 33 eggs were recorded in a single clutch during field studies.

Although males tend to grow bulkier and bigger than females, Komodo monitors have no obvious morphological differences between sexes except in the arrangement of a specific part of precloacal scales. By use of this method, Auffenberg (1981) estimated a male to female ratio of 3:1 on Komodo Island. However, the scale pattern is not always clear, and probing the cloaca for presence or absence of inverted hemipenes is troublesome because females have hemiclitoral sacs at approximately the same position of male hemipenes, and sex can often be confused. A variety of different techniques have been evaluated on captive specimens: laparoscopy (for visually examining gonads by means of an endoscope), radiography (to detect the presence of skeletal elements located inside the hemipenes), assessment of plasma testosterone concentration differences between sexes, and, more recently, ultrasound examination to image the presence or absence of ovarian follicles. A combination of the above techniques recorded a sex ratio close to 0.50 in captive clutches (Halverson and Spelman 2002). However, although accurate, these methods have not given consistent results for animals younger than 8 months of age, and their employment in field studies is hampered by the requirement for quality equipment. Thus far, DNA technology has provided the most consistent results (Halverson and Spelman 2002). Hatchlings gain about 30% of their weight in 3 months, and after about 5 years, they may weigh 25 kg and have a TL of over 2 m. Growth continues slowly throughout life. Males reach a larger size than females, and longevity of over 30 years has been suggested on the basis of captive records and field observations (Auffenberg 1981; Gepak, unpublished data).

Movement

Analysis of individual activity over a 5-week period has recorded a nonuniform distribution of distances traveled per day (Sastrawan and Ciofi 2002). Periods of low activity were often followed by days of intense movement. About 65% of movements were over distances of less than 1000 m, but a few sparsely distributed movements were longer excursions of up to 5500 m. Home ranges varied from 258 ha to 529 ha and were larger in males than in females. Males had about 35% of their home range shared with an individual of the same sex, whereas a higher proportion of the activity area (up to 99%) can overlap between specimens of opposite sex.

According to Auffenberg (1981), the feeding strategy of *V. komodoensis* is one of the main factors determining both spatial and temporal patterns of movements. Part of the monitor's diet consists of carrion. The smell released by decomposing carcasses carried by the wind can lead an adult specimen to travel a long distance (up to 3 km), well beyond the boundaries of its usual activity area.

Behavioral dominance also seems to affect individual movement patterns. During mating, feeding, and occasional encounters, a set of sustained aggressive–submissive relationships is established among individuals. This behavior results in younger specimens being routinely excluded by their dominant elders. The latter, defined as residents by Auffenberg (1981), utilize the same home range during subsequent years, whereas younger subadults (transient individuals), move randomly over large areas and are potentially important in interpopulation movements and dispersal.

Population Genetics

Genetic differentiation is most marked for the populations of the north coast of Flores and of Komodo Island. Individuals sampled on north Flores showed a significant level of genetic differentiation not only from other island populations but also (although to a minor extent) from sampling sites of the west coast of Flores. Level of genetic divergence reported for the north Flores' population, geographical distance to the west coast, and absence of suitable habitats that could form a corridor for gene exchanges between north and west coast, suggest that the known range of monitors on Flores is presently structured in at least two distinct populations (Ciofi et al. 1999). Heterozygosity values were among the highest recorded in the archipelago, highlighting the importance of the island of Flores for maintenance of intraspecific genetic diversity.

Comparisons between south Rinca and Gili Dasami populations, divided by approximately 1 km of seawater, gave genetic distance measures significantly lower than any other value reported across the archipelago. The same pattern was found, to a minor extent, between north Rinca and west Flores populations (~250 m of seawater apart), where a relatively high level of gene flow was described by Ciofi and Bruford (1999).

Comparison of historical changes of sea levels and bathymetric data suggest that Komodo joined the eastern islands during the glaciation of 135,000 years and 18,000 years BP. On the other hand, Flores and Rinca were connected for most of the last 140,000 years, until the end of the last glaciation. Long-term separation of Komodo from the rest of the archipelago and strong tidal current recorded between Komodo and Rinca may have represented a major barrier for migrant individuals and therefore may account for the observed pattern of genetic divergence. Komodo represents an important gene pool for both maintaining a substantial level of genetic and adaptive variability within the species.

Lizards from Gili Motang showed significant differences in allele frequency distribution and size from other populations; however, there is no indication of a distinct evolutionary history, as is the case for Komodo Island. Gili Motang is of great conservation interest and is at the same time of particular concern because of the small population size, low heterozygosity, and high percentage of fixed alleles.

Physiology

Water influx and metabolic rates were determined on wild specimens by Green et al. (1991). Water influx had a coefficient of variation of 63% ($\overline{X} \pm SE$, 25.2 ± 16.0 mL H_2O/kg/d). CO_2 production rates had a coefficient of variation of 29% and average values of 0.129 ± 0.038 mL CO_2/g/h. Highest influx rates were found in juveniles, whereas adults had the lowest values. Adult specimens ambush their prey, using a sit-and-wait predatory technique; in contrast, juveniles are active foragers. Lower water influx and metabolic rates were recorded for the largest individual. Individual differences in hunting strategies may in part account for this discrepancy. Adult specimens can maintain lower active body temperature and have lower food and energy requirements than growing juveniles. According to Green et al. (1991), juveniles about 2.3 kg in weight metabolize approximately 95 kJ/kg/d, equivalent to 220 kJ/d, and require about 55 g of food per day. Adult lizards weighing more than 45 kg would consume 2440 kJ/d and require 610 g of food each day.

Parasites

A number of ecto- and endoparasites were described by Auffenberg (1981). Ticks were the only ectoparasites found on wild *V. komodoensis*. Species included *Aponomma komodoense, Amblyomma robinsoni,* and *Amblyomma helvolum. A. robinsoni* was the most common. Tick density was highest at the peak of the dry season and decreased significantly at the beginning of the rainy season. Ticks were mainly found on the back and sides of the body, especially above the lateral fold and at the insertion of the legs. Larvae of cestodes of the order *Pseudophyllidea* were also found beneath the skin, although lizards were thought to be accidental hosts. A small number of amoebae of the genus *Endolimax* and cestode ova were found in feces. Intestinal parasites were described from the order *Pseudophyllidea* and *Protocephalidea,* the first represented by the genus *Duthiersia,* the second by *Acanthotaenia varia.*

6.12. *Varanus kordensis*

HANS J. JACOBS

Figure 6.22. Varanus kordensis *adult (a) and hatchling (b) (photos by Hans Jacobs).*

Nomenclature

This species was described as *Monitor kordensis* by Meyer (1874) and was classified by Mertens (1942) as a subspecies of *Varanus prasinus*. Sprackland (1991) deprived this species of its taxonomic status and claimed it to be a mere color variant of *Varanus prasinus*. In a comparative study, Jacobs (2002) documented the differences between the two taxa and reestablished *Varanus kordensis* as a valid species (Fig. 6.22).

Geographic Distribution

Until now, there is no evidence that *Varanus kordensis* occurs anywhere else other than on Biak Island (formerly Schouten with its former province capital, Kordo). All tree monitor species on the smaller islands surrounding Papua New Guinea are basically black, except for

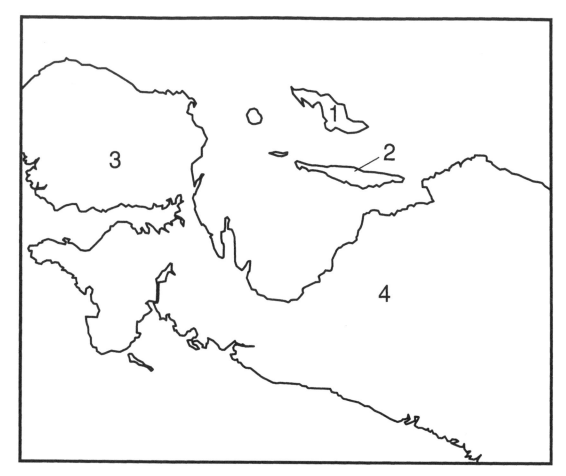

Figure 6.23. Distribution of Varanus kordensis. 1. Biak; 2. Yapen; 3. Vogelkop; 4. Irian Jaya.

V. prasinus on mainland New Guinea and *V. kordensis* (these are the only green species in the *V. prasinus* group).

Fossil Record

None.

Diagnostic Characteristics

- Slender tree monitor with an average TL of 70 to 80 cm.
- Tail nearly double the length of the body.
- Nuchal scales keeled.
- Color pattern very different from *V. prasinus*: ocelli instead of V-shaped stripes. Darker green, rather olive in color. Legs and tail patterned.
- Midbody scales average 89, 34 scales across the head in a straight line; in all aspects, decisively fewer scales than *V. prasinus*.
- Distinct bands of filled ocelli in hatchlings. Light brown bands across belly.

Description

V. kordensis differs from *V. prasinus* unmistakably in its principally olive coloration and its ocella pattern, which spreads to the legs and tail. Throat and belly are a unicolored light brown with a tinge of green. The tip of the snout is rather light. The tail—like that of all species of the *V. prasinus* group—is highly prehensile.

Size

Maximum size may reach 90 cm. The largest male I analyzed measured 850 mm in total with a head–body length of 270 mm and a tail length of 580 mm. The tail of lizards in this species can be twice as long as the body and in certain individuals even a bit longer, as in this case.

Habitat and Natural History

As well as *V. beccarii, V. prasinus,* and *V. macraei, V. kordensis* is a very specialized tree monitor and is undoubtedly diurnal. Data about its insular habitat are rare and unreliable.

The first documented successful breeding (Jacobs 2002) describes an incubation period of 190 days at about 28 to 29°C, approximately 20 days more than a clutch of *V. prasinus* incubated in the same incubator at the same time. Clutch sizes varied from three to four eggs, compared with *V. prasinus*, with up to six eggs. Eggs of *V. kordensis* are somewhat larger than those of *V. prasinus*.

6.13. *Varanus mabitang*

MAREN GAULKE

Figure 6.24. Varanus mabitang
(photos by Maren Gaulke).

Nomenclature

Varanus mabitang was described in 2001 by Gaulke and Curio. The type locality is the South Pandan Forest, northwest Panay Island, Antique Province, Philippines. Many features of its biology are similar to those of its presumed sister species, *Varanus olivaceus,* which was studied by Auffenberg (1988) (Fig. 6.24).

Geographic Distribution

At present, *Varanus mabitang* is known only from the northwest Panay Peninsula and the Western Panay Mountain Range (Gaulke and Demegillo 2001).

Fossil Record

No fossils of *Varanus mabitang* are known.

Diagnostic Characteristics

• Dorsal side black with scattering of tiny yellow dots on the posterior end of some scales of neck, back, and extremities.

- Ventral surface of head, neck, tail, extremities, and belly dark gray to blackish.

- Head scales enlarged; extremely small scales on neck, body, and tail.

- Tail triangular in cross section; upper scale crest with a well-defined, double, longitudinal keel.

- Head elongate; snout region slightly domed.

- Cranial table with well-developed bulges above temporal regions.

- Ventrals strongly keeled.

Description

Varanus mabitang is a very dark monitor lizard; the tiny yellow dots are only discernible at short range. No juveniles have been sighted by scientists so far; however, according to observations of local inhabitants, they are as dark as the adults. The eyes of *Varanus mabitang* are reddish brown, and the tongue is pinkish.

One of the characteristics of *Varanus mabitang* is the extremely large front and hind feet, with very long and curved claws. The teeth are blunted; the dentary teeth do not extend beyond the gum and are visible only as almost translucent, flat and round ovals within the gum. Body scales are very fine, and consequently, standard scale counts are high (e.g., scales from rictus to rictus 70, transverse rows of ventral scales from gular fold to a theoretical line connecting the insertion of hind legs ventrally 124, transverse rows of dorsals from gular fold to a theoretical line connecting the insertion of hind limbs dorsally 138).

Size

Only two *Varanus mabitang* have been measured so far. The holotype, a female, has a TL of 126.8 cm (52.7 cm SVL, 74.1 cm tail length, with some centimeters of tip missing). It weighed 1850 g. The other measured specimen had a TL of 175 cm (64 cm SVL) and weighed 5750 g. According to information from local people, this species can get very large, and a TL of more than 2 m was mentioned.

Habitat and Natural History

Varanus mabitang is a rain forest inhabitant. It occurs in primary and secondary lowland forest, up to an elevation of at least 450 m.

Terrestrial, Arboreal, Aquatic

On the basis of present knowledge, *Varanus mabitang* are strongly arboreal, as indicated by their long, strongly curved claws. They use tree holes as shelter sites and rest on branches high up in forest trees (Gaulke and Curio 2001; Gaulke and Demegillo 2001).

Time of Activity (Daily and Seasonal)

No studies on the activity rhythm of *Varanus mabitang* have been done so far. As inhabitants of a tropical country with no significant annual temperature changes, it can be assumed that differences in seasonal activity are minimal. Hunters have observed that *Varanus mabitang* is more active during sunny weather, when it is occasionally seen on the ground.

Thermoregulation

Nothing is known about the body temperature and thermoregulatory behavior of *Varanus mabitang*.

Foraging Behavior

As a mainly frugivorous species, it can be assumed that *Varanus mabitang* localizes fruit trees with ripe fruits olfactorily, with the help of the Jacobson's organ. Alternatively, fruit trees within the respective activity ranges might be known by their occupants and visited regularly to check the presence of ripe fruit.

Diet

Varanus mabitang has a highly specialized diet (Struck et al. 2002). It feeds on ripe fruits of some forest trees. Seeds of different types of screw palms (*Pandanus* spp.) and of a fig tree (*Ficus minahassae*) are present in its feces and gut contents. It also feeds on the fruit of palm trees (*Pinanga* sp.), and on the leaves of some screw palms and a type of shrub (with the local name "*topsi*"). Like its presumed sister species, *Varanus olivaceus*, *Varanus mabitang* possesses a well-developed cecum, an adaptation to its vegetarian diet.

Reproduction

No exact data on the reproductive cycle of this rare varanid are available. In dissected specimens, hunters counted between 6 and 12 eggs (Gaulke et al. 2002). The holotype, collected during the dry season of the area (in May), contained ovarian follicles between 5 and 7 mm in size.

Movement

Nothing is known on the activity range of this rare lizard.

Captured *Varanus mabitang* did not try to defend themselves, nor did they show any sign of threatening behavior typical of monitor lizards, such as tail coiling and uncoiling, gular extension, or hissing. While held, they let extremities, tail, and head hang down. Extended periods of time were spent completely motionless, even if disturbed by people. This behavior might be described as letisimulation. According to a hunter with a good knowledge of the life habits of *Varanus mabitang*, feigning death is a common behavior of this species after capture.

Varanus mabitang are very capable climbers. They can easily climb dipterocarps, which have smooth and long stems with a very hard bark. A large *Varanus mabitang* (TL 175 cm) was observed jumping from a branch of one tree to the branch of a neighboring tree.

Physiology

Nothing is known about the physiology of this rare lizard.

Fat Bodies, Testicular Cycles

The holotype contained well developed fat bodies (the animal was caught in May), but no other information is available.

Parasites

Two species of ticks were collected from the skin of *Varanus mabitang*, but identification of the ticks is still in progress; they are most probably *Amblyomma helvolum* and *Aponomma fimbriatum*.

6.14. *Varanus macraei*

WOLFGANG BÖHME AND HANS J. JACOBS

Figure 6.25. Varanus macraei
(photo by Hans Jacobs).

Nomenclature

A spectacular, very recent, discovery of a new species of tree monitor of the *Varanus prasinus* group was found in an import of Indonesian monitors to Germany and was described by Böhme and Jacobs (2001). The holotype is an adult female deposited in the Zoologisches Forschungsinstitut und Museum A. Koenig, Bonn, as ZFMK 74558. Five paratypes were still alive in late May 2003 (Fig. 6.25).

Geographic Distribution

As far as we currently know, *Varanus macraei* occurs only on Batanta, a small island close to the northwest coast of the Bird's Head (=Vogelkop) Peninsula, Irian Jaya, Indonesian New Guinea. Batanta is separated from the coast of the huge main island by Salawati Island where *V. prasinus* occurs. Salawati, which has far more species in common with New Guinea than Batanta (H. Meinig, personal communication), is separated from the latter by a relatively deep marine rift, the Sagewin Strait. A differently colored but otherwise similar looking population from the neighboring island of Waigeo resulted in the description of yet another new tree monitor taxon: *Varanus boehmei*

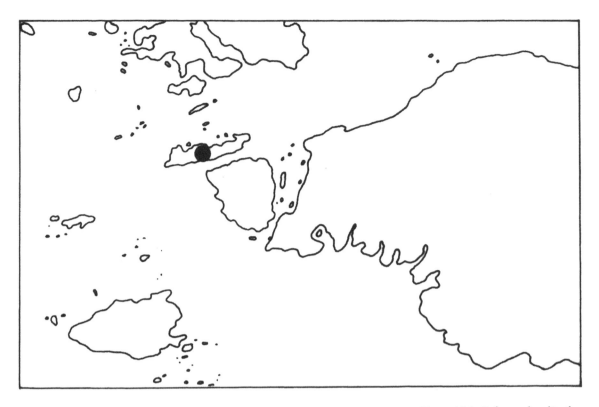

Figure 6.26. Only one locality for Varanus macraei *is known on Batanta, a small island close to the northwest coast of the Bird's Head (=Vogelkop) Peninsula, Irian Jaya, Indonesian New Guinea.*

(Jacobs 2003), which proved to be more closely related to *V. beccarii* and *V. kordensis* than to *V. macraei*. This is now the eighth taxon of the *prasinus* complex.

Fossil Record

None.

Diagnostic Characteristics

• Large, slender tree monitor, TL up to 110 cm.

• Smooth, unkeeled nuchal scales.

• Unique color pattern consisting of numerous oblique, irregular rows of blue ocelli on a black ground color on the upper side.

Description

V. macraei is unmistakable because of its blue coloration and ocellated pattern. Its nostril is much nearer to the snout than to the eye. Nuchal scales are smooth but conically convex. It possesses three to four obliquely enlarged supraocular scales. The ground color dorsally is blackish with very conspicuous, somewhat irregular crossrows of bright blue ocelli that form blue chevrons on the nape. The upper side of the head is bluish; the snout region is whitish with a distinct, light temporal stripe. The belly is a whitish gray with a bluish tinge and with some lateral brownish oblique stripes that do not meet at the midline. The tail is nearly as twice as long as head and body and highly prehensile, with 22 to 23 blue regular bands. Juveniles have dorsal ocelli without black centers, being solid dots, as in juvenile *Varanus kordensis*. The tail is completely banded and the legs intensely spotted. The belly has 12 to 14 bands. Two darker rings exist around the tip of the nose (Jacobs 2002).

Size

Maximum size is about 110 cm TL. The largest male collected measured 36 cm head–body length and 75 cm tail length. Another adult male measured 34 cm and 66 cm, respectively. The largest two females measured 31.3 cm and 28 cm head–body length and 59.9 cm and 53.3 cm tail length, respectively (MacRae, in Böhme and Jacobs 2001). Of the type series, the biggest male measured 35 cm SVL plus 65.5 cm tail, for a TL of 100.5 cm (paratype 4 in Böhme and Jacobs 2001).

Habitat and Natural History

Varanus macraei is clearly a diurnal, highly specialized tree monitor lizard, as are the other members of the *V. prasinus* species group. However, because it is a genuine new discovery, not studied before by a zoologist and therefore not hidden in the synonymy of a species already identified, virtually nothing is known about its habitat and natural history.

Reproduction

The first clutch laid in captivity consisted of 3 eggs measuring 43 by 21 mm, 45 by 20 mm, and 43 by 21 mm. The eggs weighed 9, 10, and 9 g, respectively. Incubation time is 159 days, temperature 28 to 30°C (Jacobs 2002b).

6.15. *Varanus melinus*

Thomas Ziegler and Wolfgang Böhme

Figure 6.27. Varanus melinus
(photo by Thomas Ziegler).

Nomenclature

The quince monitor lizard, *Varanus melinus*, was described by Böhme and Ziegler in 1997a (Fig. 6.27). The male holotype is deposited in the Zoologisches Forschungsinstitut und Museum Alexander Koenig (ZFMK 65737). The pet-trade-based type locality, Obi, Moluccas, Indonesia, was corrected in Ziegler and Böhme (1999) as indicated below. According to its genital morphology characteristics, *Varanus melinus* is a member of the subgenus *Euprepiosaurus* Fitzinger (see Böhme 1988; Ziegler and Böhme 1997) within the *Varanus indicus* group (Böhme and Ziegler 1997a; Ziegler and Böhme 1999). Of the currently known representatives of the *V. indicus* group, genital morphology links *V. melinus* with *V. caerulivirens*, *V. cerambonensis*, *V. indicus* sensu stricto, and *V. juxtindicus* because of the reduced and therefore derived condition of unilaterally developed hemipenial and hemiclitorial paryphasmata (Philipp et al. 1999a; Ziegler et al. 1999a, 1999b; Böhme et al. 2002). *V. melinus* is a monotypic species.

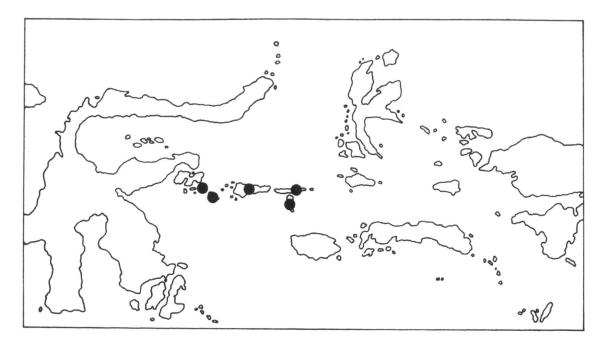

Figure 6.28. Known localities for
Varanus melinus.

Geographic Distribution

In the original description of *Varanus melinus*, Böhme and Ziegler (1997) gave the Sula Islands as well as Obi, Moluccas, Indonesia, as the distribution area of the quince monitor lizard (Fig. 6.28). Ziegler and Böhme (1999) reduced the distribution of *V. melinus* to the Sula Islands, because it was obvious that Obi was only an intermediate trading center. Occasionally Halmahera and even Sulawesi are also mentioned by Indonesian traders as distribution areas of the species, thus indicating that the species is still remarkably little known in Indonesia. The recent discovery of *Varanus melinus* can probably be correlated with the disastrous forest fires in Indonesia in 1997 (Böhme and Ziegler 1997; Bayless and Adragna 1999), because new collecting areas had to be found instead of some traditional, known collecting areas (Yuwono 1998). Because local hunters are reluctant to name the exact location of collecting areas, usually only incomplete or incorrect locality data reach exporters in Jakarta. In the Sula Archipelago, according to Sprackland (1999c), *V. melinus* seems to inhabit the islands of Mangole and Taliabu. Beside Taliabu, which is also mentioned by Lemm (1998), Bayless and Adragna (1999) further give Bowokan and Banggai island groups as distribution areas of the quince monitor lizard. Sanana Island is also sometimes reported as being inhabited by the species (Dedlmar and Böhme 2000).

Fossil Record

No fossils of *Varanus melinus* are known.

Diagnostic Characteristics

Within the subgenus *Euprepiosaurus* (see Ziegler and Böhme 1997 for diagnosis) *Varanus melinus* is distinguishable from the remaining members of the *V. indicus* group by having a yellow ground coloration with more or less developed dark marbling, the lack of any turquoise or blue coloration, a low midbody scale count (124–133), a pink-colored, light tongue, and differentiated paryphasmata only on one side of the sperm groove of the hemipenis and hemiclitoris, respectively (Böhme and Ziegler 1997a; Ziegler and Böhme 1999; Dedlmar and Böhme 2000).

Compared with other varanid taxa, *Varanus melinus* can possibly be confused with the subspecies *cumingi* of the Asiatic water monitor, *V. salvator,* because of its convergent yellow ground coloration. But *V. salvator cumingi* occurs only on the Philippines and can be distinguished from *V. melinus* by its much larger size (maximum 220 cm), having (usually) black temporal stripes, a dark colored tongue tip, and the structures of its outer genitalia (Gaulke 1991a; Ziegler and Böhme 1997; Bennett 1998).

Description

Varanus melinus is a medium-sized monitor lizard of slender habitus. Characteristic is its intensive yellow ground coloration of head, neck, body, limbs, and tail (Böhme and Ziegler 1997a). On the dorsal aspect, a dark reticulated pattern may be more or less developed mainly on the body, but also on the neck, limbs, and tail base, which may fade to increased yellow with age. The dark reticulation of the dorsum may be more or less discernible as transverse rows of blackish framed yellow ocelli. The tail usually bears black and yellow transverse bands. The tongue is uniformly pink colored.

The coloration of the distinctly darker juveniles differs considerably from that of the adults (Bayless and Adragna 1999; Dedlmar and Böhme 2000). Hatchlings are blackish with a light to yellowish pattern that consists of transverse rows of yellow ocelli on the dorsum. Later, starting at the head and neck region, the dark coloration is gradually replaced by the characteristic yellow ground coloration of the subadults and adults.

Size

The largest measured specimen of *Varanus melinus* is an adult male, depicted as paratype 1 in Böhme and Ziegler (1997a). Therein, the TL was given as 115 cm, with the SVL being 42 cm. About 2.5 years later, that same specimen measured 120 cm in total (Dedlmar and Böhme 2000). The largest known adult female has a TL of 95 cm (Dedlmar and Böhme 2000). Bayless and Adragna (1999) report a maximum size of 4 to 5 ft (1.2–1.5 m) in general. The only known hatchlings had a mean TL of 21.6 cm and a mean weight of 22 g (Dedlmar and Böhme 2000).

Habitat and Natural History

Bayless and Adragna (1999) report *Varanus melinus* to be an inhabitant of interior tropical lowland forests on Taliabu, which are characterized by a high density of dipterocarp trees. These forests are selectively logged and heavily degraded, and partly interspersed with agricultural areas. Thus, besides a population decline due to the removal of the species for the pet trade, habitat loss, and habitat destruction, the occasional killing of the species by forest workers most probably puts further pressure on the geographically restricted populations of the quince monitor lizard.

The climate of Taliabu shows little seasonal variation, with a wet season usually from November through March. Mean temperatures range from 19 to 21°C to 30 to 32°C, rainfall totals from 230 to 250–300 cm, and relative humidity remains between 82% to 90% (Bayless and Adragna 1999).

According to Lemm (1998) the quince monitor lizard appears to be shy, preferring to stay hidden under moss or logs, or high up in trees. The species seems to dislike exposure to bright sunlight without the shade of vegetation.

Terrestrial, Arboreal, Aquatic

In captivity, *Varanus melinus* shows a distinct preference for water and high humidity. Its willingness to swim and dive suggests that this species is probably a swamp dweller or an inhabitant of similar habitats, perhaps more adapted to water than its close relative *V. indicus*. In addition, because of its sharp claws and somewhat prehensile tail, the quince monitor lizard is also a good climber (Böhme and Ziegler 1997a; Ziegler et al. 1998; Lemm 1998; Bayless and Adragna 1999; Dedlmar and Böhme 2000).

Time of Activity

In Böhme and Ziegler (1997a), a tendency toward an activity at night (dusk and dawn) was stated, whereas Dedlmar and Böhme (2000) reported the species being active during the day, commencing activity in the early morning hours. Bayless and Adragna (1999) reported *Varanus melinus* to be active for only short periods of time, remaining hidden for most of the day. They further give 45 to 120 minutes daily time of activity, mainly around dusk and sunset.

Foraging Behavior and Diet

Like almost all other varanid species, *Varanus melinus* is carnivorous. Because of its size, the species certainly belongs to the top predators in its ecosystem. However, larger vertebrates as pigs or deer, as mentioned in this context by Bayless and Adragna (1999), certainly are not included in the food of the quince monitor lizard. Bayless and Adragna (1999) report that *V. melinus* is an opportunistic feeder that hunts by stealth or ambush. The intensive yellow coloration reticulated by dark might help the species to dissolve at day in the interplay of light and shade (Böhme and Ziegler 1997). As natural food items on Taliabu,

Bayless and Adragna (1999) list eggs of the Sula scrubfowl (*Megapodius bernsteinii*). According to these authors, insects such as beetles and lower vertebrates such as tree frogs are major prey items. In captivity, *V. melinus* accepted prawns, insects (crickets and grasshoppers), fish, birds, small rodents, and cat food (Böhme and Ziegler 1997a; Lemm 1998; Ziegler et al. 1998; Dedlmar and Böhme 2000). Hatchlings were reported to feed on small crickets and, within a few weeks after hatching, baby mice (Dedlmar and Böhme 2000).

Reproduction

A preserved female of *Varanus melinus*, of approximately 1 m in length, held in the herpetological collection of the Zoologisches Forschungsinstitut und Museum Alexander Koenig (ZFMK 70441), contained four large shelled eggs (23–29 by 27–31 mm). According to Lemm (1998), females have cycled in captivity both in private collections and in zoos, although eggs have been infertile. Maximum clutch sizes of up to 12 eggs are known. The first successful captive breeding of *Varanus melinus* was reported by Dedlmar and Böhme (2000): one pair (among them the largest known male, see above) yielded four clutches consisting of two to seven eggs each (mean, 4.8 eggs), which were deposited at intervals of 3.5 to 8 months (mean, 5.2 months). After about 170 days' incubation time at 28.5°C, five juveniles hatched from a clutch containing six eggs of 65 to 67 mm (mean, 66 mm) by 23 to 27.5 mm (mean, 26 mm) size. Hatchlings measured 210 to 225 mm in total, with a weight of 21 to 23 g.

Fat Bodies, Testicular Cycles

Bayless and Adragna (1999) state that fat body organs of *Varanus melinus* are usually one-fifth the total body weight of the species.

Nothing is known about thermoregulation, movements, physiology, or parasites of *Varanus melinus*.

6.16. *Varanus olivaceus*

Eric R. Pianka

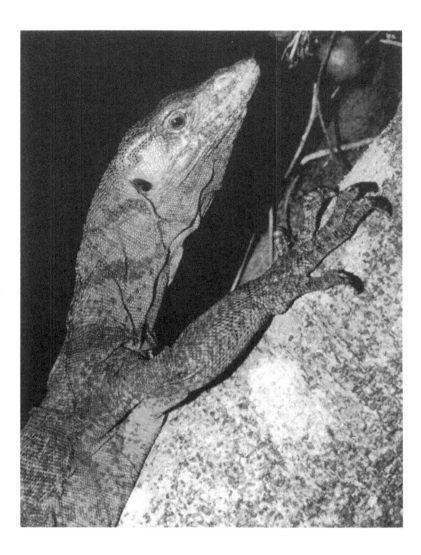

Figure 6.29. Varanus olivaceus *(photo by Daniel Bennett).*

Nomenclature

Nomenclature of this species has taken several turns. It was first described as *Varanus ornatus* by John E. Gray in 1845 from an unknown locality in the Philippine Islands. Gray's type specimen (BMNH 1946.8.30.98) is a juvenile. In 1885, Boulenger realized that the name *V. ornatus* was preoccupied, having been given to another varanid species by Daudin. Boulenger synonymized *V. ornatus* with *V. niloticus* and proposed that Gray's Philippine monitor should be called *Varanus grayi*. Walter Auffenberg discovered that Edward Hallowell had described another specimen of this species in 1857 as *V. olivaceus*. Be-

Figure 6.30. Geographic distribution of V. olivaceus.

cause Hallowell's name antedates Boulenger's, it has priority. Hallowell's type, a skin (ANSP 9916) in the collection of the Academy of Natural Sciences, Philadelphia, was "purchased in a market by Dr. Kane, U.S. Navy" (Auffenberg 1988, p. 327). Unfortunately, this type could not be located in a recent search. Another new species of Philippine monitor, *Varanus mabitang* (Gaulke and Curio 2001), has recently been discovered from the Central Philippine Island of Panay south of the geographic range of *V. olivaceus*. This new species appears to be closely related to *V. olivaceus* because it is also frugivorous (Fig. 6.29). Most of the species account for *V. olivaceus* was extracted from Auffenberg (1988).

Geographic Distribution

V. olivaceus was known from only the type and the cleaned skull of an adult specimen until 1976, when a population was rediscovered not far from Manilla. It is now known to occur on Eusbern Luzon and on the islands of Catanduanes and Polillo (Fig. 6.30).

Fossil Record

No fossils of *V. olivaceus* are known.

Diagnostic Characteristics

- Nares slits about halfway between eye and snout tip.
- Large, up to nearly 200 cm in TL.

- Greenish gray, with seven or eight darker transverse bands on body.
- Tail also banded transversely and laterally compressed with a dorsal crest.

Description

Adult *V. olivaceus* are greenish gray, with neck, back, and tail crossed by a number of darker transverse bands. The banding pattern is more pronounced in younger, smaller lizards. Heads are large, light yellow, and gray. Slitlike nasal openings exist about midway between eye and tip of snout. The laterally compressed tail has a double row of keeled scales forming a conspicuous dorsal crest. The flat lateral surface of the tail is used as a prop in climbing. Belly scales are keeled and are used for friction when climbing vertical tree trunks. Limbs are generally darker than the body, grayish to blackish brown speckled with lighter spots. Hind legs are longer than in many other monitor species. Feet are dark brown to black with gray, green, or yellow spots. Claws are black, large, sharply curved, and are used in climbing (these big lizards can hang from a single claw!). The teeth of juveniles are sharp, but those of adults are blunt, used for crushing snails.

Size

Males grow faster and reach larger sizes than females (650 mm SVL, weight 6.7 kg, vs. 509 mm SVL, weight 2.6 kg, respectively). The largest males reach over 180 to 200 cm in TL with weights in excess of 9 kg. Hatchlings are about 350 mm in TL and weigh 25 g.

Habitat and Natural History

Densely forested slopes of low mountains. These lizards rely on the camouflage provided by their disruptive coloration to avoid detection. This, coupled with their small home ranges and the fact that they live in densely vegetated rain forest areas, has allowed them to go undetected by scientists for many years.

Terrestrial, Arboreal, Aquatic

V. olivaceus are arboreal.

Time of Activity (Daily and Seasonal)

V. olivaceus are diurnal and active all year long.

Thermoregulation

Living at a tropical latitude of 14° N, *V. olivaceus* experience relatively constant warm temperatures suitable for activity over the entire course of a year. These monitors are passive thermoconformers; often they allow their body temperatures to mirror ambient conditions, although at times they do bask. Body temperatures fluctuate during the day, being near ground temperatures early in the day up until noon, but in the afternoon and early evening, body temperatures lie between ground temperatures and air temperatures. Presumably, some inertial homeothermy occurs late in the day, keeping these big lizards warm. Body temperatures range from about 28 to 38°C. Critical thermal maximum temperature is from 41.6 to 42.4°C (Auffenberg 1988).

Foraging Behavior

These monitor lizards spend most of their time in trees, descending to the ground to feed. *V. olivaceus* and *V. mabitang* are the only obligate frugivorous lizards, as well as the largest frugivores in Asian forests.

Diet

Juveniles feed largely on snails and crabs. Adults are largely frugivorous, but they also eat snails, crabs, spiders, beetles, grasshoppers, and katydids, as well as birds and bird eggs. Fruits are eaten only when they are ripe and have fallen to the forest floor (Auffenberg 1988). Fruits of different plant species are distinctly seasonal. These monitors are very choosy, eating fruits of relatively few tree species. They show a marked preference for the uncommon fruit of *Pandanus radicans*. Individual lizards may learn the positions of favored food trees and revisit them year after year. Fruits are not chewed but are eaten whole. More sugary fruits are eaten from May to July and oily fruits in February–March and from August–November. Few fruits are available during the peak of the monsoon season (November–December), during which time these varanids prey mostly on animal food.

Reproduction

Sexual maturity is reached at about 450 mm SVL at an age of 3 years. Auffenberg (1988, 1994) reports that clutches average 7.1 eggs (n = 17), ranging from 4 up to 11. Ovary weights peak in August to October. Testes are largest from May to August. Male ritualized combat occurs in late April–June. Courtship behavior is complex and ritualized. Mating occurs in June–September. Eggs are laid between July and October, probably in hollow trunks and large limbs of both standing and fallen trees. Eggs are large and make up about 18.6% of a female's body weight. Double clutches have been laid in captivity and may also occur in the wild, although most females probably lay only a single clutch per year. Hatchlings are found from May to July, suggesting a long incubation period of up to 300 days.

Movement

Males are more active than females and move throughout the day, whereas females move most during the morning hours (Auffenberg 1988). These lizards have very small home ranges.

Physiology

Alimentary tracts in this species are unlike those of other monitors. The large intestine is longer than it is among carnivorous species and possesses a cecum in which microbial endosymbionts presumably facilitate digestion.

Fat Body Cycles

Fat bodies are generally large but exhibit some seasonality, being smallest from June to August and peaking in December–January, and may constitute up to 12% of a lizard's body weight (Auffenberg 1988). Females have relatively small fat bodies while yolking up their eggs.

Parasites

Two species of ticks, *Amblyomma helvolum* commonly (96% incidence) and rarely *Aponomma fibriatum,* are ectoparasites of these lizards (Auffenberg 1988). Eleven out of 12 parasitic ticks are males. Most male ticks are lodged at the base of the lizard's claws, whereas most female ticks are found on the skin between the toes of the front feet. These lizards are also infected with an intestinal nematode *Meteterakis vaucheri,* but most lizards have few nematodes. These monitors also harbor a small undescribed malarial plasmodium in their blood (Auffenberg 1988).

V. olivaceus populations are very localized, and they are listed as endangered on the IUCN Red List of Threatened Species.

6.17. *Varanus prasinus*

Harry W. Greene

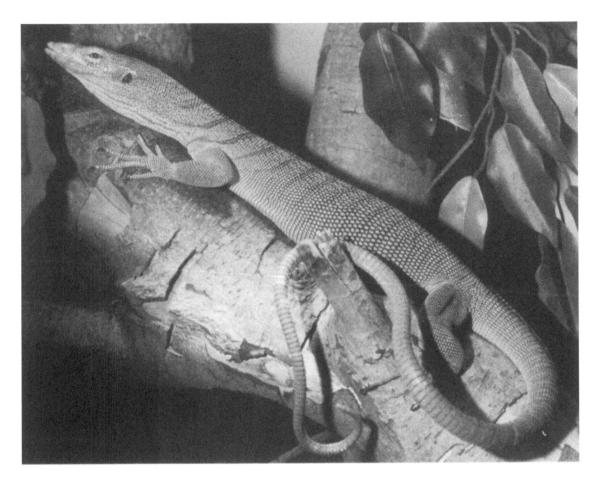

Figure 6.31. Varanus prasinus
(photo by Jeff Lemm).

Nomenclature

Varanus prasinus was named by Schlegel (1839), followed by the descriptions of *V. beccarii* (Doria 1874) and *V. kordensis* (Meyer 1874; see Mertens 1942 for synonymies and an extensive literature review). Mertens (1942) treated the latter two taxa as subspecies of *V. prasinus* along with the later described *V. p. bogerti* (Mertens 1950a). Sprackland (1991) reviewed geographic variation in this complex and synonymized *kordensis* with *V. prasinus*, elevated *V. beccarii* and *V. bogerti* to species status, and erected the names *V. telenesetes* and *V. teriae* for populations on Rossell Island off the coast of New Guinea and the Cape York Peninsula of Australia, respectively (Fig. 6.31).

Some workers have treated Sprackland's (1991) recognition of *Varanus beccari*, *V. bogerti*, *V. teriae* (in fact a synonym of *V. keith-*

Figure 6.32. Geographic distribution of Varanus prasinus, *which is found on New Guinea and on two islands between New Guinea and Australia.*

hornei, Wells and Wellington 1985), and *V. telenesetes* with skepticism, whereas others have followed him in recognizing those distinctive, allopatric populations as separate evolutionary species. Bennett (1998) followed Sprackland's (1991) arrangement, but Cogger (2000, p. 744) did not recognize *Varanus keithhornei*, of which *V. teriae* (Sprackland 1991) is a synonym, as distinct from *V. prasinus*. King and Green (1993) referred to *V. prasinus* as a species complex without citing Sprackland (1991) or mentioning his new taxa. Ziegler and Böhme (1997) examined hemipeneal morphology in specimens of *V. p. prasinus*, *V. p. kordensis*, and *V. p. beccarii* and concluded that they warranted subspecies status, whereas Jacobs (this volume) treats *V. kordensis* as a distinct species. Most recently, Böhme and Jacobs (this volume) recognized the distinctively colored population of *V. prasinus*–like monitors on Batanta Island, off the coast of New Guinea as *V. macraei*. In any case, very little is recorded about the biology of any of those taxa as separate from *V. prasinus* (but see Irwin, this volume), but future workers should consider geographic variation in habitat, diet, and other aspects of natural history in the light of the morphological variation underlying Sprackland's (1991) taxonomic treatment of the group.

Geographic Distribution

Members of the *Varanus prasinus* complex are found in appropriate habitats over much of the island of New Guinea, as well as certain offshore islands (e.g., Salawati; see Böhme and Jacobs, this volume), from sea level to at least 830 m in elevation (Greene 1986; Sprackland 1991). Populations on various islands and on the Cape York Peninsula of Australia are discussed by various authors as either *V. prasinus* or distinct species (see above, also Whittier and Moeller 1993; Cogger 2000) (Fig. 6.32).

Fossil Record and Evolution

There are no known fossils of the *Varanus prasinus* complex (Pepin 1999). Recent morphological and molecular systematic studies (Sprackland 1991; King et al. 1999; Ast 2001, and references therein) imply that these lizards are most closely related to *V. indicus* and its relatives, a clade that is in turn the sister taxon of *V. olivaceus*. Although most varanids climb well, the *V. prasinus* complex thus arose within a larger clade of substantially arboreal species. Among lizards more generally, the *V. prasinus* complex is perhaps morphologically and ecologically convergent with Old World chameleonids, the neotropical polychrotid genus *Polychrus,* and an east African lacertid, *Gastropholis prasina* (for a color photograph from life, see Spawls et al. 2002), in terms of independently derived arboreality, predation on saltatory and/or flying insects, green coloration, a prehensile tail, and perhaps foot specializations for grasping.

Diagnostic Characteristics

Lizards of the *Varanus prasinus* complex are unique among varanids in having a long (1.75 times SVL), highly prehensile tail, and mainly green or black coloration, and perhaps in specializations of the foot for grasping. Like other members of the *V. indicus* clade (Sprackland 1991), they have relatively long tails, long snouts, and long legs. *Varanus prasinus* (sensu stricto) differs from other members of the *V. prasinus* complex in having bright green coloration and in certain aspects of scalation.

Description

On mainland New Guinea *Varanus prasinus* (sensu stricto) is a brilliant green lizard, sometimes with a dorsal pattern of irregular black crossbands (for excellent color illustrations, see Bennett 1998; Cogger 2000), whereas closely related taxa on certain offshore islands and the Cape York Peninsula are almost uniformly black and *V. macraei* is black with blue ocelli (Böhme and Jacobs, this volume). As in some other species of small monitors, the claws of *V. prasinus* are exceptionally sharp for its size. Scale glands are found predominantly on the palmar, plantar, belly, thighs, and tail base regions (Andres et al. 1999).

Size

The largest recorded New Guinea *Varanus prasinus* had a SVL of 295 mm and TL of 845 mm (Loveridge 1948), and the slightly larger *V. beccarii* reaches approximately 95 cm in TL (Bennett 1998). Closely related *V. macraei* slightly exceeds 1 m in TL (Böhme and Jacobs, this volume).

Habitat and Natural History

Varanus prasinus and its closest relatives inhabit monsoon, rain, and palm forests, mangrove swamps, and cocoa plantations (Room 1974; Czechura 1980; Cogger 2000; Irwin, this volume), and it might be sympatric with at least four other species of *Varanus* in some parts of Irian Jaya (Philipp 1999a). As is the case with many cryptic, arboreal forest vertebrates, relatively little is known about the behavioral ecology of these lizards. A captive male *V. beccarii* responded to the defensive behavior of a nest-guarding female by dorsoventral flattening, hiding its head under a forelimb, and locomotor escape (Garrett and Peterson 1991). Krebs (1991, p. 227) included *V. prasinus* in a list "comparatively less specialized [monitor] species with a probably lower specialized learning ability"; this seems unlikely because emerald monitors live in a highly complex, three-dimensional habitats, exhibit diverse prey-handling tactics (see below), and might well show accompanying ethological specializations (cf. Robinson 1969; see also Horn 1999).

Terrestrial, Arboreal, Aquatic

All members of the *Varanus prasinus* complex are evidently highly arboreal (Greene 1986; Irwin, this volume).

Time of Activity (Daily and Seasonal)

Like other varanids, all members of the *Varanus prasinus* complex are presumably strictly diurnal, but little is known about their activity cycles (see Irwin, this volume).

Thermoregulation

Nothing is known about thermoregulation in free-living members of the *Varanus prasinus* complex.

Diet and Foraging Behavior

Stomachs of 29 museum specimens of the *Varanus prasinus* complex (including primarily *V. prasinus,* plus also a few *V. beccarii* and *V. bogerti*) contained 47 prey items, including 32 katydids, 3 coleopteran larvae, 3 unidentified insects, 2 grasshoppers, 2 roaches, 1 beetle, 1 centipede, 1 spider, 1 walkingstick (cf. *Eurycantha* sp.), and 1 murid rodent (cf. *Melomys moncktoni*). Modal relative prey size was about 1% of predator mass, and the relatively largest items were a 124-mm-long, 12.2-g stick insect in an adult monitor (SV 280 mm, 313 g); a 63-mm-long, 4.6-g katydid in a subadult (SV 150 mm, mass 41 g); and a 40-g rodent in an adult (SV 255 mm, 135 g). Prey typically were swallowed headfirst, and a large, spinose walkingstick evidently was dismembered before ingestion (Greene 1986). There is no evidence for plant matter in the diet of free-living *V. prasinus,* but a captive consumed bananas (Mertens 1971b). Captive members of the *V. prasinus* complex simply seize crickets and eat them, but young mice were seized by the nape, slammed against the substrate, raked with the claws, and swallowed headfirst (Greene 1986; Hartdegen et al. 1999, 2000).

Reproduction

Captive individuals of *Varanus beccarii* mated while hanging on the side of a cage (Horn 1999). In the field, two neonate *V. prasinus* with SVLs of 83 and 84 mm and obvious umbilical scars were found on October 23, and had reportedly "hatch[ed] in termite nests and fed there later to emerge"; their stomachs were empty but the colons contained several insect eggs, perhaps those of termites (Greene 1986, p. 5). Captive females of *V. beccarii* and *V. prasinus* (Greene 1986; Garrett and Peterson 1991; Horn and Visser 1991; Dedlmar 1994; Eidenmüller 1998; Bennett 1998; Bosch 1999a; Kok 2000) have laid clutches of two to four eggs, the maximum by an unusually large female (SVL, 286 mm, TL = 797 mm, 237.5 g immediately before oviposition). Captive clutches have been laid in January, March, April, November, and December; females may lay up to three clutches per year. Hatching occurs in June, September, October, and November after incubation periods of 154 to 190 days. One set of hatchlings had TLs of 205 and 210 mm and masses of 8.4 and 10.0 g; the other hatchlings weighed 10 to 11 g and were 22 cm long. Captives reach sexual maturity in about 2 years. A large captive female *V. beccarii* defended her nest box against human and conspecific intruders, before and after oviposition (Garrett and Peterson 1991).

Movement

Nothing is known about movement patterns in *Varanus prasinus*.

Physiology

Although there are no published studies on the physiology of monitors of the *Varanus prasinus* complex, three other species of arboreal varanids have significantly higher metabolic rates, irrespective of body mass, than do four terrestrial species (Thompson 1999). Because *V. prasinus* and its close relatives might be relatively slow-moving among arboreal varanids (e.g., *V. indicus*, as judged by tail prehensility in *V. prasinus*, Greene 1986), physiological studies of this species will be of great interest.

Fat Bodies, Testicular Cycles

Nothing is known about these aspects of the biology of *Varanus prasinus*.

Parasites

Bosch (1999b) reported parasite loads for a sample of 410 captive monitors from 26 species, including 46 *Varanus prasinus*, but the results are not summarized in terms of individual species (see also Upton and Freed 1990).

6.18. *Varanus rudicollis*

DANIEL BENNETT

Figure 6.33. The rough-necked monitor, Varanus rudicollis *(photo by Jeff Lemm).*

Nomenclature

Varanus rudicollis, first described as *Uaranus rudicollis* by Gray 1845 from a specimen labeled "Philippine Islands," was amended to *Varanus rudicollis* by Boulenger (1885). The species has also been known as *Varanus swarti* and *V. salvator scutigerulus*. Mertens (1942) assigned a unique subgenus, *Dendrovaranus*, to *V. rudicollis*, but Ziegler and Böhme (1997) placed it in the subgenus *Empagusia* along with *V. dumerilii* and *V. bengalensis* (Fig. 6.33).

Geographic Distribution

V. rudicollis is known to occur from southern Thailand and Burma through the Malay Pensinsula, Banka, Riau, Borneo, and Sumatra. It

Figure 6.34. Map showing geographic distribution of V. rudicollis (based on map in Bennett 1998 with permission of the publisher).

appears to be a forest dweller and has not been recorded from cultivated areas. The type locality is questionable because no further examples from the Philippines are known. A specimen labeled "Samar" in the Florida State Museum contained a north American rodent (Auffenberg 1988 and personal communication). However, it would not be at all surprising for another large monitor lizard to inhabit the Philippines (Auffenberg 1988; Gaulke and Curio 2001).

Werner (1900) recorded *V. rudicollis* from the interior of Indragiri, Sumatra. De Rooij (1915) gives the following locations: Malacca, Sumatra (Benakat in Palembang), Borneo (Matang), Kuching, Pangkalan Ampat, Sarawak, Baram River, Rejang River, Mt. Dulit (to 2000 ft), Bogon, and Samarinda. Lekagul (1969) gave its distribution in Thailand as being restricted to the south, i.e., below the Isthmus of Kra, and claimed that *V. rudicollis* was the rarest monitor in the country. Boonratana (1988) only found the species in two southern provinces, Thaleban National Park in Satun and Khlong Tom in Krabi, both areas of moist evergreen forest. Taylor (1963) recorded it from Ranong. In Burma the species is recorded from the Kamoukgyi Chaing Headwaters (Mertens 1950a). In Malaysia it is known from the Amapang Forest Reserve (Harrison and Lim 1957), the Sungai Dusun Forest Reserve, Selangor (Jasmin 1988) Pasoh Reserve, Negeri Sembilan, Bukit Lagong, Bukit Lanjan, Ulu Gombak, Sungkai, Ulu Langat, Fraser's Hill, Cameron Highlands, and the Tembeling River (Bennett and Lim 1995). Brandenburg (1983) examined material from Borneo (Samarinda, Sanga Danam, and Balikpapan) and Banka. According to Nutphand (n.d.), they are found only seldom in the southern parts of Thailand, deep in the forests and usually far away from all human habitations (Fig. 6.34).

Fossil Record

No fossils of this species are known.

Diagnostic Characteristics

- Enlarged, compressed, strongly keeled scales arranged in longitudinal rows on the neck.
- Long, narrow snout.
- Slit-shaped nares closer to the eye than the tip of the snout.
- Black body color.

Description

V. rudicollis possesses long fingers and strongly curved claws; preanal pores exist in both sexes. Body color is usually black, although green specimens are known in the pet trade. Lighter markings on the dorsal surface tend to fade with age. The skull is long and narrow with large numbers of sharp, straight teeth. The eye is large. The tail is laterally compressed with a low double keel, 127% to 161% of SVL. There are 139 to 169 scales at midbody and 79 to 90 scales from gular fold to the insertion of the hind limb.

Size

The largest recorded size of *V. rudicollis* in the wild is from a specimen in Malaysia (Ampang forest reserve) of 146 cm TL (59 cm SVL, 87 cm tail) with a weight slightly over four kilograms (Lim 1958). In Thailand, specimens are usually smaller than 1 m (Lekagul 1969); Nutphand (n.d.) gives the maximum TL as 130 cm. Mean size of 28 examined in Malaysia was 336 mm SVL, 345 g. In Sumatra, specimens of between 100 and 127 cm were seen (Werner 1900). In captivity, animals can become appreciably larger than this. Hatchlings measure about 25 cm TL and weigh about 21 g (Horn and Petters 1982).

Habitat and Natural History

V. rudicollis is only known from evergreen and mangrove forests. Ladiges (1939) found it in thick jungle and mangrove forest near a small coastal river in Sumatra. According to Lekagul (1969), the species is usually found in dense jungle rather than alongside rivers in Thailand. Bennett and Lim (1995) and Jasmin (1988) reported it from both primary and secondary forest in peninsular Malaysia.

Very little is known about the natural history of *V. rudicollis*. It is a shy animal from dense habitats that avoids human habitations. Its preferred shelter is a tree hollow. Auffenberg (1988) considered it a highly specialized insectivore, but observations from throughout its range suggest some plasticity in diet. The animal regularly feigns death when handled (Horn and Petters 1982; Nutphand, n.d.).

Terrestrial, Arboreal, Aquatic

Auffenberg (1988 and in litt.) considered *V. rudicollis* to be a highly specialized, arboreal animal. All food items in six specimens examined by him were tree-dwelling species. Other authors state that the animal shelters in trees but forages mainly on the ground. Individuals disturbed on the ground escape by climbing trees (Werner 1900; Barbour 1921; Ladiges 1939; Lim 1958; Jasmin 1988; Nutphand, n.d.; Nabhitabhata, personal communication).

Time of Activity (Daily and Seasonal)

Nabhitabhata (personal communication) was told by hunters that *V. rudicollis* was most active during the high-rainfall period of September to December in southern Thailand. Newly imported females have laid eggs in January (Horn and Petters 1982).

Foraging Behavior

V. rudicollis apparently finds most of its food by rooting in rotten plant matter, both on the ground and in trees. The long, narrow snout and slit-shaped nares evidently facilitate this behavior. Mertens (1942) claimed that it uses its tongue to lap up small insects such as ants and termites [Editors' note: this seems rather dubious to us]. Horn (personal communication) reported that captives could catch fast-moving fish from an aquarium.

Diet

Mertens (1942) believed that the diet comprised mainly ants and probably termites [Editors' note: dubious]. Werner (1900) found only insects in a stomach from Sumatra. Another from Thailand contained crabs (Nabhitabhata, in litt.). Six animals from Malaysian forests contained only large tree centipedes and phasmids up to 30 cm in length (Auffenberg 1988 and in litt.). Five stomachs from museum specimens examined by Losos and Greene (1988) contained large numbers of small prey; frogs, frog eggs, spiders, scorpions, Brachyura, Isopoda, insects (Blattoidae; Coleoptera; Orthoptera), and a mollusk. A single stomach examined by Brandenburg (1983) contained large cockroaches and grasshoppers. Jasmin (1988) reported that worms, insects, and birds eggs are believed to form part of its diet.

Reproduction

The only data are from captive individuals. Clutches average about eight eggs, with a maximum of 14, produced every 4 to 6 months (Meheffey, personal communication). Incubation takes 180 to 190 days at 29°C (Horn and Petters 1982). Successful hatching is very rarely reported in this species (Bennett 1995).

Nothing is known about thermoregulation, movements, physiology, fat body cycles, or parasites.

6.19. *Varanus salvadorii*
Hans-Georg Horn

Figure 6.35. Varanus salvadorii
(photo by Jeff Lemm).

Systematics and Taxonomy

The Papuan or crocodile monitor (*Varanus salvadorii*) was first described by Peters and Doria as *Monitor salvadorii* in 1878, and in 1885 it was renamed *Varanus salvadorii* by Boulenger (Peters and Doria 1878; Boulenger 1885; Mertens 1942, 1962, 1963). Because of several anatomical details (see below), Mertens erected a new monotypic subgenus *Papusaurus* within the genus *Varanus* of the Varanidae (Mertens 1962, 1963) containing one species, *V. salvadorii* (Fig. 6.35).

In more modern times, a new approach based on new morphological characteristics has been applied for solution of the relationships and systematics of the Varanidae. These methods used hemipeneal and lung morphology (Böhme 1988; Ziegler and Böhme 1997; Becker et al. 1989). Most of the results by hemipeneal morphology were in relative agreement with most of the subgenera erected by Mertens. As might be expected, the subgenus *Papusaurus* is part of the Indo-Australian lineage of monitors. But it is more closely related to a subclade that comprises the subgenera *Euprepiosaurus* and *V. salvator* (*Soterosaurus*) than to the Australian *Varanus* and *Odatria* (Böhme 1988). The hemipenes of *V. salvadorii* have a more primordial and in general symmetrical structure with a series of paryphasms, thus separating this species from all other species of monitors (Böhme 1988; Ziegler and Böhme 1997). Lung morphology (Becker et al. 1989) also supported the isolated position of this species among the Varanidae. But in contrast to

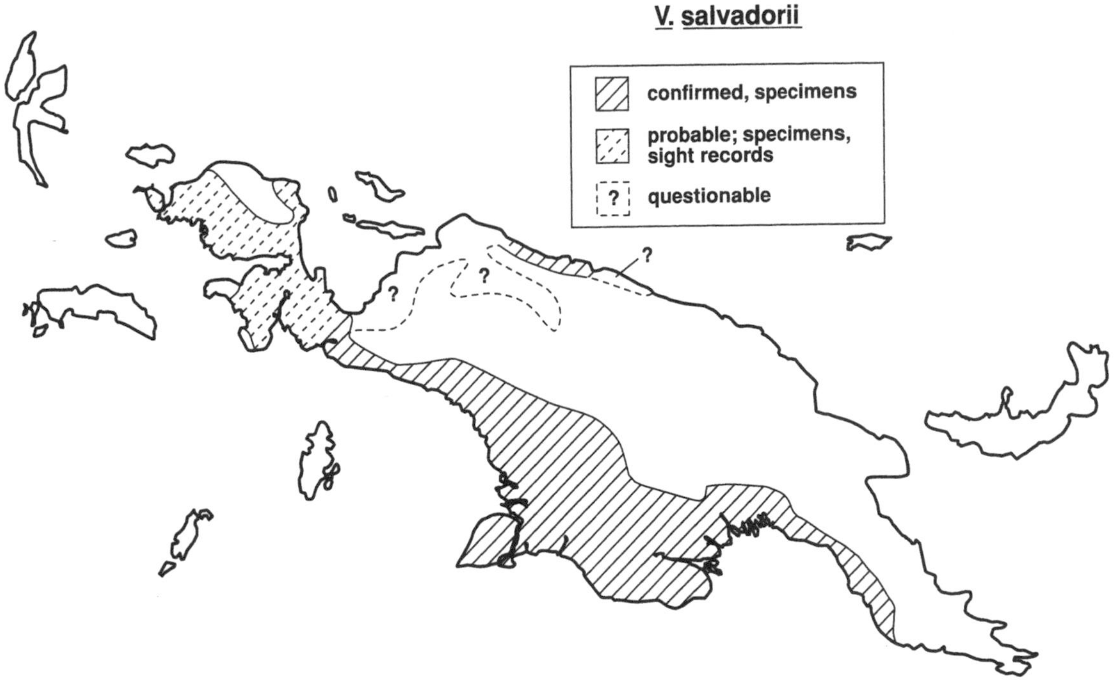

V. salvadorii

▨	confirmed, specimens
▨	probable; specimens, sight records
⌐?⌐	questionable

Figure 6.36. Geographic distribution of Varanus salvadorii *(map researched and created by Samuel S. Sweet).*

hemipeneal morphology, there is a relative close relationship to *V. varius* because of plesiomorphic characteristics.

Modern biochemical methods (microcomplement fixation analysis of serum albumins) applied by Baverstock et al. (1993) positioned *V. salvadorii* as part of the species cluster of Australian monitors of subgenus *Varanus,* including *V. komodoensis.* Although Australian monitors are closely related, *V. salvadorii* is placed outside of this cluster, but both the Australian species of the cluster and *V. salvadorii* do share a common ancestor (Baverstock et al. 1993; King et al. 1991, 1999).

Later on, the author has been informed (King and Pepin, personal communication) that one species (*V. varius*) of this cluster of Australian monitors has to be unified with *V. salvadorii,* thus modifying this part of the cladogram of the Varanidae. It seems *V. salvadorii* and *V. varius* have a common root, as was also proposed by Becker et al. (1989). On the basis of investigations of mitochondrial DNA of most of the extant species of monitors, Ast (2001) positioned *V. salvadorii* together with *V. komodoensis* and *V. varius* in a subclade of the Indo-Australian lineage. The results and conclusions based on different methods seem to be plausible and may be supported for several reasons: *V. salvadorii* inhabits mainly southern (?) (see below) New Guinea, which is separated (~90 km) only by the Torres Strait from the Australian mainland. This strait could be easily overcome by island hopping: sea level was lowered several times during glacial episodes during the last 120,000 years; thus, big and little islands could surface. Both species (*V. salva-*

dorii and *V. varius*) occupy a similar ecological niche: both are tree dwellers, but their similar appearance could be a result of evolutionary convergence.

Distribution

Unfortunately, no detailed data (census) based on field investigations are available. Mertens (1942, 1950a, 1959, 1960, 1963, 1971a) believes that this species inhabits New Guinea; according to Schultze-Westrum (1972), it is found only in southern New Guinea, and Allison (1982) believes this monitor is disjunctly distributed in New Guinean lowlands.

Varanus salvadorii has been collected or observed at the following locations in New Guinea: Dorei (northwest New Guinea), Lake Sentani, Kwawi, Jamur River (near Geelvink Bay), Setekwa River, Dinawa in Owen Stanley Range, St. Joseph River (De Rooij 1915; Mertens 1942). Near Setekwa River, this species reaches an altitude of 540 m above sea level (De Rooij 1915; Mertens 1942). Sturt Island in Fly River (Papua New Guinea), Hollandia (Dutch New Guinea, Mertens 1950a); Gulf district (southeast New Guinea, Mertens 1962); Kopi (village near Kikori, Gulf of Papua), Kikori River (Gulf of Papua), Aird Hills (Gulf of Papua, Mertens 1971a); Fosaker, Asmat District, east of Agats (West Irian), Eilanden River, Asmat district, east of Agats (West Irian; Cann 1974 and personal communication); southern Trans-Fly Western Province (O'Shea 1991) (Fig. 6.36).

Description

Detailed descriptions of *Varanus salvadorii* can be found in Mertens (1942, 1950a, 1950b, 1962), but also see De Rooij (1915). Only some major features will be given here. It is a long species of monitor, exceeding 2.5 m TL; for skull details, see Mertens (1942, 1950b). Its head is lightweight, the snout blunt and bent in profile, the nostrils nearer to the snout than to the eye and of a more rounded than slit-like form; it has several gular folds. The tail is extremely long—up to 2.7 times SVL—and is not laterally compressed; it is more rounded at the base, and more distally, it takes on a triangular shape with a small double keel (Mertens 1942, 1962; De Rooij 1915). The teeth are compressed, slightly bent, and unusually long, standing nearly vertical on jawbones, and are sharp and weakly sawn at the edge of teeth. Scales are weakly differentiated, with head scales flat and smooth, nuchals small and smooth, dorsal scales small and keeled, and tail scales larger ventrally than laterally, without rings.

Papusaurus can be separated from *Varanus* by its tail not laterally compressed, as in the subgenus, and by its many gular folds; and from *Odatria* by having a reduced parietal plate with increasing age and having nonringlike scales around the tail (Mertens 1960, 1962). *Papusaurus* can also be separated from *Odatria* by ethological differences. During ritual combats, *V. salvadorii* (and all other members of the subgenus *Varanus*) exhibit the clinch phase (see section "Behavioral

TABLE 6.1.
Morphometric data of *Varanus salvadorii.*[a]

Weight (g)	Head + Body[b] (cm)	Tail[c] (cm)	Total Length (cm)	Tail/Head + Body	Remarks
—	85	166[d]	255[e]	2.00	ZFMK 47533
—	78	153	231	1.96	ZFMK 47534
6380	74.5	149.5	224	2.01	Long-term captive
1360	51.0	116.5	167.5	2.28	Data at arrival
1250	47.8	100.8	148.6	2.11	Data at arrival
1000	44.5	120.0	164.5	2.7	Data at arrival
1440	51.8	109.5	163.3	2.11	Data at arrival
400	37.7	78.8	116.5	2.09	Data at arrival
433	36.0	75.0	111.0	2.08	Data at arrival
500	35.0	81.0	116.0	2.31	Data at arrival

[a] ZFMK, specimen of Museum König, Bonn.
[b] Tip of snout to part back of hind leg.
[c] Part back of hind leg to tip of tail.
[d] Four centimeters of tail is missing.
[e] Four centimeters added.

Observations in Captivity"), which is never taken by members of *Odatria* in such interactions (Horn et al. 1994).

Coloration is dark blackish, brown, or deep black with irregular yellow spots and ocellae and yellow rings around the tail. Color variants may occur depending on locality.

Morphometrics

Varanus salvadorii has been said to exceed eventually 4.5 m in TL (Schultze-Westrum 1972; Cann 1974). This has never been substantiated (Schmicking and Horn 1997). The largest reliable measurement of a ZFMK voucher specimen is 2.65 m (Böhme and Ziegler 1997b). Some other reliable morphometric data can be found in Table 6.1. Data are taken mainly from Schmicking and Horn (1997); compare also Horn (2002a).

Habitat, Natural History, Diet

Little is known on the habitat of *Varanus salvadorii*. Schultze-Westrum (1961) believes it a dweller of delta estuaries of big rivers (Fly, Wawoi, Kanuwe, Kikori Rivers) in southern New Guinea. This is confirmed by the origin of the first living specimen that came to Germany. This specimen was collected from Aird Hills in the Kikori River, covered by rain forest. Mertens, who later (1960) reported on this animal, denotes this area as hot-humid swampy forests of the gulf district.

Most information on the habitat of *Varanus salvadorii* mentioned above are conclusions. Fortunately, Cann, a well-experienced expert of Australian herpetofauna and an internationally recognized specialist in turtles, saw *Varanus salvadorii* during a collecting trip of reptiles (Cann 1974) in the Asmat district with its central town of Agats in Irian Jaya (West Irian).

One animal of about 60 cm in TL was caught from a sago palm (*Metroxylon sagu*) in thick rain forest, but not in a swampy vegetation. Two further specimens were encountered near the village of Fosaker when Cann (1974) and his companions were paddling along the Eilanden River. These monitors disappeared in the canopy of thick rain forest vegetation (again, not in swamplands). Cann (1974) speculates on the quickness of movements of this species, comparing it to the North American six-lined race runner (*Cnemidophorus sexlineatus*). He notes, "However, after seeing a 1.6m *V. salvadorii* leap from tree to tree, at 7.6 m from the ground, then scuttle down a tree and disappear on virtually clear ground, I believe this species may be faster" (Cann 1974). In 1996, Cann confirmed these observations with these words: "I never have seen more agile monitors. They move and leap like nothing else I saw do" (Cann, personal communication).

In general, monitors have unusual areobic properties and therefore, they even have been called mammal-like by several physiologists. In this context, *V. salvadorii* may be at the top among this series of reptiles. Very recently, staying power has been found and investigated in *V. exanthematicus* by Owerkowicz et al. (1999), who demonstrated that gular pumping operates like an injection pump to force oxygen into the lungs. All members of the subgenus *Varanus* and *Papusaurus* have the same or a very similar respiration mechanism, which explains the staying power of these animals. Energy costs of free-ranging *V. salvadorii* moving around in trees of rain forests must be very high, and gular pumping doubtlessly contributes to paying these costs.

The extreme agility can be confirmed by observations of *V. salvadorii* in terraria: I was about to feed a rat hanging from a pincer to this monitor. The distance from the monitor's head to the rat was about 1.1 m. I kept a sharp eye on this individual's head to register its first movement to catch the rat, but I could not. In a blink of an eye, the rat was gone, and I did not even see the monitor move! The muscular front and hind legs and their strong claws cooperate perfectly when these monitors move quickly. Papuan monitors are highly specialized tree dwellers.

Nothing is known precisely about the diet of free-ranging specimens because no investigation of stomach contents has yet been carried out. Villagers gave details about deer, pigs, and hunting dogs being brought down by *V. salvadorii*. O'Shea (1991) reports that even humans have been attacked. Similar information can be found in Schultze-Westrum (1972), who also claims that such prey is hauled up into the canopy and eaten there. On the other hand, Allison (1982) mentions prey more typical of monitors, e.g., insects, lizards, birds, mammals, and carrion for the crocodile monitor only. And Auffenberg (1981) mentioned chicks and eggs belonging to the diet of this species, but it remains unclear on what this information is based.

In captivity, *V. salvadorii* have been fed by 600 g meat and some eggs (specimen ~2.4 m TL) per 3 days (Mertens 1960), 10 to 20 mice, rats (?), or chicks per week (length not given; Sasse 1999), rabbits (up to 2 kg), guinea pigs, hamsters, and canned dog food (~2 m TL, Madsen 1990). Long-time keeping and breeding of the Papuan monitor is based on chicks enriched with trace elements and vitamins (up to 25 for a big specimen of 2.3 m TL per week), rats, mice, eggs, and fish (e.g., *Abramis brama*) (Schmicking and Horn 1997). First foods eaten after arrival of wild-caught specimens are noteworthy: one lizard, about 170 cm in TL sitting on a keeper's knees or legs and only held slightly by its tail, accepted chicks without force-feeding and without becoming agressive (Schmicking and Horn 1997). More details on light requirements (intensity of light) of this species can be found in Horn (2002b).

Behavioral Observations in Captivity

Nearly all behavioral observations reported here are anecdotal. Detailed and systematic investigations are missing.

Male *V. salvadorii* cannot be kept together without damaging fighting and without an interval of ritual fight, as in other species (see Horn 1994). Even in the very big cage of the rain forest house at Cologne Zoo, two males had to be separated shortly after introduction because they immediately started fighting (Pagel, personal communication). But the remaining individual tolerated four sail-tailed water lizards (~1 m TL) (*Hydrosaurus amboinensis*) without attacking them as prey over a period of 3 months up to now (Pagel, personal communication).

Keeping together males and females of approximately the same size is possible, but if a female is not in estrous and the male tries to copulate, the male's sharp claws may cause severe injuries if the female tries to escape from the male's embrace (Schmicking and Horn 1997).

Papuan monitors usually rest and bask on big branches in terraria. They sleep on the ground, but sometimes they sleep in unsual places. I observed a crocodile monitor of about 1.2 m TL submerged with its eyes closed in a water tank of the cage. Unfortunately, when I tried to take a picture of this, the specimen opened its eyes (Schmicking and Horn 1997). Such behavior may appear to be based on conditions of captivity, but Traeholt (personal communication), who did a lot of research on free-ranging *V. s. salvator* on an island in the Yellow Sea, was able to take exciting shots and even filmed *V. s. salvator* submerged on the sea bottom, sleeping there with eyes closed. Such behavior in both cases could be related to thermoregulation.

Food uptake by the Papuan monitor is slightly different in comparison with other species of similar size. These other species (e.g., *V. s. salvator* or *V. varius*) shake a mouse less violently than a rat. They seem to estimate the possible risk to them by their prey. In contrast, *V. salvadorii* shakes a rat less violently and in many cases kills it like lightning in one mouthful. They seem to be experienced with the capabilities of their sharp, long teeth, their powerful jaws, and their huge throat (Horn 1999). Normally, prey items such as rodents or insects are chased directly, but *V. salvadorii* behaves differently, sometimes demonstrating superior strategy: an adult rat was offered *V. salvadorii*. The rat was seized as usual, but it escaped from the lizard's mouth. It ran along the right side of an upright big trunk. Instead of tracking the rat directly, the monitor came along the left side of the trunk and seized the oncoming rat (Horn 1999).

Many species of monitors use lashes of the tail, e.g., *V. gouldii*, *V. salvator*, or *V. varius*, as warning reaction or for defense. In *V. salvadorii*, this behavior is less frequent. The tail is rolled up less completely and lashing is less frequent. A strange kind of warning posture is sometimes shown: the tail is rolled up like a lasso, lifted up a little, and carried behind the walking individual. The same observation has been made earlier in *V. varius* (Horn 1980). If one has seen *V. salvadorii* leaping from one branch to another—the tail extremely extended—an observer immediately understands the purpose of its extreme length (up to 2.7 times SVL): The tail helps the animal not go head over heels. It is a counterbalancing steering gear analogous to the long sticks attached to firework rockets. It is not a prehensile organ for hanging from branches.

A strange behavior—never seen by the author of this contribution of his own Papuan monitors—has been reported by Madsen (1990). If a person came near to the terrarium, the two animals turned around, showing their back and tail, and then defecated. Some time later, when they became more familiar with such situations, they only raised on their forelegs and started hissing. Madsen (1990) also reports that his two specimens—one about 2 m, the other slightly smaller and both assumed to be males—started fighting after a short time of settling down. After some time, the smaller one, being more agile, won most of these wrestling combats. At the end of such combats, it regularly mounted the bigger specimen, thus demonstrating its superior power. It also showed head weaving, like in true courtship, and in some cases, both specimens stood in an upright position, showing the clinch phase of such combats. Later on, when a female was introduced into the cage, the superior male monitor started head weaving and mounting of the

TABLE 6.2.
Data on eggs of different clutches of *Varanus salvadorii*.

No. of Eggs Laid	Date	Average Weight (g)	Average Length (cm)	Average Width (cm)	Ref.[a]
6	Oct 1989	43.3	7.5	3.4	1
12	Aug 1990	45.3	7.9	3.4	1
4	Nov 1995	60.8	8.8	4.0	2

[a] References: 1, Philippen (1994); 2, Schmicking and Horn (1997).

new female monitor but continued to mount the bigger male too (Madsen 1990). The big male could actually have been a female.

Schmicking and Horn (1997) had two breeding pairs in their terraria. Both pairs showed social interactions of a different kind. Between December and middle of February, one of the breeding pairs exhibited ritual combat, which included head weaving, the clinch phase, and a kind of wrestling. During the clinch phase, the (bigger) male continued head weaving (Schmicking and Horn 1997). Courtship behavior in the second breeding group was demonstrated first by a male rushing to a female, pushing her flanks with the tip of his snout, scratching her pelvic girdle, and head weaving. Later on, the same male mounted the female and intensively investigated her neck by darting his tongue in and out. Then the male clung by his hind legs to the female's pelvic girdle. After several attempts of this type, the female raised her tail and copulation took place (Schmicking and Horn 1997). Such behavior is very similar to that of *V. komodoensis*, as has been documentated by Auffenberg (1978). Copulation took place in a hanging position at the wall; for further details see Schmicking and Horn (1997).

In monitors, hisses are used as a warning sound. Hisses of *V. salvadorii* and a series of other monitors have been analyzed acoustically. Three different types of hisses have been established in the various species of monitors: (1) long-duration hisses with frequencies of similar amplitudes, (2) hisses with frequency gaps and low-amplitude frequencies, and (3) short-duration hisses with a broad frequency span (Young et al. 1998). Hisses of *V. salvadorii* belong to that of the third group. In this species, the frequency span of hisses is larger than in other species (Young et al. 1998).

Reproduction in Captivity

Nothing is known about reproduction of this species in the wild. Reproduction in captivity is based on an increased number of social

TABLE 6.3.
Breeding data of *Varanus salvadorii*.[a]

Eggs Laid			Incubation			Hatchlings					
Date Mating Observed	Date	No. Laid (No. Infertile)	Period (d)	Temp.(°C)	Medium	No.	Date	Length (cm)	Weight (g)	Institution (Ref.[b])
17 Jul 1991	28 Aug 1991	9 (8)	176	28.3–28.9.	Peat-vermiculite	1	18 Feb 1992	49.34	55.02	Gladys Porter Zoo(1)
10 Oct 1994	12 Nov 1994	6 (5)	203	29	Vermiculite	1	3 Jun 1995	42.5	47	(2)
Yes	22 Oct 1995	6 (5)	201	29	Vermiculite	1	10 May 1996	42.0	48	(2)
NI	24 Jul 1996	6 (0)	218–229	NI	Vermiculite	6	27 Feb–11 Mar 1997	NI	NI	Fort Worth Zoo (3)
No	17 May 1998	6 (1)	233–246	29	Perlite	5	5–8 Jan 1999	45.3–47.5	62.1–68.1	Honolulu Zoo (4)
No	12 Nov 1998	4 (0)	211–246	NI	NI	4	11–16 Jun 1999	NI	NI	Honolulu Zoo (4)

[a] NI, not indicated.

[b] 1, Hairston Adams (1996);
2, Schmicking and Horn (1997);
3, Hudson (Forth Worth Zoo, personal communication);
4, Meier (Honolulu Zoo, personal communication).

interactions. Clutches (infertile) have been deposited between August and January (Table 6.2).

Table 6.2 shows that clutch size varies between 4 to 12 eggs. The eggs show amazing differences in weight and dimensions. At present, no explanation for this is known. Three other clutches were deposited on November 12, 1994 (6 eggs), on October 22, 1995 (6 eggs), and on August 4, 1996 (10 eggs, 2 with unusual dimensions: 10 by 4.5 cm). Further dates of egg deposition are given in Table 6.3 (Schmicking and Horn 1997).

Egg deposition has been observed frequently in captivity, but most clutches contained infertile eggs. Only four successful breedings have been documented so far in the literature (Table 6.3) (Horn and Visser 1997; Schmicking and Horn 1997).

There are amazing differences in weight and less in TL of the hatchlings (Table 6.3). From data in Tables 6.2 and 6.3, egg deposition takes place in the second and hatching in the first half of the year.

Above: *Varanus acanthurus* (photo by Jeff Lemm).

Below: *Varanus baritji* (photo by Jeff Lemm).

Varanus brevicauda (photo by Eric Pianka).

Varanus caudolineatus (photo by Brad Maryan).

Varanus eremius (photo by Hal Cogger).

Varanus giganteus (photo by Eric Pianka).

Varanus gilleni (photo by Eric Pianka).

Varanus glauerti (photo by Jeff Lemm).

Varanus glebopalma (photo by Samuel S. Sweet).

Above: *Varanus gouldii*
(photo by Jeff Lemm).

Left: *V. g. flavirufus*
(photo by Eric Pianka).

Below: *Varanus keithhornei*
(photo by Jeff Lemm).

Above: *Varanus kingorum* (photo by Robert Browne-Cooper).

Below: *Varanus mertensi* (photo by Robert Browne-Cooper).

Varanus mitchelli (photo by Robert Browne-Cooper).

Varanus panoptes. (upper) *horni.* (lower) *rubidus.* (Both photos by Jeff Lemm.)

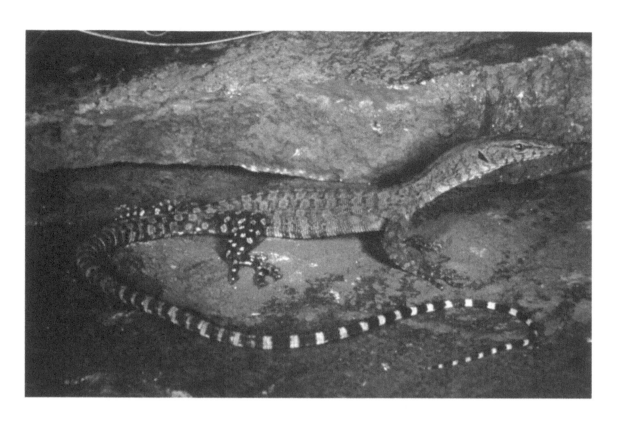

Above: *Varanus pilbarensis* (photo by Brad Maryan).

Below: *Varanus primordius* (photo by Brad Maryan).

Varanus rosenbergi (photo by Dennis King).

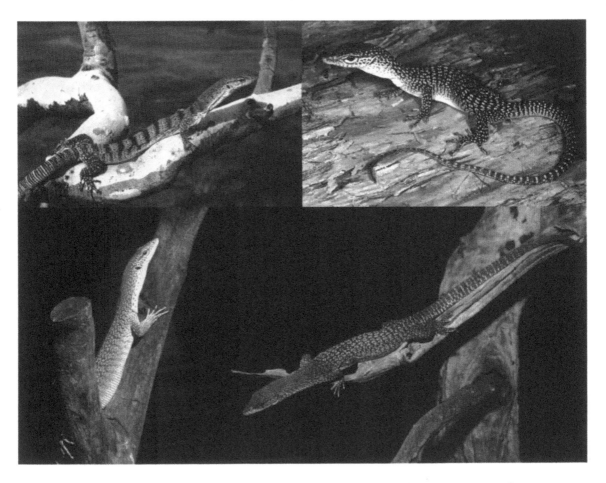

Various morphs of "*Varanus scalaris*," clockwise from upper left: *V. scalaris* (sensu stricto) from Beagle Bay (photo by David Knowles), from near Kununurra (photo by Robert Browne-Cooper), from Darwin (photo by Jeff Lemm), and the "*V. similis*" morph from Kakadu (photo by Eric Pianka).

Varanus semiremex (photo by Jeff Lemm).

Adult and juvenile *Varanus spenceri* (photo by Gavin Bedford).

Varanus storri (photo by Richard Bartlett).

Varanus tristis (photo courtesy of Dr. Harold Cogger).

Varanus varius (photo by Brian Weavers).

Heloderma horridum (photo by Dan Beck).

Heloderma suspectum
(photo by Dan Beck).

Lanthanotus borneensis (photo by Alain Compost, courtesy of Frank Yuwono).

Among all 11 clutches, clutch size varies from 4 to 11 eggs (average, 7.0 eggs). As in other species of monitors, hatchlings are more brilliant in color and they feed more on insects and small reptiles (Schmicking and Horn 1997; Allison 1982)

Parasites, Diseases

Newly obtained crocodile monitors can be heavily infested by ticks. I had to remove hundreds of them (species not determined) from breast, back, cloaca, and armpits. Application of a low concentration solution of Sebacil (Bayer) in water stopped the infestation immediately (Horn 1995; Schmicking and Horn 1997).

Fecal examinations conducted on freshly imported *V. salvadorii* resulted in identification of enteric parasites as flagellates, mite ova, cestodes, and strongyles. Normally these do not seem to cause any significant health problems (Hairston Adams 1996).

Monitors are usually infected by different species of worms. One of the animals discussed above was infested by a species of cestode (*Duthiersia*; Bosch, personal communication).

When I obtained several Papuan monitors, some of them showed severe mouth rot (*Stomatitis ulcerosa*). It could be cured relatively simply by applying of a broad-activity antibiotic (Schmicking and Horn 1997). For a detailed discussion on mouth rot in different species of reptiles including *V. salvadorii,* see Mader (1994). One of these animals died because of chronic enteritis accompanied by acute colitis (Schmicking and Horn 1997).

Review

Some time ago, an interesting review appeared (Bayless 1998). According to Bayless, New Guinea natives consider these monitor lizards to be an "evil spirit that climbs trees, walks upright, breathes fire, and kills men." One native hunter opted out of an expedition to capture one of these monitors using the excuse that he "had to go stay with his mother in law!" (Bayless 1988, p. 45).

Status

Nothing is known about the status of *V. salvadorii* in the wild. Philippen (1994) estimated the number of specimens collected for museums and/or zoos to be 30 to 50 over a period of 100 years. Thus, exploitation of this species was near zero until the 1990s. *V. salvadorii* is listed on appendix II of the Washington Convention.

Acknowledgments

Some colleagues supported this review by sharing their unpublished information and/or data with the author. Thanks therefore are due to H. Bosch, Düsseldorf; J. Cann, Sydney; R. Hudson, Fort Worth; D. Meier, Honolulu; T. Pagel, Cologne; and C. Traeholt, Copenhagen.

6.20. *Varanus salvator* (Nominate Form)

Maren Gaulke and Hans-Georg Horn

Figure 6.37. Varanus salvator salvator *(photo by Daniel Bennett).*

Systematics

The phylogenetic position of *V. salvator* within the Varanidae is still under discussion. Different conclusions are reached, depending on the method used. On the basis of external morphological features, *V. salvator* is considered a member of the subgenus *Varanus* by Mertens (1942). Later on, genital morphological investigations were carried out by Branch (1982), Böhme (1988), and Ziegler and Böhme (1997), and studies of its lung morphology by Becker et al. (1989). On the basis of his findings of an increasing complexity of the hemipeneal structure, Böhme (1988) placed *V. salvator* in a new subgenus. This subgenus was named *Soterosaurus* by Ziegler and Boehme (1997) (Fig. 6.37).

Karyological investigations of numerous members of the Varanidae, including *V. salvator*, were conducted by King and King (1975) and by Smet (1981). King and King (1975) arranged the karyotypes of the investigated varanids in six different groups, groups A to F. The *salvator* group (group C), or more precisely their karyotype, "is regarded as being the extant primordial type" (King and King 1975). On the basis of their findings, the authors prepared a map of the worldwide distribution of each of these karyotypes. The *salvator* type overlaps in the west with the *griseus* type and in the east with the *varius* type (King and King 1975). The number of chromosomes of all varanids, including *V. salvator,* is 2n = 40 (King and King 1975; Smet 1981).

Biochemical investigation methods (Holmes et al. 1975; King et al. 1991; Baverstock et al. 1993) modified these findings to some extent (changing the position of the *salvator* type between group C or B of the above chromosomal investigations).

Molecular methods (Fuller et al. 1998; King et al. 1999; Ast 2001) have also been used to clarify biogeographic and interspecific relationships. Although Fuller et al. (1998) and King et al. (1999) recognize *V. salvator* as belonging to an Indo-Asian subclade together with *V. dumerilii, V. bengalensis, V. olivaceus,* and *V. prasinus,* Ast (2001) splits this subclade into two parts (Indo-Asian A and Indo-Asian B). Indo-Asian A comprises *V. rudicollis* plus *V. salvator, s. bivittatus, s. togianus,* and *s. cumingi* (Ast 2001).

Varanus salvator Group

Water monitors, being extremely widespread throughout South and Southeast Asia and very common at least in parts of their distribution range, on first view might seem rather well known members of the varanidae. But actually they are in urgent need of a careful taxonomic reinvestigation. Up to now, the validity of some subspecies is doubtful; others might soon be elevated or reelevated to species status. The most comprehensive examination of the group was conducted by Mertens, who distinguished six subspecies in 1942 (*salvator, togianus, cumingi, nuchalis, marmoratus,* and *scutigerulus*). The only specimen of *scutigerulus,* described as *Varanus scutigerulus* by Barbour in 1932, was not seen by Mertens (1942). Reexamination showed that it actually is a *V. rudicollis* and therefore had to be removed from the *salvator* group. The form *bivittatus,* which was considered a synonym of *V. s. salvator* by Mertens (1942), was later given subspecific status (Mertens 1959). From the water monitor subspecies described or redescribed by Deraniyagala (1944, p. 47) (*andamanensis, kabaragoya, macromaculatus, nicobariensis,* and *philippinensis*), only *andamanensis* is widely considered as valid; the others were synonymized with the nominate form (*macromaculatus, kabaragoya, nicobariensis*) or *V. s. marmoratus* (*philippinensis*), respectively. Although for some subspecies (especially Philippine ones) new data were accumulated during the past decades, knowledge on others has made no or little progress since then. Besides, undescribed forms that definitely belong to the *salvator* group turn up via the pet trade once in a while, often with unknown origin. Before a satisfying state of knowledge on phylogenetic relations within the water monitor group is reached, a lot of field research and museum work will have to be completed.

Taking into account that some of the *salvator* forms might soon be elevated or reelevated to species status, it seems appropriate to present each of the widely accepted subspecies separately. However, as a result of the extremely different amount of available data for each form, the accounts will vary considerably in completeness.

Figure 6.38. Geographic
distribution of Varanus salvator
subspecies. Arrow points to the
Andaman Islands.

Legend:
- andamanensis
- bivittatus
- cumingi
- marmoratus
- nuchalis
- salvator
- togianus

Varanus salvator salvator

As a result of the enormous distribution, dozens of local names
exist for *Varanus salvator salvator*; if one is mentioned, it should be the
one of the designated terra typica, which is "Kabaragoya," but better
known in Malayan as "Biawak."

Nomenclature

Varanus salvator salvator was described by Laurenti in 1768 as
Stellio salvator. The given type locality, "America," is an obvious error;
in 1959 Mertens designated Ceylon as terra typica. A long list of
synonyms exist for this varanid (see Mertens 1942). Cantor (1847) was
the first to use the name *Varanus salvator,* and Mertens (1937) introduced the subspecies name *V. salvator salvator.*

Distribution

V. s. salvator has the largest distribution area of all recent varanids
(Fig. 6.38). It is recorded from following countries: Bangladesh, Brunei,
Burma, China, India, Indonesia, Kampuchea, Laos, Malaysia, Singa-
pore, Sri Lanka, Thailand, and Vietnam (Smith 1932). In the large
mainland countries India and China, *V. s. salvator* are restricted to the
northeastern regions (West Bengal, Orissa, Assam; main distribution
area are the Sunderbans) and the southwestern areas (Guangdong,
Guanxi, Hainan, Hong Kong, southern Yunnan), respectively. In other
countries, *V. s. salvator* are more widely distributed. Individuals reported from Taiwan are considered to have been introduced (e.g., Zhao
and Adler 1993). Bustard (1970) reports that *V. salvator* already has
reached the northern tip of Australia, but this never has been confirmed,
and Mertens (1942) thought it was impossible.

Diagnostic Characteristics

V. s. salvator are very large varanids, reaching more than 2 m TL. Narial openings are roundish to oval, much closer to the tip of the snout than to the eye. Its head is much longer (about two times) than broad; the snout is long with rounded tip, and tympanum is large. Tail is 1.36 to 1.65 times as long as SVL (tail length shows ontogenetic and sexual dimorphism: shorter in older ones and longer in males; Mertens 1942; Shine et al. 1996), laterally compressed, with a double keeled upper edge. Head scales are relatively large, flat, and smooth. Animals have four to eight well-differentiated supraoculars and 48 to 60 scales from rictus to rictus in a straight line above head. Nuchal scales are smaller than occipitals and the same size or slightly larger than dorsals. There are a total of 137 to 181 midbody scale rows and 80 to 95 ventral scale rows from the gular fold to the insertion of hind limbs. Normally, on both sides, one to two well-differentiated preanal pores exist. Medium-sized front and hind limbs have strong and curved claws (Mertens 1942).

Description

V. s. salvator have dark gray heads with some amount of whitish pattern. Neck, back, and dorsal side of tail are dark gray, with transverse rows of whitish ocelli with dark centers and irregular whitish network across back. Whitish crossbands exist on the tail. The chin is whitish with dark bands, the belly whitish with dark markings ventrolaterally, and the ventral side of the tail is whitish with dark transverse bands. The color pattern varies individually, especially the amount and distinction of the whitish markings across the backs. Geographic variation in color pattern is reported too. For example, animals from Sri Lanka are rather brightly patterned, but especially from coastal areas, very dark forms are reported (Luxmoore and Groombridge 1989). It remains to be investigated whether different forms are summarized under *V. s. salvator* at present. Juveniles show a brighter color pattern than adults and become darker during ontogeny.

Morphometrics

V. s. salvator is the largest member of the *salvator* group, and includes the longest varanid ever reported (a TL of 321 cm for a *V. s. salvator* from Sri Lanka; Randow 1932). Free-living specimens can reach a weight up to 25 kg (Jasmin 1989; Traeholt 1995a). However, individuals with a TL of around 300 cm and of weight of more than 15 kg obviously are very rare. Among 1237 *V. s. salvator* measured in Perak, Malaysia (caught for the skin trade), the longest individual was a male of 225 cm TL; the longest female had a TL of 175 cm (Khan 1969). Females reached a weight up to 7.28 kg, males up to 14.1 kg. On average, males were heavier than females of the same size class. However, the weight of females varied greatly with their reproductive condition (Khan 1969). The mean TL for 42 females measured in a skinnery in South Sumatra, Indonesia, was 149.6 cm, and for 80 males was 142.7 cm (Shine et al. 1996). The largest of the measured animals was a male with a TL of 203 cm. The mean TL for 165 individuals (males and females) measured in the field in North Sumatra, Indonesia, was

TABLE 6.4.
Morphometric data for *Varanus s. salvator.*[a]

No.	Sex	Weight (kg)	Total Length (cm)
1	?	15.5	221 (1)
2	?	7.8	180 (1)
3	?	16.0	200 (1)
4	?	19.7	213 (2)
5	?	17.5	176 (3)
6	?	12.8 (4)	195
7	?	13.5 (4)	210
8	?	3.2 (4)	134
9	?	5.8 (4)	177
10	?	12.4 (4)	195
11	?	2.8 (4)	157
12	?	7.4 (4)	177
13	?	3.0 (4)	139
14	?	15.8 (5)	180

[a] References appear after values in parentheses and are as follows: 1, Traeholt (1997a); 2, Jasmin and Amin Abdullah (1987/88); 3, Lederer (1929); 4, Traeholt (1997c); 5, Traeholt (1994b).

126 cm (Erdelen et al. 1997). The largest specimen was a male with a TL of 218 cm, and the heaviest individual weighed 13.3 kg. Average length of 55 individuals measured in a skinnery in the same area was 140 cm (Erdelen et al. 1997). The largest of 60 *V. s. salvator* measured in the field in West Kalimantan, Indonesia, had a TL of 224 cm; the average length was 143 cm (Auliya and Erdelen 1999; Erdelen et al. 1997). The heaviest individual weighed 12.7 kg. The average length of 29 *V. s. salvator* measured in a skinnery in the same area was 110 cm (Erdelen et al. 1997).

The largest male measured by Traeholt in Malaysia (1997a) had a TL of about 210 cm (SVL 92 cm) and a weight of 15.5 kg. The largest female measured 180 cm (SVL 76 cm) and weighed 7.8 kg. In the same study, Traeholt determined that individuals from Tulai Island, Malaysia, have larger average weights and body sizes than in mainland populations.

TABLE 6.5.
TABLE 6.5.
Growth in captivity over a period of 10 years.[a]

Age(y)	0[b]	0.75	2	3	4	5	6	7	8	10[c]
Total length (cm)	25	50	85	110	138	156	164	171	173	189

[a] Data from Lederer (1929).
[b] At arrival.
[c] Information by Lederer to Flower (1937).

Table 6.4 contains results of some additional measurements.

In captivity, growth development has been studied by Lederer (1929) over a period of 10 years, shown in Table 6.5.

Data from skinneries may underestimate the actual size structure of the population. Large individuals are often unfit for the skin trade because of their heavily scarred skins (Luxmoore and Groombridge 1989). However, no significant difference can be noted between data taken in skinneries (Khan 1969; Shine et al. 1996; Erdelen et al. 1997) and in the field (Erdelen et al. 1997; Auliya and Erdelen 1999). Measurements taken in skinneries have the advantage that sex determination is reliable, whereas this still is a big problem in field studies (e.g., Auliya and Erdelen 1999).

Only few long-term field data on growth rate of weight and/or length exist for *V. s. salvator*. In the course of field surveys in Sungai Tembeling, Pahang, Malaysia, three water monitors were recaptured after almost 4 years. The weight growth rate among the three animals varied significantly. Although one had an average weight growth rate of 1200 g per year during the 4 years, the second animal had an average weight growth rate of 1000 g, and the third of only 400 g per year (Jasmin and Amin Abdullah 1987/88). Only for the first animal were length measurements also provided. It increased its size from 101 cm TL to 166 cm TL within the 4 years, which means an average annual length growth rate of 22 cm (Jasmin and Amin Abdullah 1987/88).

V. s. salvator has been a frequent subject of anatomical studies. Because of restricted space here, only very brief information on this subject can be given.

Skin

The skin of *V. s. salvator* was and is of great importance to the tanning industry and leather trade. Skin of *V. s. salvator* can be easily distinguished from that of, e.g., *V. bengalensis* and *V. flavescens*; that of *V. s. salvator* shows a medium-sized grain (oval scales with a black bump), whereas *V. bengalensis* has a fine-sized grain (flat scales) and *V. flavescens* has a coarse one (Fuchs 1974). The bumps contain most of the black pigmentation, which is distributed by 55% to 60% in the corium and by 40% to 45% in the epidermis (Fuchs 1977) of *V. s. salvator*. Every scale is surrounded by numerous small granulae, and osteoderms have been noticed (Fuchs 1977).

Skull and Dentition

Most intensive and comprehensive studies on the skull of monitors including *V. s. salvator* have been carried out by Mertens (1942). He determined numerous indices of the cranium (e.g., cranium–length index, cranium length–width index, length–weight index of maxillae, width index of parietal plate) that can be used for comparison of interspecific and intraspecific differences in varanids (Mertens 1942). These data will not be repeated here.

Skulls of *V. s. salvator* are long, extended, relatively flat, and lightly built, and they have the general shape of a pyramid, as in most lizards. The skull has a broad cranial foramen. With increasing age, the parietal plate tends to be smaller backward (Mertens 1942). Part of the snout is relatively short, as is the case with the septomaxillaria in a juvenile rather than in an adult (Mertens 1942); the lower jawbone is shorter than the upper in general. The height of the cranium in *V. s. bivittatus* may be bigger than in *V. s. salvator*. Small holes (foramen maxillaria) exist in the upper and the lower jawbone (Mertens 1942).

Comparative investigations on kinesis of skulls of *Tupinambis* (*T. teguixin*) and *Varanus* (*V. salvator*) resulted in eventual passive mobility of juvenile *Tupinambis*, while the skull of *Varanus* is mesokinetic (Hofer 1960). Frazzetta (1962) defines mesokinetic skulls as having only a frontoparietal joint, whereas amphikinetic applies to skulls having both a metakinetic plus mesokinetic joint. In this way, he notes the rather complex linkage system to be amphikinetic. He developed a model where four different units of the skull (rotatory joints) interconnect to form a four-link system that then is called a quadric-crank mechanism or chain. While swallowing prey, a monitor moves its skull jerkily from side to side and/or presses the victim to the ground, allowing a new grip by the jaws. Head and neck push forward jerkily, and the inertia of the victim is used to ram the prey back into the mouth (Frazzetta 1962). The quadric-crank chain model has been applied to and discussed on the results of feeding in live specimens (*V. bengalensis*, *V. exanthematicus*, *V. albigularis*, and *V. komodoensis*) of *Varanus* by use of cinematography. Amphikinetic movements in the skull allow a stronger posterior recurvature of the maxillary teeth (Rieppel 1979). Rieppel (1978) also discussed the phylogeny of the cranial kinesis in lower vertebrates, and he concluded the primordial component part to be situated in the splanchnocranium.

Teeth are compressed laterally and serrated at the posterior edge. There are 8 to 9 premaxillary teeth and 11 to 13 maxillary and mandibular teeth. Number of teeth in many cases is not complete because of frequent replacement of dentition (Mertens 1942; Bullet 1942). Teeth are positioned with broad bases on the inner side of the jaws. With respect to the differing skull size depending on age of a specimen, the number of teeth generations during the life of a monitor must be numerous (Bullet 1942).

Tongue

All specimens of monitors, including *V. salvator*, have an extremely long, deeply forked, highly protrusible, slender tongue. The tongue is not used for manipulating food but instead became an accessory to the sense of smell. Scent particles are collected and transported to the paired Jacobson's organ, where the scent is analyzed. (This view has been criticized: according to Oelofsen and van den Heever [1979], the tips of the tongue do not enter Jacobson's organ, as was established by means of X-rays and radioactive-marked tips of the tongue in live *V. albigularis*.) The tongue is associated with a number of cartilages and bones, the combination of which is called the hyoid skeleton (Bellairs 1969; Hofer 1960).

On the basis of anatomical and functional investigations mainly on *V. exanthematicus* but also on other varanids including *V. salvator*, Smith (1986) established that the tongue has lost a roughened dorsal surface and if not in action, it is withdrawn into a sheath, which means it is not able to move freely. *Varanus* tongues are far more specialized than in any other lizard and resemble the tongue of snakes (Smith 1986). The morphology of the tongue and the hyoid apparatus is unique in varanids because of its general specialization (scent collector and analyzer) and because of its extreme protrusibility; connected with this the increased mobility and strength of the hyobranchium (Smith 1986).

Habitat

V. s. salvator occurs in primary and secondary forests, along the coast, in agricultural areas, and even in villages. They are found up to altitudes of at least 1100 m (Erdelen 1991). Preferred habitats throughout their extremely large range include fresh- and brackish water environs and their direct vicinity, mainly mangrove swamps and riparian habitats (e.g., Dryden and Wikramanayake 1991; Erdelen 1991; Gaulke and De Silva 1997; Pandav 1993; Pandav and Choudhury 1996; Traeholt 1997a; Wikramanayake 1995; Wikramanayake and Green 1989; Wikramanayake and Dryden 1993).

Terrestrial, Arboreal, Aquatic

V. s. salvator are often referred to as semiaquatic or amphibic (e.g., Mertens 1942; Wikramanayake and Dryden 1993) or even aquatic (e.g., Smith 1932) in their living habits. They can even cross seawater barriers between different islands. *V. s. salvator* was the first terrestrial vertebrate species to recolonize the Krakatau Islands, 25 years after the destructive eruption in 1883 (Rawlinson et al. 1990). But besides being excellent swimmers and divers, they spend a large part of their time on land and are good climbers as well. Juveniles are more arboreal than adults.

Nights are spent in burrows, in tree holes, on tree branches, in dense vegetation, or even in the water (e.g., Gaulke and De Silva 1997; Traeholt 1995b). The average length of seven water monitor burrows investigated in Malaysia was 9.5 m; the depth ranged between 1.2 and 2.4 m, with the entrance sloping downward in an average angle of 4.8°

(Traeholt 1995b). The entrance size ranged from 40 by 42 cm to 75 by 67 cm. Temperature inside the burrows varied between 25 and 27.8°C and was more stable than ambient temperature.

Time of Activity (Daily and Seasonal)

V. s. salvator are diurnal, as are all known varanids.

V. s. salvator are active from sunrise to sunset. In a population in the Bhitarkanika mangroves, Orissa, India, active individuals were sighted from 0600 to 1800, the general activity level (based on sightings of animals outside their burrows) was highest during 1200 to 1500 (Pandav and Choudhury 1996). Activity level, but especially conducted activity, was highly correlated with ambient temperature. Although most basking activities took place when ambient temperatures ranged between 21 and 27°C, most foraging activities took place during temperatures between 29 and 31°C. In winter (December–February, when temperatures can drop to 10°C), activities started later and ended earlier than during summer.

V. s. salvator studied at Uda Walawe National Park, Sri Lanka, showed very similar activity times. Active animals were observed from 0600 to 1800, with main activity times between 0900 and 1600 and an activity peak around 1300 (Dryden and Wikramanayake 1991). A juvenile *V. s. salvator* in North Sumatra, Indonesia, equipped with a radio transmitter, also had its peak activity times from 1200 to 1500 (Gaulke et al. 1999). In Sungai Tembeling, Pahang, Malaysia, water monitors start their activity at about 1000; only large lizards were observed even between 1800 and 2000 (Jasmin and Amin Abdullah 1987/88). Activity peak in the area was reached between 1300 and 1500 (Jasmin 1990). These data from different countries indicate that activity peaks are reached during the warmest time of the day. A seasonal change in activity rhythm most probably is much less pronounced in tropical areas, which includes the entire Sunda region, compared with more temperate regions on the Asian mainland.

Thermoregulation

Average cloacal active temperature of *V. s. salvator* measured on the Malayan peninsula and on Tulai Island, Malaysia, was 30.4°C, with a daily fluctuation between 29.5°C and 37.3°C. Correlation between ambient temperature and cloacal temperature appeared strong; thermoregulatory behavior included cooling in water and sleeping in burrows (Traeholt 1995a). Activity temperatures taken on Sri Lanka were 29.9°C (Wikramanayake and Green 1989) and 30.4 ± 2.1°C (Wikramanayake and Dryden 1993). *V. s. salvator* prefer thermally stable microhabitats during day and night to maintain relatively constant body temperatures. Their body temperatures are lower than those of terrestrial varanids (Wikramanayake and Dryden 1993, 1999). A preferred body temperature of 36 to 38°C for *V. salvator* is suggested by Meeks (1978) on the basis of experiments under artificial conditions. Experimental animals could regulate their body temperatures to fluctuations of approximately 2 to 3°C. Preferred basking sites of *V. s. salvator* are litter and sand on sun-exposed ground and tree branches. Main basking time is around 0900 (Pandav and Choudhury 1996).

Behavior

The principal hunting method of *V. s. salvator* is open pursuit (Traeholt 1993, 1994a). Areas are searched by continued abrupt forward and sideways movements with constant tongue flicking. This foraging behavior is often "area-concentrated" (Traeholt 1993), meaning that an area that has proved a successful hunting ground before is revisited. Depending on prey species detected, different catching methods are used. Insects and other small prey, like frogs, are caught with a "jump catch," a fast forward movement. Crabs and turtle eggs are usually dug out of their burrows or nests, respectively (Traeholt 1993, 1994a). Rotten carcasses can be detected up to 100 m away, but normally only carcasses found within the home range are approached (Traeholt 1994b).

Ritual combat has been observed several times both in the wild and in captivity (Hediger 1962; Horn et al. 1994). Most interesting and important: combatants exhibit the clinch phase that is characteristic for members of all subgenera, excluding members of the subgenus *Odatria*.

In captivity (a zoo), a Malayan water monitor learned to tell keeper and visitor apart: it reacted to its name, and it learned to voluntarily change between its exhibit and a reserve terrarium for, e.g., cleaning operations (Lederer 1933).

In monitors, hisses are used as a warning sound. Hisses of *V. salvator* and a series of other monitors have been analyzed acoustically. Three different types of hisses have been established in the various species of monitors: (1) long-duration hisses, (2) hisses with frequency gaps, and (3) short-duration hisses (Young et al. 1998). Hisses of *V. salvator* belong to that of the first group (Young et al. 1998).

Diet

V. s. salvator are generalized carnivores. Food varies with respect to the habitat, as different habitats shelter a different fauna. In mangrove habitats and coastal areas, different crab species comprise an important part of the food. In oil palm estates, insects and rodents are the main food source, and in the direct vicinity to human settlements, organic waste and leftovers are a regular food source (Shine et al. 1996; Traeholt 1994a). But other animals, such as birds and bird eggs, reptiles and reptile eggs, amphibians, and fish, are food sources too (e.g., Deraniyagala 1931; Harrison and Lim 1957; Jasmin 1986; Rashid and Diong 1999; Smith 1931). Carrion is eaten, but live prey is preferred (Traeholt 1994b). Cannibalism does occur (e.g., Shine et al. 1996). In a tourist area on Pulau Tulai, Malaysia, Traeholt (1994a) demonstrated a seasonal variation in diet depending on the peak tourist season (May to September). During this time, the diet of *V. s. salvator* consisted mainly of tourists' leftovers such as cooked fish and even sandwiches, and they appeared extremely well fed. This is a good example to show the high adaptability of *V. s. salvator*.

Reproduction

At least in its tropical distribution areas, *V. s. salvator* have a very long, or even circumannual reproductive season (e.g., Erdelen 1989, 1991; Shine et al. 1996), but egg-laying peaks seem to vary regionally.

In a study of Shine et al. (1996), all investigated adult males in South Sumatra, Indonesia, had large testes containing sperm, but testes were larger in April than in October. All adult females investigated during August and April were reproductively active, containing oviductal eggs or vitellogenic follicles. Only very little reproductive activity was observed in October. Most females produced multiple clutches each year, as was evident by the presence of ovarian scars from previous clutches. Clutch size ranged from 5 to 22 eggs, with a mean of 13 eggs, and was correlated with female size (data collected during different visits to a skinnery in South Sumatra). All mature males investigated in skinneries in North Sumatra were reproductively active, with testes being largest in August. Reproductively active females also were found year round, but many fewer animals were reproductively active during the drier period in the middle of the year. Clutch sizes ranged from 6 to 17 eggs (Shine et al. 1998).

Khan (1969) reports data on reproductively active females from Perak, Malaysia. Some individuals containing oviductal eggs or vitellogenic follicles were found throughout the year, but a distinct peak season was reached during September–October (with ~50% of dissected females being in reproductive condition), whereas from February to May, only very few individuals contained eggs. (Data are from dissected animals caught for the skin trade.)

In Thailand, water monitors lay eggs during the dry season (February to May), according to Ramesh Boonratana (1989); Smith (1935) reports egg laying during June. For Hainan, Schmidt (1927) reported egg laying during July; in India egg laying was observed during June–July (Biswas and Kar 1981; Biswas and Acharjyo 1977).

In a breeding group of *V. s. salvator* at the Madras Crocodile Bank, India (originating from Orissa, India), females laid one or two clutches per year, with a mean clutch size of 13.8 (Andrews and Gaulke 1990). Clutch size was independent of female size, as egg numbers laid by one female varied considerably during different ovipositions. Instead, an inverse correlation between egg size and clutch size was observed, with the smallest clutch containing the largest and heaviest eggs, and vice versa.

Clutch sizes for *V. s. salvator* all through its range vary from 5 (Shine et al. 1996) to 40 (found in the oviducts, Erdelen 1989), but it is assumed that extremely large clutches are not laid at once, but in several smaller clutches about every 8 days (Thalibsjah, personal communication, in Erdelen 1989).

V. s. salvator eggs are very variable in size, a length of 64 to 82.6 mm, a width of 32.3 to 45 mm, and a weight of 30 to 87.2 g is reported from different countries, with a high size variability occurring in one population and even in different clutches of one female (Andrews and Gaulke 1990; Anonymous 1978; Biswas and Kar 1981; Kratzer 1973; Meer Mohr 1930; Moharana and Pati 1983; Schmidt 1927; Vogel 1979a, 1979b). Much larger eggs, with a length up to 100 mm, are reported from Sri Lanka (Deraniyagala 1931).

Incubation times vary depending on incubation temperature. At a temperature of 25 to 30°C incubation took 9 months, an increase to 31 to 32°C decreased incubation time to 8 months, and a further temperature increase to 32 to 33°C reduced incubation time to 7 months in a breeding group kept at Madras Crocodile Bank, India (Andrews and Gaulke 1990). Time intervals between hatching of the first and last hatchling of one clutch ranged from 2 to 4 days (Andrews and Gaulke 1990). Throughout the literature, incubation times vary between 207 to 327 days for artificially incubated clutches (Groves 1984; Honegger 1971).

Incubation temperatures in opened nest holes in termite mounds (India) ranged between 28 and 29.5°C, while ambient temperature ranged between 27 and 32°C (Biswas and Kar 1981).

Nest holes are dug in sandy soil along riverbanks, at the base of trees or bamboo clumps (Ramesh Boonratana 1989), in termite mounds (Biswas and Kar 1981), and in rotten or hollow trees (Laidlaw 1901; De Rooij 1915). The length of the nest hole is variable, Biswas and Kar (1981) report a length of about 30 cm for four nest holes dug in termite mounds.

Behavior that can be explained as nest guarding (Anonymous 1978; Biswas and Kar 1981) and parental care (helping to release the young from their burrows during hatching time, Ramesh Boonratana 1989) is reported for *V. s. salvator*.

Data on age of maturity are scant. Male *V. s. salvator* in Sumatra, Indonesia, reach maturity at a SVL of about 40 cm (TL ~100 cm), females at a SVL of about 50 cm (TL ~125 cm) (Shine et al. 1996, 1998). This length will be reached within 2 to 3 years, depending on food availability, if compared with growth rates recorded in kept animals (e.g., Andrews and Gaulke 1990).

In general, published breeding data in captivity fit well to the above mentioned observations on egg deposition and hatching in the wild.

V. salvator has been bred several times in captivity (Horn and Visser 1989, 1997). When these data were compiled one problem arose: Authors in some cases did not always clearly describe what subspecies they had kept under their care. Yet it can be assumed in most cases that the nominate form is meant. Some data cited there as unpublished results have now appeared as full papers (e.g., Ott 1997; Herrmann 1999).

Sexual maturity in captivity is attained at about 100 cm TL or 50 cm SVL and at the end of 2 years. Egg-laying seasons are closely related with those in the wild (Andrews 1995). Numerous data (e.g., size of females at egg deposition, clutch size, clutch weight to body weight ratio in percent) support the given statements (Andrews 1995). However, the data compiled by Andrews were collected in a zoo in the tropics.

Movement

Temporary activity areas of six *V. s. salvator* (equipped with transmitters) in an oil palm estate in North Sumatra, ranged between 1.5 and 22.6 ha. Activity areas were widely overlapping, indicating a lack of territoriality. Four of the studied varanids were resident throughout the observation period of 4 weeks, one was a transient, one could not be observed long enough to determine its status (Gaulke et al. 1999). The home range of *V. s. salvator* in Sungai Tembeling, Pahang, Malaysia, does not exceed 200 m², according to observations of Jasmin and Amin Abdullah (1987/88). The home range of a radio-collared female *V. s. salvator* at Sg. Putih Forest Reserve, Malaysia, was about 9000 m² (Jasmin 1989). One large lizard at Sungai Dusun canal, Malaysia, traveled a distance of nearly 2 km per day during his foraging activities. Younger individuals have smaller home ranges than large ones, and different home ranges overlap (Jasmin 1990).

Home range size certainly varies depending on the habitat. Water monitors living in a habitat with a high nutritious value (as mangrove swamps) obviously need smaller activity areas than conspecifics in less favorable habitats. Even so, although activity areas overlap, a specific site can only feed a certain number of individuals, which explains why some are at least temporary transients, others temporary residents. Home ranges and movements of *V. s. salvator* from two different habitats did not vary significantly within a local habitat, but sizes differed between populations in the two habitats. Home range sizes varied from 0.014 to 0.317 km² (Traeholt 1997c).

Physiology

Because of their relatively low activity temperature, the metabolic rate of *V. s. salvator* is low compared with some other varanid species according to Gleeson (1981), but according to Dryden et al. (1992), water flux and metabolic rates of *V. s. salvator* are substantially higher than in varanids from semiarid and arid habitats, and higher than predicted for other tropical lizards (water influx rate, mean of 54.5 mL kg^{-1} d^{-1}; metabolic rate, mean of 0.195 mL CO_2 g^{-1} h^{-1}).

The parietal eye is of high importance for the diurnal activity rhythm in *V. s. salvator*. Five water monitors, caught on Tioman Island, Malaysia, were fitted with radio transmitters and released into a large enclosure for observations. Animals with uncovered parietal eye showed a clear diurnal rhythm, with activities restricted to daytime. After covering the parietal eye with aluminum foil, the mean locomotive activity of the water monitors decreased drastically. Their body temperature decreased as a result of a lack in thermoregulatory behavior. The animals were often observed to be active at night, and often slept outside a burrow (Traeholt 1997b).

Longevity records for this monitor were 5.1 to 8.7 years in captivity (Anonymous 1966) and 10 years, 8 months according to an older report (Flower 1937).

Fat Bodies

A difference in fat body stores among males and females of *V. s. salvator* is not evident, but seasonal variation in fat body size occurs (Shine et al. 1996), and detailed data have not been reported.

Parasites, Diseases

The tick species *Aponomma lucasi* was found in most *V. s. salvator* investigated at Taman Negara, Malaysia. Tick number per individual varied from 3 to 30, with the highest concentration on the inner folds of the fore- and hind legs (Jasmin 1986). The spiruid nematodes *Tanqua*

TABLE 6.6.
Morphometric data of *Varanus salvator salvator* (black color form).

No.	Weight (g)	Head + Body[a] (cm)	Tail (cm)[b]	Total Length (cm)	Head Length (cm)	Head Width (cm)	Front Leg (cm)	Hind Leg (cm)
1	910	38.6	63.3[c]	101.9	6.9	3.8	7.3	9.2
2	1030	40.9	62.1[c]	103.0	6.8	4.0	7.2	10.2
3	1520	45.3	74.5	119.8	7.6	4.4	8.2	11.7
4	610	39.7	58.5	98.2	6.6	4.0	7.2	9.2

[a] Tip of snout to back part of hind leg.
[b] Back part of hind leg to tip of tail.
[c] Tip of tail missing.

tiara were found in stomachs of *V. s. salvator* in northern Sumatra, with higher numbers in juveniles than in adults (Shine et al. 1998). A male Malayan water monitor that spent 5 years in captivity showed medial calcification of limited extent, with secondary thickening in the aortive arch (Finlayson and Woods 1977).

Varanus salvator salvator (Black Color Morph)

Description

In the 1980s, an unusual colored water monitor was exported from Thailand. Upper surface was totally black and also the ventral side was nearly black or contained small right angular whitish spots. Nuchalia clearly bigger than dorsalia and caudalia, tail clearly compressed and with double keel, blue-black tongue (Horn and Gaulke, unpublished data).

It could be demonstrated, however, by importation of a big male specimen (140 cm) of the normal color phase of *V. s. salvator* of nearly the same area that the black one was only a black color phase and not a new species. A comparison of the hemipeneal morphology of both color phases showed the same structure (Böhme et al., unpublished data).

Habitat, Distribution

Mangrove swamps and/or forests of western Thailand (Derwanz, personal communication)

Morphometrics

Table 6.6 contains some morphometric data (Horn and Gaulke, unpublished data).

6.21. *Varanus salvator* (Subspecies)

Hans-Georg Horn and Maren Gaulke

Figure 6.39. Varanus s. cumingi
(photo by Maren Gaulke).

Figure 6.40. Varanus s. nuchalis
(photo by Maren Gaulke).

Introduction

Currently six subspecies of *Varanus salvator* are recognized: *Varanus salvator andamanensis, V. s. bivittatus, V.s. cumingi, V.s. marmoratus, V. s. nuchalis, V. s. togianus* (Mertens 1963), besides the nominate form *V. s. salvator* (Figs. 6.39, 6.40).

Varanus salvator andamanensis

Systematics, Taxonomy

This subspecies was first described by Deraniyagala (1944). Later on Mertens (1959) considered *V. s. andamanensis* taxonomically not to be valid and he synonymized it with the nominate form of Sri Lanka. However, some time later Deraniyagala (1961) confirmed his view on the basis of some more specimens, which was also accepted by Mertens (1963).

Description

V. salvator type monitor. Its yellow dorsal pigmentation is less pronounced than in the nominate form. Spots consist of one or two weak rows across the body in half grown individuals. The snout is more acuminate than in the nominate form, the subspecies has 8 to 10 supraoculars as compared with 5 to 6 in the former (Deraniyagala 1944, 1961). Kala (1998) describes a gray background coloration and orange spots in hatchlings arranged in rings around the body. This is in contrast to hatchlings of the nominate form (Kala 1998).

Distribution

One subspecies, *V. s. andamanensis,* is restricted to the Andaman Islands in the Indian Ocean (Deraniyagala 1944, 1961).

For geographic distribution of various subspecies of *V. salvator,* see Fig. 6.38 on page 246.

Morphometrics

In general the TL of this subspecies is clearly smaller than the nominate form: 101.0 cm (*V. s. andamanensis*) as compared with 207.0 cm in *V. s. salvator* (Deraniyagala 1961). But Krammig (1977) visiting the Andaman Islands, measured a specimen of 198 cm TL, which is similar to the value given by Deraniyagala (1961) for the nominate form. Tail length of the above mentioned specimen: 58.0 cm (Deraniyagala 1961). Kala (1998) mentions specimens of 250 cm and 200 cm in TL, respectively. Other data are: Male 148 cm TL, 71 cm SVL, 9.0 kg; female: 133 cm TL, 61 cm SVL, 5.75 kg; female 104 cm TL, 56 cm SVL, 3.75 kg. Both females had missing tips of tails (Kala 1998).

Habitat, Diet

Evergreen rain forests, dried flat wetlands and littoral forests, mangrove swamps (Kala 1998); for illustration of some of these see Krammig (1977). The Andaman water monitor feeds on eggs of sea turtles and crocodiles as well as on crabs in mangrove marshes at low tide (Kala 1998). Carrion is also accepted (Krammig 1977).

Behavior

After having been caught, the monitor feigned death (Krammig 1977).

Reproduction

Hatchlings have been observed during March and April. No special behavioral activities during reproductive time were noticed in captivity. In captivity two eggs were noticed on the soil surface on August 8, 1996, but seemed to be eaten next day (Kala 1998). Instead, however, four hatchlings were found in the enclosure on April 4, 1997. Three more hatchlings plus a dead one were observed by the end of April. Hatchlings were assumed to be from one clutch and they seemed to be around 10 days old. Thus the incubation period under seminatural conditions can be estimated to be about 300 days (Kala 1998). On April 30, the seven hatchlings weighed 29.5 to 34.0 g (mean 31.4 g), they had a TL of 30.0 to 33.4 cm (mean 30.1 cm) and their SVL was 13.4 to 15.0 cm (mean 14.2 cm) (Kala 1998).

Status

"Although this subspecies is commonly hunted by tribals and settlers for meat, there is no trade for skins and population declines are not suspected" (Kala 1998).

Varanus salvator bivittatus

Systematics, Taxonomy

This subspecies is closely related to the nominate form. Therefore synonymized by Mertens (1942) with the nominate form *V. s. salvator*. Described by Mertens in 1959 and confirmed by the same author in 1963.

Description

Along the sides of the neck behind the ear *V. s. bivittatus* shows a black band und under this a yellowish band (compare pictures of a hatchling alive by Kopstein [s.a.]), which may be divided occasionally into spots. But nevertheless these markings always can be identified (Mertens 1959). This is in contrast to specimens from Sri Lanka in which these markings are missing. The dorsal spots in both forms undergo an ontogenetic change during life and may vanish totally to produce a uniform dark coloration (Mertens 1959; see also pictures of a big specimen alive by Kopstein s.a.). These bands are typical for specimens from Java and they are less pronounced in specimens from the lesser Sunda Islands.

Distribution

Java (Terra typica), Lombok, Sumbawa, Flores, Wetar (Mertens 1959, 1963).

Morphometrics

TL of *V. s. bivittatus* does not seem to differ very much if at all from that of *V. s. salvator*. Vogel (1979a) estimated size classes of this subspecies to be 30 to 59 cm (10.0%), 60 to 89 cm (25.6%), 90 to 119 cm (18.8%), 120 to 149 cm (30.0%), 150 to 179 cm (13.3.%), 180 to 210 cm (2.2%) in the nature reserve Udjung Kulon in western Java. (Percentages are calculated from original data.)

Habitat, Diet

This subspecies inhabits evergreen rain forests, clearings of these and banks of rain forest rivers (Vogel 1979a, personal communication).

No detailed investigations on stomach contents of *V. s. bivittatus* are available. Carcasses are eaten frequently (Vogel 1979a; Hoogerwerf 1970). Eggs and fledglings of birds (egrets, cormorants, ibis), which breed in colonies, are also eaten (Hoogerwerf 1951). A big specimen killed and swallowed the calf of a banteng (Hoogerwerf 1970). Crabs and sea turtles are also eaten.

Behavior

V. s. bivittatus exhibit agonistic behavior at carcasses and some-times at resting places (Vogel 1979b; Hoogerwerf 1970). In several cases, they showed the characteristic clinch phase (Vogel 1979a, 1979b; Hoogerwerf 1970; Horn et al. 1994). Detailed descriptions of other behavioral characteristics can be found elsewhere (Vogel 1979a, 1979b).

Reproduction

Kopstein (s.a.) dissected a 200-cm specimen, which contained half developed eggs in September. Copulation has been observed in September (Hoogerwerf 1970). The smallest female at Udjung Kulon involved in copulation was 140 cm, the smallest male was 170 cm (Vogel 1979a). Digging activities for preparation of nesting holes for egg deposition were observed in May. Egg deposition occurred there in May. The nesting hole contained 24 eggs. Length of eggs ranged from 6.5 to 7.2 cm, width from 4.2 to 4.5 cm, respectively (Vogel 1979a). Vogel (1979a) estimated the age of the female to be 3 to 4 years, that of the male to be 4 to 5 years. In captivity, this species mated between April 29 and May 6, a total of 6 eggs (4 infertile) were deposited on June 9 and 12 that hatched after 205 to 241 days at December 31 and February 8. Hatchling weights were 31 and 40 g (Horn and Visser 1997; Honegger 1995).

Parasites

All individuals of this subspecies were heavily infested with ticks at Udjung Kulon. One specimen died because of an extreme infestation by fly larvae: its stomach contained 10 (different species?) nematodes (Vogel 1979a).

Status

V. s. bivittatus is heavily exploited for the leather trade (Kopstein s.a., see also Hoogerwerf 1970)

Varanus salvator cumingi

Nomenclature

Varanus salvator cumingi was described by Martin in 1838 as *Varanus cumingi* (not *cumingii,* as erroneously cited in several works), with the type locality Mindanao, Philippines. The species status of this form was not doubted (e.g., Taylor 1922) until the revision of Mertens (1942), who regards it as a subspecies of *V. salvator.* The species versus subspecies status of this well-differentiated monitor lizard of the *sal-*

vator group remains controversial (e.g., Bayless and Adragna 1997). Modern investigation methods might lead to its official reelevation to species status.

Distribution

V. s. cumingi occurs on the southern and eastern islands of the Philippines, on Basilan, Bohol, Leyte, Mindanao, and Samar, and certainly on many more of the smaller offshore islands in between. This island group is known as the Greater Mindanao region, comprising of the Mindanao subregion (Mindanao, Basilano) and the East Visayan subregion (Bohol, Leyte, Samar), a clearly defined faunal subprovince supporting many endemic species and subspecies (e.g., Heaney and Regalado 1998). The occurrence of *V. s. cumingi* on Cebu (Mertens 1942) was not reconfirmed by more recent surveys (Gaulke 1991a, 1992).

Diagnostic Characteristics

V. s. cumingi are characterized by their striking, bright yellow color pattern and low scale counts (mean standard counts lower than in all other described subspecies including the nominate form).

Description

V. s. cumingi occurs in two distinctive color varieties (Gaulke 1991a, 1992). The form from Mindanao has a dominantly yellow head with differing amounts of dark markings; neck, back, and tail dark gray with large yellow crossbands, transverse rows of ocelli, and spots; chin and throat mostly uniform yellow, rarely with dark markings; ventral side whitish with irregular dark crossbands; lower side ot tail yellow with dark ocelli; upper sides of extremities mottled dark gray and yellow, lower side yellow. Individuals from the other islands are darker, with a mostly dark gray head, and much less distinctive yellow color pattern on dorsal side (Gaulke 1991a, 1992).

V. s. cumingi are exceptional amongst most varanids in so far as many juveniles are much darker and duller in coloration than adults. During ontogeny they very slowly change toward the brilliant coloration of their parents (Wicker et al. 1999).

There are no significant differences in general shape and proportions from the nominate form.

Morphometrics

The largest recorded *V. s. cumingi* has a TL of 150 cm (SVL 60 cm). Mean SVL for 10 adult specimens coming from different islands is 42.2 cm (Gaulke 1992). Mean TL of six adult specimens from Mindanao is 114.2 cm (min. 102 cm, max. 132 cm), with an average weight of 1383 g (min. 800 g, max. 2500 g) (Wicker et al. 1999).

Habitat

V. s. cumingi occurs in mangrove swamps, along vegetated riverbanks, and in agricultural areas. Whether it also occurs in inland forests and higher altitudes is not yet known.

Terrestrial, Arboreal, Aquatic

No field observations published, but probably similar to other members of the *salvator* group.

Time of Activity (Daily and Seasonal)

Diurnal. As typical for tropical species, a pronounced seasonal activity rhythm cannot be expected. However, a slight change in activity rhythm might be expected for this form, as this could be ascertained for *V. s. marmoratus*, another Philippine water monitor subspecies (see there).

Agonistic Behavior

Combat fights in *V. s. cumingi* are reported (with photo-documentation) from kept individuals (Horn 1994), and were also observed in a breeding group at the Zoological Garden Frankfurt, Germany. Combat behavior was observed between male and female, and between females. Their ritualistic fight includes the bipedal clinch phase as is typical for the subgenus *Varanus* (Horn et al. 1994).

Diet

Investigation of intestinal tracts of 10 *V. s. cumingi* showed no food differences as compared with other members of the *salvator* group. They mainly feed on small prey species such as different types of arthropods, but also vertebrates. In captive bred *V. s. cumingi*, some hatchlings refused to eat insects, but accepted earthworms and fish. Later on they also preyed on insects (Wicker et al. 1999).

Reproduction

Reproductive activities in nature probably are not highly seasonal because of the tropical nature of the climate, but actual data are very scarce. A dissected female caught on Mindanao in early November had 10 almost fully developed eggs in her oviducts. This female had a TL of 128 cm (SVL 52 cm) and a weight of 2100 g (Gaulke 1992). Other data on reproductive biology are obtained from a *V. s. cumingi* group (origin of the founder generation is Zamboanga del Norte province, Mindanao) kept at the Zoological Garden Frankfurt, Germany (Vogel 1994; Horn and Visser 1997; Wicker et al. 1999). Mating (only observed under water, even so courtship behavior was conducted on land) occurred about 4 weeks before egg laying. The female started to search for a good oviposition site up to 5 days before egg deposition. Usually the egg deposition box offered was used. Up to three clutches were laid in 1 year, with an interval of 3.5 to 5 months. Eggs were laid in the months of January, March, April, July, August, September, October, and December. Mean egg size was 3.5 by 7.6 cm, with a mean weight of 56.1 g. The nesting site was defended by the female against intruders. Eggs were guarded continuously, and if eggs were removed for artificial incubation, the nest was repaired by the female and guarding continued. Incubation temperature ranged between 27 to 28°C during the first part of incubation, and was raised to about 28 to 30°C during the second part. Incubation time varied from 190 to 220 days. Hatchlings had TLs between 28 and 31 cm (SVL, 12–14 cm), and weighed 26 to 42 g.

Parasites

Six wild caught *V. s. cumingi* brought to the Frankfurt Zoo were infected with ticks on their legs, headsides, tympanum, around the eye, and near the cloaca. In their faeces Hexamitae, Ascaridae, Strongylidae, Oxyurae, and Cestodae were determined (Wicker et al. 1999).

Varanus salvator marmoratus

Nomenclature

Varanus salvator marmoratus was described by Wiegmann in 1834 as *Hydrosaurus marmoratus*. The type locality is Luzon, Philippines. Taylor (1922) synonymized it with *V. salvator*. Mertens (1942) revised the *salvator* group and elevated it to subspecific status as *V. s. marmoratus*. Deraniyagala (1944) described the same form as *V. s. philippinensis,* which therefore is a junior synonym to *V. s. marmoratus*.

Distribution

V. s. marmoratus occurs in the northern and western parts of the Philippines. It is reported from several islands of the Palawan region (Calamian Islands, Balabac, Palawan), of the Sulu region (Bongao, Sanga Sanga, Siasi, Sibutu, Tawi Tawi), and the Luzon region (Luzon, Mindoro, Polillo), and occurs most certainly on many more of the offshore islands in between. These three regions are clearly defined faunal subprovinces of the Philippines, supporting many endemic species and subspecies (e.g., Heaney and Regalado 1998). The occurrence of *V. s. marmoratus* on Mindanao (a different faunal region), as reported in Taylor (1922), could not be reconfirmed by more recent distribution surveys (Gaulke 1991a, 1992), and is most probably erroneous. Lately (Gaulke, unpublished data), *V. s. marmoratus* was recorded on three islands of the Semirara Island group (Caluya, Panagatan III, Semirara), an island group that is included in the political boundary of Antique province, Panay (member of the West Visayan region), but actually lies on a land bridge to Mindoro (e.g., Ferner et al. 2001).

Diagnostic Characteristics

V. s. marmoratus are very similar to the nominate form, but can be differentiated by their generally darker appearance, the ventral coloration (whitish with mostly confluent, distinctive dark crossbands in *V. s. marmoratus* vs. whitish with only marginal dark markings in *V. s. salvator*) and the size of the nuchals (same size as occipitals in *V. s. marmoratus* vs. smaller than occipitals in *V. s. salvator*). With a maximum TL of about 200 cm, *V. s. marmoratus* is the largest of the Philippine water monitor subspecies, but does not get as large as the nominate form.

Description

V. s. marmoratus are dark gray, with variable amounts of irregular whitish markings on the head and indistinct transverse rows of whitish spots, seldom ocelli, across the back. Neck scales often have whitish posterior margins. The dorsal tail surface is mottled with irregular whitish marks proximally, and a differing number of whitish cross-bands distally. The chin and throat are whitish with numerous dark spots and sometimes dark crossbands. The belly is whitish with a network of irregular dark crossbands. The tail underside is whitish proximally and dark gray distally. Extremities are dark gray with numerous small whitish spots on the upper sides and whitish with irregular dark markings on the inner sides (Mertens 1942; Gaulke 1989, 1991a, 1992).

Juveniles are slighty lighter with a more pronounced color pattern than adults.

Scale counts of *V. s. marmoratus* are within the range of the nominate form (Mertens 1942; Gaulke 1989, 1991a). There are no differences in general shape and proportions from the nominate form.

Morphometrics

According to Taylor (1922), they can reach a TL of more than 200 cm. However, as he considers this form as not distinctive from *V. s. salvator*, it is unclear whether this remark refers to the nominate form or *V. s. marmoratus*. The only exact measurements he gives are from an animal of less than 100 cm TL. Among 176 animals measured (and afterward released) on Calauit Island (Calamian Islands, Palawan region), the largest individual had a TL of 193 cm (SVL 78.8 cm; the tip of its tail was missing; therefore, the TL was estimated using the ratio of tail length to SVL for large adults) and the average TL was 140.3 cm (SVL 54.8 cm). Average weight was 3.05 kg, maximum weight 7.2 kg. Average ratio of tail length to SVL was 1.56. Smaller individuals (no hatchlings were measured) had a proportionally longer tail (1.79) than larger individuals (1.45). Average SVL of males was 56.7 cm, weight 3.19 kg, whereas that of females was 53.7 cm SVL, weight 2.93 kg (data from Gaulke 1989).

Extreme sizes and size/weight relations for *V. s. marmoratus* are reported from Central Luzon by Villamor (1993). According to Villamor (1993), average TL of 80 males is 196.29 cm, but their average weight only 1.55 kg. 59 females have an average TL of 180.48 cm and an average weight of just 1.09 kg. Mean body length for all samples (107.56 cm) is much longer than their mean tail length (65.03), suggesting an error in the conversion of size and weight measures. Auffenberg (1988) reports an average SVL of 41.7 cm for *V. s. marmoratus* in southern Luzon (n = 11), and Bennett (2000) an average SVL of 39.1 cm for males (n = 30) and 34.2 cm for females (n = 6), and a mean mass of 961 g for males and 645 g for females from Polillo Island (Luzon region).

Habitat

V. s. marmoratus is widespread throughout its range. It occurs in primary and secondary forests in the island centers, along the coast, in agricultural areas, and even in villages. The prime habitats with the highest population densities are mangrove swamps and vegetated riverbanks.

Terrestrial, Arboreal, Aquatic

Habitat is as in the nominate form.

Time of Activity (Daily and Seasonal)

Diurnal. A first, slight activity peak is observed between 6 and 7 am, followed by a slight decrease with the highest activity peak during 9 and 10 am. During noon time, the hottest time of the day, activity decreases remarkably, but is followed by a third activity peak during 3 and 4 pm. First active animals can be observed during sunrise (~5 A.M.), and latest active animals around sunset (~6.30 P.M.). Nocturnal activity was once observed: the carcass of a wild pig (*Sus barbatus*) was visited between midnight and 5 A.M. by at least two different *V. s. marmoratus*. Differences in activity rhythm between dry and rainy seasons are slight. During the rainy season first activities start somewhat later, the noon time activity decrease is less pronounced, and the afternoon activity peak is about 1 hour earlier (2–3 P.M.) than during dry season. The dry season in this region spans the first half of the year, the rainy season the second half. However, rainy and dry seasons vary in intensity and duration significantly during different years (data from Gaulke 1989).

Villamor (1993) reports a higher activity of *V. s. marmoratus* from Central Luzon during the hotter and drier months (March–September) than during the cool and wet months (October–February). These data are based on trapping success, which was higher during the dry season in the area investigated.

Thermoregulation

The only available body temperature measurement is the lethal temperature of an adult *V. s. marmoratus*, who directly after death had a cloacal temperature of 43°C, while the ambient temperature during the time at this sun-exposed place was 50°C. Shortly before death this individual had its mouth wide open and panted (Gaulke 1989). This shows that *V. s. marmoratus* can actively influence body temperature, in this instance keeping it below ambient temperature. This observation was an accident (with a trapped animal), not a deliberate experiment.

Behavior

V. s. marmoratus are active foragers. More or less constantly flicking their tongues, they search for food not only in all kinds of terrestrial hiding places, but even under water (in fresh-, brackish, and even seawater), and at least younger ones also on trees. Different species of crabs, which form an important part of their prey, are dug up from soil or sand. Places that proved successful once are revisited by the same individual regularly, this behavior is called "area-concentrated search" by Traeholt (1993). Fleeing prey species are often chased. Fish caught in brackish water creeks are swallowed immediately in the water. Individuals disturbed during their search for prey run away, usually returning after a short while to continue their search at the same place (Gaulke 1989).

Combat fights in *V. s. marmoratus* were observed several times on Calauit, Palawan province (Gaulke 1989). Mostly they started at a feeding site (carcass), with both sexes being involved. The ritualistic fight of *V. s. marmoratus* includes the bipedal clinch phase as is typical for the subgenus *Varanus* (Horn et al. 1994).

Diet

V. s. marmoratus are generalized carnivores. Their main prey species are rather small animals compared with the size of the lizards. Crabs dominate, but obviously all available species of arthropods are eaten. Vertebrate prey comprises fish, anurans, reptiles, birds, and mammals. Mammalian prey are mainly small rodents, but an observation of a *V. s. marmoratus* catching a young macaque (*Macaca philippinensis*) was reported. Bennett (2000) reports *Bufo marinus* as one of the prey species; thus, *V. s. marmoratus* is the second varanid form (besides *V. s. nuchalis*) known to feed on these poisonous toads without ill effects. Bigger prey species such as mangrove crabs are dismembered with the help of the forelimbs prior to ingestion. Carrion is eaten, and a large carcass might attract several individuals. Up to nine individuals could be observed feeding and interacting around a wild boar carcass. Three animals were observed feeding at the same time on a dead dolphin. Even though heavily rotted flesh is consumed, *V. s. marmoratus* do not seem to favor it. After they tear off pieces of meat they shake them to get rid of maggots, and the snout is rubbed intensively on the ground after feeding to get rid of scraps of food. Cannibalistic behavior occurs, as remnants of *V. s. marmoratus* in the guts of a conspecific show, but is a rare exception. A dead *V. s. marmoratus* was used as bait, but during an observation period of several days, no conspecifics were attracted (Gaulke 1989, 1991b, 1992).

Reproduction

Reproductive activities probably are not strongly seasonal in *V. s. marmoratus* as a result of the tropical climate of its habitat. Whether peak breeding seasons occur remains unknown. Digging of nest holes on Calauit/Palawan province was observed in February, August, and November, courtship behavior in November. Termite mounds and holes dug in sandy soil are used as oviposition sides. One case of a mass egg-laying site is reported from Calauit (in Gaulke 1989). A wardener, who carefully opened a nest mound of megapod birds for control purposes detected between 60 to 70 eggs of *V. s. marmoratus* in a depth of about 50 cm. Mounds of megapods, who control the nest temperature carefully by adding or removing nest material, certainly are advantageous oviposition sites. Compared with termite mounds they have the advantage of having a much softer substrate. Females are very cautious while digging their nest site. If disturbed, either by conspecifics or others, they stop digging and proceed to another place. Only one clutch size of *V. s. marmoratus* (14 eggs) has so far been reported, but egg sizes were not recorded (Gaulke 1989). No data of reproduction in captivity are available.

Movement

Activity areas of 4.0, 8.2, and 8.7 ha were calculated for three adult male *V. s. marmoratus* on Calauit (Gaulke 1989, using mark-recapture methods during a 1-year study). Auffenberg (1988) conducted telemetric studies on *V. s. marmoratus* and *V. olivaceus* on Luzon, but only states that acivity areas of *V. s. marmoratus* tend to be larger than those of *V. olivaceus* (which are between 0.22 and 2.67 ha). Some individuals were observed to be resident during the observation period (~1 year), while others were transients. Activity areas of resident *V. s. marmoratus* overlap highly, and the animals do not defend specific territories (Gaulke 1989).

The maximum travel distances (bee-line) recorded between trapping site and resighting place among 36 individually marked *V. s. marmoratus* on Calauit is 1400 m (two individuals) within a time interval of 6 and 11 weeks, respectively (Gaulke 1989).

Parasites

Two tick species (*Amblyomma, Aponomma*) were recorded on *V. s. marmoratus* on Calauit. 167 investigated individuals had an average number of 17.2 ticks each. Few species were free from ticks during investigation, one individual was infested with 208 ticks. Small ticks were found in highest numbers at the tail and hind limb insertion, less often on the ventral side and the forelimb insertion, large specimens were mainly recorded at the gular fold, around the tympanum, below the snout, and along the lateral fold. Tick infestation did not differ between the sexes. Only three of the investigated monitors were heavily infested by mites. Cestodes and nematodes were recorded in the feces of one individual, but no systematic investigation was conducted (data from Gaulke 1989). Villamor (1993) reports the following ecto- and endoparasites for *V. s. marmoratus* from Central Luzon: ticks (*Aponomma lucasi*), nematodes (*Tanqua tiara, Hastospicullum macrophallus, Ophidascaris esa*), cestodes (*Acanthotaenia daeleyi*).

Varanus salvator nuchalis

Nomenclature

Varanus salvator nuchalis was described by Guenther in 1872 as *Hydrosaurus nuchalis*. The type locality was given as "Philippines." Boulenger (1885) listed it as *Varanus nuchalis,* and Taylor (1922) still considered it as valid species in his monograph on Philippine lizards, but mentions its close relation to *V. salvator.* Because of this close relation, Mertens (1942) regarded it as a subspecies of *V. salvator.* Until now the name *V. s. nuchalis* is valid. However, the last word on the species/subspecies status of this well differentiated monitor lizard of the *salvator* group is not spoken, modern investigation methods might lead to its reelevation to species status.

Distribution

V. s. nuchalis occurs on the islands of Boracay, Cebu, Guimaras, Masbate, Negros, Panay, Sibuyan, Siquijor, Ticao, and most probably on many more of the smaller offshore islands in between. This island group in the Central Philippines is known as the West Visayan region, a clearly defined faunal subprovince supporting many endemic species and subspecies (e.g., Heaney and Regalado 1998). The occurrence of *V. s. nuchalis* on Luzon and Mindoro (Mertens 1942 erroneously mentions Mindanao instead of Mindoro as one of the possible localities), as given in Taylor (1922), was not reconfirmed in more recent surveys (Gaulke 1991a, 1992).

Diagnostic Characteristics

V. s. nuchalis are characterized by their strongly enlarged, rather prominent nuchal scales, significantly larger than in any other form of the *salvator* group. Coloration is highly variable, but a light longitudinal middorsal stripe is present in most individuals. Taylor's remark that little variation is evident (Taylor 1922) was due to the small sample size available to him.

Description

V. s. nuchalis occurs in two distinctive color varieties (Gaulke 1991a, 1992). One form is quite colorful: head with variable amounts of black and white coloration; back and tail dark with indistinct to distinct transverse rows of whitish or yellow spots or ocelli and a median light stripe; upper sides of extremities dark with numerous small, bright yellow spots; throat completely whitish, whitish with dark banding, yellow with dark banding, or almost uniform dark gray; underside of belly and extremities either whitish or yellow, with variable amounts of irregular, dark, transverse banding. The second color variety is a very dark, in some cases almost a melanistic form. Dark specimens occur sympatrically with the lighter color form on Panay and Negros, but the brightly colored form dominates. The dark form dominates on Masbate, Ticao, and Boracay. Actually, so far no lighter-colored specimens were recorded on Masbate and Ticao.

There is no distinctive color difference between adults and juveniles.

Scale counts of *V. s. nuchalis* are within the range of the nominate form (Mertens 1942, Gaulke 1991a). There are no differences in general shape and proportions from the nominate form.

Morphometrics

With an average TL of about 100 cm and a measured maximal length of about 145 cm, *V. s. nuchalis* is the smallest of the described water monitor subspecies, but single individuals might grow larger. The size of one sighted animal in northwest Panay was estimated between 160 to 170 cm, but no measurements were taken. There is no evident average size difference between males and females; however, the longest measured animal was a male. The average weight is about 1.12 kg, the heaviest animal, a male, weighed 2.8 kg (data from Gaulke 1992; Gaulke and Reiter 2001).

Habitat and Natural History

V. s. nuchalis is widespread throughout its range. It occurs in primary and secondary forests in the island centers, along the coast, in agricultural areas, and even in villages. Preferred habitats with the highest population densities are mangrove swamps and vegetated riverbanks.

Terrestrial, Arboreal, Aquatic

Habitat is as in the nominate form.

Time of Activity (Daily and Seasonal)

Activity is diurnal. As typical for tropical species, a pronounced seasonal activity rhythm cannot be expected. However, a slight change in activity rhythm might be expected for this form, as this could be ascertained for *V. s. marmoratus*, another Philippine water monitor subspecies (see there).

Thermoregulation

No body temperature measurements have been taken in *V. s. nuchalis*. Especially in the morning, after leaving their nocturnal retreat, they can be seen basking. Basking places include sun exposed tree branches and sun exposed places on the ground.

Behavior

See nominate form.

Diet

V. s. nuchalis was the first varanid known to ingest the highly poisonous cane toad *Bufo marinus* without ill effects from its toxin (Alcala 1957; Gaulke 1992). In the stomach of one individual from Negros, several lumps of hair were found, which might have been regurgitated as gastric pellets later on (Gaulke 1992). Otherwise, no differences in diet compared with the other subspecies are reported.

Reproduction

Reproductive activities probably are not highly seasonal as a result of the tropical climate, but peak seasons might occur. On Negros, gravid females are most often caught from July to December according to the local Negritos, which are traditional monitor hunters (Gaulke 1986). In mid-July, the digging of a nest hole was observed on Negros (Gaulke 1989). The nest hole was horizontally dug in a sandy creek bank. It was about 60 cm long and the entrance 20 cm wide, widening somewhat in the inner parts. The nine eggs laid had an average width of 30.5 mm, an average length of 66.5 mm, and an average weight of 33.8 g directly after deposition. Ten days later, the average width was 31.9 mm, the average length 66.6 mm, and the average weight 37.9 g. The eggs are significantly smaller than the eggs known of the nominate form (see there). The egg-laying female had a SVL of 45 cm (TL 110 cm) and a weight of 1.95 kg after egg deposition (Gaulke 1989).

A female caught on Negros in December (SVL 48.5 cm, TL 117 cm, weight 2.45 kg) had 12 fully developed eggs in her oviducts (Gaulke 1992). Observations on incubation times, sexual maturity, or courtship behavior in this subspecies are still missing. No data of reproduction in captivity are available.

Movement

Information on movement patterns of *V. s. nuchalis* are anecdotal at present. A semiadult *V. s. nuchalis* (estimated TL 50 cm) used the same tree hole in a garden near Dumaguete City/Negros as a night retreat and as a hiding place during the daytime for several months, as observed by the owner and Gaulke (1989). Another individual (estimated TL during first sighting 60 to 70 cm, but considerably larger in the meantime) used the same tree hole as a night retreat close to a research station in the forest of northwest Panay Peninsula since August 1999. It leaves the tree hole at about 9 or 10 in the morning and spends about an hour basking on the same tree before climbing down to start foraging (Gaulke and Reiter 2001). An activity area of 15.3 ha (convex-polygon method) was ascertained for an adult male (TL 134 cm) in the same area using radiotelemetry. Distances between 70 and 457 m were traveled daily (Gaulke and Reiter 2001). As in other forms of the water monitor, home range size and sedentarity presumably vary highly depending on the habitat.

Parasites

V. s. nuchalis are regularly infested by at least two species of ticks. Cestodes and nematodes were found in the gastrointestinal tract (Gaulke, unpublished data).

Varanus salvator togianus

Description

Varanus salvator togianus is a uniform dark brown form of *V. s. salvator*. A recent color photograph (Rummel, personal communication) shows a deep black individual. Nuchals are similar in size to dorsals. A total of 160 scales exist around the midbody, with 74 ventrals from the gular fold to the beginning of the hind legs (Mertens 1942, 1959, 1963).

Morphometrics

TL of the type specimen is 111 cm (Mertens 1942, 1959, 1963).

6.22. *Varanus spinulosus*

KAI M. PHILIPP, THOMAS ZIEGLER,
AND WOLFGANG BÖHME

Nomenclature

In 1941, Robert Mertens described the Solomon's keeled monitor lizard, *Varanus indicus spinulosus,* from a single specimen from San Jorge Island ("Georgs Island"), Solomon Islands. The type specimen, deposited in the Natural History Museum, Vienna (NMW 23387), remained the single known representative until 1989, when five more specimens showed up in the pet trade (Sprackland 1993a, 1994, 1997). Sprackland (1994) elevated *V. indicus spinulosus* to full specific rank on the basis of its distinct phenotype and its sympatry with typical *V. indicus* on Santa Ysabel Island.

The specific name refers to the diagnostic spikelike nuchal scales.

V. spinulosus is generally listed as a member of the subgenus *Euprepiosaurus* Fitzinger, within which it is placed in the *V. indicus* group (see Böhme et al. 1994). However, its exact systematic position is still unsolved, because up to now, no molecular or genital morphology studies have been carried out (e.g., Ziegler and Böhme 1997).

No subspecies of *V. spinulosus* have been recorded.

Geographic Distribution

V. spinulosus is only known from two neighboring Solomon Islands (Fig. 6.41): San Jorge and Santa Ysabel (Mertens 1941; Sprackland 1993a). On Santa Ysabel, *V. spinulosus* lives in sympatry with *V. indicus* (McCoy 1980; Sprackland 1993a).

Fossil Record

No fossils of *V. spinulosus* are reported.

Diagnostic Characteristics

- Moderately large, up to about 100 cm in TL.
- Dorsum dark brown with rows of yellow solid spots.
- Tongue pink.
- Nuchals conical and pointed.
- High scale count around midbody.

Description

V. spinulosus has a deep chocolate brown to black dorsal surface, becoming tan below. Solid spots of lime green or yellowish form four

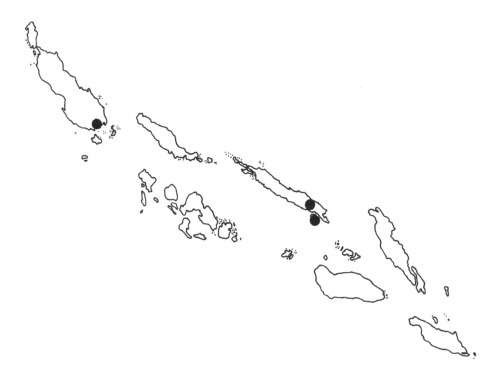

Figure 6.41. Known localities for Varanus spinulosus in the Solomon Islands.

broad transverse bands on the dorsum, from shoulders to hips. Each band consists of four spots, and the most anterior pair of vertebral spots touch middorsally. Between these bands are numerous yellowish speckles, also arranged in transverse rows, forming distinct ocelli. The head is dorsally and laterally uniform dark, lacking any light markings. The tongue is pink for its entire length. Limbs are dark brown, slightly speckled with yellow. Tail has light, thin bands, with those on the distal two-thirds only about two scales wide.

Coloration and pattern of juvenile *V. spinulosus* are unknown.

The snout of *V. spinulosus* is distinctly shorter, broader, and higher than in other members of the *V. indicus* group. The head is 1.56 times longer than broad and 2.03 times longer than high (e.g., *V. indicus*, 1.7–2.2 and 2.4–3.2, respectively) (Mertens 1941; Philipp 1999b).

The tail of *V. spinulosus* is laterally not as strongly compressed as in other members of the *V. indicus* group; it has a double keel along the dorsal apex.

The nuchals and anterior dorsal scales are spiked and hull-shaped, widely separated from each other. Compared with other members of the *V. indicus* group (sensu Böhme et al. 1994), *V. spinulosus* is very small scaled. Mertens (1941) counted about 210 scales around the midbody.

Size

SVL of the type specimen is 31.2 cm, and its tail length is 55.0 cm (Mertens 1941).

The size of only one more captive specimen of the Baltimore Zoo is recorded. This specimen had a TL of about 100 cm (30 cm SVL), weighting 841 g (Kirschner et al. 1996).

Habitat and Natural History

Reptile traders from the Solomon Islands reported that *V. spinulosus* is not particularly rare on Santa Ysabel and that it inhabits mountainous, forested parts of the island. The coloration of specimens from Santa Ysabel is said to differ from those of San Jorge (Sprackland 1993a, 1994).

Terrestrial, Arboreal, Aquatic

Published pictures of captive specimens indicate that *V. spinulosus* is an excellent climber (e.g., Kirschner et al. 1996).

Diet

No data on the diet of wild ranging animals are known. Some specimens in captivity feed regularly on rodents, whereas others prefer fish, refusing insects and rodents (Sprackland 1993a, 1994; Kirschner et al. 1996).

Reproduction

Observations at the Baltimore Zoo indicate that reproductive reciprocity between *V. spinulosus* and *V. indicus* is rare or nonexistent and argues, besides the phenotypic differences, for specific separation of *V. spinulosus* and *V. indicus* (Sprackland 1993a, 1994).

6.23. *Varanus timorensis*

Dennis R. King and Lawrence A. Smith

Figure 6.42. Varanus timorensis (upper photo by James Murphy, lower photo from Roti by Ron Johnstone).

Nomenclature

The Timor monitor was described by Gray in 1831. No type has been designated (Cogger et al. 1983). Mertens described two subspecies of *V. timorensis*: *scalaris* from west Kimberley Western Australia in 1941 and *similis* from Groote Island in 1958. In doing so, he introduced the concept that *V. timorensis* was a species that ranged from the Lesser Sundas of Indonesia and across northern Australia, a concept that persists in many publications to this day (Fig. 6.42).

Mertens' visit to the Western Australian Museum in 1957 was a watershed event for him in beginning to understand the complexities of his *V. timorensis* group in northern Australia, although clinging to the idea that *V. glauerti* (which he described in 1957) and *V. scalaris* were subspecies of *V. timorensis* obviously caused him problems. In 1957, to avoid having two sympatric subspecies of *V. timorensis* in west Kimberley, he elevated *V. t. scalaris* to a full species. Soon after (Mertens (1958), he reversed his decision, relegating *scalaris* to a subspecies and elevating *glauerti* to a species.

The most helpful character in distinguishing *V. scalaris* from *V. timorensis* in all its forms is the presence of enlarged spiny scales at the base of the tail in males. These enlarged scales are absent in true *V. timorensis*. Mertens (1958) discusses this character, but only from the point of distinguishing species, rather than species groups. Storr et al. (1983) used the name "*scalaris*" for all northern Australian and southern New Guinea *V.* "*timorensis.*" *V. scalaris* is now considered a species complex containing several as yet undescribed species (L. A. Smith, personal observation). Mitochondrial DNA has confirmed that *V. timorensis* of Indonesia are not conspecific with *V.* "*timorensis*" of southern New Guinea and the Northern Territory (Ast 2001).

The four insular populations of true *V. timorensis* exhibit some significant variations. The Roti population has been formally described (*Varanus auffenbergi*) (Sprackland 1999). A review of variation in all four populations is in preparation (Smith et al., in preparation)

Geographic Distribution

V. timorensis is found on Timor and the nearby smaller islands of Savu, Roti, and Semau in Indonesia (Fig. 6.43).

Fossil Record

No fossils are known.

Diagnostic Characteristics

• Dorsal pattern more or less distinct regular transverse rows of ocelli on a gray, blackish-gray, or brownish-gray background.
• White on ventral surface.
• Tail black with white rings.
• Small head scales.
• Round tail.

Size

TL is about 60 cm and tail length is from 137% to 176% of SVL (Bennett 1998). Maximum weight of 290 g (King 1993). One juvenile was collected that had a TL of 158 mm and a weight of 3.6 g. Total hatchling lengths are approximately 140 to 150 mm (Eidenmüller 1986).

Figure 6.43. Known localities for Varanus timorensis.

Habitat and Natural History

The original habitats of this species were monsoon forests, much of which have been cleared for farming, rain forests, and grass palm thickets (Sautereau and de Bitter 1980). Most specimens were captured on rocky coastal areas up to 50 m in elevation, although two locations reach altitudes of up to 670 and 700 m (Schmutz and Horn 1986).

Terrestrial, Arboreal, Aquatic

Varanus timorensis is terrestrial and arboreal.

Time of Activity (Daily and Seasonal)

During hot periods in the middle of the day, they usually shelter (Schmutz and Horn 1986).

Thermoregulation

Nothing is known about thermoregulation in V. timorensis.

Foraging Behavior

Stone walls and rocky areas are often used for basking sites and hunting areas for invertebrates (Schmutz and Horn 1986). V. timorensis also forage in remnant thickets (Sautereau and de Bitter 1980).

Diet

Stomachs of 2 of 4 specimens contained a scorpion and a typhlopid snake (Losos and Greene 1988), whereas those for 22 of 47 specimens contained grasshoppers (n = 6), spiders (n = 3), scorpions (n = 2), cockroaches (n = 2), geckos, and other individual species of invertebrates; the 25 other stomachs were empty (King 1993).

Reproduction

Testes and ovaries of lizards collected during May were significantly larger than those captured in October and September. No oviductal eggs were found. It seems that *V. timorensis* breed early in the dry season (May to July) and incubate clutches of 7 to 11 eggs for between 93 to 186 days in captivity (Debitter 1981; Eidenmüller 1986; Horn and Visser 1989). At hatching, young are 55 to 70 mm SVL, weighing 4.5 to 6 g, with a TL of 163 to 174 mm (Bennett 1998; Eidenmüller 1986).

There was no significant difference between numbers of males and females captured.

Movement

Nothing is known of the movements of *V. timorensis*.

Testicular Cycle

Testes collected in May are larger than those collected in September–October.

Physiology, Fat Bodies

Nothing is known about the physiology or fat bodies of *V. timorensis*.

Parasites

Ticks were collected from 17 of 47 specimens, and all those found were *Aponomma soambawensis*.

6.24. *Varanus yuwonoi*

KAI M. PHILIPP, THOMAS ZIEGLER,
AND WOLFGANG BÖHME

Figure 6.44. *Varanus yuwonoi*
(photo by Jeff Lemm).

Nomenclature

Harvey and Barker described the black-backed mangrove monitor lizard, *Varanus yuwonoi,* in 1998. The type locality is Jailolo (1°N, 127.5°E), Halmahera Island, Moluccan Islands, Indonesia. The female holotype specimen (UTA R-41281) is stored at the University of Texas at Arlington (Fig. 6.44).

The species is named after Frank Bambang Yuwono, Jakarta, Indonesia.

V. yuwonoi was placed within the *V. indicus* group by Harvey and Barker (1998) because of external features, and this placement was subsequently corroborated by genital morphology studies by Ziegler and Böhme (1999).

No subspecies of *V. yuwonoi* are recorded.

Geographic Distribution

V. yuwonoi is only known from around Jailolo and Tanah Putih. Local inhabitants around Paca (northeastern Halmahera) have never seen a blue-tailed monitor species (Harvey and Barker 1998). Whether *V. yuwonoi* is restricted to this small area of Halmahera or has a larger distribution, even including other Moluccan Islands, is unknown (Fig. 6.45).

Fossil Record

No fossils of *V. yuwonoi* are reported.

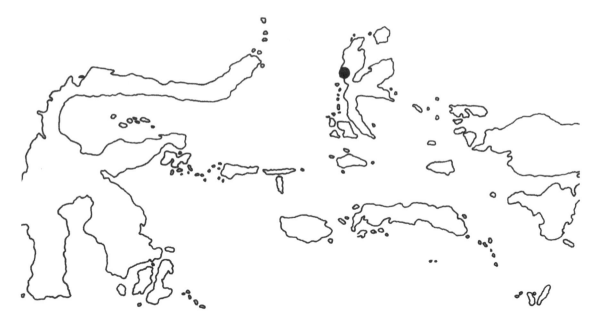

Figure 6.45. Known locality for
Varanus yuwonoi.

Diagnostic Characteristics

• Large monitor of slender habitus.

• Dorsum melanistic anteriorly, posteriorly yellow-brown.

• Tail laterally compressed, blue banded with a double keel on the apex.

• Tongue body dark violet, tines light.

• Hemiclitores strongly asymmetrical, having paryphasmata stretching to both lobes.

Description

The upper side of head and neck as well as the anterior third of dorsum is black. The black blends to a yellow-brown pattern with some green that covers the caudal two-thirds of dorsum and the first third of the tail. The green gets more and more intensive at the tail and turns to blue for the last two-thirds of tail. Venter is pale sulfur yellow, brightest in the gular region. The tips of the tongue are light; the body of the tongue on the dorsal side is dark purple, and the ventral side and base of the tongue are pink. Juveniles are more intensely colored. The black pattern of the head and neck is arranged in distinct ocelli on a yellow-turquoise background color. Separate black blotches fuse more and more as the animals grow older (Harvey and Barker 1998; Ziegler et al. 1998; Sprackland 1999c). Scales of *V. yuwonoi* are very small, e.g., 174 to 196 around midbody (Harvey and Barker 1998; Ziegler and Böhme 1999).

Size

V. *yuwonoi* is probably the largest representative of the V. *indicus* group. The longest measured individual is the female holotype, with a TL of 137.7 cm (SVL, 53.2 cm; tail length, 84.5 cm). Tail is about 160% of SVL (Harvey and Barker 1998; Ziegler and Böhme 1999). We have pictures from pet trade specimens available proving that V. *yuwonoi* can grow to about 1.5 m (see also Ziegler et al. 1998). Lemm (1998) reported that males may reach 6 ft.

Habitat and Natural History

Only a few data on this large species have come out of Halmahera. V. *yuwonoi* is said to inhabit the rain forest of the steep slopes of the mountain range. During dry periods, this species is often encountered on the ground and frequently near stream courses at the base of the hills. In wet periods, they are rarely encountered because they may move up from the streams into the densely forested steep slopes. Known specimens were collected between 50 and 300 m in elevation. Disturbed V. *yuwonoi* seek refuge by climbing trees (Harvey and Barker 1998). In captivity, this species was observed to be partially arboreal (Lemm 1998).

Diet

Nothing is known of the diet of V. *yuwonoi* in the wild. In this context, it seems interesting to note that in Halmahera, a specimen was found trapped in a large trash barrel, probably attracted by the dead fish in the barrel (Harvey and Barker 1998).

In captivity, these lizards readily feed on large insects, fish, and mice (Kirschner and Koschorke 1998). Ziegler et al. (1998) stress that the one specimen in their care would only feed at night.

References

Ahl, E. 1932. Eine neue Eidechse und zwei neue Frösche von der Insel Jobi. *Mitt. Zool. Mus. Berlin* 17:892–899.

Alcala, A. C. 1957. Philippine notes of the ecology of the giant marine toad. *Silliman J.* 4(2):90–96.

Allison, A. 1982. Distribution and ecology of New Guinea lizards. In *Biogeography and ecology of New Guinea*, ed. J. L. Gressit, 2:803–813. The Hague: Junk.

Andres, K. H., M. von Düring, and H.-G. Horn. 1999. Fine structure of scale glands and preanal glands of monitors Varanidae. *Mertensiella* 11:277–290.

Andrews, H. V. 1995. Sexual maturation in *Varanus salvator* (Laurenti, 1768), with notes on growth and reproductive effort. *Herpetol. J.* 5:189–194.

Andrews, H. V., and M. Gaulke. 1990. Observations on the reproductive biology and growth of the water monitor (*Varanus salvator*) at the Madras Crocodile Bank. *Hamadryad* 15(1):1–5.

Anonymous. 1966. A survey of recent longevity records for reptiles and amphibians in zoos. *Int. Zoo Yrbk.* 6:487–499.

———. 1978. *Varanus salvator* breeding at Madras Snake Park. *Hamadryad* 3(2):4.

Ast, J. C. 2001. Mitochondrial DNA evidence and evolution in Varanoidea (Squamata). *Cladistics* 17:211–226.

Auffenberg, W. 1978. Social and feeding behavior in *Varanus komodoensis.* In *Behavior and neurology of lizards,* ed. N. Greenberg and P. D. McLean, 301–331. Rockville, Md.: U.S. Department of Health, Education, and Welfare, National Institute of Mental Health.

———. 1979. Intersexual differences in behavior of captive *Varanus bengalensis* (Reptilia: Lacertilia, Varanidae). *J. Herpetol.* 13:313–315.

———. 1981. *The behavioral ecology of the Komodo monitor.* Gainesville, Fla.: University Presses of Florida.

———. 1984. Notes on the feeding behavior of *Varanus bengalensis* (Sauria: Varanidae). *J. Bombay Nat. Hist. Soc.* 80:286–302.

———. 1988. *Gray's monitor lizard.* Gainesville, Fla.: University Presses of Florida.

———. 1994. *The Bengal monitor.* Gainesville, Fla.: University Presses of Florida.

Auffenberg, W., H. Rahman, F. Iffat, and Z. Perveen. 1989. A study of *Varanus flavescens* (Hardwicke and Gray) (Sauria: Varanidae). *J. Bombay Nat. Hist. Soc.* 6(3):286–307.

Auliya, M. A., and W. Erdelen. 1999. A field study of the water monitor lizard (*Varanus salvator*) in West Kalimantan, Indonesia—New methods and old problems. *Mertensiella* 11:247–266.

Barbour, T. 1921. Aquatic skinks and arboreal monitors. *Copeia* 1921: 42–44.

———. 1932. A new Bornean monitor. *Proc. New Engl. Zool. Club* 13:1–2>.

Bartholomew, G. A., and V. A. Tucker. 1964. Size, body temperature, thermal conductance, oxygen consumption, and heart rate in Australian varanid lizards. *Physiol. Zool.* 37:341–354.

Baverstock, P. R., D. King, M. King, J. Birell, and M. Krieg. 1993. The evolution of species of the Varanidae: Microcomplement fixation analysis of serum albumins. *Aust. J. Zool.* 41:621–638.

Bayless, M. K. 1998. The artrellia: Dragon of the trees. *Reptiles* 6:32–47.

Bayless, M. K., and Q. Dwyer. 1997. Notes on the peach-throated monitor. *Reptile Amphib. Mag.* 26–30.

Bayless, M. K., and J. A. Adragna. 1997. Monitor lizards in the Philippine Islands: A historical perspective (Sauria: Varanidae). *Asia Life Sci.* 6(1/2):39–50.

———. 1999. The Banggai Island monitor. Notes on distribution, ecology, and diet of *Varanus melinus. Vivarium* 10(4):38–40.

Becker, H.-O., W. Böhme, and S. F. Perry. 1989. Die Lungenmorphologie der Warane (Reptilia: Varanidae) und ihre systematisch stammesgeschichtliche Bedeutung. *Bonn. Zool. Beitr.* 40(1):27–56.

Bedford, G., and K. A. Christian. 1996. Tail morphology related to habitat of varanid lizards and some other reptiles. *Amphibia-Reptilia* 17:131–140.

Behrmann, H. J. 1981. Haltung und Nachtzucht von *Varanus t. timorensis. Salamandra* 17:198–201.

Bellairs, A. 1969. *The life of reptiles.* Vol. 1. New York: Universe Books.

Bennett, D. 1995. *A little book of monitor lizards.* Aberdeen: Viper Press.

———. 1998. *Monitor lizards: Natural history, biology and husbandry.* Frankfurt am Main: Edition Chimaira.

————. 2000. Notes on *Varanus salvator marmoratus* on Polillo Island, Philippines. In *Wildlife of Polillo Island, Philippines,* ed. D. Bennett, 29–32. Aberdeen: Viper Press.

Bennett, D., and B. L. Lim. 1995. A note on the distribution of *Varanus dumerilii* and *V. rudicollis* in Peninsular Malaysia. *Malayan Nat. J.* 49:113–116.

Biebl, H. 1995. Haltungsrekord bei der Pflege eines Dumeril-Warans (*Varanus dumerilii*). *Monitor (Frankfurt)* 4(1):24–29.

Biswas, S., and L. N. Acharjyo. 1977. Notes on ecology and biology of some reptiles occuring in and around Nandankanan Biological Park, Orissa. *Rec. Zool. Surv. India* 73:95–109.

Biswas, S., and S. Kar. 1981. Some observations on nesting habits and biology of *Varanus salvator* (Laurenti) of Bhitarkanika Sanctuary, Orissa. *J. Bombay Nat. Hist. Soc.* 78:303–308.

Böhme, W. 1988. Zur Genitalmorphologie Der Sauria: Funktionelle Und Stammesgeschichtliche Aspekte. *Bonn. Zool. Monogr.* 27:5–176.

————. 1991. New findings on the hemipenial morphology of monitor lizards and their systematic implications. *Mertensiella* 2:42–49.

Böhme, W., H.-G. Horn, and T. Ziegler. 1994. Zur Taxonomie der Pazifikwarane (*Varanus indicus*–Komplex): Revalidierung von *Varanus doreanus* (A. B. Meyer, 1874) mit Beschreibung einer neuen Unterart. *Salamandra* 30(2):119–142.

Böhme, W., and T. Ziegler. 1997a. *Varanus melinus* sp. n., ein neuer Waran aus der *V. indicus*—Gruppe von den Molukken, Indonesien. *Herpetofauna* 19(111):26–34.

————. 1997b. Großwarane im Museum Koenig, mit Bemerkungen zu Afrikas größer Echse. *Tier. Mus.* 5(3):65–74.

————. 1997c. On the synonymy and taxonomy of the Bengal monitor lizard *Varanus bengalensis* (Daudin, 1802) complex (Sauria: Varanidae). *Amphibia-Reptilia* 18(2):207–211.

Böhme, W., and H. Jacobs. 2001. *Varanus macraei* sp. n., eine neue Waranart der *V. prasinus*—Gruppe aus West Irian, Indonesien. *Herpetofauna* 23(133):5–10.

Böhme, W., K. M. Philipp, and T. Ziegler. 2002. Another new member of the *Varanus* (*Euprepiosaurus*) *indicus* group (Sauria, Varanidae): An undescribed species from Rennell Island, Solomon Islands. *Salamandra* 38(1):15–26.

Boonratana, R. 1988. Distributional survey and trade of *Varanus* species in Thailand. Consultancy report to CITES.

Boonratana, R. 1989. Distributional survey and trade of *Varanus* spp. in Thailand. In *Asian monitor lizards: A review of distribution, status, exploitation and trade in four selected species,* ed. R. Luxmoore and B. Groombridge. Report to the CITES Secretariat. Cambridge: World Conservation Monitoring Centre, Annex 7.

Bosch, H. 1999a. Successful breeding of the emerald monitor (*Varanus p. prasinus*) in the Löbbecke Museum + Aquazoo, Düsseldorf (Germany). *Mertensiella* 11:225–226.

————. 1999b. Parasite burdens of monitors in captivity. *Mertensiella* 11:189–192.

Boulenger, G. A. 1885. *Catalogue of the lizards in the British Museum.* Vol. 2. London: Taylor and Francis.

Branch, W. R. 1982. Hemipeneal morphology of Platynotan lizards. *J. Herpetol.* 16(1):16–38.

Brandenburg, T. 1983. *Monitors in the Indo-Australian archipelago.* Leiden: Brill.

Bullet, P. 1942. Beiträge zur Kenntnis des Gebisses von *Varanus salvator*. *Laur. Vjschr. Naturf. Ges. Zürich* 87:139–192.

Burt and Burt. 1932. Herpetological results of the Whitney South Sea Expedition VI. *Bull. Am. Nat. Hist.* 63:461–597.

Bustard, H. R. 1970. *Australian lizards.* Sydney: Collins.

Cann, J. 1974. Collecting in Irian Jaya (West New Guinea) during 1972. *Bull. Herp. R. Zool. Soc N. S. W.* 1(3):4–14.

Cantor, T. E. 1847. A catalog of reptiles inhabiting the Malayan Peninsula and Islands. *J. Asiat. Soc. Bengal* 16:608–1078. Reprint, 1966, Asher, Amsterdam.

Ciofi, C., and M. W. Bruford. 1999. Genetic structure and gene flow among Komodo dragon populations inferred by microsatellite loci analysis. *Mol. Ecol.* 8:S17–S30.

Ciofi, C., M. Beaumont, I. R. Swingland, and M. W. Bruford. 1999. Genetic divergence and units for conservation in the Komodo dragon *Varanus komodoensis. Proc. R. Soc. Lond.* B 266:2269–2274.

Cogger, H. G. 1981. A biogeographical study of the Arnhem Land herpetofauna. In *Proceedings of the Melbourne Herpetological Symposium,* ed. Banks and Martins, 148–155. Melbourne: Zoological Board of Melbourne.

———. 1992. *Reptiles and amphibians of Australia.* Chatswood, Australia: Reed Books.

———. 2000. *Reptiles and amphibians of Australia.* 6th ed. Sydney: Reed New Holland.

Cogger, H. G., E. E. Cameron, and H. M. Cogger. 1983. *Zoological catalogue of Australia.* Vol. 1, *Amphibia and Reptilia.* Canberra: Australian Government Publishing Service, Bureau of Flora and Fauna.

Czechura, G. V. 1980. The emerald monitor *Varanus prasinus* (Schlegel): An addition to the Australian mainland herpetofauna. *Mem. Queensl. Mus.* 20:103–109.

Daudin, F. M. 1802–1803. Histoire naturelle générale et particuliere des reptiles. Vol. 8. Paris: Mus. Nat. History.

Debitter, P. M. 1981. *Varanus timorensis timorensis. Lacerta* 40:48–49.

De Lisle, H. F. 1996. *The natural history of monitor lizards.* Malabar, Fla.: Krieger.

De Rooij, N. 1915. *The reptiles of the Indo-Australian archipelago.* Leiden: Brill.

Dedlmar, A. 1994. Haltung und Nachzucht des Smaragdwarans (*Varanus (Odatria) prasinus*). *Salamandra* 30:234–240.

Dedlmar, A., and W. Böhme. 2000. Erster Nachzuchterfolg beim Quittenwaran, *Varanus melinus* Böhme and Ziegler, 1997. *Herpetofauna* 22(127):29–34.

Deraniyagala, P. E. P. 1931. Some Ceylon lizards. *Spolia Zeylanica* 16:139–180.

———. 1944. Four new races of the "Kabaragoya" lizard, *Varanus salvator. Spol. Zeyl.* 24:59–65 (plates 10–12).

———. 1947. The names of water monitors of Ceylon, the Nicobars and Malaya. In *Proceedings of the 3rd Annual Session of the Ceylon Government Press,* 2:12.

———. 1961. The water monitor of the Andaman Islands: A distinct subspecies. *Spol. Zeyl.* 29:203–204 (plates 1 and 2).

Dryden, G. L. 1965. The food and feeding habits of *Varanus indicus* on Guam. *Micronesica* 2(1):73–76.

Dryden, G. L., and E. D. Wikramanayake. 1991. Space and time sharing by

Varanus salvator and *V. bengalensis* in Sri Lanka. *Mertensiella* 2:111–119.

Dryden, G. L., B. Green, E. D. Wikramanayake, and K. G. Dryden. 1992. Energy and water turnover in two tropical varanid lizards, *Varanus bengalensis* and *V. salvator*. *Copeia* 1992(1):102–107.

Earll, C. R. 1982. Heating, cooling and oxygen consumption rates in *Varanus bengalensis*. *Comp. Biochem. Physiol.* 72A:377–381.

Eidenmüller, B. 1986. Observations on the care and a recent breeding of *Varanus (Odatria) timorensis timorensis* Gray 1831. *Salamandra* 22:197–181.

———. 1998. Bemerkungen zur Haltung und Nachtzucht von *Varanus p. prasinus* (Schlegel, 1839) und *V. p. beccarii* (Doria, 1874). *Herpetofauna* 20:8–13.

Engelmann, W. E., and H.-G. 2003. Erstmalige Nachzucht von Karl Schmidt's Waren, *Varanus jobiensis*, im Zoo Leibig. *Zool. Garten N.F.* 73(6):353–358.

Erdelen, W. 1989. Survey of the status of the water monitor lizard (*Varanus salvator*; Reptilia: Varanidae) in South Sumatra. In *Asian monitor lizards: A review of distribution, status, exploitation and trade in four selected species*, ed. R. Luxmoore and B. Groombridge. Report to the CITES Secretariat. Cambridge: World Conservation Monitoring Centre, Annex 3.

———. 1991. Conservation and population ecology of monitor lizards: The water monitor *Varanus salvator* (Laurenti, 1768) in south Sumatra. *Mertensiella* 2:120–135.

Erdelen, W., F. Abel, and M. Riquier. 1997. *Status, Populationsbiologie und Schutz von Bindenwaran (Varanus salvator), Netzpython (Python reticulatus) und Blutpython (Python curtus) in Sumatra und Kalimantan, Indonesien*. Bonn: Abschlussbericht an das Bundesamt für Umwelt, Naturschutz und Reaktorsicherheit.

Ferner, J. W., R. M. Brown, R. V. Sison, and R. S. Kennedy. 2001. The amphibians and reptiles of Panay Island, Philippines. *Asiatic Herpetol. Res.* 9:34–70.

Finlayson, R., and S. J. Woods. 1977. Arterial disease of reptiles. *J. Zool. Lond.* 183:397–410.

Fisher, H. I. 1948. Locality records of Pacific Island reptiles and amphibians. *Copeia* 1948; 1:69.

Flower, S. S. 1937. Further notes on the duration of life in animals. III. Reptiles. *Proc. Zool. Soc. Lond. Ser. A* 1–39.

Frazzetta, T. H. 1962. A functional consideration of cranial kinesis in lizards. *J. Morphol.* 111:287–319.

Frost, M. 1995. Dumeril's monitor hatched at Zoo Atlanta. *AZA Communique*, July 1995:26.

Fuchs, Kh. 1974. Die asiatischen Reptilienhäute. *Leder* 25(1):1–13.

———. 1977. Histologie und mikroskopische Anatomie der Haut des Bindenwarans (*Varanus salvator*). *Stuttg. Beitr. Naturk. Ser. A*, Nr. 299:1–16.

Fuller, S., P. Baverstock, and D. King. 1998. Biogeographic origins of goannas (Varanidae): A molecular perspective. *Mol. Phylogen. Evol.* 9(2):294–307.

Garrett, C. M., and M. C. Peterson. 1991. *Varanus prasinus beccarii* (NCN) behavior. *Herpetol. Rev.* 22:99–100.

Gaulke, M. 1986. Ueber die Situation des Bindenwarans (*Varanus salvator nuchalis*) auf Negros, Philippinen. *Herpetofauna* 8(44):16–18.

————. 1989. Zur Biologie des Bindenwaranes, unter Berücksichtigung der paläogeographischen Verbreitung und der phylogenetischen Entwicklung der Varanidae. *Cour. Forsch. Inst. Senck.* 112:1–242.

————. 1991a. Systematic relationship of the Philippine water monitors as compared with *Varanus s. salvator,* with a discussion of dispersal routes. *Mertensiella* 2:154–167.

————. 1991b. On the diet of the water monitor, *Varanus salvator,* in the Philippines. *Mertensiella* 2:143–153.

————. 1992. Taxonomy and biology of Philippine water monitors (*Varanus salvator*). *Phil. J. Sci.* 121(4):345–381.

Gaulke, M., and A. De Silva. 1997. Monitor lizards of Sri Lanka: Preliminary investigations on their population structure. *Lyriocephalus* 3(1):1–5.

Gaulke, M., W. Erdelen, and F. Abel. 1999. A radio-telemetric study of the water monitor lizard (*Varanus salvator*) in North Sumatra, Indonesia. *Mertensiella* 11:63–78.

Gaulke, M., and E. Curio. 2001. A new monitor lizard from Panay Island, Philippines (Reptilia, Sauria, Varanidae). *Spixiana* 24(3):275–286.

Gaulke, M., and A. Demegillo. 2001. Gut verstekt in den Baeumen: Der Panay Waren. *Mitt. Zool. Gesellschaft Frankfurt* 4:4–6.

Gaulke, M., and J. Reiter. 2001. *Varanus salvator nuchalis,* eine wenig bekannte Unterart des Bindenwarans. *Draco* 7:42–49.

Gaulke, M., E. Curio, A. Demegillo, and N. Paulino. 2002. *Varanus mabitang,* a rare monitor lizard from Panay Island and a new conservation target. *Silliman J.* 43:24–41.

Gleeson, T. 1981. Preferred body temperature, aerobic scope and activity capacity in the water monitor *Varanus salvator. Physiol. Zool.* 54(4): 423–429.

Gray, J. E. 1831. A synopsis of the species of the class Reptilia. In *The animal kingdom,* ed. G. Griffith. London: Francis and Taylor.

————. 1845. Catalogue of lizards in the British Museum.

Green, B., D. King, M. Braysher, and A. Saim. 1991. Thermoregulation, water turnover and energetics of free-living Komodo dragons, *Varanus komodoensis. Comp. Biochem. Physiol. A Comp. Physiol.* 99:97–101.

Greene, H. W. 1986. Diet and arboreality in the emerald monitor, *Varanus prasinus,* with comments on the study of adaptation. *Fieldiana (Zoology),* n.s. 31:1–12.

Groves, J. 1984. Water monitor hatched. *Chicago Herpetol. Soc. Newsletter,* October 1984.

Guenther, A. 1872. On two species of *Hydrosaurus* from the Philippine Islands. *Proc. Zool. Soc. Lond.* 1872:145–146.

Hairston Adams, C. 1996. Crocodile or Papuan monitor. In *Taxon mangagement accounts,* ed. H. S. Hammack. Fort Worth: Fort Worth Zool. Park, Am. Zoo Aqu. Ass. (AZA).

Hallowell, E. 1857. Notes on the reptiles in the collection of the Museum of the Academy of Natural Sciences, Philadelphia, pp. 146–153.

Halverson, J., and L. H. Spelman. 2002. Sex determination. In *Biology and conservation of Komodo dragons,* ed. J. B. Murphy, C. Ciofi, C. de La Panouse, and T. Walsh. Washington, D.C.: Smithsonian Institution Press.

Hardwicke, T., and J. E. Gray. 1827. A synopsis of the species of the saurian reptiles, collected in India by Major-General Hardwicke. *Zool. J.* 3:213–229.

Harrison, J. L., and B. L. Lim. 1957. Monitors of Malaya. *Malayan Nat. J.* 12(1):1–10.

Hartdegen, R. W., D. Chiszar, and J. B. Murphy. 1999. Observations on the feeding behavior of captive black tree monitors, *Varanus beccari. Amphibia-Reptilia* 20:330–332.

Hartdegen, R. W., D. T. Roberts, and D. Chiszar. 2000. Laceration of prey integument by *Varanus prasinus* (Schlegel, 1839) and *V. beccarii* (Doria, 1874). *Hamadryad* 25:196–198.

Harvey, M. B., and D. G. Barker. 1998. A new species of blue-tailed monitor lizard (genus *Varanus*) from Halmahera Island, Indonesia. *Herpetologica* 54(1):34–44.

Hauschild, A. 1998. Haltung und Nachzucht des Dumeril-Warans, *Varanus dumerilii. Herpetofauna* 115:27–34.

Heaney, L. R., and J. C. Regalado. 1998. *Vanishing treasures of the Philippine rain forest*. Chicago: Field Museum.

Hediger, H. 1934. Beitrage zur Herpetologie und Zoogeographie Neu Britanniens und einiger umliegender Gebiete. *Zool. Jahrb. Syst.* 65: 441–582.

———. 1962. Tierpsychologische Beobachtungen aus dem Terrarium des Züricher Zoos. *Rev. Suisse Zool.* 69:317–324.

Herrmann, H.-W. 1999. Husbandry and captive breeding of the water monitor, *Varanus salvator* (Reptilia: Sauria: Varanidae) at the Cologne Aquarium (Cologne Zoo). *Mertensiella* 11:95–103.

Hofer, H. 1960. Vergleichende Untersuchungen am Schädel von *Tupinambis* und *Varanus* mit besonderer Berücksichtigung ihrer Kinetik. *Morph. Jahrb.* 100(4):706–746.

Holmes, R. S., M. King, and D. King. 1975. Phenetic relationships among varanid lizards based upon comparative electrophoretic data and karyotypic analyses. *Biochem. System. Ecol.* 3:257–262.

Honegger, R. E. 1971. Zoo breeding and crocodile bank. *Proceedings of the First Working Meeting of Crocodilians Special* 1:86–97.

———. 1995. Bindenwaran. *Irbis* 12, Bull. 1. Zürich, Tiergarten-Ges.

Hoogerwerf, A. 1951. Warnemingen gij voedel zoekende watervogels en varanen. *Limosa* 24(1/2):55–59.

———. 1970. *Udjung Kulon: The land of the last Javan rhinoceros.* Leiden: Brill.

Horn, H.-G. 1977. Notizen zur Systematik, Fundortangaben und Haltung von *Varanus* (*Varanus*) *karlschmidti* (Reptilia: Sauria: Varanidae). *Salamandra* 13(2):78–88.

———. 1980. Bisher unbekannte Details zur Kenntnis von *Varanus varius* auf Grund Von feldherpetologischen und terraristischen Beobachtungen. *Salamandra* 16(1):1–18.

———. 1985. Beiträge zum Verhalten von Waranen: Die Ritualkämpfe von *Varanus komodoensis* Ouwens, 1912 und *V. semiremex* Peters, 1869 sowie die Imponierphasen der Ritualkämpfe von *V. timorensis timorensis* Gray, 1931 und *V. t. similis* Mertens, 1958 Sauria: Varanidae. *Salamandra* 21:169–179.

———. 1994. Der Ritualkampf von *Varanus salvator cumingi. Herpetofauna* 16(92):27–30.

———. 1995. Odyssee und Rettungsversuch geschmuggelter Reptilien mit Anmerkungen zu Sinn und Unsinn praktischen Naturschutzes. International Symposium Vivaristik Dokumentation, 35–37.

———. 1999. Evolutionary efficiency and success in monitors: A survey on behavior and behavioral strategies and some comments. *Mertensiella* 11:167–180.

———. 2002a. Die Echsen der Platynota: Biologie und Terrarienhaltung. In *Zootierhaltung-Tiere in menschlicher Obhut: Reptilien, Amphibien, Fische, Wirbellose,* ed. W.-E. Engelmann. Frankfurt: Vlg. H. Deutsch.

———. 2002b. Beleuchtung im Virarium (Paludarium, Terrarium). In *Praxis Ratgeber: Vivariumbeleuchtung,* ed. Kh. Sauer, B. Steck. H. Schuchardt, and H.-G. Horn. Frankfurt am Main: Edition Chimaira (in press).

Horn, H.-G., and B. Schulz. 1977. *Varanus dumerilii,* wie ihn nicht jeder kennt. *Aquarium* 11(9):37–38.

Horn, H.-G., and G. Petters. 1982. Beitrage zur Biologie des Rauhnackwarens, *Varanus (Dendrovaranus) rudicollis. Salamandra* 18(1–2):29–40.

Horn, H.-G., and G. J. Visser. 1989. Review of reproduction of monitor lizards *Varanus* spp. in captivity. *Int. Zoo Yrbk.* 28:140–150.

———. 1991. Basic data on the biology of monitors. *Mertensiella* 2:176–187.

———. 1997. Review of reproduction of monitor lizards *Varanus* spp in captivity. II. *Int. Zoo Yrbk.* 35:227–246.

Horn, H.-G., M. Gaulke, and W. Böhme. 1994. New data on ritualized combats in monitor lizards (Sauria: Varanidae) with remarks on their function and phylogenetic implications. *Zool. Garten* N.F. 64(5):265–280.

Jacobs, H. J. 2002a. Zur morphologischen Variabilität der nominellen Smaragdwaran-Taxa *Varanus prasinus* (Schlegel, 1839) und *Varanus kordensis* (A. B. Meyer, 1874) mit Bemerkungen zur Erstzucht des letzteren. *Herpetofauna* 24(137):21–34.

———. 2002b. Erstnachzucht von *Varanus macraei. Herpetofauna* 24(141):29–33.

———. 2003. A further new emerald tree monitor lizard of the *Varanus prasinus* species group from Waigeo, West Irian (Squamata: Sauria: Varanidae). *Salamandra* 39(2):65–74.

James, C. D., J. B. Losos, and D. R. King. 1992. Reproductive cycles and diets of goannas (Reptilia: Varanidae) from Australia. *J. Herpetol.* 26:128–136.

Jasmin, A. 1986. Preliminary study of growth rate and movement of water monitor lizard (*V. salvator*) at Taman Negara. *J. Wildlife Parks* 5:63–78.

———. 1988. Report of licenses, kills and exports of *Varanus* species in peninsular Malaysia. Consultancy report to CITES.

———. 1989. Report of licences, kills and exports of *Varanus* species in Peninsular Malaysia. In *Asian monitor lizards: A review of distribution, status, exploitation and trade in four selected species,* ed. R. Luxmoore and B. Groombridge. Report to the CITES Secretariat. Cambridge: World Conservation Monitoring Centre, Annex 4.

———. 1990. The monitor lizard of Sungai Tembeling, Taman Negara. *J. Wildlife Parks* 10:109–115.

Jasmin, A., and M. Amin Abdullah. 1987/88. Growth rate and behaviour of water monitor lizard (*Varanus salvator*) at Sg. Tembeling, Taman Negara. *J. Wildlife Parks* 6/7:58–66.

Kala, N. 1998. Captive Breeding of *Varanus salvator andamanensis* Deraniyagala, 1944. *Hamadryad* 22:122–123.

Khan, M. 1969. A preliminary study of the water monitor, *Varanus salvator. Malayan Nat. J.* 22(2):64–68.

King, D. R. 1993. The diet and reproductive condition of free-ranging *Varanus timorensis. W. Austr. Naturalist* 19:189–194.

King, M., and D. King. 1975. Chromosomal evolution in the lizard genus *Varanus* (Reptilia). *Aust. J. Biol. Sci.* 28:89–108.

King, D., M. King, and P. Baverstock. 1991. A new phylogeny of the Varanidae. *Mertensiella* 2:211–219.

King, D., and B. Green. 1993. *Monitors: The biology of varanid lizards.* Malabar, Fla.: Kreiger.

King, D. R., and J. E. Keirans. 1997. Ticks (Acari: Ixodidae) from varanid lizards in eastern Indonesia. *Rec. W. Austr. Mus.* 18:229–230.

King, D., S. Fuller, and P. Baverstock. 1999. The biogeographic origins of varanid lizards. *Mertensiella* 11:43–49.

Kirschner, A., T. Müller, and H. Seufer. 1996. *Faszination Warane.* Kirschner and Seufer Verlag.

Kirschner, A., and H. Koschorke. 1998. Das Portrait: *Varanus yuwonoi* Harvey and Barker 1998. *Sauria* 20(4):1.

Kok, R. 2000. Ervaringen en geslaagde kweek met de smaragdvaraan, *Varanus prasinus* (Schlegel, 1839). *Lacerta* 58:109–112.

Kopstein, F. (s.a.) *Zoologische Tropenreise.* Leiden: Vlg. Kolff.

Krammig, D. 1977. Die Andamanen—Eine Inselwelt voller Geheimnisse. *Tier* 1977(1):12–15.

Kratzer, H. 1973. Beobachtungen über die Zeitigungsdauer eines Eigeleges von *Varanus salvator. Salamandra* 9(1):27–33.

Krebs, U. 1979. Der Dumeril-Waran (*Varanus dumerilii*), ein spezialisierter Krabbenfresser? *Salamandra* 15(3):146–157.

———. 1991. Ethology and learning: From observation to semi-natural experiment. *Mertensiella* 2:220–232.

Ladiges, W. 1939. Herpetologische Beobachtungen auf Sumatra. *Zool. Anz.* 128:235–249.

Laidlaw, F. F. 1901. On a collection of lizards from the Malay Peninsula, made by members of the "Skeat Expedition" 1899–1900. *Proc. Zool. Soc. Lond.* 1901:301–310.

Laurenti, J. N. 1768. *Specimen medicum eshibens synopsin reptilium emendatam cum experimentis circa venema et antidota reptilium austriacorum.* Vienna: Syn. Rept.

Lederer, G. 1929. Ein zahmer Bindenwaran. *Lacerta, Zeitschr. Vivar. Kde.* 26:19–20.

———. 1933. Beobachtungen an Waranen im Frankfurter Zoo. *Zool. Garten N.F.* 6:118–126.

Lekagul, B. 1969. Monitors of Thailand. *Conserv. News SE Asia* 8:31–32.

Lemm, J. 1998. Year of the monitor: A look at some recently discovered varanids. *Reptiles* 6(9):70–81.

Lim, B. L. 1958. The harlequin monitor lizard. *Malay Nat. J.* 13:70–72.

Losos, J. B., and H. W. Greene. 1988. Ecological and evolutionary implications of diet in monitor lizards. *Biol. J. Linn. Soc.* 35:379–407.

Loveridge, A. 1948. New Guinean reptiles and amphibians in the Museum of Comparative Zoology and United States National Museum. *Bull. Mus. Comp. Zool.* 101:304–430.

Luxmoore, R., B. Goombridge, and S. Broad, eds. 1988. Significant trade in wildlife: A review of selected species. In CITES appendix II, Reptiles and vertebrates. Cambridge: IUCN.

Luxmoore, R., and B. Groombridge, eds. 1989. *Asian monitor lizards: A* review of distribution, status, exploitation and trade in four selected species. Report to the CITES Secretariat. Cambridge: World Conservation Monitoring Centre.

Lydekker, R. 1886. Fauna of the Karnul caves. *Palaeontol. Ind.* (10)4:1–58.

Mader, D. R. 1994. Reptile medicine. *Reptiles* 1(3):24–27.

Madsen, F. 1990. Rapport fra Malmö Reptil center: Papua-varanen (*Varanus salvadorii*). *Nordic Herp. Soc.* 33:70–76.

Martin, W. 1838. [Description of *Varanus cumingi.*] *Proc. Zool. Soc. Lond.* 1838:69–70.

McCoid, M. J. 1993. Reproductive output in captive and wild mangrove monitors (*Varanus indicus*). *Varanews* 3(3):4.

McCoid, M. J., and R. A. Hensley. 1991. Mating and combat in *Varanus indicus*. *Herpetol. Rev.* 22:16–17.

———. 1993. Observations of *Varanus indicus* in the Mariana Islands. *Varanews* 3(6):4–5.

McCoid, M. J., and G. J. Witteman. 1993. *Varanus indicus* diet (mangrove monitor). *Herpetol. Rev.* 24:105.

———. 1994. Factors in the decline of *Varanus indicus* on Guam, Mariana Islands. *Herpetol. Rev.* 25(2):60–61.

McCoy, M. 1980. Reptiles and amphibians in the Solomon Islands. *Wau Ecol. Inst. Handbook* 7:1–80.

Meeks, R. 1978. On the thermal relations of two oriental varanids: *Varanus bengalensis nebulosus* and *Varanus salvator*. Cotswold Herpetological Symposium, 32–47.

Meer Mohr, J. C. 1930. Over eieren van *Varanus salvator* en van *Python curtus*. *Trop. Nat.* 19:156–157.

Mertens, R. 1926. Über die Rassen einiger indo-australischer Reptilien. *Senckenbergiana* 8:272–279.

———. 1937. Zwei Bemerkungen über Warane (Varanidae). *Senckenbergiana* 19:177–181.

———. 1941. Zwei neue Warane des australischen Faunengebietes. *Senckenbergiana* 23(4/6):261–272.

Mertens, R. 1942. Die Familie der Warane (Varanidae). *Abh. Senck. Naturf. Ges.* 462, 466 467:1–391.

———. 1950a. Notes on some Indo-Australian monitors (Sauria: Varanidae). *Am. Mus. Novitates* 1456:1–7.

———. 1950b. Der Schädel von *Varanus salvadorii*. *Neue Ergebn. Probl.* (Klatt-Festschrift, Leipzig), 561–566.

———. 1951. A new lizard of the genus *Varanus* from New Guinea. *Fieldiana* 31(43):467–471.

———. 1957. Two new goannas from Australia. *Western Aust. Nat.* 5:183–185.

———. 1958. Bemerkungen über die Warane Australiens. *Senckenbergia Biol.* 39:229–264.

———. 1959. Liste der Warane Asiens und der Indo-australischen Inselwelt mit systematischen Bemerkungen. *Senck. Biol.* 40:221–240.

———. 1960. Der Papua-Waran. *Kosmos* 56:547–549.

———. 1962. Papusaurus, eine neue Untergattung von *Varanus*. *Senck. Biol.* 43(5):331–333.

———. 1963. Liste der rezenten Amphibien und Reptilien, Helodermatidae, Varanidae, Lanthanotidae. *Tierreich Lief.* 79:v–x, 1–26.

———. 1971a. Über eine Waransammlung aus dem östlichen Neuguinea. *Senck. Biol.* 52(1/2):1–5.

———. 1971b. Unerwartete Bananenfresser unter Reptilien. *Salamandra* 7:39–40.

Meyer, A. W. 1874. Übersicht über die von mir auf Neu-Guinea und den Inseln Jobi, Mydore und Mafoor im Jahre 1873 gesammelten Amphibien. *Mber. K. Akad. Wiss. Berlin* 1874:128–138.

Moehn, L. D. 1984. Courtship and copulation in the Timor monitor, *Varanus timorensis. Herpetol. Rev.* 15:14–16.

Moharana, S., and S. Pati. 1983. Het eierleggen van *Varanus salvator* in het Nandan Kanan Zoological Park in India. *Lacerta* 41(4):67–68.

Morwood, M. J., P. B. O'Sullivan, F. Aziz, and A. Raza. 1998. Fission-track ages of stone tools and fossils on the east Indonesian island of Flores. *Nature* 392:172–176.

Murphy, J. B. 1972. Notes on Indo-Australian varanids in captivity. *Int. Zoo Yrbk.* 12:199–202.

Neill, W. T. 1958. The occurrence of reptiles and amphibians in saltwater areas and a bibliography. *Bull. Mar. Sci. Gulf Carrib.* 8(1):1–97.

Neugebauer, W. 1973. Familie Warane. In *Grzimeks Tierleben VI (Kriechtiere),* ed. B. Grzimek, 324–337. Zürich: Kindler Verlag.

Nutphand, W. N.d. *The monitors of Thailand* (in Thai). Bangkok: Mitphadung Publishing Office.

Oelofsen, B. W., and J. A. van den Heever. 1979. Role of the tongue during olfaction in varanids and snakes. *S. Afr. J. Sci.* 75(8):365–366.

O'Shea, M. 1991. The reptiles of Papua New Guinea. *Bull. Br. Herpetol. Soc.* 37:15–31.

Ott, T. 1997. Nachzuchtergebnisse von *Varanus salvator. Elaphe* 5(4):6–11.

Owerkowicz, T., C. G. Farmer, J. W. Hicks, and E. Brainerd. 1999. Contribution of gular pumping to lung ventilation in monitor lizards. *Science* 284:1661–1663.

Pandav, B. 1993. A preliminary survey of the water monitor (*Varanus salvator*) in Bhitarkanika Wildlife Sanctuary, Orissa. *Hamadryad* 18:49–51.

Pandav, B., and B. C. Choudhury. 1996. Diurnal and seasonal activity patterns of water monitor (*Varanus salvator*) in the Bhitarkanika mangroves, Orissa, India. *Hamadryad* 21:4–12.

Pepin, D. J. 1999. The origin of monitor lizards based on a review of the fossil evidence. *Mertensiella* 11:11–42.

Peters, W., and G. Doria. 1878. Catalogo dei reptili e dei batraci racolti de O. Beccari, L. M. D'Albertis e A. A. Bruijn nella sotto regione austromalese. *Ann. Mus. Civ. Stor. Nat. Genova* 13:323–450.

Philipp, K. M. 1999a. Niche partitioning of *Varanus doreanus, V. indicus* and *V. jobiensis* in Irian Jaya: Preliminary results. *Mertensiella* 11:307–316.

———. 1999b. Zur Systematik, Zoogeographie und Ökologie des Pazifikwaran-Artkomplexes (Reptilia: Squamata: Varanidae). M.A. thesis. Zoologische Staatssammlung.

Philipp, K. M., W. Böhme, and T. Ziegler. 1999a. The identity of *Varanus indicus*: redefinition and description of a sibling species coexisting at the type locality (Sauria: Varanidae, *Varanus indicus* group). *Spixiana* 22:273–287.

Philipp, K. M., T. Ziegler, and W. Böhme. 1999b. Der Türkiswaran *Varanus caerulivirens* Ziegler, Böhme und Philipp, 1999. *Herpetofauna* 21(122):10–11.

Philippen, H.-D. 1994. Der Papuawaran, *Varanus [Papusaurus] salvadorii* (Peters and Doria). *Monitor* 3(1):17–18.

Pitman, C. R. S. 1962. More snake and lizard predators of birds. *Bull. Br. Ornithol. Club* 82(3):45–55.

Prasad, K. N., and P. Yadagiri. 1986. Pleistocene cave fauna, Karnool District, Andhra Pradesh. *Rec. Geol. Surv. India* 115:71–77.

Radford, L., and F. L. Payne. 1989. The reproduction and management of *Varanus dumerilii. Int. Zoo Yrbk.* 28:153–155.

Randow, H. 1932. Fauna und Flora von Dehiwala auf Ceylon. *Wschr. Aquar. Terrar. Kunde* 29:471–473.

Rao, R. J. N.d. Survey of the conservation status of monitor lizards (*Varanus* sp.) in India. School of Studies in Zoology. Jiwaji University, Gwalior, 1994–97. Available at: http://envfor.nic.in/divisions/re/ta5p5.html. Accessed October 1, 2003.

Rashid, S. M. A., and C. H. Diong. 1999. Observations on *Varanus salvator* feeding on *Oligodon octolineatus. Hamadryad* 24(1):48–49.

Raven, H. C. 1946. Predators eating green turtle eggs in the East Indies. *Copeia* 1946(1):48.

Rawlinson, P. A., A. H. T. Widjoya, M. N. Hutchinson, and G. W. Brown. 1990. The terrestrial vertebrates of Krakatau Islands, Sunda Strait, 1883–1986. *Phil. Trans. R. Soc. Lond. B* 328:2–28.

Rese, R. 1983. Der Timorwaran *Varanus timorensis timorensis*: Haltung und Zucht. *Sauria* 4:13–15.

Rieppel, O. 1978. The phylogeny of cranial kinesis in lower vertebrates with special reference to the Lacertilia. *N. Jb. Geol. Paläont. Abh.* 156:335–370.

———. 1979. A functional interpretation of the varanid dentition (Reptilia, Lacertilia, Varanidae). *Geigenbaurs Morph. Jahrb.* 125:797–817.

Robinson, M. H. 1969. Defenses against visually hunting predators. *Evol. Biol.* 3:225–259.

Rooij, N. de. 1915. *The reptiles of the Indo-Australian archipelago.* Leiden.

Room, P. M. 1974. Lizards and snakes from the Northern District of Papua, New Guina. *Br. J. Herpetol.* 5:438–446.

Rotter, J. 1963. *Die Warane (Varanidae).* Die Neue Brehm Bacherei, Heft 325. Wittenberg: A. Ziemsen Verlag.

Sarker, S. U. 1987. Crocodiles and lizards of Siragonj (Pabna) Bangladesh. *Tigerpaper* 14(4):30–32.

Sasse, A. 1999. Der Papuawaran—*Varanus salvadorii* Peters and Doria, 1878. *Jahrbuch Herpetol. Terrarist. Vereinig. Österreich* 1999:55–56.

Sastrawan, P., and C. Ciofi. 2002. Distribution and activity pattern of *Varanus komodoensis.* In *Biology and conservation of Komodo dragons,* ed. J. B. Murphy, C. Ciofi, C. de La Panouse, and T. Walsh, 42–77. Washington, D.C.: Smithsonian Institution Press.

Sautereau, L., and P. de Bitter. 1980. Notes sur l'élevage et la reproduction en captivité du varan de Timor Sauria-Varanidae. *Bull. Soc. Herpetol. Française* 15:4–9.

Schlegel, H. 1839. *Abbildungen neuer oder unvollständig bekannter Amphibien, nach der Natur oder dem Leben entworfen und mit einem erläuternden Texte begleitet.* Düsseldorf: Arne.

Schmicking, T., and H.-G. Horn. 1997. Beobachtungen bei der Pflege und Nachtzucht des Papuawarans, *Varanus salvadorii* (Peters and Doria 1878). *Herpetofauna* 19(106):14–23.

Schmidt, K. P. 1927. The reptiles of Hainan. *Bull. Am. Mus. Nat. Hist.* 54:395–465.

Schmutz, E., and H.-G. Horn. 1986. The habitat of *Varanus Odatria t. timorensis* Gray 1831 Sauria: Varanidae. *Salamandra* 22:147–152.

Schultze-Westrum, Th. 1961. Wasser- und Baumreptilien auf Neuguinea. *Kosmos* 57:247–252.

———. 1972. *Neuguinea*. Bern: Kümmerly and Frey.

Shine, R., P. S. Harlow, J. S. Keogh, and Boeadi. 1996. Commercial harvesting of giant lizards: The biology of water monitors *Varanus salvator* in Southern Sumatra. *Biol. Conserv.* 77:125–134.

Shine, R., Ambariyanto, P. S. Harlow, and Mumpuni. 1998. Ecological traits of commercially harvested water monitors, *Varanus salvator,* in northern Sumatra. *Wildlife Res.* 25:437–447.

Sights, W. P. 1949. Annotated list of reptiles taken in Western Bengal. *Herpetologica, Chicago* 5(4):81–83.

Smet, W. H. O. de. 1981. Description of the orcein stained karyotypes of 36 lizard species (Lacertilia, Reptilia) belonging to the families Teiidae, Scincidae, Lacertidae, Cordylidae and Varanidae (Autarchoglossa). *Acta Zool. Pathol. Antverp* 76:73–118.

Smith, H. C. 1931. The monitor lizards of Burma. *J. Bombay Nat. Hist. Soc.* 34:367–373.

Smith, M. A. 1932. Some notes on monitors. *J. Bombay Nat. Hist. Soc.* 35:615–619.

———. 1935. The fauna of British India. In *Reptiles and amphibians*. Vol. 2, *Sauria*. London: Taylor and Francis.

Smith, K. K. 1986. Morphology and function of the tongue and hyoid apparatus in *Varanus* (Varanidae, Lacertilia). *J. Morphol.* 187:261–287.

Spawls, S., K. Howell, R. Drewes, and J. Ashe. 2002. *A field guide to the reptiles of East Africa: Kenya, Tanzania, Uganda, Rwanda and Burundi*. San Diego: Academic Press.

Sprackland, R. G. 1991. Taxonomic review of the *Varanus prasinus* group with descriptions of two new species. *Mem. Queensl. Mus.* 30:561–576.

———. 1993a. Rediscovery of a Solomon Islands monitor lizard (*Varanus indicus spinulosus*) Mertens, 1941. *Vivarium* 4(5):25–27.

———. 1993b. The taxonomic status of the monitor lizard *Varanus dumerilli heteropholis* Boulenger 1892. *Sarawak Mus. J.* 44:113–121.

———. 1994. Rediscovery and taxonomic review of *Varanus indicus spinulosus* Mertens, 1941. *Herpetofauna* 24(2):33–39.

———. 1997. Mangrove monitor lizards. *Reptiles* 1997(3):48–63.

———. 1999a. Character displacement among sympatric varanid lizard species: A study of natural selection and systematics in action (Reptilia: Squamata: Varanidae). *Mertensiella* 11:113–120.

———. 1999b. A new species of monitor from Indonesia. *Reptile Hobbyist*, February:20–27.

———. 1999c. Species, species everywhere: Where do new species come from? *Vivarium* 10(3):7–8, 36–38.

Storr, G. M., L. A. Smith, and R. E. Johnstone. 1983. *Lizards of Western Australia II: Dragons and monitors*. Perth: Western Australian Museum.

Struck, U. A. V. Altenbach, M. Gaulke, and F. Glaw. 2002. Tracing the diet of the monitor lizard *Varanus mabitang* by stable isotope analyses (Δ^{15}N, Δ^{13}C). *Naturwissenschaften* 89:470–473.

Subba Rao, M. V., and K. Rao. 1984. Feeding ecology of the Indian common monitor, *Varanus monitor*. In *Sixth Annual Symposium on Captive Propagation and Husbandry*, ed. D. Marcellilni, 197–204. Thumond, Md.: Zoological Consortium.

Taylor, E. H. 1922. The lizards of the Philippine Islands. *Phil. J. Sci. Monogr.* 17:1–269.

———. 1963. Lizards of Thailand. *Univ. Kans. Sci. Bull.* 44(14).

Thompson, G. 1999. Goanna metabolism: Different to other lizards, and if so, what are the ecological consequences? *Mertensiella* 11:79–90.

Traeholt, C. 1993. Notes of the feeding behaviour of the water monitor, *Varanus salvator. Malayan Nat. J.* 46:229–241.

———. 1994a. The food and feeding behaviour of the water monitor, *Varanus salvator*, in Malaysia. *Malayan Nat. J.* 44:331–343.

———. 1994b. Notes on the water monitor *Varanus salvator* as a scavenger. *Malayan Nat. J.* 47:345–353.

———. 1995a. A radio-telemetric study of the thermoregulation of free living water monitor lizards, *Varanus s. salvator. J. Comp. Physiol. B* 165:125–131.

———. 1995b. Notes on the burrows of the water monitor lizard, *Varanus salvator. Malayan Nat. J.* 49:103–112.

———. 1997a. Population dynamics and status of water monitor lizards, *Varanus salvator*, in two different habitats of Malaysia. In *Wetlands biodiversity and development: Proceedings of Workshop 2 of the International Conference on Wetlands and Development*, ed. W. Giesen, 147–160. Kualalumpur: Wetlands International.

———. 1997b. Effect of masking the parietal eye on the diurnal activity and body temperature of two sympatric species of monitor lizards, *Varanus s. salvator* and *Varanus b. nebulosus. J. Comp. Physiol. B* 167:177–184.

———. 1997c. Ranging behavior of the water monitor lizard *Varanus salvator. Malayan Nat. J.* 50:317–329.

Uchida, T. 1967. Observations on the monitor lizard as a rat control agent on Ifaluk, Western Caroline Islands. *Micronesica* 3(1):17–18.

Upton, S. J., and P. S. Freed. 1990. Description of the oocysts of *Eimeria beccari* n. sp. (Apicomplexa: Eimeriidae) from *Varanus prasinus beccari* (Reptilia: Varanidae). *Syst. Parasitol.* 16:181–184.

Villamor, C. I. 1993. Morphometry and conservation status of water monitor lizard (*Varanus salvator*) in the Philippines. *Asia Life Sci.* 2(2):113–120.

Visser, G. J. 1985. Notizen zur Brutbiologie des Gelbwarans *Varanaus (Empagusia) flavescens* (Hardwicke and Gray, 1827) im Zoo Rotterdam. *Salamandra* 21(2/3):161–168.

Vogel, P. 1979a. Zur Biologic des Bindenwarans (*Varanus salvator*) im westjavanischen Naturschutzgebiet Ujung Kulon. Ph.D. diss., University of Basel.

———. 1979b. Innerartliche Auseinandersetzungen bei freilebenden Bindenwaranen (*Varanus salvator*). *Salamandra* 15(2):65–83.

Vogel, D. 1994. Erstmalige Nachzucht des Mindanao Bindenwarans (*Varanus* [*Varanus*] *salvator cumingi*) im Zoologischen Garten Frankfurt. *Monitor* 3(1):41. (Internal circular of the GDHT.)

Walsh, T., D. Chiszar, F. G. F. Birchard, and K. M. T. A. Tirtodiningrat. 2002. Captive management and growth. In *Komodo Dragons: Biology and conservation*, ed. J. B. Murphy, C. Ciofi, C. de La Panouse, and T. Walsh. Washington, D.C.: Smithsonian Institution Press, 178–195.

Wells, R. W., and R. C. Wellington. 1985. A classification of the Amphibia and Reptilia in Australia. *J. Herpetol., Suppl. Ser.* 1:1–61.

Werner, F. 1900. Reptilien und Batrachier aus Sumatra, gesammelt Herrn. Gustav Schieder jnr. im Jahre 1897–1898. *Zool. Jharbuch (Systematik)* 13:479–508.

Whittier, J. M., and D. R. Moeller. 1993. *Varanus prasinus* (the emerald goanna) on Moa Island, Torres Strait, Australia. *Mem. Queensl. Mus.* 34:130.

Wicker, R., M. Gaulke, and H.-G. Horn. 1999. Contributions to the biology, keeping and breeding of the Mindanao Water Monitor (*Varanus s. cumingi*). *Mertensiella* 11:213–223.

Wiegmann, A. F. A. 1834. Beitraege zur Zoologie, gesammelt auf einer Reise um die Erde, von F. J. F. Meyen. Siebente Abhandlung. *Nova Acta Physico-Medica* 9:183–268.

Wikramanayake, E. D. 1995. Activity and thermal ecophysiology of two sympatric monitor lizards in Sri Lanka. *J. South Asian Nat. Hist.* 1:213–224.

Wikramanayake, E. D., and G. L. Dryden. 1988. The reproductive ecology of *Varanus indicus* on Guam. *Herpetologica* 44:338–344.

———. 1993. Thermal ecology of habitat and microhabitat use by sympatric *Varanus bengalensis* and *V. salvator* in Sri Lanka. *Copeia* 1993(3):709–714.

———. 1999. Implications of heating and cooling rates in *Varanus bengalensis* and *V. salvator. Mertensiella* 11:149–156.

Wikramanayake, E. D., and B. Green. 1989. Thermoregulatory influence on the ecology of two sympatric varanids in Sri Lanka. *Biotropica* 21:74–79.

Wikramanayake, E., W. Ridwan, and D. Marcellini. 1999. The thermal ecology of free-ranging Komodo dragons, *Varanus komodoensis,* on Komodo Island, Indonesia. *Mertensiella* 11:157–166.

Wilson, S. K., and D. G. Knowles. 1988. Monitors or goannas—Family Varanidae. In *Australia's reptiles: A photographic reference to the terrestrial reptiles of Australia,* 353–361. Sydney: Collins.

Young, B. A., G. Abishahin, M. Bruther, C. Kinney, and J. Segroi. 1998. Acoustic analysis of the defensive sounds of *Varanus salvator* with notes on sound production in other varanid species. *Hamadryad* 23(1):1–14.

Yuwono, F. B. 1998. The trade of live reptiles in Indonesia. *Mertensiella* 9:9–15.

Zhao, E., and K. Adler. 1993. *Herpetology of China.* Society for the Study of Amphibians and Reptiles.

Ziegler, T., and W. Böhme. 1996. Über das Beutespektrum von *Varanus dumerilii* (Schlegel 1839). *Salamandra* 32(3):203–210.

———. 1997. Genitalstrukturen und Paarungsbiologie bei squamaten Reptilien, speziell der Platynota, mit Bemerkungen zur Systematik. *Mertensiella* 8:3–207.

———. 1999. Genital morphology and systematics of two recently described monitor lizards of the *Varanus (Euprepiosaurus) indicus* group. *Mertensiella* 11:121–128.

Ziegler, T., W. Böhme, and U. Schweers. 1998. Spektakuläre Neuentdeckungen innerhalb der Pazifikwaran-Gruppe. *Reptilia* 3(6):14–16.

Ziegler, T., W. Böhme, and K. M. Philipp. 1999a. *Varanus caerulivirens* sp. n., a new monitor lizard of the *V. indicus* group from Halmahera,

Moluccas, Indonesia (Squamata: Sauria: Varanidae). *Herpetozoa* 12(1/2):45–56.

Ziegler, T., K. M. Philipp, and W. Böhme. 1999b. Zum Artstatus und zur Genitalmorphologie von *Varanus finschi* Böhme, Horn und Ziegler, 1994, mit neuen Verbreitungsangaben für *V. finschi* und *V. doreanus* (A. B. Meyer, 1874) (Reptilia. Sauria: Varanidae). *Zool. Abh. Dresden* 50(2):267–279.

Ziegler, T., W. Böhme, B. Eidenmüller, and K. M. Philipp. 2001. A note on the coexistence of three species of Pacific monitor lizards in Australia (Sauria, Varanidae, *Varanus indicus* group). *Bonner Zool. Beiträge* 50(1–2):27–30.

7. Australian Varanid Species

Twenty-four named varanid species are endemic to Australia. One, *Varanus panoptes*, has reached southern New Guinea, where it is given distinct subspecies status, *V. panoptes horni*. A member of the *V. scalaris* species complex is also found in New Guinea. There are two distinct clades of Australian monitors: pygmy monitors belong to the subgenus *Odatria* (16 described species, plus many more that remain undescribed), whereas all but one (*V. varius*) of the larger species belong to the subgenus *Varanus*, which includes *V. giganteus, V. gouldii, V. mertensi, V. panoptes, V. rosenbergi,* and *V. spenceri*. *V. varius* is a member of another small clade that includes *V. salvadorii* and *V. komodoensis*. Several Asian monitor species, including members of the *indicus* (*doreanus, finschi,* and *indicus,* sensu stricto) and *prasinus* (*keithhornei*) species complexes, have reached northern tropical Australia. Many other new species, including members of the *gouldii, scalaris,* and *tristis* species groups, remain to be described. *V. "glauerti"* from Kakadu is an as yet undescribed new *Odatria* species, as is *V. "storri"* from central Queensland. Another undescribed new *Odatria* species from Western Australia is related to *V. caudolineatus* and *V. gilleni*.

Accurate range maps convey vital information about where a species occurs. We were fortunate that all the Australian museums allowed us to use and print their records. We culled some of the most obvious outlying records, which were clearly erroneous, but include the actual museum numbers and localities so that future workers can check doubtful records. Some dot maps and some localities are probably wrong, but these are the best records available. Records from non-Australian museum collections are not shown, but these will be scant compared with those from Australia.

Abbreviations used for these localities are as follows: HS, Homestead; Hwy., Highway; Is., Island; PO, Post Office; Mt., Mountain; Ra., Range(s); RR, railroad; SF, State Forest; Stn., Station; TO, Turnoff; AM, Australian Museum; NMV, National Museum of Victoria; NTM, Northern Territory Museum; QM, Queensland Museum; SAM, South Australian Museum; WAM, Western Australian Museum.

7.1. *Varanus acanthurus*

GIL DRYDEN

Figure 7.1. Varanus acanthurus *(photo by Jeff Lemm).*

The spiny-tailed monitor was described in 1885 by Boulenger. Its taxonomy was reviewed by Storr (1980) and is now considered a species complex within the subgenus *Odatria* (Cogger 1986; King and Horner 1987). The holotype from the northwest coast of Australia is in the British Museum (BMNH 1945.8.30.97) (Cogger et al. 1983) (Fig. 7.1).

Geographic Distribution

Varanus acanthurus is an arid-adapted form extending over the tropical and subtropical portions of Western Australia, the Northern Territory, northwest Queensland, and associated offshore islands (Storr 1980). Although Cogger (1992) claimed that they occur in northern South Australia, Houston and Hutchinson (1998) assert that they are not found in that state. One record from southern Queensland is very close to the South Australian border (Fig. 7.2).

QM R2143 Gregory Downs, via Burketown—18.65 139.25; QM R2979 Stapleton —13.18 131.033; QM R10200 Frewena, near—19.43 135.4; QM R10790, R10792 Doomadgee Mission Stn.—17.93 138.817; QM R24890 Dajarra, 19 km S—21.93 139.583; QM R31983 Mary Kathleen, 20 km N—20.63 140.05; QM R41850, R41858 Oorida area, Diamantina Lakes—23.77 141.133; QM R43753 Dajarra, ~20 km E, on Duchess Rd.—21.58 139.633; QM R47081 Riversleigh Stn.—19.03 138.75; QM R52714, R52730 Lawn Hill Stn., Century Project Site—18.75 138.583; QM R52763 Beagle Bay, 71 km S—17.55 122.417; QM R52786 Cloncurry, 120 km

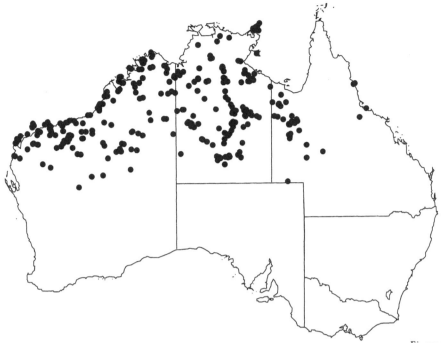

Figure 7.2. Known localities of collection of Varanus acanthurus.

NW—19.92 139.75; QM R52795, R52797 Beagle Bay, 71 km S—17.55 122.417; QM R52852 Quamby, 9.6 km N—20.25 140.25; QM R53285 Lawn Hill Stn., Century Site—18.75 138.617; QM R54183 Barkly Hwy., 80 km W Mt. Isa—20.25 139; QM R58878 Corella Rd. crossing, 36.6 km N Cloncurry on Normanton Rd.— 20.45 140.317; WAM R13718 between Wave Hill and Inverway Stns.—17.05 130.15; WAM R75520 2 km W Fitzroy Crossing—18.2 125.55; WAM R94844 5 km E Roebuck Plains HS—17.92 122.5; AM R83 Derby—17.3 123.617; AM R219 25 mi inland from Cairns—17 145.667; AM R754 Cairns District—16.92 145.767; AM R9911, R9912 Roper River N.A.—14.73 134.733; AM R11037, R11038 Groote Eylandt, Gulf of Carpentaria—13.98 136.467; AM R11905 Alice Springs— 23.7 133.867; AM R14063 The Granites—20.58 130.35; AM R15100, R16634, R18858, R19185 Mt. Isa—20.73 139.483; AM R21031 Charters Towers—20.08 146.267; AM R38713 11 M. W Springvale, 40 mi N Hall's Creek—17.75 127.5; AM R40281 Soudan, near Barrie Caves; AM R49242, R49243 Alice Springs—23.7 133.867; AM R49376, R49377 25 km NW Refrigerator Bore—20.9 130.617; AM R49417, R49418; R49419, R49475, R49474, R49560 Mt. Doreen—22.03 131.333; AM R49476, R49477 25 km NW Refrigerator Bore—20.9 130.617; AM R49685, R49768, R53777 Alice Springs—23.7 133.867; AM R51911 2 km NW of Pullya Pullya Dam, Ringwood Stn.—23.73 134.933; AM R51913 Hills near Green-leaves Caravan Park, Alice—23.67 133.883; AM R51914 near Flynns Memorial, Alice Springs—23.7 133.85; AM R54257 N Mt. Isa—20.67 139.483; AM R55054, R55055, R55056. R55057 12 km N Borroloola—16 136.283; AM R60196 5 km S Renner Springs on Stuart Hwy.—18.4 133.8; R60197, R60198 Barrow Creek Roadhouse—21.53 133.883; AM R60199 27 km SW of Barrow Creek on Stuart Hwy.—21.7 133.75; AM R60220, R60221 29 km N of Barkly Hwy. on Borroloola Rd.—19.07 135.183; AM R60221 29 km N of Barkly Hwy. on Borroloola Rd.— 19.07 135.183; AM R64804 Mt. Isa—20.73 139.483; AM R65961 54.5 km W Middleton—22.63 141.167; AM R65967 2 km E Mt. Isa—20.73 139.517; AM

R68825 Camooweal—19.92 138.117; AM R72002 135 km S Mt. Isa on Boulia Rd.—21.7 139.533; AM R72019, R72020, R72020, R72037 106 km N Boulia—22.05 139.583; AM R72163, R72164, R72185, R72186 38 km E Camooweal—19.97 138.467; AM R72191, R72192 Wickham Rd. 41 km SW Victoria River Downs—16.68 130.667; AM R72690 Bullo River on Crossing of Katherine—Kunu—15.7 129.633; AM R73081 Wickham River Waterhole—16.68 130.667; AM R80250 90 km S of Alice Springs on Main Hwy.—24.35 133.45; AM R81390 near Lyndon River, base of Exmouth Peninsula—23.65 114.267; AM R81391 Fortesque River Crossing—21.3 116.133; AM R81399 Warroora Stn.—23.48 113.8; AM R90857, R90858, R90859, R90860 Mt. Doreen—22.03 131.333; AM R90870, R90872, R90873 Mt. Doreen—22.1 131.417; AM R95320 25 km S Larrimah on Stuart Hwy.—15.73 133.367; AM R95464, R95465 Elliot Dump—17.55 133.533; AM R95465 Elliot Dump—17.55 133.533; AM R95468 30 km N of Barry Caves on Barkley Hwy.—19.95 136.467; AM R100578 26.9 km N of the Wittenoom-Newman Rd. via the Port Hedland Rd.—22.2 118.75; AM R100668 Rubbish Tip on South Hedland—20.4 118.633; AM R100901 New Rubbish Tip 5.5 km NNE of Broome—17.92 122.25; AM R101431, R101432 Rubbish Tip on Sandfire Flat Roadhouse; AM R101657, R101658 "Yardie Creek"—22.3 113.833; AM R111206 Mt. Isa—20.73 139.483; AM R128355 Forrest River Mission—15.18 127.85; AM R128861 Broome—17.97 122.233; AM R140444 Wallal (Rubbish Tip)—19.77 120.667; AM R141688 5 km S of Tennant Creek—19.7 134.35; AM R141728 Tennant Creek Tip—19.65 134.183; AM R142917 64 km N of Tea-Tree—21.47 133.417; AM R143881 17.7 km N Barkly Hwy. via Kajabbi Rd.—20.57 139.733; AM R143882 19 km N Barkly Hwy. via Kajabbi Rd.—20.57 139.733; AM R147228 198 km W Camooweal on Barkly Hwy. (Old Frewina Roadhouse—19.88 135.333; AM R147243 13.3 km S Wauchope on Stuart Hwy.—20.92 134.217; AM R147248, R147249, R147250 19.9 km N Wauchope on Stuart Hwy. (=1 km N Dixon's Creek Crossing)—20.48 134.25; AM R147252, R147253, R147254 49 km S Wauchope on Stuart Hwy.—21.22 134.117; ANWC R0516 5 mi NW of Argyle Downs—16.2 128.742; ANWC R0941 ~3 km N of Amelia Spring, Macarthur Rd.—16.58 136.167; ANWC R1142 Nourlangie Rock, S of Jabiru [Editors' note: this is probably a *V. baritji*]—12.85 132.783; ANWC R2227 Opalton, S of Winton—23.27 142.7; ANWC R2776 Lake Auld, Great Sandy Desert—22.42 123.833; ANWC R2904 Townsville—19.27 146.817; ANWC R3056 White Springs, S of Port Hedland—21.78 118.8; ANWC R5016, R5017, R5018, R5019 Alice Springs—23.7 133.867; NMV T1096, T1098, T1100 Mt. Doreen Stn.—22.08 131.417; NMV D 4974 Barrow Creek—21.53 133.883; NMV T 1097 Mt. Doreen Stn.—22.08 131.417; NMV D 7992 Kuridala, Cloncurry—21.28 140.5; NMV T 1099 Mt. Doreen Stn.—22.08 131.417; NMV D 5632 Tennant Creek—19.65 134.183; NMV R 919 Broome—17.97 122.233; NMV D 44855 Birdsville—25.9 139.35; NMV D 2931, D2943, D2944 Tennant Creek—19.65 134.183; NMV D 12786 Ormiston Gorge—23.62 132.717; NMV D 2932 Tennant Creek—19.65 134.183; NMV D 2929 Tennant Creek—19.65 134.183; NMV D 5619 Barrow Creek—21.53 133.883; NMV D 3556 Alice Springs—23.7 133.867; NMV D 2933 Tennant Creek—19.65 134.183; NMV T 1101 Labbi Labbi—21.52 128.633; NMV D 5618 Barrow Creek—21.53 133.883; NMV D 4913 Oenpelli—12.32 133.05 [Editors' note: probably a *V. baritji*]; NMV D 56417 Alice Springs—23.7 133.867; NMV D5631, D5634, D5635 Tennant Creek—19.65 134.183; NMV T1102, T1103, T1104 Roper River—14.68 134.733; NTM R25459 Tanami—20.53 130.032; NTM R25460 Wollogorang Stn., Echo Gorge Camp; NTM R25461 Mt. Isa—20.73 139.483; NTM R25463 Wauchope—20.65 134.217; NTM R25464 26.8 km S Tea-Tree—22.43 133.4; NTM R25465 Kimberleys—26.32 119.783; NTM R25466 Victoria River Downs Stn.—16.4 131.017; NTM R25467, R25470 Barrow Creek—21.52 133.883; NTM R25469 Rabbit Flat—20.43 130.05; NTM R25472 Prowse Gap, N of Aileron—22.57 132.333; NTM R25473 Manbuloo Stn., Katherine—14.52 132.2; NTM R25474 16.8 km S Barrow Creek—21.63 133.75; NTM R25475 3.9 km W Victoria River Downs—16.37 130.983; NTM R25476 11 mi N Alice—23.55 133.85; NTM R25477 Devils Marbles—20.57 134.267; NTM R25478 Wauchope—20.65 134.217; NTM R25479, XXX—19.42 135.4; NTM R25480 Elliott—17.55 133.533; NTM R25481 Red Point Marchinbar Is.—11.28

136.583 [Editors' note: probably a *V. baritji*]; NTM R25482 Nita Down Stn.—
19.13 121.7; NTM R25483 50 km N Broome—17.97 122.317; NTM R25484
Devils Marbles—20.53 134.25; NTM R25485, R25486 Nita Downs—19.13 121.7;
NTM R25487, R25493, R25494, R25509 Frewena—19.43 135.4; NTM R25488,
R25489 Barrow Creek—21.53 133.9; NTM R25490 Wycliffe Well—20.8 134.233;
NTM R25491 Elliott—17.55 133.533; NTM R25492 Mt. Gillen, Macdonnell
Ranges, Alice Springs—23.72 133.8; NTM R25495 Barradale [Editors' note: no
latitude and longitude; if Mt. Borradaile, it's probably a *V. baritji*]; NTM R25497
Manbulloo Stn.—14.52 132.2; NTM R25498, R25568, R25569—19.92 135.417;
NTM R25499 Rabbit Flat—20.43 130.05; NTM R25500 Three Ways—19.45
134.217; NTM R25501 Wauchope—20.65 134.217; NTM R25502 Wollogorang
Stn., Echo Gorge; NTM R25503. R25511 Horden Hill, Granites—20.65 130.317;
NTM R25504 Centre Is. (Scrub), Survey Bay—Hill; NTM R25505, R25507 Nita
Downs—19.13 121.7; NTM R25506 26.6. km E Three Ways, on Barkly Hwy.—
19.43 134.45; NTM R25510 Heartbreak Hotel Area—16.68 135.717; NTM R25512
The Granites—20.58 130.35; NTM R25513 Barrow Creek—21.52 133.883; NTM
R25514 4.7 km E Boroloola TO, Barkly Hwy.—19.58 135.867; NTM R25515 1.2
km S Dunmarra—16.37 133.417; NTM R25516 3.9 km W Victoria River Downs—
16.37 130.983; NTM R25517, R25518 2 mi WNW Victoria River Downs—16.37
130.983; NTM R25519 Dunmarra—16.68 133.7; NTM R25520 Cape Crawford
Area—16.59 135.885; NTM R25521 Cape Crawford Area—16.62 135.867; NTM
R25522, R25523 Cape Crawford Area—16.13 135.717; NTM R25524, R25525
Macarthur River Stn., Barney Hill—16.42 136.1; NTM R25526 Nita Down Stn.—
19.13 121.7; NTM R25527 Barkly Hwy. 21 km E Rockhampton Downs TO—
19.38 135.283; NTM R25529 Bullita Stn.; NTM R25530 Alice Valley, Giles Spring
Yard—23.67 132.9; NTM R25531 Thozet Box Camp, Loves Creek Stn.—23.48
134.717; NTM R25532 Inturtupa Waterhole, Ellery Creek—23.82 133.067; NTM
R25533 17 km E Top Springs—16.62 131.933; NTM R25534 4.5 km S Dun-
marra—16.73 133.4; NTM R25535 32 km E Three Ways, Barkly Hwy.—19.4
134.5; NTM R25536 4.5 km S Dunmarra—16.32 133.383; NTM R25537 32 km
E Three Ways, Barkly Hwy.—19.4 134.5; NTM R25538 Manbuloo Stn., Katherine
—14.52 132.2; NTM R25539 17 km E Top Springs—16.62 131.933; NTM R25540,
R25541 4.5 km S Dunmarra—16.32 133.383; NTM R25542 215 km W Borro-
loola—16.75 135.05; NTM R25543 4.5 km S Dunmarra—16.73 133.4; NTM
R25544 32 km E Three Ways, Barkly Hwy.—19.4 134.5; NTM R25545, R25546
Old Wave Hill Dump, 13 km E Police Stn.—17.48 130.936; NTM R25547
Wauchope—20.65 134.217; NTM R25548 Alyawarre Desert Area—20.7 135.6;
NTM R25549 Green Swamp Well—19.28 132.667; NTM R25550 Supplejack Stn.,
Blue Bush Bore—19.28 129.6; NTM R25551 Ruins of Lod Dalmore Hmstd.—
19.75 135.983; NTM R25552 No 15 Bore Rockhampton Downs—19.32 135.517;
NTM R25553 No 14 Bore Rockhampton Downs—19.32 135.517; NTM R25554
178 km S Broom by Frazier Downs T-Off—18.78 121.817; NTM R25555 112 km
W Katherine—18.23 131.6; NTM R25556 11 km S Top Springs—16.65 131.767;
NTM R25557, R25558, R25559, R25560 Wave Hill Stn. Old Homestead Dump—
17.48 130.95; NTM R25561 112 km W Katherine—18.23 131.6; NTM R25562,
R25563 15 km E on Edith Falls Rd.—14.17 132.1; NTM R25564 2 Island Bay—
11.08 136.733; NTM R25565 11 km S Top Springs—16.65 131.767; NTM R25566
Frewena—19.43 135.4; NTM R25567 Adelaide River, Town. 20 km S—15.42
131.15 [Editors' note: perhaps a *V. baritji*]; NTM R25570 Duncan Hwy., 41 km N
Wave Hill TO—17.7 128.817; NTM R25571 Hanson River Crossing, Willowra
Rd.—22 133.433; NTM R25572 Barrow Creek—21.53 133.9; NTM R25573
Supplejack Stn., 7 km W Blue Bush Bore—19.27 129.583; NTM R25574 239 km
W Victoria Crossing Victoria Hwy.—16.05 129.267; NTM R25575 1 km SSE of
Mt. Windajong—21.05 132.883; NTM R25576 ~1 km SSE of Mt. Windajong—
21.05 132.883; NTM R25577 Guluwuru Is.—11.52 136.417; NTM R25578 Jirgari
Is.—11.8 136.133; NTM R25579 Raragala Is. (South)—11.64 136.168; NTM
R25580 Raragala Is. (South)—11.65 136.184; NTM R25581 Alyawarre Desert
Area—19.85 136; NTM R25582 Cape Crawford Area—16.6 135.884; NTM
R25583 Cape Crawford Area—16.62 135.785; NTM R25584, R25585, R25586
Heartbreak Hotel Area—16.68 135.717; NTM R25587 Sangsters Bore, 3 mi S—

20.88 130.35; NTM R25588 Tanami Desert—19.38 129; NTM R25589 Barkly Hwy.—19.73 135.9; NTM R25590 Elliott, 65 km S—18.13 133.55; NTM R25591 Tennant Creek—19.07 134.2; NTM R25592 Kurundi Stn.—20.6 134.083; NTM R25593 Barkly Hwy.—19.73 135.9; NTM R25594 Tanami—20.63 130.417; NTM R25595 Kurundi Branch Creek—20.47 134.433; NTM R25596 Top Springs, 80 mi NW; NTM R25597 Mt. Isa; NTM R25598 Tanami—19.85 130.517; NTM R25599 Tennant Creek—19.07 134.2; NTM R25600 Tanami—20.58 130.008; NTM R25601 Musselbrook Res, Border Waterhole—18.6 137.984; NTM R25602 Musselbrook Res, Track to Lawn Hill—18.86 138.567; NTM R25603 Stuart Hwy., 49 km S Wauchope—21.23 134.13; NTM R25604 Stuart Hwy., 19.9 km N Wauchope —20.49 134.253; NTM R25605 Keep River National Park—15.75 129.083; NTM R25606 Bauhinia Downs Stn.—15.92 135.317; NTM R25607 Limmen Gate National Park—15.78 135.333; NTM R25608 Wadamunga Lagoon, Roper River— 14.8 134.943; NTM R25609 English Company Isles, Pobasso Is.—11.9 136.454 [Editors' note: probably a *V. baritji*]; NTM R25610 Thozet Gum Camp—23.5 134.717; NTM R25611 Big Bromby Is.—11.84 136.673 [Editors' note: probably a *V. baritji*]; NTM R25612 Tanami—20.6 131.002; WAM R953 Pago—14.13 126.717; WAM R4291 La Grange—18.67 122.017; WAM R11202, R11203, R11204 Wotjulum, June 1954—16.18 123.617; WAM R11824, R11825, R11826, R11827, R11828, R11829, R11830, R11831, R11832 Wotjulum—16.18 123.617; WAM R12130 Roebourne—20.78 117.15; WAM R12334, R12335, R12336 Wotjulum—16.18 123.617; WAM R12897 Lowendall Is.—20.65 115.567; WAM R13083 Woodstock—21.62 118.95; WAM R13665 Beverley Springs. W Kimberley—16.72 125.467; WAM R14363 Legendre Is. Dampier Archipelago—20.38 116.867; WAM R14590 Braeside—21.2 121.017; WAM R14902 Mt. Edgar—21.3 120.067; WAM R14903, R14904, R14905 Mundabullangana. Near Homestead— 20.52 118.067; WAM R14906 Mt. Edga—21.3 120.067; WAM R20002, R20003, R20003 Mt. Herbert—21.33 117.217; WAM R20824 36 km W of Alice Springs— 23.7 133.517; WAM R21222 Fossil Downs Stn.—18.15 125.783; WAM R22936 Mardie—21.18 115.983; WAM R23802 32 km N of Larrimah—15.28 133.217; WAM R23803 16 km N of Larrimah—15.43 133.217; WAM R24067, R24068, R24080, R24081, R24082 Mt. Hart. Within 6 km of HS—16.82 124.917; WAM R24295 23 km N of Wauchope—20.43 134.217; WAM R24736 Mt. Hart. Within 6 km of HS—16.82 124.917; WAM R26345 Katherine—14.47 132.267; WAM R26700 Presumably West Kimberley District—17.5 123.5; WAM R26927 64 km N of Windy Corner. Gibson Desert (254 km)—22.98 125.183; WAM R27018 33 km E of Well 24. Canning Stock Route—23.12 123.667; WAM R28019, R28020 Mt. Phire—19.32 121.7; WAM R28021, R28022 Yardie Creek. Tantabiddi Well— 21.93 113.967; WAM R28024, R28025, R28026, R28034 Kalumburu.—14.3 126.633; WAM R29061 Mundabullangana Stn.—20.52 118.05; WAM R29121 Wittenoom Gorge. "9.6 km up the gorge"—22.33 118.333; WAM R31210 Woodstock—21.62 118.85; WAM R31414 Exmouth Townsite—21.93 114.117; WAM R31559 near Wyndham.—15.48 128.117; WAM R36592 White Springs Ruins. 109 km N of Wittenoom—21.25 118.333; WAM R37245 South Muiron Is.—21.68 114.317; WAM R37339, R37340 West Lewis Is. Dampier Archipelago—20.6 116.617; WAM R37442 Hermite Is. Montebello Is.—20.48 115.517; WAM R37982 Nullagine—21.9 120.1; WAM R40475 Kunmunya Mission. 23 km NE of Kuri Bay—15.43 124.667; WAM R41301 Augustus Is. Bonaparte Archipelago—15.32 124.533; WAM R42965 Augustus Is. Bonaparte Archipelago—15.4 124.567; WAM R43134 Surveyor's Pool. Mitchell Plateau—14.67 125.733; WAM R43165, R43201, R43320 Mitchell Plateau—14.87 125.833; WAM R44057 Sir Graham Moore Is., Bonaparte Archipelago—13.88 126.567; WAM R45205 Mt. Leisler —23.35 129.35; WAM R46074 Frazier Downs—18.8 121.717; WAM R46175 Meetheena HS—21.3 120.467; WAM R46435 La Grange—18.68 121.767; WAM R46688, R46691, R46692, R46693 Prince Regent River Reserve—15.47 125.483; WAM R46972 Prince Regent River Reserve—15.47 125.667; WAM R50690 Drysdale River National Park. on Slope of Ashton Range—15.27 126.717; WAM R50988 Old Doongan HS—15.33 126.533; WAM R51026, R53323 Yardie Creek —21.88 114.017; WAM R51215 S of Lake Betty—19.58 126.367; WAM R51714

5 km E of Nullagine—21.88 120.167; WAM R51944 Durba Springs—23.75 122.517; WAM R52711 Marandoo Minesite—22.62 118.133; WAM R54151, R54160, R54161 near Mchughes Bore Dampier Downs (Edgar Ranges)—18.38 123.05; WAM R55451 45 km NE of Mt. Isa, 71 km S of Julius Dam TO—20.43 139.783; WAM R56366 Port Warrender—14.5 125.833; WAM R56703, R58764 Barrow Is. SE corner—20.8 115.4; WAM R58765 near Town Point Barrow Is.—20.78 115.467; WAM R58766, R58767 near Shark Point Barrow Is.—20.87 115.417; WAM R58768 near Shark Point Barrow Is.—20.87 115.417; WAM R58769 Lake Argyle, 16 km S of Main Ord Dam—16.3 128.8; WAM R58770, R58771 Lake Argyle—16.25 128.75; WAM R58772 La Grange—18.67 122.017; WAM R58773, R59024 Kalumburu—14.3 126.633; WAM R58774 La Grange. Injudinah Creek—18.63 121.867; WAM R58775, R59025 Woodstock Stn.—21.62 118.95; WAM R58949, R58950 Mt. Herbert—21.33 117.217; WAM R58992 Mulyie HS—20.47 119.517; WAM R59976 Lake Argyle, on Main Dam Site—16.12 128.733; WAM R59977 Mannerie Swamp Nita Downs—19.08 121.583; WAM R60107 Argyle Downs HS—16.28 125.8; WAM R60551 Mitchell Plateau—14.8 125.817; WAM R60564 Lawley River—14.67 125.933; WAM R60675 Mitchell Plateau—14.58 125.75; WAM R61073 Yardie Creek The Watercourse—22.33 113.817; WAM R62444 Mulyie HS—20.47 119.517; WAM R63160 Shay Gap—20.5 120.15; WAM R63291 Balgo Hill—20.12 127.8; WAM R63419 Twin Heads—20.25 126.533; WAM R63469, R63498 Tobin Lake—21.68 125.75; WAM R63585, R63586, R63587 Anketell Ridge—20.68 122.833; WAM R63884 2 km SW Karara Well No 24 Canning Stock Route—23.13 123.3; WAM R63952, R63965, R63966 12 km 21 degree Well No. 29 Canning Stock Route—22.45 123.917; WAM R68101 4.5 km 182 degree Mt. Bruce—22.64 118.147; WAM R69505 Ankatell Ridge—20.43 122.183; WAM R70760 30.2 km 238° Marillana HS—22.78 119.16; WAM R73530 8 km E Woodstock HS—21.62 119.033; WAM R73856 Harding River Area 25 km S Roebourne—21 117.067; WAM R74863 5 km W of Roebourne—20.78 117.1; WAM R74875 Mulga Downs HS—22.1 118.467; WAM R75507 2 km NW of Fitzroy Crossing—18.17 125.583; WAM R75807 Nita Downs—19.08 121.683; WAM R76415 10 km SSW Cooya Pooya HS—21.12 117.117; WAM R76568 Tom Price—22.75 117.767; WAM R76584 11 km NE Warroora HS—23.43 113.683; WAM R77199 Walsh Point—14.57 125.833; WAM R77200 Mitchell Plateau—14.87 125.8; WAM R77334, R77376 Mitchell Plateau—14.82 125.839; WAM R77411, R77412 Mitchell Plateau—14.83 125.825; WAM R77418 Mitchell Plateau—14.73 125.733; WAM R77534, R77607, R77609 Port Warrender—14.59 125.763; WAM R77560 Mitchell Plateau—14.82 125.839; WAM R77599 Port Warrender—14.57 125.8; WAM R77608, R77611 Mitchell Plateau—14.63 125.804; WAM R78238 Barrys Caves—20.05 136.667; WAM R78929 26 km SE of Mt. Augustus—24.53 117.167; WAM R79108, R79129 1.3 km NNW of Tunnel Creek Cave—17.6 125.133; WAM R79150, R79151 Roebuck Plains—17.93 122.517; WAM R80424 15 km N of Mt. Murray—22.35 115.55; WAM R81441 51 km SSE Well 31 Canning Stock Route—22.92 124.667; WAM R81467 Rudall River—22.58 122.183; WAM R82603 Lower Carawine Pool—21.48 121.017; WAM R83455 34 km NE of Napier Downs—17.15 125.067; WAM R83769 12 km NE of Well 26, Canning Stock Route—22.88 123.583; WAM R87072 13 km NNW Mt. Murray—22.38 115.517; WAM R87098 18 km SW of Glen Hill Homestead—16.67 128.267; WAM R87751, R87752 3 km W of Cobra HS—24.2 116.433; WAM R89560 Varanus Is., Lowendal Is.—20.65 115.567; WAM R89562 Lachlan Is., Buccaneer Archipelago—16.62 123.517; WAM R89977 Lowendal Is.—20.65 115.567; WAM R90574 Woodstock—21.61 119.04; R90627 Woodstock—21.61 119.031; WAM R90786 3.4 km S Woodstock HS—21.65 118.95; WAM R90857 Woodstock—21.61 119.031; WAM R90858 200 Metres S Gallery Hill—21.67 119.041; WAM R94142 Varanus Is., Lowendal Is.—20.65 115.567; WAM R94373 Balgo Mission—20.15 127.967; WAM R94753 Ca 80 km S Telfer—22.31 122.055; WAM R94952 29 km W Thompson Hills—21.33 124.75; WAM R94977 22 km W Thompson Hills—21.32 124.817; WAM R95241 Glenayle Stn.—25.25 122; WAM R95311 Shoesmith Cliffs—21.5 125.35; WAM R96236, R96237 Mitchell Plateau—14.73 125.733; WAM R96243 54 km SW Port Hedland

[Yule River Crossing on NW Coastal Hwy.]—20.72 118.3; WAM R97617 Kalumburu—14.3 126.633; WAM R99196 Woodstock Stn.—21.61 119.031; WAM R99816, 99817 3 km NW Millstream Headquarters—21.57 117.067; WAM R100813 Yarraloola Stn.—21.57 115.867; WAM R101362 Wyndham—16.75 123.95; WAM R103102 Bungle Bungle National Park.—17.57 128.517; WAM R103182 Bungle Bungle National Park.—17.73 128.15; WAM R104956 Mandora HS—19.73 120.85; WAM R106031, R106032, R106033, R115715 Koolan Is.—16.15 123.75; WAM R106040 South Hedland—20.4 118.6; WAM R108887 12.5 km at 145 Degrees to Lake Hancock Gibson—24.78 124.6; WAM R113071 Lesley Salt Works—20.27 118.897; WAM R113611 42 km NNE Auski Roadhouse—21.98 118.833; WAM R114760, R114761, R114762, R114763 Cleaverville—20.65 117; WAM R117063 18 km W of Mt. Stuart—22.43 115.883; WAM R117268, R117269 Hermite Is., Montebellos—20.48 115.517; WAM R119340 near Cape Malouet, Barrow Is.—20.72 115.4; WAM R119667, R119668 Emma Gorge Cockburn Range—15.83 128.033; WAM R121201, R121202, R121203 Thevenard Is.—21.47 114.983; WAM R123943 Cape Range 10 km S Exmouth—22 114.083; WAM R125099 15 km E of Newman—23.37 119.9; WAM R100736 Woodstock, Site 2—21.61 118.956; WAM R100742 Woodstock, Site 7—21.61 119.031; WAM R100779 Woodstock, Site 3—21.61 118.962; WAM R100781 Woodstock, Site 24—21.59 119.078; WAM R100782 Woodstock, Site 22—21.59 119.084; WAM R104078 Woodstock, Site 22—21.59 119.084; WAM R104094 Woodstock, Site 0—21.66 119; WAM R104137 Woodstock, Site 22—21.59 119.084; WAM R104189 Woodstock, Site 4—21.61 118.974; WAM R131726 Vicinity of Newman—23.35 119.733; WAM R135022 Mt. Whaleback—23.33 119.669; WAM R132554 Degrey River Stn.—20.27 119.333; SAM R01754 Well No. 5—25.37 121; SAM R01855 Amac Donald Downs—22.42 135.117; SAM R01855 Bmac Donald Downs 22.42 135.117; SAM R04644 near top of Mt. Herbert—21.32 117.217; SAM R04645, R04646 Tambrey HS—21.63 117.6; SAM R18719 NE of Alice Springs—23.27 134.817; SAM R34210 7 km E of Mt. Isa—20.72 139.55; SAM R34211, R34212, R34213, R34233 6 km E of Camooweal—19.92 138.167; SAM R34256, R34257 Macarthur River Stn.—16.67 135.85; SAM R34264 Redbank Mine—17.18 137.767; SAM R38816, R38817, R38818 Barrow Creek—21.53 133.883; SAM R38828 Hatches Creek Mine—20.95 135.217; SAM R42677 138 km N of Boulia—21.73 139.55; SAM R42680 Dajarra—21.68 139.5; SAM R42688 7 km NE of Mt. Isa Telecom Repeater Stn.—20.72 139.55; SAM R42694 37 km NNE of Mt. Isa—20.57 139.733; SAM R42700 Quamby—20.37 140.283; SAM R29309 Barradale—22.87 114.95; SAM R02683 Banka Banka—18.8 134.033; SAM R03530 Mandora Stn.—19.73 120.833; SAM R03899 Yuendumu—22.25 131.8; SAM R05134A, R05134B, R05135A, R05135B, R05135C, R05135D, R05857, R05859 Doomadgee Mission—17.93 138.817; SAM R06012A, R06012B, R06012C, R06124A, R06124B, R06124C Tennant Crk—19.65 134.183; SAM R13541 Bing Bong Stn.—15.62 135.35; SAM R13787 Finke River—23.95 132.767; SAM R13788 Tennant Creek—19.65 134.183; SAM R13789 Barrow Creek Stn.—21.53 133.883; SAM R15715 81 mi N of Kulgara—22.45 115.983; AM R147239 5 km N Daly Waters on Stuart Hwy.—16.27 133.383.

Fossil Record

None known.

Diagnostic Characteristics

Diagnostic characteristics are as follows (see Storr 1980; Cogger 1992 for variations):

• Small- to medium-sized, about 70 cm total length (TL).

• Tail nearly round without dorsal crest; spiny dorsal and lateral scales, 1.3 to 2.3 times as long as snout–vent length (SVL).

- Head scales small and smooth; interorbitals largest.

- Dark brown dorsally with yellowish ocelli on trunk; head striped and tail banded with same colors; white or cream belly.

- Scales lateral to vent more spiny in males (Thissen 1992).

The *V. acanthurus* complex exhibits considerable color variation on a north–south cline (Storr 1980).

    V. baritji has been elevated to species status, and *V. a. insulanicus* seem larger and much darker than previously examined forms (King and Horner 1987). Application of molecular analyses may well reveal forms of taxonomic distinction.

Size

    SVLs for 80 specimens from the Northern Territory Museum collection ranged from 66 to 241 mm for males (n = 39) and 95 to 210 mm (n = 41) for females.

    SVLs of 221 specimens in the Western Australian Museum collection ranged from 63 to 250 mm for males (n = 99) and 57 to 205 mm for females (n = 122) (King, in preparation).

    Sixteen *V. acanthurus acanthurus* from the Northern Territory mainland had SVLs of 120 to 240 mm, with a mean of 180 mm.

    Three *V. acanthurus insulanicus* had SVLs of 226 to 250 mm with a mean of 237 mm (King and Horner 1987).

    Three small animals from Barrow Island were suspected to be dwarves (Case and Schwaner 1993).

Habitat and Natural History

    These secretive monitors are usually associated with stony ridges and rocky outcrops, where they live under slabs or in crevices (Cogger 1992). In the absence of rocks, they shelter in burrows associated with spinifex (Swanson 1979) or in trees (Stammer 1970).

Terrestrial, Arboreal, Aquatic

See above.

    Wild animals are usually taken from refugia, suggesting minimal daytime exposure. They have, however, been caught in mammal traps, been seen active, or been found dead on the road, suggesting some movement between shelter sites (King and Rhodes 1982). One radio-tracked animal ventured 71 m in 45 days (Dryden et al. 1990). Captive spiny tails emerge to feed (Thissen 1992) and bask extensively under visible-spectrum lamps (Krebs 1999).

Foraging Behavior

    The paucity of sightings of foraging animals suggests that *V. acanthurus* is a sit-and-wait predator. This is indirectly supported by relatively low metabolism indices in captive (Thompson and Withers 1994) and free-ranging (Dryden et al. 1990) animals. Captive lizards have been observed using their tails to poke prey from hiding places

(Horn 1999). These small monitors are doubtlessly preyed upon by carnivorous mammals, birds, and other reptiles. A pygmy python (*Antaresia perthensis*) swallowed a 45-cm *V. acanthurus* (Browne-Cooper 1998).

Diet

Two series of museum specimens give us the best picture of prey taken. Stomach contents examined from 21 specimens in American museums suggest that the diet consists of orthopterans (predominantly), beetles, cockroaches, and lizards—agamids, geckos, and skinks (Losos and Greene 1988). Although the stomachs of larger monitors usually held more items (Losos and Greene 1988), there was no correlation between predator and prey size (King, in preparation).

Stomach contents of 269 animals in the Western Australian Museum and 84 in the Northern Territory Museum were examined, and stomachs of 167 of these contained food. The most frequent items were grasshoppers (n = 78), beetles (n = 25), lizards (geckos, agamids, and skinks; n = 19), cockroaches (n = 13), spiders (n = 6), snails (n = 4), and stick insects (n = 4). One animal had eaten a small dasyurid. Unfortunately, 25 stomachs contained invertebrates that could not be identified because of advanced digestion or poor preservation.

A captive hatchling kept for 30 months ate a wide variety of reptiles as well as bits of meat and mice (Husband 1980).

Reproduction

Sex ratio approaches 1M:1F. All males 120 mm SVL or greater were spermatogenic (histologically determined), as was one 89 mm SVL. The smallest female with enlarged ova was 102 mm SVL, and all those with oviductal eggs were at least 140 mm SVL. Reproduction seems seasonal, with ovulation between August and November (late dry season) and a single yearly clutch (King and Rhodes 1982).

Clutch size ranged from 2 eggs in wild animals (Bustard 1970) to 18 eggs in captivity (Krebs 1999). The average ± SE size of 14 clutches laid in captivity was 7.9 ± 0.9 eggs (Horn and Visser 1989). In captivity, 30 eggs averaged 4.5 g, and 21 hatchlings averaged 15 cm TL. Incubation varies between 3 and 5 months at different temperatures (Horn and Visser 1989; Krebs 1999). One nest of hatchlings was excavated by Husband (1980). It was at the end of a tunnel dug some 40 cm deep in bulldozed earth. Eight hatchlings averaged 66 mm SVL.

Movement

Movement is presumed to be limited. In a study on water and energy turnover in *V. acanthurus,* conducted in central Australia, maximal distance moved by a *V. acanthurus* between refuge sites was only 71 m. Most animals stayed at the same site for several days without moving during the day (Dryden et al. 1990).

Physiology

Mean metabolic rates during summer near Alice Springs were low in comparison with those of other varanids. Approximately 70% of their estimated water requirement came from food. The rest was presumably provided by the moist environment of the refugial atmosphere (Dryden et al. 1990). Oxygen consumption in captivity is also low (Thompson and Withers 1994).

Parasites

The incidence of ticks on *Varanus acanthurus* is very low. Only 13 of 269 specimens in the reptile collection of the Western Australian Museum were infested with ticks (Sharrad and King 1981; and King, in preparation). Twelve lizards were infested by *Amblyomma limbatum,* and one specimen was infested by *Aponomma fimbriatum.*

A sample of 122 specimens of *Varanus acanthurus* in the Western Australian Museum, 86 in the Northern Territory Museum, and 4 in the Queensland Museum was examined by Jones (1983b), who found that 74% of the specimens were infested by nematodes (mainly gastric nematodes), with 65% infested with *Abbreviata hastaspicula,* 8% with *Abbreviata antarctica,* 4.7% with *Abbreviata confusa,* and 1.0% with *Abbreviata* sp. in their stomachs. These nematodes are also found in other species of reptiles in Western Australia. Nematodes were also found in 66 of 147 other specimens of *V. acanthurus* in the reptile collection of the Western Australian Museum, but these have not yet been identified.

7.2. *Varanus baritji*

Max King

Figure 7.3. Varanus baritji *(photo by Jeff Lemm).*

Nomenclature

This spiny-tailed monitor from the subgenus *Odatria* Gray 1838 was described by King and Horner (1987). The holotype, held in the Northern Territory Museum, specimen NTM R13192 (female), was collected in 1985 from a rock outcrop named Mirrngadja in the Arafura swamp 12°39'S, 135°12'E, in Australia's Northern Territory. This species was named after Dr. N. G. White, who collected the holotype. "*Baritj*" means white in the aboriginal Ritharrju-Wagilak language group, speakers of which occupy the area where the lizard was originally found (Fig. 7.3).

Geographic Distribution

Specimens of *V. baritji* live on rock outcrops and in stony country in the northeastern area of the Northern Territory, north of 15°S. Populations of *Varanus acanthurus* occur to the south and west of the *V. baritji* distribution and come into contact with it at Katherine River and Adelaide River. Populations of *V. acanthurus insulanicus* occur to the east of *V. baritji* and are found on three islands in the northwestern Gulf of Carpentaria (King and Horner 1987) (Fig. 7.4).

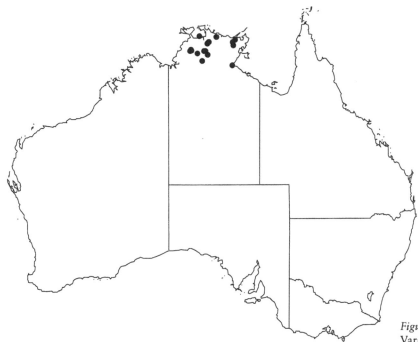

Figure 7.4. *Known localities for* Varanus baritji.

AM R51912 Daly River TO, Stuart Hwy.—13.5 131.17; AM R88844 Jabiluka Project Area. See topographical map 5472.1:100000 SE—12.6 132.92; AM R133283 Doyndji, Eastern Arnhem Land—12.4 135.47; AM R133286 Doyndji, Eastern Arnhem Land—12.4 135.47; AM R133348 Doyndji, Eastern Arnhem Land—12.4 135.47; AM R133349 Doyndji, Eastern Arnhem Land—12.4 135.47; NTM R25613 Kakadu National Park—12.8 132.77; NTM R25614 Kakadu National Park, Stage 3—13.5 132.45 NTM R25615 Port Roper, 30 km W—14.9 135.13; NTM R25616 8 km N Pine Creek—13.8 131.78; NTM R25617 Kakadu National Park—Stage 3—13.9 132.78; NTM R25618 7 mi S Adelaide River Town—13.4 131.15; NTM R25619 7 mi W Stuart Hwy. Daly River Rd.—13.5 131.05; NTM R25620 7 mi S Adelaide River Town—13.4 131.15; NTM R25621 7 mi S Adelaide River Town—13.4 131.15; NTM R25622 20 km S Adelaide River Town—13.5 131.18; NTM R25623 Katherine—14.5 132.27; NTM R25624 4.3 km E Goomadeer River Crossing, Arnhem Land—12.1 133.62; NTM R25625 Kakadu National Park, Stage 3—12 132; NTM R25626 4.3 km E Goomadeer River Crossing, Arnhem Land—12.1 133.62; NTM R25627 Kakadu National Park—Stage 3—13.5 132.58; NTM R25628 Mirrngadja, Arafura Swamp, Arnhem Land—12.7 135.2; NTM R25629, R25630, R25631, R25632, R25633, R25634 Doyndji (Arnhem Land)—12.9 135.32.

Fossil Record

No fossils of *V. baritji* are known.

Diagnostic Characteristics

Varanus baritji is a moderate-sized member of the *V. acanthurus* species complex of spiny-tailed monitors, which includes *V. acanthurus* Boulenger 1885, *V. acanthurus insulanicus* Mertens 1958, *V. primordius* Mertens 1942, *V. storri* Mertens 1966, and *V. storri ocreatus* Storr 1980. *Varanus baritji* is distinguished from other members of this complex and other varanids by a combination of size, scalation, coloration, and back pattern.

Description

A moderate-sized spiny-tailed monitor with an elongate and slender body, *V. baritji* has a characteristic tail that is round in section anteriorly and is triangular in section at the midlength. Dorsal and lateral caudal scales have strong spinose keels, and the tail has two keeled crests.

The dorsal surface of the neck and head are mid brown and generally immaculate, although some specimens have black spots. The back color is a uniform ochre with occasional black spots. In some specimens, the black spots coalesce to form a reticulate pattern or rings. However, the absence of yellow and black striping on the neck, and the absence of yellow and black ocelli on the back, distinguish this species from *V. acanthurus,* its closest relative. Note that specimens of *V. baritji* have a distinctive lemon-yellow gular region. The tail has alternating bands of black and light brown scales. The ventral tail surface and abdomen are buff colored.

Size

SVLs of 12 *V. baritji* specimens measured ranged from 121 to 252 mm with a mean of 171 mm. Tail lengths ranged from 206 to 472 mm on these specimens with a mean of 297 mm. The SVL to tail-length ratio had a mean of 0.58. Legs were relatively short, with forelimbs ranging from 33 to 70 mm, with a mean of 48 mm. Hind limbs measured from 45 to 95 mm with a mean of 63 mm. The head had ear-to-snout lengths ranging 21 to 39 mm, with a mean of 27 mm.

Habitat and Natural History

These rock dwelling monitors live in cracks in rock outcrops or massifs, and they are also found beneath rock exfoliations and sometimes (though not always—see below) in rudimentary burrows beneath rocks lying on soil. *Varanus baritji* have been collected in sandstone, granitic, and limestone outcrops (King and Horner 1987). Sweet (1999) found that the species constructed regularly used burrows beneath rocks on shallow slopes. *V. baritji* have very good burrows up to a meter deep, usually under a large embedded rock. They also have other smaller, "day" burrows that can also be used overnight in warm weather (Bedford, personal observation).

Specimens of *V. baritji* live in the tropical north of Australia and are found in a variety of habitats ranging from stony hills and slopes to dissected ranges and escarpments. The predominant vegetation type is tropical savanna woodland dominated by *Eucalyptus* with a dense understory of spear grass. Sweet (1999) saw and trapped *V. baritji* in *Allosyncarpia* woodland and dry woodland in Kakadu National Park. He noted they were sympatric with *V. glauerti, V. glebopalma* and *V. gouldii* in these areas (Sweet 1999). Large males can have a sentry post from which they can see the areas of other *V. baritji*. They tend to remain dormant after habitat is burned each year (July–November), waiting for rain (Bedford, personal observation).

Terrestrial, Arboreal, Aquatic

Varanus baritji is a specialized rock-dwelling species. Specimens of this monitor use the spines on their tails to maneuver themselves over steep rocky surfaces and within rock cracks. When attacked by predators, these animals inflate their bodies to wedge themselves into cracks and use their tails as anchors in this defensive position.

Foraging Behavior

These monitors use rock outcrops as a shelter and a safe place to find food, and they also move into adjacent grasslands to forage for insects and other prey. They actively hunt large therophosid spiders (*Selenocosmia* sp.) in their burrows and beat them on the ground, removing most appendages and setae before swallowing them (Sweet 1999).

Reproduction

The only published information about reproduction for *V. baritji* is the observation of King and Horner (1987) that the holotype of *V. baritji* laid three eggs in captivity in early June 1985. Egg laying may occur in the dry season in this species. *V. baritji* breed in July, and females lay eggs during the dry season in August to September (Bedford, personal observation). Clutch sizes of up to nine eggs are known and eggs hatch after 95 to 110 days (Bedford, personal observation).

V. baritji are active in the early morning and late afternoon. They are sit-and-wait ambush foragers and will eat anything they can subdue (Bedford, personal communication).

Unfortunately, to my knowledge, no published information exists on activity patterns, thermoregulation, diet, physiology, water balance, energetics, fat bodies, or parasites found on this species.

7.3. *Varanus brevicauda*

ERIC R. PIANKA

Figure 7.5. Varanus brevicauda *hatchling (a) and adult (b)(photos by Eric Pianka).*

Nomenclature

Commonly called the short-tailed monitor, *Varanus brevicauda* was described by Boulenger in 1898. Type specimens (syntypes) in the British Museum of Natural History (catalog specimens 1946.8.30.46–47) came from Sherlock River, Nickol (as Nicol) Bay, Western Australia (Cogger et al. 1983) (Fig. 7.5).

Geographic Distribution

Varanus brevicauda is widely distributed in arid regions of Western Australia, the southern half of the Northern Territory, northernmost South Australia, and remote western Queensland (Fig. 7.6).

138.5; QM R52766 Deep Creek, E of Broome—17.77 122.75; AM R49089, R49090 Roebourne—20.78 117.15; AM R100985, R101549 CA 6.6 km N of Sandfire Flat Roadhouse via the Great Northern Hwy.—19.32 121.267; AM R110614, R111909 8 km N of Mirrica Bore, Ethabuka Stn., NW of Bedourie—23.73 138.467; AM R125964 Debessa Stn. rubbish tip—17.8 124.083; AM R134751 Ewaninga, 40 km S Alice Springs—24 133.9; AM R138498 Ethabuka Stn., near Bedourie—25.97 139.1; NTM R25635—20.78 130.333; NTM R25636 NT—22.25 131.8; NTM R25637, R25638, R25639, R25642, R25643, R25644, R25645 Sangsters Bore—12 km SW—20.87 130.267; NTM R25640, R25649, R25650, R25651, R25652 Sandfire—19.68 121.267; NTM R25641 Atartinga, 29 km SW—22.57 133.933; NTM R25646 Lake Surprise, 24 km SE—20.42 131.983; NTM R25647 Site 2, Mt. Riddock Stn., Dulcie Range—22.52 135.417; NTM R25648 Alyawarre Desert Area—19.62 135.617; NTM R25653 Lake Surprise, 24.5 km SE; NTM R25654 Alyawarre Desert Area—20.93 137.067; NTM R25655

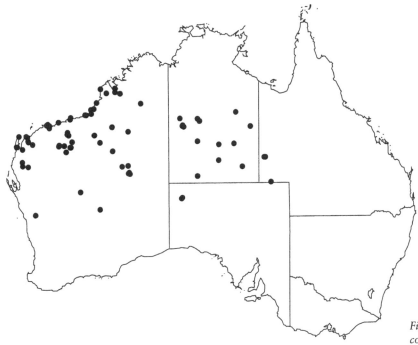

Figure 7.6. Known localities of collection for Varanus brevicauda.

Rabbit Flat—20.18 130; NTM R25656 Salt Beef Lake—20.95 130.417; NTM R25657 Tanami—20.22 131.767; NTM R25658, R25659, R25660, R25661, R25662 Ewaninga—24 133.9; NTM R25663 Lake Surprise—20.18 131.801; NTM R25664 Illogwa Creek, 2 km E—24.57 136.25; WAM R1023, R1024, R1025 Wallal—19.78 120.633; WAM R2124 De Grey Stn.—20.18 119.183; WAM R5330 Marrilla—22.97 114.45; WAM R12159 Tardun—28.78 115.75; WAM R12622 Abydos Stn.—21.42 118.917; WAM R13837 Derby—17.3 123.617; WAM R14915 Mundabullangana—20.52 118.067; WAM R16876 Ningaloo on campsite—22.7 113.967; WAM R16884 29 km E of Ningaloo—22.7 113.967; WAM R20350 32 km S of Derby—17.6 123.633; WAM R28029 La Grange—18.68 121.767; WAM R29118 Roebourne. Old Rifle Range—20.78 117.15; WAM R29768 16 km SE of Urala HS—21.87 114.983; WAM R40273, R40274 Coulomb Point—17.37 122.15; WAM R40616 24 km E of Mt. Madley—24.5 124.25; WAM R41058 Roebourne—20.78 117.15; WAM R44329 Coulomb Point—17.37 122.15; WAM R45806 Wallal—19.78 120.633; WAM R45807 Wallal—19.78 120.633; WAM R46168 Anna Plains—19.25 121.483; WAM R51301 34 km W of Sandfire Flat Roadhouse—19.78 120.767; WAM R54347 Marandoo—22.62 118.133; WAM R60433 16 km ENE of Pardoo Palm Spring—20.07 119.233; WAM R60693 Karratha (Presumably)—20.73 116.85; WAM R60874 Marandoo—22.62 118.133; WAM R62440 Vlaming Head—21.8 114.1; WAM R62864 7 km NNW of Erlistoun—28.28 122.117; WAM R63913 4.5 km NNE Karara well no. 24 Canning Stock Route—23.12 123.367; WAM R64236 38 km 210° Mctavish Claypan—20.93 123.25; WAM R66326 31 km 138° Mt. Meharry—23.18 118.792; WAM R67937 36 km 137° Mt. Meharry—23.21 118.825; WAM R69778 NE Slope Mt. Bruce—22.58 118.167; WAM R70773 22.4 km 222° Marillana HS—22.79 119.263; WAM R73136 30.2 km 238° Marillana HS—22.78 119.16; WAM R73137 30.2 km 238° Marillana HS—22.78 119.16; WAM R75108 3 km NE of Willare Bridge Roadhouse—17.7 123.633; WAM R81191, R81307, R85385.

R85386, R85387, R85388, R85389 Koordarrie HS—22.3 115.033; WAM R81307 Koordarrie HS—22.3 115.033; WAM R81347 21 km WSW Marillana HS—22.25 119.4; WAM R84051 1 km W of Nanutarra Homestead—22.5 115.5; WAM R90571, R90578 Woodstock—21.61 118.988; WAM R90600, R90898 Woodstock—21.61 118.956; WAM R90893, R90904 200 m S Gallery Hill—21.67 119.041; WAM R91135, R91136 Harding River Dam—21 117.167; WAM R96248 Ca 80 km S Telfer—22.32 122.076; WAM R99145, R99177, R99199 Woodstock Stn.—21.67 119.041; WAM R99836 2 km S Charlies Knob—25.07 125; WAM R99838 10 km S Charlies Knob—25.15 125.083; WAM R100303 Mt. Lawrence Wells—26.8 120.2; WAM R100943 9 km SSW Charlies Knob—25.12 124.933; WAM R101345 19 km 324° Bohemia Downs HS—18.75 126.15; WAM R102069 Urala Stn.—22.2 115.033; WAM R102157 7.5 km ENE Mt. Windell—22.63 118.614; WAM R103967 Peron Peninsula Shark Bay—26.1 113.633; WAM R104076, R104136, R104209, R104219 Woodstock Stn.—21.62 118.95; WAM R106055 Percival Lakes—21.37 124.867; WAM R108879 18.0 km at 170 Degrees to Lake Hancock, Gibson Desert Nature Reserve—24.46 124.833; WAM R114758 Cleaverville—20.65 117; WAM R121356 11.8 km W of Mardathuna Homestead—24.07 114.473; WAM R127164 Nifty Mine—21.67 121.583; WAM R129008 Urala Stn.—21.78 114.863; WAM R100734 Woodstock, Site 5—21.61 118.988; WAM R121138, R121139, R121351 Mr3—24.43 114.5; WAM R121352 Ke2—24.5 115.018; WAM R121354 Ke1—24.49 115.031; WAM R124913 Ke2—24.5 115.018; SAM R36239 8 km SSE Curtin Springs HS on Mulga Pk Rd.—25.39 131.767; SAM R48810 9.3 km NNW of Mt. Cheesman Pi10301—27.33 130.293; SAM R48822 12.2 km NW of Mt. Cheesman Pi10201—27.34 130.238.

Fossil Record

No fossils of *Varanus brevicauda* are known.

Diagnostic Characteristics

• Diminutive size, maximal TL of only 230 mm (118 mm SVL).

• Tail round and short, about the same length as the body.

• Short legs.

Description

This very distinctive little monitor should not be confused with any other species. Dorsal surfaces are brown or reddish brown, flecked with darker brown spots, and ventral areas are pale white or pale brown. There is a faint temporal stripe. The tail is prehensile, round, and short, only about 87% to 116% of SVL with keeled scales. Legs and tail are quite short.

Size

Varanus brevicauda is the smallest of all monitor lizards, reaching adult size at 70 to 110 mm SVL. These dwarf monitors range from 39 to 118 mm SVL (maximum TL ~230 mm). Relative to other species of *Varanus, V. brevicauda* have a shorter head, neck, limbs, and tail (Thompson and Withers 1997c). However, their bodies are relatively longer, and they have from 31 to 35 presacral vertebrae, whereas other monitors have only 28 to 31 (Greer 1989), suggesting that natural selection has favored a longer body. Hind limbs are shorter than in most monitor lizards, being only slightly longer than the forelimbs (Christian and Garland 1996). Heads of males are relatively longer than those of females at the same SVL (Pianka 1994a). In the Great Victoria Desert, females are slightly larger than males (Pianka 1994a). However, in central Australia, males are longer and heavier than females (James 1996). Body weights of adults vary from 8 to 17 g. Recapture studies show that *V. brevicauda* grow slowly (James 1996).

Habitat and Natural History

Varanus brevicauda are found in hummock grasslands in the red sand deserts of interior Australia. This secretive terrestrial pygmy monitor is seldom seen but can be locally common. The vast majority of specimens are collected in pit traps. Dozens have pit trapped on flat sand plains covered with large, long unburned, clumps of spinifex, possibly the preferred habitat of *V. brevicauda* in the Great Victoria Desert (Pianka 1994a, 1996). At this site, 90% of lizards were found on flat sand plains, and only two lizards were trapped on sandridges. However, at another spinifex dominated sand dune site in central Australia, *V. brevicauda* are associated with sandridges, where they are quite abundant, with densities estimated at about 20 adults per hectare (James 1994). The typical monitor lizard threat posture and behavior has been conserved in the evolution of these diminutive monitors, which hiss and lunge with their throat inflated as if they are a serious threat (Pianka, personal observation).

Terrestrial, Arboreal, Aquatic

Varanus brevicauda are terrestrial but they may well climb around within spinifex tussocks. Their tails are very muscular and prehensile, and they "hang on for dear life" when inside a spinifex grass tussock, using their legs as well as their tail (Pianka, personal observation). They appear to rely on spinifex for protection. One was found in the stomach of a *Varanus gouldii flavirufus* (Pianka 1994a).

Time of Activity (Daily and Seasonal)

In central Australia, *Varanus brevicauda* are most active during the spring, becoming inactive and possibly brumating (the ectothermic equivalent of hibernation) during autumn and winter (James 1996). Similarly, in the Great Victoria Desert, they were captured from August to January.

Thermoregulation

Two *V. brevicauda* were dug up in shallow burrows during August. One must have been active immediately before being exhumed because crisp, fresh tail lash marks were at the burrow's entrance and the lizard had a body temperature of 35.4°C, 10°C above ambient air temperature (Pianka 1970a).

Foraging Behavior

Because these sedentary lizards seldom leave the protective cover of spinifex tussocks, they must forage within tussocks.

Diet

Stomach contents show that these small monitor lizards eat reptile eggs, centipedes, beetles, grasshoppers, other small lizards (especially skinks), spiders, and cockroaches, as well as the larvae of Lepidoptera and other insects. One female weighing 9.1 g contained an adult 1.5-mL *Ctenotus calurus;* this prey item constituted 16.5% of the *brevicauda*'s body weight (Pianka 1994a).

Reproduction

In the Great Victoria Desert, the smallest male with enlarged testes was 82 mm SVL, and the smallest gravid female was 94 mm SVL (Pianka 1994a). In central Australia, sexual maturity is reached in males at about 70 mm SVL and in females at about 83 mm SVL (James 1996). Males may become reproductive at an age of about 10 months, but females probably do not mature until their second spring, at an age of about 22 months (James 1996). A male fell into the same pit only hours after a female was removed from that pit trap, suggesting that males may follow scent trails to find females (Pianka, personal observation). Clutch size is usually two to three eggs, although larger clutches have been reported. Relative clutch masses of two females with oviductal eggs was 16.7 g (Pianka 1994a). Mating occurs in the spring (September–October), and eggs are laid in November. Hatchlings emerge in late January to February and are about 42 to 45 mm SVL and weigh only about 2 g. In captivity, incubation periods of 70 to 84 days and 18 to 25°C have been reported (Schmida 1974).

Movement

Mark–recapture studies have shown that *V. brevicauda* do not move very far (James 1996). Average movements (19 individuals recaptured with 24 recaptures total) ranged only from 14 to 25 m over periods of up to 751 days. One lizard, however, did disperse 400 m in 700 days. Because the probability of recapture diminishes rapidly with distance, these lizards could be more mobile than it appears.

Physiology

Thompson and Withers (1997a) report standard evaporative water loss rates (mg H_2O/g/h) at 35°C of 0.472 ± 0.038. Standard metabolic rates of *V. brevicauda* ($\dot{V}o_2$ and $\dot{V}co_2$, both measured in mL/h) at 25°C are 1.12 ± 0.117 and 0.88 ± 0.08, respectively, and at 35°C are 2.71 ± 0.254 and 2.02 ± 0.200 (Thompson and Withers 1997b). Maximal metabolic rates at 35°C ($\dot{V}o_2$ and $\dot{V}co_2$, both measured in mL/h) are 56.4 ± 4.44 and 74.3 ± 6.28 (Thompson and Withers 1997b).

Factorial aerobic scope of *V. brevicauda* is 20.95. Metabolic rates are relatively low among these small monitors but well above those measured for other small species of lizards (Thompson and Withers 1997b).

Fat Bodies, Testicular Cycles

In central Australia, testes are largest from July through November and regress during winter (James 1996).

Parasites

V. brevicauda stomachs often contain many small, as yet unidentified, nematodes (they are probably a species of *Abbreviata*).

7.4. *Varanus caudolineatus*
GRAHAM THOMPSON

Figure 7.7. Varanus caudolineatus *(photo by Brad Maryan).*

Nomenclature

Boulenger (1885) describes a single specimen caught by Mr. Duboulay at Champion Bay, Western Australia. Champion Bay is on the north side of the port of Geraldton (28°46'S, 114°36'E), the western end of the current geographic distribution of *V. caudolineatus*. The holotype is in the British Museum (BMNH 64.7.22.3; Cogger et al. 1983). Although Fuller et al. (1998) did not include *V. caudolineatus* in their analysis of the phylogeny of varanids, there is little doubt *V. caudolineatus* is closely related to *Varanus gilleni*, as it is morphologically and ecologically very similar, with the most significant difference being the larger size of *V. gilleni*. *Varanus caudolineatus* (Fig. 7.7) is a member of the subgenus *Odatria*.

Distribution

Varanus caudolineatus is found in the western central area of Western Australia (Fig. 7.8). Its distribution abuts that of *Varanus gilleni* on the eastern side (Pianka 1969). It appears to have a southern boundary at about latitude 30°44'S and an eastern boundary at about longitude 123°E. Northern and northeastern geographic boundaries are not as clearly defined. The geographical distribution of an unde-

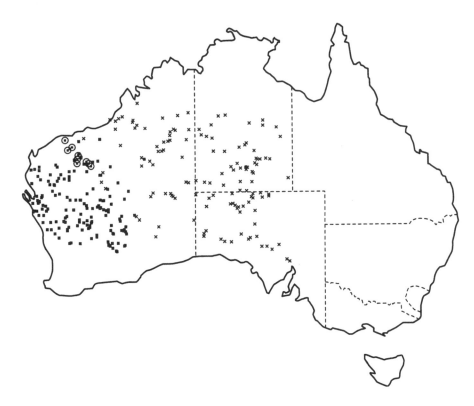

Figure 7.8. Geographical distribution of V. caudolineatus, V. gilleni, and the undescribed Varanus sp., from voucher specimens in Australian museums. Solid squares, V. caudolineatus; crosses, V. gilleni; circles with a dot, undescribed species.

scribed *Varanus* sp. lies between that of *V. gilleni* and *V. caudolineatus* in the Pilbara of Western Australia (Fig. 7.8). The apparent sympatry between these three species indicated in Fig. 7.8 probably represents specimens that have been misidentified in Australian museum collections, which is highly likely because few people are aware of the third undescribed species.

AM R49092, R49093, R49094: Ajana, ~3 mi Ex Coastal Hwy.—27.9 114.72; AM R130698 Wooramel River Crossing on Carnarvon-Mullewa Rd.—25.7 115.97; ANWC R2040 Carnegie Stn., NE of Wiluna—25.8 122.97; ANWC R2171, R2172, R2173, R2174: Millbillillie Stn., E of Wiluna—26.6 120.3; ANWC R2176 Lake Violet Stn., E of Wiluna—26.6 120.62; ANWC R2208 Millbillillie Stn., E of Wiluna—26.6 120.58; ANWC R2215 Meekatharra—26.5 118.5; NMV D 3558 Minilya—23.8 113.97; NMV D 7622 Mingenew—29.2 115.43; NMV D 1777, 1783 Middalya, 129 km inland from Carnarvon—23.9 114.77; NTM R25665 Meekatharra—26.6 118.47; NTM R25666 Youanmi, 5 km WSW—28.6 118.78; NTM R25667, R25668: Albion Downs—27.2 120.38; WAM R330, R331: Quinns —27.0 118.63; WAM R1191 Warriedar Stn. Yalgoo—29.1 117.18; WAM R3415, R3416, R3423: Laverton—28.6 122.4; WAM R3818 Narloo Stn. Wurarga—28.3 116.18; WAM R3881 Well 5, Canning Stock Route—25.3 121; WAM R3903 Well 29. C.S.R.—22.5 123.88; WAM R4288 Tambrey Stn.—21.6 117.6; WAM R4732, R4733; Gullewa. Yalgoo—28.6 116.32; WAM R4933, R4934: Murchison Downs —26.8 118.65; WAM R5048 Marilla—22.9 114.47; WAM R7279 Caron—29.5 116.32; WAM R7378, R7379: Belele Stn. Meekatharra—26.3 117.63; WAM R8167, R8168: Warroora Stn.—23.4 113.8; WAM R8257 Grants Patch—30.4 121.12; WAM R10611 Minilya Stn.—23.8 113.97; WAM R10812 Quinns. via Nanni NE—27.0 118.63; WAM R12278 Mundiwindi—23.8 120.25; WAM R12407 Kathleen Valley. Near Leonora—27.4 120.65; WAM R13362 Jiggalong—

23.3 120.78; WAM R13711 Overlander Roadhouse, Shark Bay TO—26.4 114.47; WAM R13857 Cosmo Newbery—28 122.9; WAM R14240 Kalgoorlie—30.7 121.47; WAM R14917 27 km E of Marillana—22.6 119.67; WAM R14918 11 km S of Cue—27.5 117.9; WAM R15785, R15786, R15787, R15788: Mileura. Ejah Paddock—26.3 117.33; WAM R15789 Mileura. Charlies Creek—26.3 117.33; WAM R17681 Mt. Margaret—28.8 122.18; WAM R19600 Cosmo Newbery Mission—28 122.9; WAM R19771 Kathleen Valley—27.4 120.65; WAM R19789, R19790: Albion Downs—27.2 120.38; WAM R20241 near Kangiangi HS—21.6 117.28; WAM R21100 32 km NE of Yelma HS—26.3 121.93; WAM R21137 24 km SW of Wiluna—26.7 120.05; WAM R21138 70 km SW of Wiluna—27.0 119.72; WAM R21139 70 km SW of Wiluna—27.0 119.72; WAM R21149 35 km N of Sandstone—27.6 119.3; WAM R21150 51 km SW of Sandstone—28.3 118.93; WAM R21178, R21179, R21180, R21181: 11 km SW of Youanmi—28.6 118.73; WAM R22652 Fields Find—29.0 117.25; WAM R22870 Yalgoo—28.3 116.68; WAM R22995 Ajana—27.9 114.63; WAM R23908, R23909, R23910: Laverton—28.6 122.4; WAM R25149 2 mi SE Turee Creek—23.6 118.67; WAM R25883 Ajana—27.9 114.63; WAM R26068, R26069, R26070: Jiggalong. Vicinity of Mission—23.3 120.78; WAM R26320 Greenough (Probably Eastern Goldfields)—30.7 121.5; WAM R27231 Kathleen Valley. Wanjarri Stn.—27.3 120.55; WAM R28028 the Well Spring—25.0 121.58; WAM R28290 Albion Downs—27.2 120.38; WAM R28394, R28955, R28956: Coordewany Stn.—25.6 115.97; WAM R29111 32 km S of Mt. Magnet—28.3 117.85; WAM R30967, R30968, R30969: Albion Downs. Within 13 km of HS—27.2 120.38; WAM R31381, R31382, R31383, R31384, R31385: 35 km NE of Mingenew—28.9 115.7; WAM R31681 14 km S of Menzies—29.8 121.03; WAM R34563, R34564: Callagiddy Stn. 32 km SE of Carnarvon—25.0 114.03; WAM R34685 64 km N of Beacon.-29.8 117.87; WAM R37897 Callagiddy.—25.0 114.03; WAM R39043 Youanmi—28.6 118.83; WAM R39765 Callagiddy—25.0 114.03; WAM R41793 Kalli near Cue—26.9 117.12; WAM R44529 Woodleigh—26.1 114.55; WAM R46618 Yarrie—20.6 120.2; WAM R46621 Linden—29.3 122.42; WAM R47347 Byro Stn.—26.0 116.15; WAM R47374 Sandstone—27.9 119.3; WAM R47630 Mileura. Macquarie Mill—26.3 117.33; WAM R47793 30 km SE of Bulloo Downs—24.2 119.78; WAM R47794 23 km Nw of Mt. Bruce—22.5 117.97; WAM R48151 5 km Nw of Kirkalocka HS—28.5 117.72; WAM R48152 16 km S of Mt. Magnet—28.2 117.85; WAM R49943 Woodleigh—26.1 114.55; WAM R49992, R49993: Wilroy Reserve. 19 km S of Mullewa—28.7 115.5; WAM R51092 60 km S of Leonora. 175 km N of Kalgoorlie—29.4 121.33; WAM R51157 5 km S of Warriedar HS—29.2 117.18; WAM R51189 Orabanda, 1 km N—30.3 121.07; WAM R52893, R52894, R52895, R52896: Mt. Augustus—24.3 116.92; WAM R54230 Marandoo—22.6 118.13; WAM R54594 15 km E of Hamelin Pool—26.3 114.33; WAM R54595 Wooramel HS—25.7 114.28; WAM R56834 Marandoo—22.6 118.13; WAM R57377, R57378:Woodleigh Stn.—26.1 114.55; WAM R59607 20 km ENE of Meadow Stn. HS—26.6 114.8; WAM R59620 34 km SSE of Nerren Nerren—27.3 114.85; WAM R60130 Youanmi—28.6 118.83; WAM R60429 2 km S of Overlander—26.4 114.47; WAM R60642 13 km W of Mungawolagudi Claypan—26.8 115.32; WAM R62171 Marandoo—22.6 118.13; WAM R62865, R62866, R62866 21–22 km SE of Mt. Keith—27.3 120.67; WAM R63655 25 km NNW Winning HS—22.9 114.45; WAM R64436 Cooloomia HS—26.9 114.3; WAM R65820 12.0 km ENE Comet Vale—29.9 121.24; WAM R65883 8.3 km SSE Mt. Linden—29.3 122.47; WAM R65886 8.00 km SSE Mt. Linden—29.3 122.46; WAM R65961 2.5 km N Mt. Linden—29.3 122.42; WAM R66004 7.75 km SSE Mt. Linden—29.3 122.47; WAM R69118 18.5 km ENE Yuinmery HS—28.5 119.19; WAM R69293 9.5 km SSE Banjiwarn HS—27.7 121.66; WAM R69295 12.5 km SSE Banjiwarn HS—27.8 121.68; WAM R69315 12.5 km SSE Banjiwarn HS—27.8 121.68; WAM R70831 24 km 333° Mt. Windarra—28.3 122.12; WAM R70895 3.5 km 005° Yowie Rockhole—30.4 122.35; WAM R71062 Billabong Roadhouse—26.8 114.62; WAM R72653, R72654: Comet Vale—29.9 121.58; WAM R72749 12.25 km ENE Comet Vale—29.9 121.24; WAM R72830 7.75 km SE Mt. Linden—29.3 122.45; WAM R72892, WAM R72893: 7.75 km SSE Mt. Linden—29.3 122.47; WAM

R73142 26.6 km 225° Marillana HS—22.8 119.23; WAM R73143 26.6 km 225° Marillana HS—22.8 119.23; WAM R73223 2 km 015° Yowie Rockhole—30.4 122.35; WAM R73309 5.5 km 137° Black Flag—30.5 121.28; WAM R73433 9 km 190° Mt. Elvire HS—29.4 119.58; WAM R73624 Ophthalmia Range—23.2 119.12; WAM R74671 18.5 km ENE Yuinmery—28.5 119.19; WAM R74689 8 km ENE Yuinmery—28.5 119.09; WAM R75834 Weano Gorge—22.3 118.25; WAM R75858 Coomalbidgup, Lort River; WAM R78488 7 km NE of Yowie Rockhole—30.4 122.37; WAM R78545, R78546: 21 km SE of Mt. Keith—27.3 120.67; WAM R78594 22 km S of Mt. Elvire Homestead—29.5 119.6; WAM R81917 Jibberding—29.8 116.97; WAM R81922 6 km W of Jibberding on White Well—29.8 116.92; WAM R84028 37 km SE of Ashburton Downs Homestead—23.6 117.3; WAM R85239, R85240, R85241:Dead Horse Rocks—29.3 121.28; WAM R87637 12 km SW of Yinnetharra HS—24.7 116.1; WAM R87661 13 km SSW of Mt. Phillip HS—24.5 116.23; WAM R87754 4 km NNE of Mt. Phillip HS—24.3 116.32; WAM R87770 12 km SW Yinnietharra—24.7 116.1; WAM R88098 11 km N of Nerren Nerren HS—27.0 114.63; WAM R91684 15 km N Karalund'e—26 118.68; WAM R92900 14 km SW Hamelin HS—26.5 114.1; WAM R94715 Agnew—28.0 120.52; WAM R95287 7 km N Goolthan Goolthan Hill—28.0 116.73; WAM R95527 3 km S Nannowtharra Hill—28.3 117; WAM R96122, R96123, R96124, R96125: Millrose Area—26.1 120.72; WAM R96679 17 km WNW Wadina HS—27.9 115.47; WAM R97786 Yoothapinna Stn.—26.5 118.5; WAM R100314 Mt. Lawrence Wells—26.8 120.2; WAM R101298 Wanjarri Nature Reserve, 2 km N of Kathleen Valley—27.3 120.65; WAM R102016 5.4 km N Joy Helen Mine—23.2 115.77; WAM R102098 Wongida Well, Barlee Range—22.9 115.85; WAM R108999 Marandoo—22.6 118.13; WAM R115204 Eurardy Stn.—27.5 114.67; WAM R116670 Black Range, 15 km WSW Marangaroon—23.7 115.47; WAM R117117 Upper Gascoyne—24.7 116.07; WAM R125540 Randall Well, 210 km NNE Meekatharra—25.3 119.4; WAM R104100 Woodstock, Site 5—21.6 118.99; WAM R122576 Md4—25.6 114.62; WAM R122622 Md5—25.7 114.6; WAM R122854 Mr2—24.4 114.51; WAM R123783 Gj3—25.1 115.43; WAM R125186 Ke3—24.5 114.97; WAM R125889 Ke3—24.5 114.97; WAM R132507 Jundee 0 [Editors' note: no latitude] 120.8; SAM R03448 Yandeearra Stn.—21.2 118.4; SAM R22808, R22809: Wannoo—26.8 114.62; SAM R29254, R29255, R29255: 57 km S Leonora—29.3 121.27; SAM R29401 1 km S Wannoo—26.8 114.62; SAM R31202, R31203: Leonora—28.8 121.33; SAM R35574 Mileura Stn.—26.3 117.33: SAM R42428 King of Creation Mine—28.0 122.32.

Diagnostic Characteristics

Varanus caudolineatus can be confused with *V. gilleni*, *V. eremius*, *V. kingorum*, and an undescribed *Varanus* sp. *Varanus caudolineatus* is distinguishable from *V. eremius* by the small, raised, and smooth head scales compared with the keeled head scales of *V. eremius*. Most often, *V. eremius* are gray or reddish brown, with blackish-brown spots adjacent to pale spots on the dorsal surface of the abdomen. Striping on the tail of *V. eremius* extends almost the full length. *Varanus caudolineatus* is distinguishable from *V. gilleni* and the undescribed species mostly by the head pattern. *Varanus caudolineatus* has a number of irregularly shaped small dark blotches on the dorsal surface of the head that are not arranged in a pattern. *Varanus gilleni* has a number of longitudinal streaks, and the undescribed species has both streaks and blotches or spots on the dorsal surface of the head. The dorsal surface of the abdomen of *V. caudolineatus* has circular black dots with a lighter color surrounding the dark spot. The dorsal surface of the abdomen of *V. gilleni* has a series of transverse bars, broken bars, or rows of dots. The dorsal surface of the undescribed species is speckled, and there are often transverse bands of dark speckles across the back. The distal two-thirds of the tail of *V. caudolineatus* contains four to six stripes; the distal third of the tail of *V. gilleni* contains four stripes, and sometimes these are broken. The anterior end of the tail of *V. gilleni* has transverse bars or rows of dots. The anterior half of the undescribed species has transverse bars or dots, and the distal half has four to six stripes. The ventral surface of the head and throat of *V. caudolineatus* has gray dots, and these extend to the anterior end of the abdomen and in some specimens most of the abdomen. *Varanus gilleni* has fewer dots on the ventral surface of the head and throat than *V. caudolineatus*. The undescribed species has dots on the ventral surface of the throat and anterior section of the abdomen, sometimes extending to the entire abdomen.

Description

This small semiarboreal goanna is morphologically and ecologically similar to *V. gilleni* and to the undescribed species.

Coloration

Body color varies across its range. Scales are strongly keeled. At either end of the continuum, there are two color morphs. For the gray morph, the dorsal surface of the head is gray with dark, irregularly shaped blotches and a dark stripe running through the eye to above the tympanum. The reddish-brown morph generally has more dark markings on the dorsal surface of the head, and the background color is gray-brown. For the gray morph, the background color of the dorsal surface of the abdomen is gray with dark brown-black spots, surrounded by a paler circular area. In the reddish-gray morph, the gray background is replaced with a reddish-brown background. The distal two-thirds of the tail has four to six longitudinal stripes, from which it derives its name. The ventral surface is off-white, gray, or brownish-white with less well defined spots than on the head and throat. These spots sometimes extend to the abdomen, legs, and base of the tail.

Neonates and juveniles are more distinctly colored replicates of adults.

Size

Neonates *V. caudolineatus* hatch at about 1.9 g, with a SVL of 54 mm and a tail length of 61 mm. Adult females grow to a SVL of 118 mm, and males grow to a SVL of 123 mm. The heaviest nongravid *V. caudolineatus* I have measured was 27.2 g. Most are less than 20 g, although Smith (1988) reports a postoviposition female with a body mass of 28 g (Table 7.1; Figs. 7.9–7.11).

Table 7.1 provides a comparison of the relative appendage lengths for *V. caudolineatus, V. gilleni,* and the undescribed species. These data show very little difference in body shape among *Varanus* sp., *V. caudo-*

TABLE 7.1.
Limb dimensions as a percentage of snout–vent length (SVL) for *V. caudolineatus,*
V. gilleni, and *Varanus* sp., an undescribed species similar to the other two goannas.[a]

Characteristic	V. caudolineatus		V. gilleni		Varanus sp.	
	Male	Female	Male	Female	Male	Female
No. of specimens	45	22	13	9	10	13
Max. SVL (mm)	125	118	175	167	148	140
Forelimb length	24.6 ± 0.27	23.8 ± 0.25	23.3 ± 0.25	23.0 ± 0.50	23.7 ± 0.52	23.9 ± 0.37
Hind limb length	32.0 ± 0.26	30.6 ± 0.63	29.0 ± 0.39	29.0 ± 0.11	31.1 ± 0.68	30.7 ± 0.52
Head length	18.5 ± 0.12	17.4 ± 0.19	17.6 ± 0.16	16.7 ± 0.32	17.5 ± 0.32	17.6 ± 0.49
Head width	9.9 ± 0.12	9.7 ± 0.19	9.2 ± 0.12	8.9 ± 0.16	9.5 ± 0.19	9.3 ± 0.22
Head depth	6.4 ± 0.08	6.1 ± 0.13	5.6 ± 0.11	5.3 ± 0.11	6.1 ± 0.11	6.0 ± 0.14

[a] Values are expressed as percentages ± 1 SE.

lineatus, and *V. gilleni,* with *Varanus* sp. being placed between the other two species.

Figure 7.9 indicates the extent of overlap in the morphology of *V. caudolineatus, V. gilleni,* and *Varanus* sp. in morphological space based on a canonical discriminant analysis of body appendage lengths. If the sexes for each species are not separated in a discriminant analysis, it is difficult to detect a difference between *Varanus* sp. and the other two species. Differences between sexes are almost as great as differences among species.

When SVL data for *V. caudolineatus* from Western Australian Museum specimens were combined with my unpublished data, the maximum size of adults increases as you move east (Fig. 7.10). There is no obvious change in size with latitude. The relationship between SVL and body mass on a log–log plot is linear (Fig. 7.11) and can be represented by the equation

$$\text{SVL (mm)} = 45.60m^{0.309} \ (n = 38, r^2 = 0.96),$$

where m = body mass in g.

Branch (1982) describes the hemipenes of *V. caudolineatus* as asymmetrical and club shaped, with a simple sulcus that runs to the crotch and drains into a flattened nude apical. *Varanus caudolineatus* has six to seven shallow papillose stiffened frills that traverse the distal asulcal region, with two frills that are noticeably narrower. The most distal frill fuses with a papillose ridge that demarcates the flattened nude apical region. The right horn is three to four times the size of the left; both are blunt-tipped and straplike.

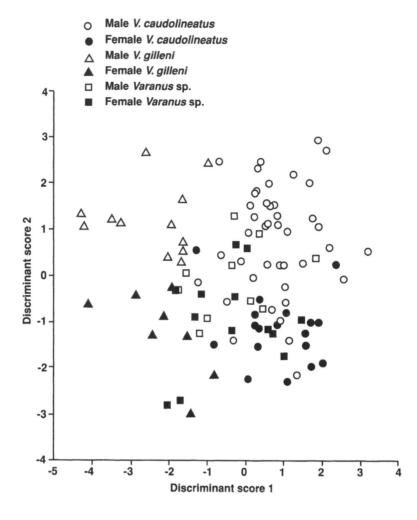

Figure 7.9. Canonical variates
1 and 2 for male and female
V. caudolineatus, V. gilleni, *and*
Varanus *sp. based on a range of*
morphological measures
displaying the overlap in
morphology among species.

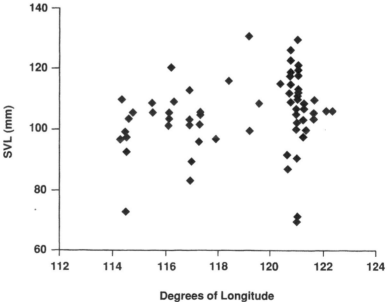

Figure 7.10. SVL *of* Varanus
caudolineatus *plotted against*
longitude of capture site to show
the increase in size for the more
easterly located specimens.

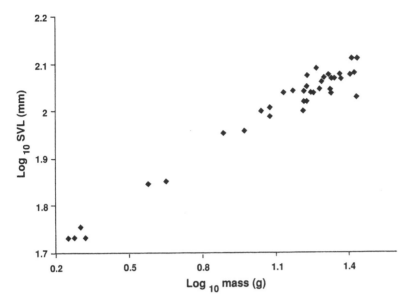

Figure 7.11. Linear relationship between logarithmically transformed SVL and body mass for V. caudolineatus.

Habitat and Natural History

This small semiarboreal species is almost always found in tree hollows or under bark. Its distribution reasonably closely matches that of the western geographical distribution of *Acacia aneura,* except for the coastal area from Shark Bay to Geraldton. The distribution of *V. gilleni* matches the more eastern geographical distribution of *Acacia aneura.* Around Dead Horse Rocks, Western Australia (29°22'S, 121°17'E), *V. caudolineatus* is mostly found in rock crevices. Seldom is this wary goanna encountered on the ground; and they have been observed to locate themselves in tree hollows with head protruding, observing their surroundings.

Diet

On the basis of stomach contents of Western Australian Museum specimens, Pianka (1969) reports their diet to consist of invertebrates (mostly roaches, grasshoppers, and spiders) and geckos (*Gehyra variegata* and *Rhynchoedura ornata*). Thompson and King (1995) report the diet of *V. caudolineatus* at Atley Station, Western Australia (28°25'S, 119°07'E), to be considerably different from those in the Western Australian Museum collection. Stomachs of 88 museum specimens examined contained mostly spiders, grasshoppers, and other unidentified invertebrates, whereas *V. caudolineatus* from Atley Station contained predominantly scorpions and spiders. *Varanus caudolineatus* from Atley Station had also devoured a *Menetia greyii* and a *Ctenotus schomburgkii.* Most spiders and scorpions eaten were ground-burrowing species, and given there is no evidence that *V. caudolineatus* is active at night (Thompson 1993; Thompson and King 1995) and *V. caudolineatus* tracks were found leading to entrances of scorpion burrows, these lizards probably enter burrows in search of prey. Average size of prey items was 0.55 mL (Pianka 1969); however, *V. caudolineatus* can subdue and devour relatively large prey. For example, Thompson et al. (1997) report a captive 11-g *V. caudolineatus* eating a 4-g gecko.

Varanus caudolineatus is seasonally active. Pianka (1969, 1994a) caught them between November and March in the southern part of their geographic distribution. Pianka (1994a) reports a mean field active body temperature of 37.8°C (SD = 3.45, n = 10). Pit-trapping programs within their distribution catch more specimens on hot days (>35°C; S. Thompson, personal communication), suggesting that these goannas prefer a relative high ambient temperature.

Thompson (1993) studied the spatial ecology of *V. caudolineatus* with a radioactive tracer and reported the mean maximum linear distance traveled by *V. caudolineatus* in 1 day was 33.9 m and ranged from 14 to 156 m. They were not active at night even through ambient temperature is in the high 30s to low 40s (°C). *Varanus caudolineatus* use the same tree as a retreat for many consecutive days and use only a small number of trees within a confined activity area. Thompson (1993) reports finding the same *V. caudolineatus* within 35 m of its initial capture after 74 days at Atley Station, and another *V. caudolineatus* was caught within 30 m of its initial capture site after 9 months at Ora Banda (30°22'S, 121°04'E; S. Thompson, personal communication), suggesting that these lizards have small home ranges. Searches for *V. caudolineatus* (unpublished data) suggest that they may live in loose colonies because their distribution does not appear to be evenly spaced over apparently suitable habitat.

Combat Ritual

Thompson et al. (1992) report the male combat ritual for *V. caudolineatus* to be similar to that of *V. gilleni* (Carpenter et al. 1976) and most other varanids in the *Odatria* group. This involves wrestling, longitudinal rolls while embraced with ventral surfaces adjacent, lateral twisting and flexing, and occasional biting on the flank, limbs, and tail. While embraced with both fore- and hind limbs, two adversaries will form an arch from their snout to their tail.

Reproduction

Observations of captive *V. caudolineatus* suggest that mating behavior is similar to that in other varanids. A male will nudge and tongue-flick a female, particularly around the head and neck. Mating occurs when the male is able to lie along the top of a female, using a hind limb and tail to expose the cloaca, enabling the male to insert a hemipene.

Male testes were largest during July and August, and a female with enlarged oviducts was collected on December 30 (Pianka 1969). Smith (1988) reports a female *V. caudolineatus* with a postoviposition mass of 28 g, laying four eggs (total mass of 9 g) on October 23, 1986. Unpublished data (G. Thompson) suggests *V. caudolineatus* oviposits over an extended period (wild-caught gravid females laid three eggs on

September 22, 1992; three eggs on October 12, 1993; four eggs on December 16, 2000; and three eggs on January 15, 1995). Four eggs laid in 2000 were incubated between 29 to 31°C and hatched after 77 days. Postoviposition body mass of this female was 10.6 g and the four eggs weighed 9.6 g, a contribution of ~47.5% of its total mass to four eggs, a very large relative clutch mass (Thompson and Thompson, 2002). Pianka (1994a) reports mean clutch size of V. caudolineatus to be 4.3 (n = 6). Mean mass, SVL, and tail length of neonates hatched in captivity were 1.95 g, 54.7 mm, and 61.5 mm, respectively (n = 4, Thompson and Thompson, 2002).

Physiology

Standard metabolic rate for V. caudolineatus is within the range expected for other similar-sized lizards and is represented by the equation

$$\log_{10} \dot{V}_{O_2} \text{ (mL h}^{-1}) = -1.46 \pm 0.43 \log \text{ mass (g)} \pm 0.38 \, T_b \, (°C)$$

(Thompson and Withers 1994, 1997b). However, maximal metabolic rate for V. caudolineatus is the highest recorded for any lizard:

$$\dot{V}_{O_{2max}} \text{ mL h}^{-1} = 13.2 m^{0.72},$$

where m = mass in g at 35°C, with the consequence that factorial aerobic scope (maximal/standard metabolic rate) at 35°C is also the highest recorded for any lizard (>>35). Field metabolic rate (FMR) for V. caudolineatus is ~0.46 mL CO_2 g^{-1} h^{-1} during summer when active (Thompson et al. 1997).

When active during summer, the daily water requirement for V. caudolineatus is ~31.6 mL kg^{-1} d^{-1} (Thompson et al. 1997). Body water content was slightly higher than for most other goannas—about 80% of total mass (Thompson et al. 1997). Evaporative water loss at 35°C is H_2O (mg h^{-1}) = 1.48$m^{0.41}$, where m = mass in g (Thompson and Withers 1997a). Thompson and Withers (1997d) partitioned evaporative water loss for three small varanids, V. caudolineatus, V. brevicauda, and V. eremius, and report at 14, 20, and 25°C, evaporation across pulmonary surfaces was 4.7%, 2.4%, and 5.9% of total water loss, respectively. Like other small goannas, at low body temperatures, V. caudolineatus has an irregular ventilation pattern, including extended periods with no ventilation. At 14, 20, and 25°C, the mean longest nonventilatory periods for V. caudolineatus were 61.0, 16.8, and 10.2 minutes, with the longest nonventilatory period of 137 minutes at 14°C (Thompson and Withers 1997d).

Parasites

Jones (1992) reports the nematode Abbreviata levicauda (Nematoda: Physalopteroidea) occurs in low numbers (4.3%) in the gut of V. caudolineatus. Infected specimens are widespread within the known geographical distribution. Jones (1992) suggests that this nematode is probably the same species that is found in Varanus tristis, another semiarboreal species.

7.5. *Varanus eremius*

ERIC R. PIANKA

Figure 7.12. Varanus eremius
(photo by Hal Cogger).

Nomenclature

Lucas and Frost (1895) described this species and later illustrated it in color (Lucas and Frost 1896). Type specimen (holotype) is held at the National Museum of Victoria (as NMV D9136); the type locality is Idracowra, Northern Territory (Fig. 7.12).

Geographic Distribution

Varanus eremius occurs in the red sand deserts of Western Australia, the southern half of the Northern Territory, northern South Australia, and remote southwestern Queensland. This widespread terrestrial species occurs throughout the Great Victoria Desert, the Gibson Desert, the Little Sandy Desert, the Great Sandy Desert, the Tanami Desert, and the deserts of the central ranges, as well as in the Simpson Desert (Fig. 7.13).

AM R7635 407 mi EW Line—30.5 132.15; AM R10854, R10855, R10856 Hermannsberg—23.95 132.77; AM R11903, R11904 Coniston Stn., 169 mi W Alice Springs—22.15 132.52; AM R33497 Port Hedland—20.3 118.58; AM R49709 38.1 mi N Neale Junction—27.72 125.82; AM R90866 62.1 km NW Chilla Well via Rabbit Flat—Yueudumu Rd.—20.95 130.65; AM R90875 64.4 km W Ayers Rock (by road)—25.18 130.5; AM R101499 32.2 km SE of Onslow PO by road—21.9 115.17; AM R105694 0.5 km N of Cooralya TO on NW Coastal Hwy.—24.38 114.0; AM R110613 8 km N of Mirrica Bore, Ethabuka Stn., NW of Bedourie—23.73 138.47; AM R111316 Denham, Shark Bay—25.93 113.53; AM R113137 Simpson Desert, 37 km W of Muncoonie Homestead, NW of Birdsville—25.18 138.28; AM R125075, R125076, R125077, R125078, R125079, R125080,

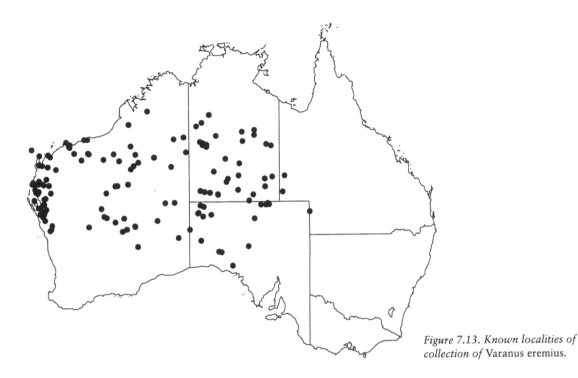

Figure 7.13. Known localities of collection of Varanus eremius.

R125081, R125082 Mt. Liebig—23.3 131.37; AM R125083 Ooldea—30.45 131.83; AM R140522 Denham (Rubbish Tip)—25.92 113.53; AM R147247 19.9 km N Wauchope on Stuart Hwy. (=1 km N Dixons Creek Crossing)—20.48 134.25; ANWC R2203 Desert Farm, Wiluna—26.58 120.3; NMV D 9136 Idracowra—25.0 133.78; NMV D 478 Illamurta—24.3 132.68; NMV D 2947 Tennant Creek—19.65 134.18; NMV T 109 Labbi Labbi—21.52 128.63; NMV D 527 Charlotte Waters—29.98 134.83; NMV D 159 Alice Springs—23.7 133.87; NMV D 1787, D 1772 Middalya, 129 km Inland from Carnarvon—23.9 114.77; NMV D 99 Oodnadatta—27.55 135.45; NMV D 160 Alice Springs—23.7 133.87; NTM R25669 213.2 km W Yuendumu—20.83 130.67; NTM R25670 Horden Hill, Granites—20.65 130.32; NTM R25671 E Bank Armstrong Creek 100 km W Ayres Rock—25.08 130.05; NTM R25672 Atartinga Stn.—22.57 133.93; NTM R25673 Sangsters Bore—12 km SW—20.87 130.27; NTM R25674 Frewena—19.42 135.4; NTM R25675 Frewena—19.43 135.4; NTM R25676 Frewena—19.92 135.42; NTM R25677 Alyawarre Desert Area—20.77 136.58; NTM R25678 Wave Hill Stn., 15 km SSW Flora Bore—17.97 130.92; NTM R25679 Alyawarre Desert Area—20.93 137.07; NTM R25680 Simpson Desert, Lake Caroline—23.78 137.2; NTM R25681 Tanami—20.53 130.03; NTM R25682 Sangsters Bore—20.88 130.42; NTM R25683 Tanami—20 131.62; NTM R25684 Western Tanami—19.05 129.65; NTM R25685 Andado Stn.—25.42 135.28; NTM R25686 Simpson Desert—23.95 136.45; NTM R25687—20.53 130.03; NTM R25688 Winnecke Creek—18.75 130.22; WAM R11353 32 km E of Mullewa—28.55 115.17; WAM R11609 Laverton—28.63 122.4; WAM R12386 near Wannoo—26.82 114.62; WAM R14658 Warburton Range—26.13 126.58; WAM R14916 Maloora Rock-hole—29.2 127.93; WAM R17081 5 km E of Roebourne—20.77 117.2; WAM R20082, R20083, R20083 Tambrey—21.63 117.6; WAM R22001, R22020 War-burton Range—26.13 126.58; WAM R27230 Wanjarri Stn. Kathleen Valley—27.32 120.55; WAM R28015 Boologooro—24.35 114.03; WAM R33597 29 km N

of Ajana—27.68 114.63; WAM R34145 93 km E of Warburton Range Mission—
26.13 127.52; WAM R36593 White Springs. 109 km N of Wittenoom—21.25
113.33; WAM R37570 Lockwood Spring. Kalbarri National Park—27.32 114.47;
WAM R37571 Lockwood Spring. Kalbarri National Park—27.32 114.47; WAM
R39144 Well 23. C.S.R.—23.08 123.22; WAM R40529 Wanjarri Stn. 16 km E of
Kathleen Valley—27.41 20.82; WAM R40603, R40614 145 km N of Carnegie—
24.47 122.97; WAM R40677 Callagiddy Stn. South Bore—25.05 114.03; WAM
R40749 Se Carnarvon Range—25.22 120.75; WAM R40873 48 km W of Balgo
Hill Mission—20.15 128.42; WAM R42298 Well 12 Canning Stock Route—24.58
121.7; WAM R42307 Well 12 C.S.R.—24.57 121.87; WAM R45126 58 km E of
Jupiter Well—22.88 127.17; WAM R45261 Well 38. C.S.R.—21.95 125.53; WAM
R48117 E Yuna Reserve. 30 km ESE of Yuna—28.43 115.32; WAM R55167 Peron
HS. Peron Peninsula—25.83 113.55; WAM R56859 16 km S of Gascoyne Junction—25.21 15.22; WAM R56971 44 km NE of Yuna. Proposed Yuna Reserve—
28.05 115.33; WAM R59621 14 km ENE of Meadow Stn. HS—26.63 114.77;
WAM R60406, R60407, R60408 Boologooro Stn. 80 km N of Carnarvon—24.35
114.03; WAM R60981 Giralia HS—22.68 114.37; WAM R61076 Vlaming Head—
21.8 114.1; WAM R63377 10 km 224° Lens Bore—20.33 127.43; WAM R63941
12 km 21° Well 29 Canning Stock Route—22.45 123.92; WAM R64273 38 km
209° Mctavish Claypan—20.93 123.22; WAM R64414, R64415 16 km SW
Billabong Roadhouse—26.92 114.5; WAM R69289, R69323, R69324 12.5 km
SSE Banjiwarn HS—27.8 121.67; WAM R69539 18 km 236° Cooloomia—27.04
114.15; WAM R70028 5.3 km 269° Mt. Percy—17.62 124.87; WAM R70929 46.5
km 251° Goorda Tower—18.89 123.04; WAM R73857 Harding River Area 25 km
S Roebourne—21 117.07; WAM R73858 Harding River Area 30 km S Roebourne—
21.05 117.12; WAM R73995 5 km ENE Kurrana Well—22.15 120.48; WAM
R74768 12.5 km SSE Banjiwarn—27.8 121.67; WAM R80254 4 km W of Barradale
Roadhouse—22.85 114.92; WAM R81311 Old Onslow—21.72 114.95; WAM
R81758 3 km E Greenough Point—25.25 113.88; WAM R83760 Well 26 Canning
Stock Route—22.77 123.5; WAM R88607 13 km SW Manberry HS—24.07
114.05; WAM R88608 3 km NW Bullara HS—22.65 114.03; WAM R90603,
R90716 Woodstock—21.61 118.97; WAM R90891 Gallery Hill—21.67 119.04;
WAM R91250 58 km NNE Queen Victoria Spring—29.98 123.87; WAM R92838
15 km S Hamelin HS—26.56 114.2; WAM R95004, R95007, R95017 Lake Auld—
21.73 123.67; WAM R95338 Burrup Peninsula—20.57 116.8; WAM R96249 Ca
80 km S Telfer—22.32 122.08; WAM R102398, R102408 Barlee Range Nature
Reserve—23.06 115.79; WAM R108531 7–8 km WSW Point Salvation—28.2
123.6; WAM R108533, R108534, R108535, R108548, R108550, R108552,
R108554, R108555, R108558, R108567, R108570 7–8 km WSW Point Salva-
tion—28.47 122.83; WAM R108546, R108551, R108553, R108557, R108560,
R108561, R108568, R108571 7–8 km WSW Point Salvation—28.25 123.6; WAM
R113070 Lesley Salt Works—20.32 118.9; WAM R114618 Carnarvon—24.91
13.65; WAM R114780, R114781, R114781 Atley Stn.—28.23 119.07; WAM
R114783 2 km W Wittenoom—22.23 118.32; WAM R121357 2.5 km W of
Overlander—Denham Rd.—26.49 114.06; WAM R127047 Nifty Mine—21.67
121.58; WAM R129102 De La Poer Range Nature Reserve—27.47 122.58; WAM
R121253 Mr4—24.41 114.47; WAM R121348 Ne2—27.06 114.59; WAM
R121349 Bb1—25.13 113.82; WAM R121350 Ke1—24.49 115.03; WAM R121353
Cu—24.4 113.4; WAM R121355 Bb5—25.13 113.77; WAM R121411 Cu5—
24.19 113.46; WAM R122452 Na4—26.55 113.96; WAM R122492 Na2—26.49
114.06; WAM R122651 Md1—25.63 114.7; WAM R122693 Md5—25.71 114.6;
WAM R122702 Md2—25.62 114.69; WAM R122862 Mr3—24.43 114.5; WAM
R123247 Ke2—24.51 15.02; WAM R123613 Pe4—25.84 113.61; WAM R123741
Bo1—24.41 113.66; WAM R125771 Wo1—26.22 114.6; WAM R125807 Wo3—
26.21 14.54; WAM R125820 Wo2—26.21 114.58; WAM R125858 Md2—25.62
114.69; WAM R126719 Pe2—25.88 113.55; WAM R126732 Bo4—24.41 113.75;
WAM R126735, R126743 Bo1—24.41 113.66; WAM R126747 Mr2—24.44
114.51; WAM R126766 Bo2—24.41 113.68; WAM R126825 Bo4—24.41 113.75;
WAM R126885 Bb4—25.12 113.73; WAM R126899 Bo3—24.41 113.71; WAM

R126912 Bb5—25.13 113.77; WAM R126927 Bb2—25.13 113.81; SAM R00076, R00077, R00078, R00079 Hermannsburg—23.95 132.77; SAM R00532 Charlotte Waters—25.92 134.9; SAM R06090 Tennant Creek—19.65 134.18; SAM R11171 4 mi W Refrigerator Bore—21.05 130.7; SAM R22236 SE Alice Springs—24.67 136.5; SAM R29875 7 km S along Mulga Pk Rd. SSE Curtin Springs HS—25.38 131.77; SAM R32243 39 km SW Halinor Lake—29.45 130.17; SAM R32292 50 km SW Halinor Lake—29.53 130.14; SAM R33810, R33811 12 km S Bloodwood Well—26.95 140.95; SAM R35970 2 km W of Purni Bore Campsite—26.28 136.08; SAM R36238 Yulara Townsite—25.23 131.02; SAM R36272 13.6 km Along Mulga Pk Rd. from Lasseter H/Way—25.43 131.78; SAM R43805 Serpentine Lakes Edge—28.5 129.02; SAM R43840 Sand Dunes 1–2 km E of Serpentine Lakes Access Rd.—28.5 129.03; SAM R44402 8.5 km NW of Mt. Kintore—26.5 130.43; SAM R44449 12.5 km ENE of Mt. Cooparinna—26.35 130.09; SAM R45395 9.7 km S of Ampeinna Hills—27.16 131.13; SAM R45496 16 km NNE of Inila Rockwaters; SAM R45497 23.5 km NW of Inila Rockwaters—31.62 133.26; SAM R48609 4.7 km NNE of Mt. Cheesman—27.37 130.35; SAM R48762 3.1 km WNW of Mt. Lindsay—27.02 129.85; SAM R48779 6.6 km WNW of Mt. Lindsay—27.06 129.82; SAM R49902 44.7 km E of Purni Bore—26.32 136.54; SAM R49942 29.4 km ENE of Purni Bore—26.23 136.38; SAM R49961 38 km E of Purni Bore—26.23 136.47; SAM R50057 77 km E Purni Bore—26.33 136.87; SAM R50071 71.9 km E Purni Bore—26.31 136.82; SAM R50075 68.1 km E Purni Bore—26.19 136.77.

Fossil Record

No fossils of *Varanus eremius* are known.

Diagnostic Characteristics

- Small size.
- Rusty red above.
- Pale below except for gray-mottled throat.
- Round, longitudinally striped tail.
- Keeled head scales.

Description

A very distinctive, small terrestrial monitor lizard. Rusty red above, with numerous scattered dark brown spots mixed with smaller creamy or yellowish flecks. Tail round in cross section, with six cream and dark brown to black stripes. Caudal scales keeled but not spinose. Black stripe from snout to eye extending to temporal region. Thin cream stripe extends from the jaw to the base of the front leg. Pale below except for gray mottling on the throat. Small keeled head scales. Total of 85 to 110 midbody scale rows. Digits moderate, slightly compressed, with nearly straight claws. Nostrils round, on the sides of the head, much closer to the tip of the snout than to the orbit of the eye.

In some parts of its geographic range, lizards are more brown or more yellowish than they are in the red sand deserts. No sexual dimorphism in color is evident. Juveniles look like adults but their color pattern is crisper.

Size

V. eremius is a dwarf monitor with a SVL ranging from 59 to 160 mm (TL ~140–500 mm). Heads of males are relatively longer than those of females at the same SVL (Pianka 1994a). Tail lengths are about 1.5 times SVL (regression equation: tail length = 24.6 ± 1.47 SVL). Average body weight of 33 adult lizards was 41.5 g.

Habitat and Natural History

Varanus eremius are associated with hummock grasslands in the red sand deserts of interior Australia. This terrestrial pygmy monitor is common, but this beautiful little red *Varanus* is extremely wary and hence seldom seen. Nevertheless, their unique and conspicuous tracks allow substantial inferences to be drawn about their activities (Pianka 1968, 1994a).

The following statements about *Varanus eremius* are based on impressions gained while following literally hundreds of kilometers of *V. eremius* tracks on foot. Individuals usually move over great distances when foraging. Fresh tracks often cover distances of up to a kilometer. Because these lizards show little tendency to stay within a delimited area, home ranges would appear to be extremely large. These monitor lizards are attracted to fresh holes and diggings of any sort and will often visit any man-made digging within a few days after it is made. While moving, *V. eremius* drag their tails straight behind them, frequently changing direction abruptly, leaving characteristic looped tail track patterns in the sand. *V. eremius* do not sway their heads from side to side like most larger terrestrial monitors, and they do not appear to use their tongues to detect prey via vomeronasal olfaction like larger species do. Instead, these monitors hunt visually. In contrast to *V. gouldi flavirufus*, *V. eremius* do not dig for their prey, but rather rely on catching it above ground. More than once, I have noted an *V. eremius* track intercept the track of another smaller lizard with evidence of an ensuing tussle. One such pair of tracks came together, rolled down the side of a sandridge leaving a trail of big and little tail lash marks, and finally become one track, dragging away a fat belly!

Terrestrial, Arboreal, Aquatic

Varanus eremius are strictly terrestrial. I have never seen one even attempt to climb. Lucas and Frost (1896) noted its straight claws as compared with those of *Varanus gilleni,* which is arboreal and has strongly recurved claws. Moreover, *Varanus eremius* is the smallest widely foraging terrestrial monitor lizard. *V. brevicauda* are much more sedentary (James 1996).

Time of Activity (Daily and Seasonal)

Unlike larger species of monitors, *Varanus eremius* are active all year long, although they appear to be most active during late winter and early spring (Pianka 1968, 1994a).

Thermoregulation

Mean ± SD body temperature of 75 active individuals was 37.3 ± 3.24°C. The slope of a least-squares linear regression of body temperature on ambient air temperature is 0.22, suggesting that these lizards actively thermoregulate.

Foraging Behavior

In a typical foraging run, an individual *V. eremius* often visits and goes down into several burrows belonging to other lizard species, especially the complex burrow systems of the nocturnal skink *Egernia striata*. These activities could be in search of prey, related to thermoregulatory activities, and/or simply involved with escape responses. Certainly a *V. eremius* remembers the exact positions of burrows it has visited, because it almost inevitably runs directly to the nearest one when confronted with the emergency of a lizard collector.

One was once observed to attack a small skink from ambush. On this occasion, a large *V. eremius* jumped out of a loose spinifex (*Triodia basedowi*) tussock when a small blue-tailed skink (*Ctenotus calurus*) came within a few centimeters of the edge of the tussock (Pianka 1968).

Unlike larger monitors such as *V. giganteus*, *V. gouldii flavirufus*, and *V. panoptes*, which walk with their legs below them and their bodies and tails elevated above ground, *V. eremius* walk in the typical primitive tetrapod stance with their bodies close to the ground, dragging their tail straight behind. As stated above, they leave a very distinctive tail drag track mark, which frequently turns back on itself.

Diet

W. E. Schevill reported that a Hermannsburg specimen had eaten a mouse (Loveridge 1934). A survey of the stomach contents of 130 of these pygmy monitor lizards (82 of which contained prey items; Pianka 1994a) revealed that 76% of the diet of *V. eremius* by volume consists of other lizards, with large grasshoppers plus an occasional large cockroach or scorpion constituting most of the remainder. Nearly any other lizard species small enough to be subdued is eaten: stomachs with food contained 53 individual lizard prey items representing some 14 other species in addition to other items. Prey species eaten by *V. eremius* include the following skinks: *Ctenotus calurus, C. dux, C. grandis, C. lesueuri, C. piankai, C. quattuordecimlineatus, C. schomburgkii, Lerista bipes, Menetia greyi, Morethia butleri,* and the pygopodid *Delma,* plus the agamids *Ctenophorus inermis, C. isolepis,* and *Gemmatophora longirostris.*

One *V. eremius,* weighing 42.5 g, had eaten a *Ctenophorus inermis* with a body mass of 12.2 mL (28.7 % of the *V. eremius*'s body mass).

Reproduction

Male ritualized combat has never been reported for this species but undoubtedly occurs. The smallest gravid female *V. eremius* had a SVL of 110 mm, whereas the smallest male with enlarged testes had an SVL of 116 mm, suggesting that females may reach sexual maturity earlier than males (Pianka 1994a). Mating occurs in the austral spring, during October–November. Clutch size varies from 2 to 6 eggs, with a mean of 3.6 eggs (n = 14 eggs). Relative clutch mass of three females with eggs in their oviducts averaged 15.3% of female body weight. Eggs are laid during November to December and hatchlings emerge in mid-January to early February (incubation time is short, probably ~60 to 90 days). Hatchling SVLs range from 59 to 64 mm (weights 1.9–3.3 g).

Movement

Home ranges are large, as explained above. This species is active all year long but exhibits distinct seasonal changes in activity, being most active during late winter and early spring.

Physiology

Thompson and Withers (1997a) report standard evaporative water loss rates (mg H_2O/g/h) of 0.242 ± 0.037 at 25°C and of 0.474 ± 0.050 at 35°C. Standard metabolic rates of *V. eremius* ($\dot{V}o_2$ and $\dot{V}co_2$, both measured in mL/h) at 25°C are 2.38 ± 0.324 and 1.91 ± 0.271, respectively, and at 35°C are 6.16 ± 0.906 and 4.74 ± 0.697 (Thompson and Withers 1997b). Maximal metabolic rates at 35°C ($\dot{V}o_2$ and $\dot{V}co_2$, both measured in mL/h) are 91.9 ± 17.0 and 112.9 ± 25.1 (Thompson and Withers 1997b). Factorial aerobic scope of *V. eremius* is 12.98. Metabolic rates are typical for small monitors but well above those measured for other small species of lizards (Thompson and Withers 1997b).

Fat Bodies, Testicular Cycles

Fat bodies show no clear seasonal cycle and can be large at any time of year, probably depending on immediate past conditions for feeding. Testes regress during May and June and are largest during the mating season October and November.

Parasites

All 22 stomachs of *Varanus eremius* from two study sites examined by Jones (1995) contained encysted larval physalopterid nematodes.

7.6. *Varanus giganteus*

HANS-GEORG HORN AND DENNIS R. KING

Figure 7.14. Varanus giganteus (*photo by Eric Pianka*).

Systematics, Taxonomy

Varanus giganteus, the perentie, was described in 1845 by Gray in the catalog of the specimens of lizards on Barrow Island, Western Australia, in the collection of the British Museum. The scientific name given at the time was *Hydrosaurus giganteus*. Later, Boulenger (1885) renamed it *Varanus giganteus*. In a subsequent volume of this catalog, these descriptions were based on external morphology, e.g., numbers and shape of scales, length of the head plus body, length of tail, and length of front and hind legs (Mertens 1942, 1958, 1963). Because of the fine scalation, Mertens believed that the perentie is closely related to *V. varius* and places this species between *V. varius* and *V. komodoensis* (Mertens 1942, 1958) (Fig. 7.14).

Hemipeneal morphology is a more modern approach to taxonomic assignment than classical systematic morphology. On the basis of such investigations, Böhme (1988b) and Ziegler and Böhme (1997) found that *V. giganteus*, with its relatively plesiomorphic hemipeneal structure, had to be member of the *gouldii* group that comprises only Australian members *V. g. gouldii* and subspecies, *V. spenceri*, *V. mertensi*, and *V. giganteus*. *V. komodoensis* and *V. varius* belong to this subclade of the Indo-Australian radiation of monitors too, but they do not belong to the *gouldii* group. Members of the odatrian group are related to this group but form another subclade.

Molecular biology has also been applied to monitors. King and King (1975) used karyotyping and Holmes et al. (1975) employed comparative electrophoretic measurements of four proteins to elucidate evolutionary lineages of this group of reptiles. They established six different groups, A to F, of karyotypes of which the *salvator* group

(group C) formed the most primordial one and from which the authors believed the *Odatria* group (group D) evolved, and from this, the *gouldii* group (group A), including *giganteus,* originated (King and King 1975). The remaining three groups need not be considered here. The electrophoretic measurements (Holmes et al. 1975) resulted in a similar phylogeny to King and King (1975).

An evolutionary series such as this (*salvator-Odatria-gouldii*) conflicts with Dollo's law of irreversibility, which says that complex characteristics, once lost during evolution, are renewed for a second time in the same manner as before very rarely. Monitors of the *salvator* group exhibit the clinch phase during ritual combats, which means the combatants stand upright like wrestlers. The same ethological character is shown by monitors of the *gouldii* group, including *V. giganteus* (for illustration of the clinch phase of *Varanus giganteus,* see Card 1994), whereas members of the odatrian group lack this character (Horn et al. 1994).

Therefore, the odatrian group of monitors cannot precede the *gouldii* group. Also, Böhme (1988b) did not find hemipeneal relationships of *V. salvator* to any other group of monitor, but because of basic phylogenetic principles, he does not see any relation of *Odatria* as predecessor to the *gouldii* group, including *V. giganteus.* The lung character of *V. gouldii* is primordial too. Thus all these characteristics fail to support any odatrian ancestry for the *gouldii* group (Becker et al. 1989). The debate ended when King (1990) rediscussed the origin of Australia's reptiles on the basis of older chromosomal and immunogenetic data, and of more varanid taxa.

By use of the technique of microcomplement fixation of serum albumins combined with chromosomal data, King et al. (1991) and Baverstock et al. (1993) erected a phylogeny that showed a separation of *Odatria* from the *salvator* group and a younger separation of a southeast Asian-Australian subclade containing *V. varius* and *V. komodoensis* from which the purely Australian subgenus *Varanus* (including *V. giganteus*) finally split off.

Phylogentic relationships of monitors were further clarified by DNA sequencing (King at al. 1999; Fuller et al. 1998), resulting in two monophyletic clades of monitors: Australian (*Odatria*) and Indo-Australian (subgenus *Varanus* with *mertensi, gouldii, giganteus,* and *varius, komodoensis, salvadorii*), which do have a common root. Ast (2001) used mitochondrial DNA sequences and demonstrated that the Varanoidea form a monophyletic group of anguimorph lizards, which comprise the odatrian monitors and the *gouldii-giganteus* group (subgenus *Varanus*) as sister subclades with a common ancestor.

On the basis of modern phylogenetic systematics, comparative evolutionary methods were used to reconstruct the probable course of evolution of size in monitors (Pianka 1995). Small body size evolved only twice (in the subgenera *Odatria* and *Euprepiosaurus*), whereas large body size evolved at least three times in (1) *V. niloticus,* (2) *V. salvadorii,* and (3) *V. giganteus, V. komodoensis,* and the (extinct) *Varanus* (formerly *Megalania*) *priscus.*

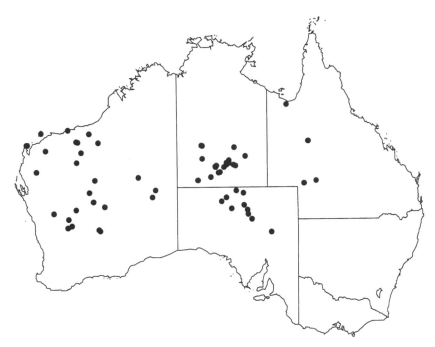

Figure 7.15. Known localities for Varanus giganteus.

Distribution

V. giganteus is found only in the arid interior of Australia. Cogger (1992) and Vincent and Wilson (1999) present distribution maps that comprise far western Queensland through central Australia to the coast of Western Australia. For local distribution in South Australia, see distribution map by Houston (1978), and for distribution in Western Australia, see Storr (1980) and Storr et al. (1983). The terra typica has been given as the north coast of Australia by Mertens (1963), whereas Storr (1980) identifies north coast of New Holland with Barrow Island (Western Australia). For a detailed discussion on this subject and information on distribution, see Horn and Visser (1988). Recent findings by Thompson et al. (2002) revealed that *V. giganteus* is extending its range down the coast southward in Western Australia (Fig. 7.15).

QM R17565 Kynuna District—21.58 141.92; QM R49821, R49822, R49823, R49824, R49825, R49826, R49827, R49829: Cuddapan Stn.—25.65 141.5; QM R51749 Windorah—25.42 142.65; AM R8257 Norseman S of Kalgoorlie—32.20 121.78; AM R17964 Hermannsburg Mission—23.95 132.77; AM R21574 Todd River Stn., Alice Springs—23.85 134.50; AM R49682 20 km S Finke—24.65 133.05; AM R60218 Greenleaves Caravan Park, Alice Springs—23.70 133.87; AM R65227 Etopia Stn.—22.23 134.57; AM R111731 Cycad Gorge, Palm Valley—24.07 132.75; AM R111732 Mt. Doreen Camp—22.08 131.42; AM R125073 Mt. Liebig—23.30 131.37; AM R128640 Utopia Stn., via Alice Springs—22.23 134.57; ANWC R1782 North Pool, N of Wiluna—26.43 120.15; NMV D 68026 Palm Valley, Finke River—24.05 132.71; NTM R25689 1.25 mi SE Meteor Craters, Alice Springs—24.58 133.18; NTM R25690 47 5 km S Alice Springs—24.07 133.6; NTM R25691 Catholic School, Alice Springs—23.7 133.87; NTM R25692 Stuart Arms Hotel, Alice Springs—23.7 133.87; NTM R25693 Angas Downs—25.05 132.28; NTM R25695 14 mi S Alice Springs—23.92 133.75; NTM R25696 Residency Yard, Parson St., Alice Springs—23.7 133.87; NTM R25697 Alice Springs—

23.7 133.87; NTM R25698 Opp Alice Springs Drive-In—23.7 133.87; NTM R25699 Alice Springs, General Cemetary—23.7 133.92; NTM R25700 Wolfram Hill, Mt. Doreen Stn.—22.03 131.33; NTM R25701 Lambina Stn.—26.95 133.77; NTM R25702 Arltunga Mines—23.43 134.07; NTM R25703 Todd River Stn.—24.0 134.7; NTM R25704 Indiana Stn.—23.05 135.67; NTM R25705 Jessie Gap—23.75 134.02; NTM R25706 Ayers Rock—25.35 131.03; WAM R756 Anketell—28.03 118.85; WAM R9029, 9030, 9031, 9032: Yardie Creek—21.88 114.02; WAM R12637 Bamboo Creek—20.87 120.17; WAM R12923 Yalgoo—28.35 116.68; WAM R13454 Woodstock Stn.—21.62 118.95; WAM R13708 Mt. Wardiacco, 60 km NE of Paynes Find—28.88 118.12; WAM R22710 between Turee Creek and Prairie Downs—23.58 118.92; WAM R26262 Mundabullangana—20.52 118.05; WAM R27232 Wanjarri Stn., Kathleen Valley—27.32 120.55; WAM R28004 Tantabiddi Well. Yardie Creek—21.93 113.97; WAM R28280 Mariltana—22.63 119.4; WAM R31441 near Exmouth—21.93 114.12; WAM R48729 5 km ESE of Miss Gibson Hill. In Valley—26.87 126.38; WAM R52114 Woodstock—21.62 118.95; WAM R53596 Charles Knob. Young Range—25.05 124.98; WAM R53608 Carnarvon Range—25.27 120.68; WAM R60128 Woodstock Stn.—21.62 118.95; WAM R65660 10 km ENE of Comet Vale. Granite Dam—29.91 121.24; WAM R74094 4 km NNW Eel Pool Davis River—21.73 121.03; WAM R74766 13 km Se Banjiwarn—27.79 121.7; WAM R78177 15 km S of Menzies—29.82 121.08; WAM R81668 west face of Kennedy Range—24.52 114.97; WAM R84044 16 km W of Mt. Stuart—22.43 115.9; WAM R87388 47 km S of Dromedary Hill—29.47 118.45; WAM R96597 8 km NNE Bimbijy—29.63 118.07; WAM R28005, R28006, R48959, R82925, R82926, R97578, R97584, R97586: Barrow Is.—20.8 115.4; WAM R100760 Woodstock, Site 11—21.67 119.04; SAM R00075 Arrabury Stn.—30.07 138.28; SAM R00094 William Creek—28.92 136.33; SAM R01524 Hermannsburg—23.95 132.77; SAM R14780 ~6 km E Nilpinna HS—28.48 135.98; SAM R20578 19 km E Mt. Willoughby—27.95 134.32; SAM R20988 Peake Ruins near Oodnadatta—28.08 135.9; SAM R27034 Mt. Liebig—23.3 131.37; SAM R28227 Dalhousie Ruins—26.52 135.47; SAM R31801 Yutjiri Gorge—26.23 126.7; SAM R33352 49 km S Burketown—18.07 139.77; SAM R34342 Mintabie Opalfield—27.32 133.3; SAM R40461 Eringa Ruin—26.3 134.73; SAM R43786 Allandale HS Waterhole—27.63 135.58.

Diagnostic Characteristics

• Head scales smooth and very small, nostrils lateral, close to tip of snout.

• Tail strongly compressed.

• Dorsally dark brown to black, stippled with yellow in adults.

• Spots on the back large, usually black centered and edged.

• Whitish underneath.

Description

The perentie is the largest Australian monitor. Although it can attain a relatively high weight, it has a slender body and a medium-sized tail (tail length is less than 1.5 SVL). Tail is compressed with small double keel. Head and neck are relatively long with small scales; smooth dorsal scales are arranged in 150 to 155 rows around the body from gular fold to beginning of the hind legs. Scales on tail are keeled. The nostril is round to oval, and the ear is relatively large. The canthus rostralis is strong. Legs have strong, sharp claws (Mertens 1942, 1958; Cogger 1992; Storr 1980; Storr et al. 1983; Worrell 1963). A number

TABLE 7.2.
Morphometric data of *Varanus giganteus*.

Weight (g)	Snout–Vent Length (cm)	Tail Length (cm)	Total Length (cm)	Tail/Head + Body	Ref.[p]
Not indicated[a]	81.0[o]	96.6[m]	177.6	1.19	1
Not indicated[b]	64.8[o]	84.0[o]	148.8	1.30	1
5350[c]	73.6	95.4[n]	169.0	1.30	2
2000[d]	55.0	75.0[n]	130.0	1.36	2
5090[e]	72.8	94.2[n]	167.0	1.29	2
2500[f]	59.1	80.9[n]	140.0	1.37	2
3523[g]	64.50	84.89[n]	149.39	1.32	2
5171[h]	72.46	93.43[n]	165.89	1.29	2
1819[i]	54.69	75.64[n]	130.33	1.38	2
2059[j]	57.60	76.74[n]	134.34	1.33	2
77[k]	21.38	34.60[n]	55.98	1.62	2
40[l]	15.02	22.52	37.53	1.5	3

[a,c,e] Males.
[b,d,f] Females.
[a,b,c,d,e,f] Breeding pairs in the wild.
[g] Total of 28 males combined (average)
[h] Total of 14 sexually mature males combined (average).
[i] Total of 27 females combined (average).
[j] Total of 21 sexually mature females combined (average).
[k] Total of four juveniles in the wild combined (average).
[l] Total of six captive-born hatchlings combined (average).
[m] Missing 8–10 cm of tail.
[n] Length of tail calculated from the author's data
[o] Rather than snout–vent length, the tip of the snout to part back of hind legs and part back of hind leg to tip of tail was measured.
[p] References: 1, Horn and Visser (1988); 2, Heger (2000); 3, Bredl and Horn (1987).

of 30 presacral and 130 postsacral vertebrae have been determined in this species (Greer 1989).

The mineralized bones in perentie hemipenes makes it a member of group C of a classification based on variation of shape and number of hemibacula (Card and Kluge 1995).

Basic coloration is whitish or creamy, tending to brown or dark above with age, sometimes reddish; legs are dark brown, reticulated around neck and throat. Bright yellowish spots edged by black rings form transverse rows from the neck to two-thirds of the tail. The end of tail has uniform yellowish-white patches on the ventral surface (Cogger 1992; Worrell 1963; Storr 1980; Storr et al. 1983).

Morphometrics

Different maximum sizes have been reported for this impressive lizard. Bustard (1970) mentions a specimen from near Alice Springs that measured 2.13 m. Butler (1970) recorded the biggest perentie on Barrow Island (Western Australia) to be 1.93 m long and weigh 17.24 kg. Cogger (1992) reports one exceeding 2 m; Houston (1978) gives 2.4 m. Stammer (1970) even reported on a perentie of 2.59 m. Stirling (1912) describes a specimen of 2.28 m TL; Storr (1980) and Storr et al. (1983) state up to 2 m, Wilson and Knowles (1988) mention 2.5 m, and Worrell (1963) believes it reaches 2.43 m.

A selection of some more detailed data of different specimens is listed in Table 7.2.

The smallest reproductive female found together with a much bigger male in the wild had a TL of 130 cm and a head–body length of 55.0 cm; the smallest reproductive male had a TL of 150 cm and a head–body length of 66.6 cm (Heger 2000)

Thus, it can be estimated that male *V. giganteus* reach sexual maturity at about a TL of 140 to 160 cm, whereas breeding females may exceed 110 to 130 cm. The data given in Table 7.2 all exceed this estimation, which means all these specimens could reproduce. The largest monitor listed in Table 7.2 has a TL of 177.6 cm, missing 8 to 10 cm of the tail. This means the maximum size of 200 to 250 cm must be a rare exception. Captive-born hatchlings have an average TL of about 37 cm and an average weight of 40 g; wild-caught juveniles (55.8 cm TL, 77 g) must have been already several weeks old. Some additional data on SVL and head and hind leg length of *V. giganteus* are given by Pianka (1986, 1994a).

King et al. (1989) investigated this species intensively on Barrow Island (Western Australia). These investigators recorded weights from 150 to 11,700 g with an average of 4460 g and a SVL of 23.0 to 88.0 cm. TL was 36.0 to 194.0 cm (King et al. 1989). In a further publication, King and Green (1993c) described the relationship of 84 captured *V. giganteus* concerning SVL as a function of body mass and TL as a function of body mass. In both cases, the resulting diagrams demonstrated a strong increase in SVL and TL depending on a body mass between 0 and 4 kg, although later the growth function becomes an asymptotic to maximal size (SVL, 1 m; TL, 2 m). The ratio of tail length to SVL fits well to a decrease with increasing body mass, which means an increase in size and age (King and Green 1993c).

An attempt to understand the demographic (size, age) situation of the *V. giganteus* from Cape Range National Park (Western Australia) was unsuccessful. Sizes of captured females and males, respectively, are arranged in a bar diagram according to their SVLs. But the result was inconclusive with respect to clearly defined size groups or year classes (Heger and Heger 1992).

Thompson and Withers (1997c) investigated the comparative morphology of 17 Western Australian monitors. Although a commonly held view is that varanids do not vary morphologically (e.g., in shape, except body mass), a significant and nonisometric variation in the relative appendage dimensions was found in this study, which included *V. giganteus*. Numerous morphometric data (e.g., SVL, thorax–abdomen, length dimensions of head, legs, tail) were determined and treated mathematically. These data showed this group of reptiles is not morphologically conservative (Thompson and Withers 1997c). Only a few results can be mentioned here. For example, if the average thorax–abdomen length of all species is taken as a standardized measure, relative tail lengths subdivide species examined into three groups: long-tailed (e.g., *V. glauerti*, *V. pilbarensis*), short-tailed (e.g., *V. brevicauda*, *V. caudolineatus*), and tails of medium length (e.g., *V. giganteus*, *V. mertensi*). Head length was (relatively) largest for *V. giganteus* and smallest for *V. mertensi*. *V. giganteus*, along with *V. glebopalma*, *V. glauerti*, and *V. pilbarensis*, has an appreciably longer neck than other species (e.g., *V. gilleni*). Forelimb lengths of *V. giganteus* and *V. glauerti*, *V. pilbarensis*, and *V. glebopalma* are relatively longer than in *V. mertensi* or *V. gilleni* (Thompson and Withers 1997c). Sexual dimorphism is also demonstrated by these data; males are generally longer than females, and their tails are longer. The authors discriminated four broad categories of habitat use and morphology of species on the basis of relative appendage length. *V. giganteus* and *V. eremius* are members of the widely foraging terrestrial varanids (Thompson and Withers 1997c).

Fossil Record

Pleistocene specimens from Australia have been attributed to "perhaps *V. giganteus*" (Lydekker 1888). Nopsca (1908) referred them to *V. giganteus*, and Fejerváry (1918) placed them as *Varanus*, cf. *V. giganteus*, whereas Estes (1983) suggests the material should be restudied. (All citations after Estes 1983.)

Habitat, Natural History, Diet

V. giganteus is a terrestrial lizard, but if approached, it sometimes takes refuge in trees. Many observations about this animal are anecdotal. Most reporters on this species agree with Stirling (1912) that the perentie is encountered mainly in hilly landscapes where rocky outcrops, gorges with big boulders, and deep crevices or caves useful as burrows are abundant. Such areas are preferred if sandy plains, deserts, and clay pans interrupt the rocky formations and thus allow foraging (Waite 1929; Worrell 1963; Brunn 1981, 1982; Horn and Visser 1988; Wilson and Knowles 1988; Visser and Horn 1989; Cogger 1992). Detailed illustrations are presented by Horn and Visser (1988, 1990a, 1990b). Stammer (1970) observed this species in western Queensland on gray "bull dust" country and limestone flags, on red sand loam soil under big slabs, and in dry creek beds. Perenties can be also encountered in totally flat, sandy country. Much information on *V. giganteus* can be gained by careful study of the spoor in sandy areas (Pianka 1982, 1986, 1994a). Pianka (1982, 1994a, 1994b), who investigated the ecology of lizard communities in the Great Victoria Desert at Red Sands and the L area, at first could not track down any spoor of a perentie but encountered this species relatively frequently some years later. This may be due to an increasing rabbit population. He found perenties in large, deep burrows (1 m deep, 7 to 8 m long) in the banks of sandridges. The depth may be necessary for cooling purposes of perenties in this hot habitat. The following observation is worth mentioning. Pianka dug out a specimen from its burrow and released it 5 to 7 km away. It wandered around several kilometers, as could be seen from its tracks, but then turned directly toward the very few large burrow systems in the area (Pianka 1982). Either it was guided by its memory from earlier searches for food, by scents, or by both signals.

Brunn (1978) found this species in an unusual habitat: a large perentie was first seen basking on the road. It allowed the observer to get within 2 m, but eventually climbed a tree. No rocky outcrop was in sight there, just a few mulga (*Acacia aneura*) trees, which were thinly covering the dry plains. In search for food, perenties will wander long distances in open plains far away from rocky outcrops—for example, on gibber plains with a few trees (Brunn 1981, 1982). A specimen was observed by Pianka (1982) in an even more unusual habitat. It exploited the muddy water of the Reetz Creek Billabong (Western Australia)! It seemed very familiar with the lukewarm water and was "acting more like an aquatic species." A similar unusual observation was made in western Queensland by Brunn (1980). He and his companions saw something in a drainpipe of about 15 m length and 1 m in diameter, filled with knee-deep water. After being disturbed, a little perentie ran out of the pipe and dived into the water, but it was soon captured. After release, it took refuge in a tree.

Perenties not only take shelter in natural crevices and holes, but they also use burrows dug out by burrowing bettongs (*Bettongia lesueur*) (Robinson 1992). This can only happen on Barrow Island and not for continental Australia because this species of bettong is only extant on Barrow Island (Western Australia) and is extinct on the mainland. Hopping mice (*Notomys* spp.) excavate similar tunnel systems in sandy areas, which may also be used by *V. giganteus* (Wilson and Knowles 1988). *V. giganteus* hibernate from May to August in the above-mentioned crevices, holes, and burrows (Stirling 1912).

According to Stirling (1912), a perentie is supposed to have broken a dog's forelegs with a blow of its powerful tail, and a native woman was knocked down by a large perentie's tail. Such observations have not been confirmed in recent times, although they have been repeated several times. Pianka (1982) reports that a medium-sized perentie, which first behaved docilely in his hands, became extremely pugnacious when standing on its own feet. When harassed and cornered, perenties may raise their body on their hind limbs and hiss loudly with throat extended (Wilson and Knowles 1988). Brunn (1981, 1982) describes this species as a shy reptile, which, if danger threatens, will lie motionless on the ground or may look "uninterested" in the observer. On the other hand, Horn and Visser (1989, 1990a, 1990b) encountered a large female that behaved pugnaciously, hissing loudly with extended throat and finally threatening the observer with mouth wide open. A very large specimen behaved uninterested, or trusted in its powerful jaws, although the photographer was lying on the ground at a distance of only 1.2 m (Horn and Visser 1989). If a juvenile specimen is accidentally encountered close by in the field, it sometimes shows its extreme fear by staring into space (Horn, unpublished data).

Fyfe (personal communication) found most large perenties (over 1.2 m) can be approached easily if you move slowly and if the lizard sees you coming. However, if surprised, they generally run away quickly (even plunging off 6-m-high rock walls). Sometimes they crouch down close to the substrate when approached. A foraging perentie can be followed without being unduly worried (Fyfe, personal communication).

Perenties are carnivorous lizards and feed on any prey they can overpower. Additionally, they feed on carrion (Waite 1928; Bustard 1970; Stammer 1970; Brunn 1981, 1982; Cogger 1992; King 1999). Further detailed and direct observations include those of Pianka (1994a, 1994b), who obtained a half-digested *V. gouldii* from a perentie by stomach flushing. In the same area, perenties were feeding on road-killed kangaroos. Brunn (1981, 1982) reported on perenties feeding on rabbits, a dog, and small kangaroos. Stirling (1912) and Waite (1928) believe echidnas are immune to a perentie's attack. The diet of perenties consists of lizards, snakes, turtles and turtle eggs, small mammals, and carrion (King 1999). King reports that a 1.5-m perentie swallowed another perentie of 1.2 m TL. After 30 minutes, it walked away with half a meter of the victim's tail hanging out of its mouth (King 1999). Thus, this species is cannibalistic occasionally. Fyfe (personal communication) saw a juvenile perentie about 0.6 m long devouring an agamid ("*Physignathus,*" now placed in *Lophognathus longirostris*) nearly 0.5 m long. The majority of the tail of the victim hung down from the mouth of the small perentie over the next 24 hours. Fyfe also reported a curious observation of a field herpetologist (Gavin Bedford) who radiotracked central carpet pythons (*Morelia bredli*) near the Alice Springs airport. One python emitted its signal from an old rabbit burrow. When Bedford dug up the burrow, he found a 1.8-m male perentie that disgorged a large mulga snake (*Pseudechis australis*). When the mulga snake was cut open, it contained the dead carpet python (Fyfe, personal communication). With binoculars, Frauca (1973) observed a perentie tearing pieces out of the carcass of a kangaroo.

A detailed investigation of stomach contents (King et al. 1989) gave the following results: 13 perenties in the Western Australian Museum, Perth, had eaten lizards, snakes, a bird, and a bat. Juvenile specimens had eaten a spider and an undetermined number of grasshoppers. Six stomachs were empty (King et al. 1989). The diet based on feces (25 samples) collected on beaches consisted mainly of eggs or hatchlings of turtles (15 of 18 samples) and of small- to medium-sized mammals (13 of 25 samples) (King et al. 1989). A large perentie can consume up to 30% of its body weight (King 1999). Another investigation of stomach contents of museum specimens also revealed that *V. giganteus* prey mainly on vertebrates (James at al. 1992).

Six specimens investigated contained 50% lizards, 31.1% orthopterans, and 7.1% chilopods, with mammals making up the remainder. Some further information on prey captured by perenties (Losos and Greene 1988) also reveals that this species is more carnivorous than insectivorous: a 0.896-kg specimen ran down a 21-g *Lophognathus longirostris*. Perenties patrol beaches in search for turtles and their eggs, and they lie in ambush under vehicles for scavenging seagulls, which are captured in short sprints. A 5-kg specimen of 0.72 m SVL was found to have eaten a 100-g agamid, *Pogona barbata* (Losos and Greene 1988). Other unpublished observations mentioned there were later published (King et al. 1989).

For photographic illustrations of a perentie swallowing a rabbit, see Visser and Horn (1989); and for illustration of a perentie devouring a wallaby, see King (1999).

Behavioral Observations

Several less important behavioral observations on *V. giganteus* have already been mentioned. Only two more unusual behavioral traits will be described here.

Several authors (Stirling 1912; Waite 1928; McPhee 1979; Strimple 1988; Christian et al. 1994; Strimple and Strimple 1996; and others) have reported the unusual swiftness of this monitor when alarmed and trying to escape, and when moving at their top speed, the forelimbs are raised from the ground: they run bipedally. Christian et al. (1994) reviewed bipedalism in reptiles, including three species of varanids (*V. giganteus, V. gouldii, V. mertensi*) and discussed the physical background of this kind of movement.

But none of these authors had ever witnessed this fascinating event. We are happy to present very recent observations on this subject by Fyfe. He reports,

> I chased a 1.8 meter Perentie over flat ground at Ayers Rock one day. The goanna was very fast and easily ran away from me, even though I ran as fast as I could. The Perentie started running on all four legs but then raised up onto just the rear legs with no apparent reduction in speed. There were grasses to about 30 cm high and scattered bushes and dead timber to 1.0–2.0 metres high. The Perentie ran more or less in a straight line for rocks about 50 meters away. When a bush or dead timber was in its path, the lizard kept running straight at it and slid up and over it without slowing down. Maybe the raised and inclined body enabled the Perentie to "ski jump" or toboggan over these obstacles that I had to run around. (Fyfe, personal communication)

Perenties exhibit another interesting behavioral trait mainly during the reproductive season: ritual combat as a competition for females. Such social interactions, which also can be carried out as competition for food or basking sites, comprise several different phases: epigamic display, encompassing phase, clinch phase, and catch and subpressing phase (Horn et al. 1994).

Waite (1928) published a picture of the clinch phase (both individuals are standing upright with mutual embrace) of two monitors that were thought to be two *V. giganteus,* although closer examination showed that this picture was of two *V. spenceri* (Horn 1981). The clinch phase of *V. giganteus* was first documented photographically by Card (1994). The clinch phase exhibited by *V. giganteus* (subgenus *Varanus*) separates this species clearly from members of the subgenus *Odatria.* The encompassing phase (both individuals are standing head to back of the opponent, neck bowed, throat extended) has also been documented photographically by King (1999). The section on reproduction documents behavioral observations during courtship.

Ecology

Activity Pattern

Heger (2000) carried out the most comprehensive investigation of activity patterns and thermal requirements in perenties in Cape Range National Park (Western Australia). However, there could be local differences within the broad distribution area of this species.

Generally, Heger observed that perentie activity patterns changed depending on season. During winter (June), activity peaked broadly (55 minutes per hour of activity) between 1100 and 1200 hours (starting at 0900 hours and reaching zero at 1700 hours), whereas in late summer (April), a smaller peak was registered at 0900 hours (10 min/h) followed by a minimum between 1100 to 1200 hours and a broad peak (35 min/h) between 1400 to 1600 hours. In December, which is Australian midsummer, four activity peaks of low intensity (10–15 min/h) reflected the hot season (Heger 2000). Activity was generally interrupted during the middle of the day. Further, time of maximal activity, going to earth, and hauling out were influenced by season and environmental conditions (temperature, sun, rain). In winter (June), maximal activity coincided roughly, as expected, with maximal temperature, and in contrast to this, maximal activity in summer coincided roughly with cooler temperatures. Influence of day length remains unclear (Heger 2000).

Body Temperature, Thermoregulation

Pianka (1982) measured body temperatures of 38.8°C and 37.6°C at air temperatures of 28.3°C and 24.8°C, respectively, in two perenties. Mean body temperature of 36.7°C was recorded for three further specimens (Pianka 1994b). Heatwole and Taylor (1987) mention a mean body temperature (T_B) of 36.8°C for *V. giganteus.* Heger (2000), on the other hand, collected four types of different body temperature data: (1) at initial capture, (2) during routine radiotracking, (3) during all-day remote measurements, and (4) during behavioral observations. An average individual body temperature was calculated from these

TABLE 7.3.
Grand mean body temperature (T_b).[a]

Sex	T_b (°C)	Range (°C)	n	Size[b]	T_b	Range	n
Male	35.14	28.15–39.80	28	H	33.6	—	1
Female	35.16	31.30–39.20	19	S	34.78	34.10–35.93	3
				M	35.59	28.15–39.80	30
				L	34.27	32.23–39.50	14

[a] Data from Heger (2000). Standard error not indicated.
[b] Size classes are as follows: H, hatchling (18.0–30.0 cm); S, juveniles (30.1–45.0 cm);
M, semiadults (45.1–65.0 cm); L, adults (65.1–85.0 cm).

data for each lizard in order to compare the mean values for sex and size of those lizards. Results are tabulated in Table 7.3.

These data show no differences in mean body temperatures (T_b), either between sexes or among size classes. This is also demonstrated if mean body temperatures (T_b) are plotted as a function of either SVL or mass; again, no differences were discernible (Heger 2000). If average body temperatures (T_b) were studied as a function of month, some differences depending on month arose. Body temperatures varied most among individuals during May to August but remained high within a relative stable range from September to April (Heger 2000). On the other hand, if mean individual body temperatures are compared, no differences between sexes were revealed, but there were clear differences in size classes and month (Heger 2000). An interesting trend was also noticed in individual lizards. These showed similar high body temperatures between morning and sunset during summer, but some individuals exhibited a kind of oscillating pattern (medium-sized lizards of around 36–37°C during the whole day), whereas a large individual peaked around 0900 to 1100 hours and again between 1500 to 1800 hours (36–37°C also) but had a plateau of mean body temperature of only 34°C between roughly 1100 to 1500 hours (Heger 2000).

These data and observations indicate that thermoregulation in perenties takes place by basking, posturing or body orientation, shuttling, microhabitat choice, conduction, and eventual convection (Heger 2000). Basking takes place by presenting the whole body to the sun; they may first present only their head to the sun after emerging from their burrow. After some time, they emerge a little further, or they may shuttle between sun and shade, or they may lift the body to avoid contact with hot substrate and even may climb shrubs and trees to be cooled by strong winds (convection) (Heger 2000). Fluctuations in body temperatures could be followed directly and may be illustrated by an individual perentie: after emerging from its burrow, T_b increased during activity. After eating a lizard, it moved into shade, and T_b (=36.4°C) stabilized for some time, then increased again. As T_b reached 38°C, the animal moved 1 m, seeking better shade. T_b then diminished slightly and increased slowly again. When its body temperature was 38.8°C, the animal put its upper body on a tree trunk, thus having no ventral contact to the soil. Cooling became rapid (T_b = 36.6°C), and the perentie again resumed activity, which increased T_b (to 39°C). It sought shade, and T_b decreased. Near the end of the day, T_b decreased more rapidly, which caused basking and a small increase of T_b (0.5°C). After this, the animal showed some activity and rested on warm soil at the end of the day (T_b = 34.9°C). Sometimes perenties are very selective before going to earth for nighttime shelters. A small female investigated half a dozen crevices before going to earth. Another female hauled in, and its T_b decreased rapidly inside the burrow. It moved out of that burrow and went to earth again in another rocky crevice. There, T_b stabilized and heat loss clearly diminished (Heger 2000).

Fat Bodies

Like many lizards, *V. giganteus* possess a dorsoventrally compressed bright yellow mass of fat, which lies free enclosed in the extraperitoneal cavity. Weight may be up to 12% of the total body weight. Knowledge of these fat bodies is incomplete. This mass of fat seems to be related to the fat bodies in amphibians and seems to be related to sexual activity. This is confirmed by the observation that they reach their maximal weight in spring when the breeding season starts. These fat bodies may also serve as a reserve during hibernation (Stirling 1912). Another speculation on their use is that they may be used as a source of material the mother reptile can draw on to yolk up its eggs.

Home Ranges

Perenties are not territorial. Instead, they inhabit extended home ranges, which are consistent for individuals from year to year. Home ranges overlap between and within sexes (Heger 2000). However, home ranges vary significantly between sexes (Heger and Heger 1992; Heger 2000), as can be seen in Table 7.4.

Home range size varies considerably between individuals and between adult males and mature females during breeding season. This seems to be mainly due to an increased migratory instinct of the males at this time. More detailed information on this subject is shown in Table 7.5.

Metabolic Rates, Turnover Experiments

Six *V. giganteus* (one juvenile, two females, three males) from Cape Range National Park (Western Australia) were used to determine standard metabolic rates at 25°C and 35°C under calibrated experimental conditions.

TABLE 7.4.
Home range size for male and female perenties.[a]

Sex	n	Mean ± SE (ha)	Range (ha)
Male	9	325.6 ± 127.0	28.5–1147.5
Female	11	47.5 ± 9.1	8.1–97.9

[a] Data from Heger (2000).

TABLE 7.5.
Home range sizes for male and female perenties during breeding and nonbreeding seasons.[a]

Season	n	Mean ± SE (ha)	Range (ha)
Breeding male	6	288.30 ± 140.39	4.78–884.24
Nonbreeding male	9	120.57 ± 22.40	24.93–226.40
Breeding female	6	48.67 ± 12.21	8.06–93.73
Nonbreeding female	10	35.01 ± 4.67	21.43–72.75

[a] Data from Heger (2000).

The specimens showed mean mass-specific standard metabolic rates (mL/g h) $\dot{V}o_2$ of 0.043 and $\dot{V}co_2$ 0.030 at 25°C and $\dot{V}co_2$ 0.095 and $\dot{V}o_2$ 0.076 at 35°C, respectively. Respiratory quotients calculated from these data are 0.71 at 25°C and 0.80 at 35°C. (Thompson et al. 1995). The intraspecific relationships of standard metabolic rate as a function of body mass $\dot{V}o_2$ (mL/h) are $0.0896g^{0.90}$ at 25°C and $0.126g^{0.96}$ at 35°C.

The analogous relation for $\dot{V}co_2$ (mL/h) reveals $0.052g^{0.92}$ at 25°C and $0.094g^{0.97}$ at 35°C (Thompson et al. 1995). The mean intraspecific mass exponent for other species of reptiles has been reported to be 0.67, ranging from 0.51 to 0.80. For *V. giganteus*, the data are 0.93 ($\dot{V}o_2$) and 0.95 ($\dot{V}co_2$) at the temperature under discussion. If analogous data for other varanids are included, the intraspecific mass exponent for varanids varies from 0.9 to 1.1, in contrast to the lower values mentioned above for other reptiles. Thus, the intraspecific mass-specific metabolism for any given species of varanid is nearly independent of body mass. At present, no explanation for this unusually high intraspecific mass exponent can be given (Thompson et al. 1995). King and Green

TABLE 7.6.

Data obtained from water and sodium input and energy
output in *Varanus giganteus.*[a]

Quality	Season 1		Season 2	
	Mar 1983, ± SE	n	Jan–Mar 1984, ± SE	n
Water input (mL kg^{-1} d^{-1})	5.77 ± 2.21	6	29.98 ± 7.20	13
Sodium input (mmol kg^{-1}d^{-1})	0.61 ± 0.22	5	2.93 ± 0.55	11
Prey input (g kg^{-1} d^{-1})	7.3	—	36.4	—
Energy assimilated (kJ kg^{-1} d^{-1})	38.9	—	194.0	—
CO$_2$ output (l kg^{-1} d^{-1})	2.26 ± 0.77	4	4.73 ± 0.89	10
Energy output (kJ kg^{-1} d^{-1})	56.5 ± 19.3	—	118.3 ± 22.3	—

[a] Data from Green et al. (1986). Long-time measurements not indicated.

(1999) mention a FMR of 0.17 mg/h of CO$_2$ production for a 7.7-kg
perentie in summer. The energy used calculated from this was 101 kJ
kg/d.

Water, sodium, and thus energy influx or input and efflux or output
in *V. giganteus* has been determined by turnover experiments using
radioactive-labeled water ^{3}H$_2$O and H$_2$^{18}O (with hydrogen isotope
tritium and oxygen isotope ^{18}O) and with sodium chloride ^{22}NaCl (with
sodium isotope ^{22}Na). In two different seasons on Barrow Island (Western Australia), perenties were captured, measured, and marked, and
separate intraperitoneal injections of known concentrations of the
labeled chemicals were given (Green et al. 1986). Animals were recaptured between 6 to 39 days after release and blood samples taken from
the caudal vein. In some additional cases, recapture was successful 1
year after release, and even after that length of time detectable amounts
of ^{3}H$_2$O could be found. The goannas acquired water mainly by feeding
on turtle hatchlings (*Chelonia mydas*) and lost water by defecation,
through skin, and by exhalation through the lungs. Several additional
determinations of water content in monitors and their main diet were
made that led to data shown in Table 7.6.

One result should be mentioned first: interestingly, no significant
correlation between isotope turnover rates and body weight could be
found. Table 7.6 shows strong differences between two seasons (1 vs. 2)
despite prevailing climatic conditions. Water input was highly correlated with sodium input and carbon dioxide output despite general
differences in the individuals. The mean prey input of 7.3 g kg^{-1} d^{-1}
during season 1 leads to an amount of 38.9 (kJ kg^{-1} d^{-1}) of assimilated
energy, which is below the mean of 57 kJ kg^{-1} d^{-1} metabolized by

perenties; thus, the specimens were in a negative energy balance (Green et al. 1986). In contrast, the analogous prey data for season 2 of input are (mean) 36.4 g kg^{-1} d^{-1} and the value of assimilated energy is 194.0 kJ kg^{-1} d^{-1}. The perenties were in a positive energy balance compared to the mean metabolic rate of 118 kJ kg^{-1} d^{-1} (Greenet al. 1986). An explanation for the strong differences of metabolic rates and estimated feeding rates are, according to Green et al. (1986), the very different activity patterns in perenties, which also may comprise changes in foraging strategies. Water turnover of 22.3 mL kg^{-1} d^{-1} of a single specimen of *V. giganteus* of 7.7-kg weight containing 68.4% body water during summer has been tabulated (King and Green 1999). A mean body water content of 68.4% is not valid for all reptiles. For example, *V. gouldii* has a body water content of 77%, *V. griseus* of 78% (Green at al. 1986).

Reproduction

Little is known about reproduction of perenties in the wild. King et al. (1989) examined specimens from the Western Australian Museum collection and found that individual males (SVL, 45.0 and 88.0 cm) had enlarged testes when collected in October or November; two smaller specimens collected in September (SVL, 25.5 and 24.0 cm) and April had small testes. Furthermore, a gravid female (49.0 cm SVL) with nine oviductal eggs was collected in November; two other females (35.0 and 38.0 cm SVL) collected in March and July were obviously inactive. From these individual data, King et al. (1989) concluded *V. giganteus* to be reproductive in (Australian) spring and early summer.

Therefore, if two wild perenties of appropriate size, a mature female and a male, were seen together during fieldwork at Cape Range National Park, they were assumed to be a reproductive couple. Three observations of this type occurred between December 8 and January 4, 1991, which is during the assumed breeding season. These pairs were not aggressive toward each other, and they attempted to remain close to each other (Heger and Heger 1992; Heger 2000). In central Australia south of Kulgera, another couple (female 148.8 cm, male 177.6 cm TL, respectively) was encountered in early October. They both were lying in a small cave. For comparison of head size of this couple, see picture reproduced in Horn and Visser (1988).

In captivity, *V. giganteus* has been bred several times in Australia (Bredl 1987; Bredl and Horn 1987; Horn and Visser 1997; Irwin 1996b, 1997).

Irwin (1996b) gave a very impressive and detailed description of the courtship of four perenties kept in a semioutdoor enclosure at Queensland Reptile and Fauna Park, Beerwah, Queensland. The youngest female (1.16 m TL) was not engaged in any of the courtship activities described below. The largest male 1 (1.8 m, wild caught), normally dominant in the enclosure, played a more passive role, whereas the younger male 2 (1.4 m, captive bred) was the sexually more successful partner during courtship (Irwin 1996b). Only a few points of the courtship and mating activities can be mentioned here.

The animals were inactive and refused to eat during winter (June and July 1994) but started activities (basking, feeding) at the beginning of August 1994. Test hole digging started at 1400 hours on September 17, 1994, and was continued on October 4, 17, and 18, 1994. Males 1 and 2 became very agitated, and they moved from one place to another at 0800 hours on October 19, 1994. Then the female chased the larger male 1. Although the dominant male 1 first showed no interest in her, he suddenly mounted her. His tongue flicks were very rapid; he licked her eyes and snout, and he tried to get her tail raised. This was not successful, and she escaped. The younger male 2 tried the same. During the course of events, males 1 and 2, interestingly enough, showed no aggressive behavior toward each other at all. On the other hand, the female suddenly started pressing her cloaca to the ground while walking, and she slid over this for some meters. Mating became very vigorous over a period of 9 days until October 28, and the female could only rest if she was out of sight of the males. On this date, male 2 mounted her twice and she lifted her tail, which allowed intromission of the hemipenes. Male 1 also could mount her, but she always remained passive. On November 1, mating interactions stopped totally. In contrast to her previous behavior, the female started biting the tail of male 2. Male 1 regained its dominance (Irwin 1996b). A detailed description of egg deposition is given by Irwin (1996b) and is excellently illustrated in Irwin (1997).

Two further interesting observations have to be reported. During filling up the nesting hole, the female was extremely alert, and at any noise, she stopped digging and started tongue flicking. After that, she again used her head like a shovel to push more sand forward. During this activity "she exerted so much force packing down with her snout that her legs and feet came off the ground" (Irwin 1996b). After some hours of back filling the sand and closing the entrance of the hole by raking with her forelegs, she went away and rested. The entrance of the hole could not be detected. When she noticed male 2, she chased and seized him and finally wrestled. He overpowered her and the same started again. In the end, she left the place and rested under a rock with male 1. Strangely enough, male 1 seemed to identify her with three tongue flicks and appeared to stand guard over her. Male 1 became very aggressive and attacked male 2 (Irwin 1996b). Irwin et al. (1996) also videotaped nocturnal nesting by *V. giganteus*. Some data of different egg depositions are listed in Table 7.7.

Unfortunately, Irwin (1996b, 1997) did not report number of eggs, number of hatchlings, or incubation temperature.

These data indicate a time difference of about 4 weeks between mating and egg deposition (Horn and Visser 1989). Incubation time is similar to that of other large monitors. Number of eggs of clutch 5 was 13; an average weight of eggs calculated from these data (Irwin 1996, 1997) was 95.62 g, average length 8.73 cm, and average width 4.45 cm. Time of mating, of egg deposition, and hatching coincided approximately with the scant data from wild lizards.

Incubation times from other sources (Bredl 1987; Bredl and Horn 1987; Horn and Visser 1997), and assuming a roughly similar medium

TABLE 7.7.
Data of five different clutches of *V. giganteus* deposited at Queensland
Reptile and Fauna Park, Beerwah.[a]

Reproductive Behavior	Clutch 1	Clutch 2	Clutch 3	Clutch 4	Clutch 5
Mating observed	22 Nov 1986	2 Nov 1988	5 Nov 1991	3 Nov 1992	20 Oct 1994
Eggs deposited	31 Dec 1986	30 Nov 1988	7 Dec 1991	28 Nov 1992	20 Nov 1994
Hatching	22 Aug 1987	18 Jul 1989	13 Aug 1992	30 Jul 1993 4 Aug 1993	NI
Incubation period (days)	234	200	249	228–233	NI

[a] Data from Irwin (1996b, 1997). NI, not indicated.

incubation temperature of 29 to 31°C, suggest the average incubation period in *V. giganteus* is about 219 days.

Incubation time is correlated positively with egg mass, neonate mass, and maximal adult SVL, which also means a positive correlation with incubation period (Thompson and Pianka 2001). Larger eggs need no longer incubation than do smaller eggs; however, their allometric relationship is negatively correlated. Consequently, members of the subgenus *Varanus*, e.g., *V. giganteus*, with longer incubation times lay their eggs later in summer; these incubate through the cooler wintertime, and hatching occurs in the next spring or summer. This reveals a general difference between the much smaller members of the subgenus *Odatria* and the larger of subgenus *Varanus* (Thompson and Pianka 2001). The same authors also point out that the egg mass of a single egg of *V. giganteus* (87.7 g) is approximately 28 times that of egg of *V. caudolineatus* (3.1 g), but the assumed maximal SVL of an adult *V. giganteus* (79.5 cm) is only six times that of *V. caudolineatus* (13.2 cm) (Thompson and Pianka 2001).

Hatching data of *V. giganteus*, e.g., measurement of hatchling length and weight, are scarce in the literature. Some of these data are collated in Table 7.8.

These data show no unusual incubation periods or unusual lengths or weights of this species compared with analogous data of other large species of monitors, such as *V. komodoensis*, *V. salvadorii*, or *V. varius*.

Thompson and Pianka (2001) demonstrated positive correlation of incubation times for varanid eggs with maximum adult and neonate SVL, and egg mass for 33 species of varanids, including *V. giganteus*. A

TABLE 7.8.
Hatching data of *V. giganteus*.

| Mating Observed | Eggs Laid | | | Incubation | | | | Hatchlings | | | | Ref.[b] |
	Date	Eggs[a]	Period (d)	Temp. (°C)	Medium	No.	Date	Length (cm)	Weight (g)	
5 Nov 1991	7 Dec 1991	5 (2)	249.	30	Vermiculite	3	13 Aug 1992	NI[c]	NI	1
3 Nov 1992	28 Nov 1992	11 (2)	197–218	30	Vermiculite	9	14 Jul–4 Aug 1993	37.5–41.8	NI	1
NI	25–28 Nov 1990	10 (8)	220.	30–32	Vermiculite	2	3 Jul 1991	~40	NI	1
22–23 Dec 1985	26 Jan 1986	11 (5)	231.5	30–32	Sphagnum	6	10–17 Sep 1986	37.5	40	2, 3

[a] Data are expressed as number of eggs (number of eggs infertile or damaged).
[b] References: 1, Horn and Visser (1997); 2, Bredl (1987); 3, Bredl and Horn (1987).
[c] NI, not indicated.

positive correlation was also found for SVL of the adults with egg mass, clutch size, neonate body mass, and neonate SVL in these species. Results for, e.g., *V. giganteus* can be compared with all species examined using a bar diagram. Authors emphasize that "body mass has a much greater influence on reproductive output of *Varanus* than phylogeny." Data were taken mainly from literature (Thompson and Pianka 2001).

Parasites

Parasitic infestations of reptiles are of ecological importance, and they can sometimes be used to characterize the distribution of a reptile and vice versa.

Jones (1985, 1991) investigated infestation of 13 perenties with gastric parasites of the family Physalopteridae. These parasites are nematodes of the genus *Abbreviata*. He found that besides an infestation with six other species of nematodes, the species *Abbreviata perenticola* is very specific for *V. giganteus*, and their distribution coincides with that of this species of *Varanus* (for distribution map, see Jones 1991), whereas other species of *Abbreviata* can be detected in several species of *Varanus*. In the more southern Western Australian distribution area of *V. giganteus*, *Abbreviata perenticola* was detected only as larvae, not as adults. Arthropods are the probable intermediate hosts of *A. perenticola*, and thus the perentie is less infested with this parasite than *V. gouldii* because perenties eat more vertebrate prey than the latter (Jones 1985).

Reviews

Two reviews have appeared on this species, by Strimple (1988) and Strimple and Strimple (1996).

Status

V. giganteus is listed in appendix II of the Washington Convention. In the wild it does not seem be endangered because of its broad distribution in Australia's arid regions (Pianka 1982). However, feral animals such as dogs, cats, foxes, and dingoes could impact wild populations (Irwin 1997).

7.7. *Varanus gilleni*

HANS-GEORG HORN

Figure 7.16. Varanus gilleni *(photo by Eric Pianka).*

Systematics and Taxonomy

Gillen's pygmy monitor (*Varanus gilleni*) was described by Lucas and Frost (1895). On the basis of external morphology (Mertens 1942), *V. gilleni* clearly belongs to the subgenus *Odatria,* which comprises all dwarf monitors because of their size and rounded tail (Mertens 1942, 1958, 1963) (Fig. 7.16). More detailed investigation by classical methods (external morphology) supported the older results (Mertens 1942). Hemipeneal morphology (Branch 1982; Böhme 1988b; Ziegler and Böhme 1997) again demonstrated clearly that *V. gilleni* is a member of the *Odatria* group. Ossifications have been found in the hemipeneal structure of *V. gilleni* (Shea and Reddacliff 1986; Card and Kluge 1995), but see criticism by Werner (1988). For an X-ray photograph of the hemibacula of a male *V. gilleni,* see Greer (1989).

Karyotyping a series of varanids resulted in six different groups, roughly fitting Mertens' subgenera. *V. gilleni* falls in the subgenus *Odatria.* The number of chromosomes was constant (2n = 40) for all species (including *V. gilleni*) investigated (King and King 1975; King and Green 1999).

Biochemical methods consistently demonstrate a monophyletic clade for all pygmy monitors including *V. gilleni,* but excepting *V. eremius* within the Indo-Australian group of Varanidae (Holmes et al. 1975; King et al. 1991, 1999; Baverstock et al. 1993). Findings made via molecular biology support this categorization (Ast 2001).

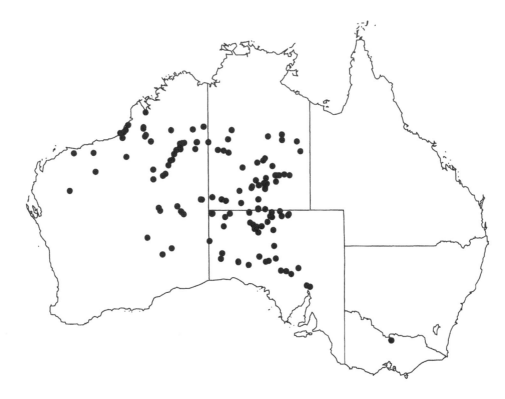

Figure 7.17. Known localities for Varanus gilleni.

Size of recent monitors has also been used to discuss evolution of the Varanidae. There are two hypotheses: dwarfism versus medium-sized lizards as a very ancient character. On the basis of those assumptions, different possibilities are discussed (Pianka 1995). *Odatria,* including *V. gilleni,* is positioned at one end of a phenogram, which is ultimately related to a cladogram based on microcomplement fixation (Baverstock et al. 1993). Losos and Greene (1988) are more convinced that medium-sized varanids were predecessors of extant monitors. For discussion of differences of the *Odatria* and *Varanus* clade from an ethological point of view, see Horn et al. (1994).

Distribution

Gillen's pygmy monitor is found only in Australia. Cogger (1992) shows a distribution map for this monitor that comprises northern parts of South Australia, southern parts of Northern Territory, and middle parts of Western Australia (Fig. 7.17; cf. Husband and Vincent 1999). Some more detailed records have been listed in the literature: Mertens (1942, 1963; total area of distribution), Pianka (1969; northern, western, southern, and central Australia), Storr (1980; Western Australia), Storr et al. (1983; Western Australia), Houston (1978; South Australia). The record from Victoria is most likely an error.

AM R2141, R2142 Tennant Creek—19.65 134.183; AM R10857 Hermannsberg—23.95 132.767; AM R11906 Alice Springs—23.7 133.867; AM R16013 Palm Valley Creek—24.07 132.75; AM R17532, R17533 Tomkinson Ras.; AM R17566, R17567, R17568, R17608, R17652; Mt. Davies, Tomkinson Ras.—26.22 129.233; AM R17940, R17941, R18257, R18493, R19609, R20574 Mt. Davies, Tomkinson Ra.—26.22 129.267; AM R26394 Petermann Ranges (vicinity), 18 mi from WA Border—24.9 129.283; AM R41206 Central Mt. Stuart—21.93 133.25; AM R41509 Gordon Downs Stn.—18.75 128.583; AM R41795, R41796 Emily Gaps, Amoonguna Sett. Alice Springs—23.77 133.933; AM R51917 Arid Zone Research Inst. Farm, Alice Springs—23.77 133.883; AM R51918, R51919 Ellery Creek, near Hermannsburg—23.92 133.917; AM R60200 27 km SW of Barrow Creek on Stuart Hwy.—21.7 133.75; AM R60201, R60202 Kulgera—27.83 133.3; AM R64810, R64811 Alice Springs—23.7 133.867; AM R84542 2 km S of Kulgera on Stuart Hwy.—25.85 133.3; AM R84543 on Marla Bore Tank on Stuart Hwy.—27.32 133.55; AM R90499 2 km N Kulgera—25.8 133.3; AM R90867 64.4 km W Ayers Rock (by road)—25.18 130.5; AM R90868 103.1 km S Neale Jct. via Connie Sue Hwy.—29.08 125.733; AM R90876 65.6 km W Ayers Rock (by road)—25.18 130.5; AM R95791 5 km E of Barry Caves—20.05 136.7; AM R100986, R100987, R100988, R100989, R100990, R100991, V, R100992, R100993, R100994, R100995, R100996, R100997, R100998, R100999, R101000, R101416, R101417, R101418, R101419, R101420, R101421, R101422, R101423, R101424, R101425, R101426, R101427, R101428, R101429 ~6.6 km N of Sandfire Flat Roadhouse via the Great Northern Hwy.—19.32 121.267; AM R104826 Alice Springs—23.7 133.867; AM R123454 5 km E of Barry Caves—20.05 136.733; AM R128641 Utopia Stn., via Alice Springs—22.23 134.567; AM R130659 40 km S Alice Springs on Old South Hwy.—24 133.933; AM R134370 4.5 km from W end of Mt. Augustus Scenic Loop Rd., Mt. Augustus Stn.—24.32 116.75; AM R147264, R147265, R147266 18 km from Stuart Hwy. on Ross River Rd.—23.75 134.017; NMV D38, D39, D41, D42, D43, D44, D46, D71, D72, D73, D74, D75, D76, D2919, D2942, D5638, D5639, D5640, D5641, D2940, Tennant Creek—19.65 134.183; NMV D549, D2927, D4920, D4921, D4923, D4973, D4975, D5622, D5624, D5625 Barrow Creek—21.53 133.883; NMV D51965, D56403, D56404 Kulgera—25.83 133.3; NMV D67457 Kathleen Creek, Watarrka Site 2—24.35 131.678; NMV D161 Alice Springs—23.7 133.867; NMV D45 Tennant Creek—36.32 145.05; NMV D2706 Charlotte Waters—29.98 134.833; NMV D 11760 near Charlotte Waters—29.98 134.833; NTM R25707 Mt. Gillen Alice Springs—23.72 133.8; NTM R25708 NW Stock Route—21.07 130.75; NTM R25709, R25713, R25715, R25716, R25728 Jesse Gap, Alice Springs—23.75 134.017; NTM R25710 Alice Springs—23.7 133.867; NTM R25714 Barrowck—21.53 133.883; NTM R25717 5 km E Harts Range—22.98 134.95; NTM R25719 Bower Bird [Editors' note: this extralimital record requires verification]; NTM R25718, R25720, R25722 Kulgera—25.83 133.3; NTM R25724, R25725, R25726, R25727 Frewena—19.43 135.4; NTM R25729 Rockhole Bore Camp, Loves Creek Stn.—23.53 134.867; NTM R25730 Alice Springs—23.75 133.95; NTM R25731 Sangsters Bore, 12 km SW—20.87 130.267; NTM R25732, R25733, R25734, R25735, R25736 Alice Springs—23.75 133.95; NTM R25737 E Bank Armstrong Creek 100 km W Ayres Rock—25.08 130.05; NTM R25738 A I B Alice Springs—23.7 133.867; NTM R25739 Sangsters Bore, 12 km SW—20.87 130.267; NTM R25741 Plenty Hwy., 4 km E Jinka Stn. TO—22.97 135.65; NTM R25742 Frewena—19.92 135.417; NTM R25743 Plenty Hwy., 7 km E Jinka Stn. TO—23 136.067; NTM R25744 Alyawarre Desert Area—20.93 137.067; NTM R25745, R25746, R25747 Kulgera, 2 km S—25.87 133.3; NTM R25748 Mongrel Downs—20.78 129.817; NTM R25749 Kulgera, 2 km S—25.87 133.3; NTM R25750 Tanami—19.07 131.006; NTM R25751 Kulgera—25.08 133.3; NTM R25752 Lambina Stn.—26.93 134.1; NTM R25753 Tanami—19.9 130.683; NTM R25754 Alice Springs—23.7 133.883; NTM R25755 Harts Range, near Entire Creek—22.98 135.167; NTM R25756 Harts Range, Painted Canyon—23.03 134.85; NTM R25757 Ewaninga, 40 km S Alice Springs—24 133.9; NTM R25758 Kulgera, 2 km S—25.87 133.883; NTM R25759 Hermannsburg (near)—23.87 133.383; WAM

R3970 Well 37. C.S.R.—22.15 125.45; WAM R3995 Well 49. C.S.R.—20.17 126.683; WAM R3998 Wells 49 and 50, Canning Stock Route—20.2 126.833; WAM R4020 Wells 39–51, Canning Stock Route—20.83 126.167; WAM R8715 Well 43. C.S.R.—21.2 125.983; WAM R14653, R14654, R14655, R14656 War-burton Range—26.1126.583; WAM R14657 45 km NW of Warburton Mission. on Newbore—25.67 126.25; WAM R15179 8 km NNW of Warburton Range. Elder Creek—26.05 126.55; WAM R15706, R15707, R15708 32 km SE of Warburton Range—26.33 126.8; WAM R19598, R19599, R20608 Warburton Range Mis-sion—26.13126.583; WAM R21000, R21001 10 km SE of Warburton Range Mission—26.2 126.65; WAM R22007, R22011, R22012, R22221, R22022 War-burton Range Mission—26.13 126.583; WAM R24437 50 km SSE of Alice Springs. 16 km S of Ewaninga R.S.—24.13 133.933; WAM R26714 No locality but presum-ably SW Kimberley—17.5 123.5; WAM R28027 Injudinah Creek. La Grange—18.63 121.867; WAM R28814 14 km NW of Mt. Beadell—25.78 124.633; WAM R28864 77 km SW of Mt. Beadell. Sutherland Range—26.02 124.733; WAM R40111 60 mi E of No. 24 Well Canning Stock. Route—23.37 124.133; WAM R40887 Mt. Romilly; C.S.R.—20.47 126.5; WAM R45262 Well 40. C.S.R.—21.67 125.783; WAM R45768, R45769 near Alice Springs, between Emily Gap and Amoonguna—23.23 133.983; WAM R46123 32 km SW of Christmas Creek HS—19.08 125.717; WAM R46165, R46166, R46167 Anna Plains—19.25 121.483; WAM R47674 Point Massie. C.S.R.—20.72 126.5; WAM R53787, R53788, R53789 near Alice Springs—23.7 133.867; WAM R54073 Edgar Ranges Reserve—18.92 123.25; WAM R54125 Edgar Ranges Reserve—18.82 123.283; WAM R57047 30 km NNE of Stretch Range—20.73 127.85; WAM R57302, R57303, R57304 Mclarty Hills—19.48 123.467; WAM R60135 Injudinah Creek La Grange Stn.—18.63 121.867; WAM R63421 Twin Heads—20.25 126.533; WAM R63959, R63960 Well No 30 Canning Stock Route—22.5 124.133; WAM R64202 2 km 90 degree Murguga Well No. 39 Canning Stock Route—21.75 125.667; WAM R67578, R67579 Balgo Mission—20.15 128.95; WAM R75791 24 km W of Joanna Spring—20.08 123.917; WAM R75792 Dragon Tree Soak—19.65 123.383; WAM R75808 Nita Downs—19.08 121.683; WAM R76453 10 km SSW Cooya Pooya HS—21.12 117.117; WAM R81443 Wildlife Well—22.88 125.167; WAM R82606, R82607 Ca 45 km NW of Mt. Crofton—21.42 121.75; WAM R88546 55 km S Anna Plains HS—19.73 121.467; WAM R94371 Balgo Mission—20.15 127.967; WAM R100593 15 km SSW Giles—25.08 128.333; WAM R100859 10 km N Charlies Knob—23.08 125; WAM R101434 Plumridge Lakes Nature Reserve—29.6 124.9; WAM R108565 7–8 km WSW Point Salvation—28.25 123.6; WAM R125105 Yandicoogina—22.72 119.052; SAM R00956 Stuarts Rgs—28.92 134.667; SAM R00996 Kingoonya—30.92 135.317; SAM R01313 Mt. Arden—32.25 137.833; SAM R02686 Alice Springs—23.7 133.867; SAM R03257 Burt Plain 5 mi N of Mt. Hay—23.4 133.083; SAM R03439 Pilgangoora Well—21.05 118.867; SAM R03574 Wolf Creek—18.97 127.633; SAM R05297, R05298 Giles—25.03 128.3; SAM R06013, R06092 Tennant Creek—19.65 134.183; SAM R07596, R07597 Musgrave Park—26.27 130.833; SAM R10339 10 mi S Yuendumu—22.33 131.917; SAM R13697 2 mi NE Mulgathing Rocks—30.22 134; SAM R13786 Tennant Creek—19.65 134.183; SAM R14856 Bates 19 km W Barton—30.49 132.467; SAM R16745 Commonwealth Hill Stn.—30.13 134.2; SAM R18090, R19260, R24240 Uro Bluff—32.12 137.583; SAM R18223 Serpentine Lakes—28.5 129.017; SAM R19973 Indooroopilly Outstn Commonwealth Hill—29.82 133.35; SAM R21166 S Olympic Dam Roxby Downs—30.75 136.867; SAM R25533 10 km S Alice Springs on W Side Stuart Hwy.—23.8 133.85; SAM R25561, R25562 Muckera Campsite No 2—29.97 130.017; SAM R25862 Mt. Dare—26.07 135.25; SAM R32164 8.5 km SW Maralinga—30.22 131.523; SAM R32295 44 km SW Halinor Lake—29.49 130.164; SAM R32417, R32418 11 km SSW Maralinga—30.26 131.55; SAM R35961 Alka Seltzer Bore—26.3 136.017; SAM R36269 8 km Along Mt. Connor Track—25.38 131.767; SAM R38110 18 km SW of Purni Bore—26.4 135.967; SAM R40460 3.5 km SW of Mt. Crispe—26.45 135.367; SAM R41858 5 km S of Blue Hills Bore—27.18 132.868; SAM R41863 6.8 km S of Blue Hills Bore—27.19 132.869; SAM R41888 12.5 km W of Mimili—27.03

132.593; SAM R44150 1 km NE of Glendambo Roadhouse—30.97 135.75; SAM R44431 6 km (Air) WNW of Mt. Kintore—26.55 130.429; SAM R44651 6.5 km (Air) SW of Gypsum Bore—27.66 134.656; SAM R44782 Wirraminna Stn.—31.2 136.233; SAM R46855 3.5 km NW of Bitchera Waterhole—26.56 134.505; SAM R46864 7 km WNW of Brush Hole Bore—26.49 134.395; SAM R46991, R46992 5.6 km SSE Mosquito Camp Dam—26.16 134.514; SAM R47438 5.6 km NW of Phoenix Bore—27.38 133.208; SAM R47442 2.2 km W of Chameleon Bore—27.52 132.972; SAM R48618 12 km NNW of Mt. Cheesman—27.31 130.265; SAM R50196 14 km SE of Sentinel Hill—26.17 132.548; SAM R50200 8.2 km ESE of Sentinel Hill—26.12 132.52.

Diagnostic Characteristics

• Small, up to 38 cm in TL.

• Rounded to oval nostrils, equidistant between tip of snout and eye.

• Slightly compressed tail near base, rounded at the posterior part and nonkeeled.

• Proportion of tail to head plus body is 1.2 to 1.3.

• Scales around tail consist of regular rings.

• Head relatively flat, broad and short.

Description

Gillen's pygmy monitor is a small species belonging to the subgenus *Odatria*. More detailed descriptions can be found elsewhere (Waite 1929; Mertens 1942; Cogger 1992). Only a few important points are given here.

A small monitor up to 38 cm (see morphometrics) with rounded to oval nostrils, equidistant between tip of snout and eye, supraoculars not differentiated, smooth scales on upper thighs and a slightly compressed tail near base, rounded at the posterior part and nonkeeled. Proportion of tail to head plus body 1.2 to 1.3. Scales around tail consist of regular rings. Head relatively flat, broad, and short (Mertens 1942, Cogger 1992).

Mertens (1942) discusses the skull in detail. Hind legs are shorter than forelegs, and claws are strongly curved (Mertens 1942). For details of head length, SVL, and length of hind legs, see Pianka (1986, 1994a).

Dorsal coloration is rich gray-brown, laterally tending to be pale gray, with a series of reddish-brown crossbands consisting of spots and short bars on the back. The tail has dark brown stripes (Mertens 1942, Cogger 1992).

Morphometrics

Some morphometric data of Gillen's pygmy monitor are listed in Table 7.9, but see also Pianka (1986, 1994a). For growth and development of captive-born *Varanus gilleni*, see Horn and Visser (1991). Maximum length of *V. gilleni* is 38 cm according to Storr (1980). For upper extremes of SVL of males and females, see James et al. (1992).

TABLE 7.9.
Morphometric data of *Varanus gilleni*.

Weight (g)	Head + Body Length (cm)	Tail Length (cm)	Total Length (cm)	Tail/Head + Body	Ref.[b]
—	14.4	19.7	34.1	1.37	1
—	13.2[c]	—	—	—	2
—	13.2[d]	16.0	29.2	1.21	2
—	13.3[c]	17.7	31.0	1.33	2
65.0	18.6[e]	—	—	—	3
—	15.2[f]	22.4[f]	37.6[g]	1.47[g]	4
21.2	11.3[a]	12.2[b]	23.5	1.08	5

[a] Tip of snout to part back of hind leg.
[b] Part back of hind leg to tip of tail.
[c] Male.
[d] Female.
[e] Snout–vent length.
[f] Number rounded down from author's data.
[g] Calculated from author's data.
[h] References: 1, Waite (1929); 2, Mertens (1942c); 3, Pianka (1982); 4, Pianka (1986); 5, Horn and Visser (1991).

The most detailed investigation on morphometrics of Western Australian monitors, including *V. gilleni*, with respect to body mass, body shape, and especially relative size of appendages, gives some insight to interspecific and intraspecific relationships, phylogenies, ecology, and sexual dimorphism in these species (Thompson and Withers 1997c). For example, the *Odatria* clade differs from the *Varanus* clade not only because of its size, but also by morphological differences in shape. Further, males are generally larger than females. This is especially true for *V. gilleni* and the related *V. caudolineatus*, in which males have longer tails and deeper heads. Females of these species appear to be morphologically more similar to each other than to their own conspecific males. Thompson and Withers (1997c) subdivide goannas of Western Australia into four groups: (1) widely foraging terrestrial, (2) sedentary terrestrial, (3) arboreal and rock inhabiting, and (4) semiaquatic species. The third group comprises a broad group of typically arboreal (*V. gilleni, V. scalaris*) to typically saxicolous species (*V. pilbarensis, V. glebopalma*) and species found in both habitats (*V. caudolineatus, V. tristis, V. glauerti, V. mitchelli*). Saxicolous and species found in both habitats have relatively long appendages and flattened heads that contrast with the relatively short appendages of

V. gilleni and *V. caudolineatus*. Both these species can be separated because *V. gilleni* has a shallower head and slightly shorter forelegs and hind limbs than *V. caudolineatus* (Thompson and Withers 1997c).

Habitat, Natural History, Ecology, Diet

This monitor ranges throughout semiarid to arid timbered regions covered with sheoaks (*Casuarina*) and/or marble gums (*Eucalyptus gongylocarpa*) and in myall forests (*Acacia*). It is a specialized tree dweller and can be found there under bark of dead trees; it is rarely seen in the open (Waite 1929; Mertens 1942, 1958; Horn, unpublished data; Houston 1978; Pianka 1969, 1982, 1994a; Vincent and Wilson 1999; Husband and Vincent 1999).

Although mostly solitary, sometimes large congregations can be encountered in the field. This may depend either on breeding behavior or lack of shelter (Vincent and Wilson 1999; Greer 1989; Schmida 1975).

On sandy plains, monitor tracks can be used to find and identify these lizards. *V. gilleni* and *V. tristis* produce very similar tracks, but they can be discriminated by the length of their stride (Farlow and Pianka 2000).

Gillen's pygmy monitors forage both in trees and on the ground, where they prey on small reptiles such as geckos, large insects, and grasshoppers (Houston 1978). Hunting begins during the cooler hours of the morning, and the monitor returns beneath the loose bark as the sun rises and the temperature increases (Swanson 1987). They also eat bird eggs and small mammals, thus being a desert reptile with a most diverse diet (James et al. 1992; Vincent and Wilson 1999). Most precise information on the diet of this species in nature can be obtained from Pianka (1969) and James et al. (1992). Stomachs of 20 specimens were empty; 5 of them contained three grasshoppers, four *Gehyra variegata*, and one *Heteronotia binoei* (Pianka 1969; see also Pianka 1994a). Another *V. gilleni* contained parts of a large scorpion (Pianka 1982). For discussion on frequency distribution of estimated sizes and volumes of prey eaten by six species of monitors, including *V. gilleni*, see Pianka (1982). Predator–prey interactions of different desert goanna species and their prey are discussed by Pianka (1994a). An interesting defensive strategy against this predator has been developed by a species of gecko, *Diplodactylus williamsi*. According to Cloudsley-Thompson (1994), these geckos can eject a repugnant fluid over distances of about 50 cm. "This readily forms cobweb-like filaments which are difficult to remove and which are effective in deterring Gillen's pygmy monitor" (Cloudsley-Thompson 1994).

A comparison (percentage utilization of various microhabitats) was made between lizard communities of deserts (Kalahari in Africa, North America, Australia). *V. gilleni* was terrestrial in open sun (50%) and arboreal in high shade (50%) (Pianka 1986). The food intake consisted of scorpions (0.9%), grasshoppers and crickets (13.9%), and all vertebrate material including sloughed lizard skins (85.2%). From these data, a microhabitat niche breadth and a food niche breadth have been calculated (Pianka 1986). If a gecko is too large to be swallowed, Gillen's pygmy monitors "prey" on the tails of these geckos (Pianka 1969, 1994a). Results of an investigation of stomach contents of 149 museum specimens are listed in a more generalized, way plus some unidentified lizards (James et al. 1992). Diet of most species of monitors discussed in a phylogenetic context show that *V. gilleni* is exceptional because of an unusual contribution of vertebrate prey to its diet (Losos and Greene 1988).

Body temperatures of three wild-caught pygmy goannas have been measured to be 36.4, and 38.4, (Pianka 1982) and 37.4°C, respectively (Pianka 1994a). Activity temperatures differ slightly in the wild (37.4°C) and in captivity (37.1°C) (King and Green 1999). Preferred body temperature in a laboratory thermal gradient (mean) is 37°C (Bickler and Anderson 1986). Basking has been observed at temperatures up to 52°C (Vincent and Wilson 1999). Water loss from the skin is 0.06 mg·H_2O/cm²/h (King and Green 1999). Evaporative water loss as discussed by Thompson and Withers (1998) is 0.081 mg·H_2O/cm²/h; this value is below most other Australian monitors. Schmida (1975) found them under bark only during the austral summer, never in austral winter. Therefore, he assumes that this goanna must be dormant somewhere else during winter. The animals were only able to move their eyelids. If they were placed in the sun, they slowly resumed activity (Schmida 1975).

Crows and small birds of prey may occasionally catch Gillen's pygmy monitor, as do snakes. Pianka (1969, 1994b) once encountered a small (20 cm SVL) *V. gouldii* in the field that had a *V. gilleni* by the nape of its neck, with the two lizards wrapped around each other. The intruder's appearance disturbed the predator, and it ran away without letting its victim loose (Pianka 1969, 1994b).

In captivity, a broad variety of insects of suitable size have been fed (crickets—*Achaeta domesticus, Gryllus bimaculatus*; locusts—*Locusta migratoria*; a moth—*Galleria mellonella*; meat; mealworms—*Tenebrio molitor*; cockroaches—*Blabitica dubia*; and spiders of different kinds) to Gillen's pygmy monitor. It also consumes newly born mice or pinkys and European lizards (*Lacerta* ssp.) and small pieces of chopped chicken (Horn 1978; Eidenmüller 1994; Vincent and Wilson 1999). Longevity record for *V. gilleni* is 4.5 years (Greer 1989).

Physiology

V. gilleni has been investigated physiologically, e.g., standard rates of oxygen consumption at temperatures between 25 and 37°C. The mean value of this was 22% lower than that predicted by a regression equation for lizards as a group. The way of lung ventilation means that arterial partial pressure of carbon dioxide and pH value are constant with rising temperatures (Bickler and Anderson 1986). This situation is encountered in large varanids too and it is in contrast to that in other reptiles (Bickler and Anderson 1986). Discussions of the reasons of the sustaining power of varanids can be found elsewhere (Greer 1989; Horn and Visser 1997).

Behavioral Observations in Captivity

As already mentioned, Gillen's pygmy monitor is a secretive species that lives mainly under bark. With few exceptions, little is known of this species' behavior in the wild. But in captivity, several interesting observations have been made (Horn 1978; Eidenmüller 1994; Vincent and Wilson 1999). Couples can be kept together, but separation at times may be favorable because a second male may stimulate reproductive activities (Eidenmüller 1994; Vincent and Wilson 1999; Husband and Vincent 1999).

V. gilleni has been observed hanging down its head and body from a branch, suspended only by the tail (Horn, unpublished data). *V. prasinus* has also been described as having a prehensile tail. In my opinion, this is incorrect: the extremely long and thin tail is used as a steering gear, and it helps *V. prasinus* and similar long-tailed species not to go head over heels while jumping. In contrast, the relatively thick tail of *V. gilleni* can be used in the manner described. Greer (1989) says that *V. gilleni* has a prehensile tail.

Most interesting are ritual combats in this species. Males, but also females, may engage in these ritual fights. Combatants twist and roll on the ground, engaged in a wrestling fight, and eventually they display a quadrupedal arching embrace, with bodies parallel and lifted from the ground (Murphy and Mitchell 1974; Carpenter et al. 1976; Horn 1985; Vincent and Wilson 1999). A clinch phase, which is shown by most members of the subgenus *Varanus*, is missing, thus clearly positioning *V. gilleni* in the subgenus *Odatria* (Horn et al. 1994). Once copulation was observed. During the time of heavy rainfall (July), which caused high humidity, a younger male (22 cm TL) persecuted a female (27 cm TL) for 2 days. The male fixed the female at the neck/throat area with the snout and with the legs dorsally. Then the male's hind body was put under that of the female, cloacae pressing together. Copulation was carried out several times, and it took up to half an hour (Schmida 1975; Polleck 1980). Fertile eggs were deposited about 3 weeks later (compare time interval between copulation and egg deposition in different species by Horn and Visser 1989) in the humid sand of the terrarium. The female dug out a hole, deposited the eggs, and closed the egg chamber with sand (Schmida 1975).

Reproduction in Captivity

For behavioral activities before egg deposition, see the preceding section. Reproductive activities in this species have been observed during different times: July (Schmida 1975), October (Horn 1978), January (Eidenmüller 1994), September (Gow 1982), and March (Polleck 2001). This anecdotal information has been strongly confirmed by James et al. (1992), who investigated testes size of males and oviductal eggs of females in 149 museum specimens. These data show that males are reproductive from June to November, with a maximum in October. In contrast, females are reproductive through January to March and from September to November, with a maximum in October, as in males.

TABLE 7.10.
TABLE 7.10.
Data of eggs of different clutches.

No. Eggs Laid	Date	Weight (g)	Length (cm)	Width (cm)	Ref.[a]
3	Aug 1977	NI[b]	3.0	1.4	1
5[c]	NI	3.8–4.1	2.6–2.8	1.4–1.7	2
1[d]	Apr 2000	NI	2.8	1.6	3
2[e]	Apr 2000	NI	1.8	1.0	3

[a] References: 1, Schmida (1975); 2, Horn and Visser (1991); 3, Polleck (2001).
[b] NI, no information.
[c] Three eggs were infertile; two hatched.
[d] Fertile and hatched.
[e] Infertile.

Reproductive males have a SVL of 10.0 cm, whereas females are only 9.5 cm. Estimated clutch size is four, with small deviations above and below this value (James et al. 1992; Horn and Visser 1991).

Eggs deposited in terraria had the dimensions listed in Table 7.10. For additional information on egg size, see Husband and Vincent (1999). For data on eggs of wild-caught females, see Gow (1982).

Clutch size has been reported to vary between four to seven in wild-caught gravid females (Gow 1982; Greer 1989). Gillen's pygmy monitor has been bred successfully many times in captivity, but only a few detailed descriptions of this event have been published (Horn 1978; Broer and Horn 1985; Eidenmüller 1994; Eidenmüller and Wicker 1997; Horn and Visser 1997; Polleck 2001; see also Vincent and Wilson 1990). Some data published on hatchlings are listed in Table 7.11. For data on hatchlings of wild-caught females, see Gow (1982).

Given a constant incubation temperature, incubation time is correlated with size, not only in *V. gilleni* but in all species of monitor (Horn 1978; Horn and Visser 1989; Thompson and Pianka 2001). Other (positive) correlations (e.g., incubation time and egg mass or neonate SVL) have also been found (Thompson and Pianka 2001).

Status

V. gilleni is listed in appendix II of Washington Convention (CITES). In the wild, it seems not to be endangered. But Gillen's pygmy monitor needs *Acacia, Casuarina,* and *Eucalyptus* woodland habitats. Domestic stock grazing and debarking trees may cause a lack of new tree recruitment and further loss of habitat. Seen on a long timescale, this may cause localized extinctions of some populations of this monitor (Vincent and Wilson 1999).

TABLE 7.11.
Breeding data of *Varanus gilleni*.

Mating Observed	Eggs Laid Date	No.[a]	Period (d)	Temp. (°C)	Medium	Incubation No.	Date	Hatchlings Length (cm)	Weight (g)	Ref.[b]
No	20–23 Oct 1977	3	92 ± 3	29–30	Wire net[c]	1	19 Jan 1978	11.0	2.2	1
1–3 Jan 1993	30 Jan 1993	2	124–131	26–29	Vermiculite	2	NI	14	3.4	2, 3
No	7 Apr 1993	2	101–111	26–29	Vermiculite	2	NI	14	3.1	3
No	23 Jun 1993	3	106–112	26–29	Vermiculite	3	NI	14	3.4	3
No[e]	28 Jan 1994	3	84–88	27–32	Vermiculite	3	22–26 Apr 1994	11.6–14.2	2.2–3.8	4
17 Jun 1994[e]	16 Jul 1994	6 (5)	102	27–32	Vermiculite	1	27 Oct 1994	—	—	4
No[e]	13 Jan 1995	6	84–89	27–32	Vermiculite	6	7–12 Apr 1995	11.6–14.2	2.2–3.8	4
Jan–Apr 1995	5 May 1995	3 (2)	121	28	Sand	1	6 Sep 1995	11	—	4
14–28 Mar 2000	19 Apr 2000	3 (2)	110	28	Vermiculite	1	6 Aug 2000	12	—	5

[a] Number in parentheses indicates the number of eggs that were not fertile.
[b] References: 1, Horn (1978);
   2, Eidenmüller (1994);
   3, Eidenmüller and Wicker (1997);
   4, Horn and Visser (1997);
   5, Polleck (2001).
[c] See Broer and Horn (1985) and Horn (1978).
[d] NI, not indicated.
[e] Unpublished data by Banks and Vincent, Melbourne Zoo. See Horn and Visser (1997).

7.8. *Varanus glauerti*

SAMUEL S. SWEET

Figure 7.18. Varanus glauerti *(photo by Jeff Lemm).*

Nomenclature

Initially described as *Varanus (Odatria) timorensis glauerti* by Mertens (1957b) and elevated to species status by Storr (1980). Holotype: Western Australian Museum, catalog WAM R12337, from Wotjulum, West Kimberley, Western Australia (Fig. 7.18).

Geographic Distribution

Restricted to monsoonal northern Australia from the western Kimberley in Western Australia to the extreme northwestern Northern Territory; also an isolated population along the northwest margin of the Arnhem Land Plateau and its outliers in the Northern Territory. Recorded from several islands off the Kimberley coast. The Kimberley population is probably a different species than the Arnhem Land population, which remains to be described (Fig. 7.19).

AM R59658 Leichhardt Mural, Death Adder Gorge—13.05 132.867; AM R123850 Mitchell Plateau, upstream from Mitchell Falls—14.83 125.683; NTM R25760 Kalumburu Mission—14.3 126.633; NTM R25761 Flying Fox Billabong Kununurra —15.78 128.733; NTM R25762 Kununurra—15.78 128.733; WAM R11207, 12337 Wotjulum—16.18 123.617; WAM R13470 Carlton Reach Bore—15.8 128.75; WAM R13790 Kalumburu, 4 km NE of Mission—14.25 126.667; WAM R40448 Augustus Is., Bonaparte Archipelago—15.32 124.533; WAM R40449 The Largest of the Heywood Is., Bonaparte Archipelago—15.33 124.333; WAM R41302, R41303, R41304 Augustus Is. Bonaparte Archipelago—15.32 124.533WAM R41505 Champagny Is., Bonaparte Archipelago—15.3 124.25; WAM R41510 Kuri Bay—15.48 124.5; WAM R43044 Crystal Creek. Near Crystal

Figure 7.19. Known localities of collection of Varanus glauerti.

Head—14.47 125.85; WAM R43140 Surveyor's Pool. Mitchell Plateau—14.67 125.733; WAM R43325 Port Warrender. Admiralty Gulf—14.5 125.833; WAM R44016 Careening Bay. Near Port Nelson—15.08 125; WAM R44076 Sir Graham Moore Is., Bonaparte Archipelago—13.88 126.567; WAM R44122 SW Osborne Is., Bonaparte Archipelago—14.35 125.95; WAM R44137 Uwins Is. Entrance to St. George Basin—15.28 124.833; WAM R44139 St. Andrew Is., St. George Basin—15.37 125.017; WAM R44149 Byam Martin Is., Bonaparte Archipelago—15.37 124.35; WAM R46811, R46812, R46813, R46835 Prince Regent River Reserve—15.33 124.933; WAM R46858 Prince Regent River Reserve—15.53 125.233; WAM R46872 Prince Regent River Reserve—15.12 125.55; WAM R50532 Drysdale River National Park—15.05 126.75; WAM R50571 Drysdale River National Park—14.77 127.083; WAM R50603 Drysdale River National Park—14.73 126.933; WAM R50773, R50774 Drysdale River National Park—14.67 127; WAM R50887 Drysdale River National Park—15.03 126.817; WAM R56234 Mitchell Plateau, eastern face. In Watercourse—14.68 125.867; WAM R57144 20 km SSE of Barton Plains Outcamp. Carson Escarpment—14.33 126.95; WAM R60529 Mitchell Plateau—14.58 125.75; WAM R61723, R61724, R61725 14 km SSE Walsh Point Port Warrender—14.67 125.883; WAM R77010 Mitchell Plateau—14.59 125.763; WAM R77266 Camp Creek—14.89 125.75; WAM R77413 Mitchell Plateau—14.59 125.761; WAM R80042 Sunday Is., Buccaneer Archipelago—16.42 123.183; WAM R80055 Hidden Is., Buccaneer Archipelago—16.25 123.517; WAM R80074 Gibbings Is., Buccaneer Archipelago—16.17 123.517; WAM R80100 Sir Frederick Is., Buccaneer Archipelago—16.13 123.4; WAM R80505, R80520 Long Is., Buccaneer Archipelago—16.58 123.383; WAM R80524 Pasco Is., Buccaneer Archipelago—16.52 123.4; WAM R83454 30 km NW of Mt. Elizabeth HS—16.15 125.967; WAM R83689 34 km NE of Napier Downs—17.15 125.067; WAM R94839 Ca 17 km N Mt. Lacy, Mt. Elizabeth Stn.—16.13 125.983; WAM R99101 5.6 km W Evelyn Is.—14.12 127.519; WAM R99231 13.5 km NE Crystal Head on

SW Osborne Is.—14.38 125.95; WAM R100165 SW Osborne Is.—14.35 125.95; WAM R103371 Bungle Bungle National Park—17.27 128.25; WAM R103399 Bungle Bungle National Park—17.32 128.417; WAM R103400, R103401 Bungle Bungle National Park—17.25 128.3.

Fossil Record

No fossils of *Varanus glauerti* are known.

Diagnostic Characteristics

- Medium size, to 80 cm TL.
- Gracile, long neck; tail may exceed 1.8 times body length.
- Tail black with brilliant white or bluish-white rings to tip.
- Neck and shoulders yellowish to rusty, body with bands of gray to turquoise oval spots outlined by rusty or black scales.
- Prominent black temporal stripe.
- Throat yellow or white, unmarked.
- Palms and soles with enlarged, rubbery black scales.

Description

This thin and very elongate monitor has a long neck, long legs, and a whiplike, prominently ringed tail. Hatchlings resemble adults, whose coloration and pattern varies geographically. Dorsal ground color is gray to tan (West Kimberley) or yellowish to rusty (East Kimberley, Arnhem Land) on the neck and shoulders, becoming blue-gray posteriorly on the trunk, grading to black about halfway down the tail. Five to eight distinct or broken crossbands of light gray to turquoise oval spots are aligned transversely, with the color intensifying posteriorly; these spots merge into bands on the tail base and become paler, white or bluish-white rings that contrast sharply with the black distal tail. Black scales may border or outline the bands of spots posteriorly on the trunk. East Kimberley animals tend to have confluent pale markings and so appear banded, whereas West Kimberley and Arnhem Land animals retain rows of distinct spots. Limbs are dark gray or black with rows of pale yellow or white spots. The throat is white or yellow, venter pale gray, sometimes with indistinct crossbands. A very prominent dark temporal streak is bordered above and below by yellow or white. Irises of eyes are brown, and the tongue is pink. The tail is rounded in cross section and slightly depressed near the base. Nostrils are lateral, slightly less than halfway between tip of snout and eye. Dorsal scales are smooth in 122 to 151 rows (mean, 140 rows) at midbody, are uniform on trunk, and become slightly spinose on the tail base (Storr 1980; Storr et al. 1983). Scales of the undersides of the digits, palms, and soles are slightly enlarged, blackened, and rubbery in life. Claws are strongly curved. Males are larger than females, with brilliant yellow throats. Males have a patch of seven to nine enlarged and very spinose postcloacal scales that are white with black bases; these scales are smaller and white in females (illustrated by Sweet 1999).

Size

SVLs range from about 70 mm at hatching to about 250 mm (TLs, 170–800 mm). Both sexes mature at about 150 mm SVL. Average SVL and mass are 215 mm and 95 g in adult males, and 180 mm and 60 g in adult females (data from Storr 1980; James et al. 1992; and Sweet 1999). Tail length averages 1.80 times SVL (Thompson and Withers 1997c).

Habitat and Natural History

Varanus glauerti is saxicolous or arboreal. Throughout the Kimberley, it occupies gorges, escarpments, and other areas where vertical rock surfaces with deep crevices occur. Nearly all firsthand accounts of the species in the Kimberley region describe it as wholly saxicolous, with only Rokylle (1989) mentioning arboreal habits. By contrast, the isolated population in northwestern Arnhem Land is strongly arboreal: 56% of 527 observations were in trees, compared with 37% in rock outcrops (Sweet 1999). In this area, *V. glauerti* strongly favors the endemic myrtaceous tree *Allosyncarpia ternata,* which dominates the canopy in escarpment monsoon forests. Radiotracked animals foraged within the hollow trunks and major branches and spent considerable periods resting with head exposed in the broken ends of small hollow branches 12 to 20 m high in the canopy. Animals also used hollow trees of other species when ranging beyond the *Allosyncarpia* woodland and forest.

Varanus glauerti occupies rocky habitats differently from the sympatric *V. glebopalma,* favoring larger outcrops with tall vertical sides and vertical crevices. When horizontal crevices are used, they tend to be near the tops of outcrops rather than near the bases (Sweet 1999). *Varanus glauerti* have stable home ranges of 1.25 to 7.37 ha, with extensive overlap among individuals. Males and pairs often shared the same tree, and even the same branch stub, with no evidence of agonistic behavior outside of the breeding season.

Terrestrial, Arboreal, Aquatic

Varanus glauerti is terrestrial and saxicolous in the Kimberley, and primarily arboreal in Arnhem Land. Animals will forage terrestrially, especially in dense vine thickets, and appear to travel on the ground among trees and outcrops.

Time of Activity (Daily and Seasonal)

Varanus glauerti seems to be inactive during the latter part of the dry season (late July to October) in Arnhem Land, but most museum specimens from the Kimberley surveyed by James et al. (1992) were collected between June and August. Whether this reflects a real difference between populations or is an artifact of the distribution of fieldwork is not known. In Arnhem Land, *V. glauerti* becomes active in trees with the first rains of the wet season but does not often venture onto the forest floor until a dense herbaceous layer develops (December–January). Activity continues into the early dry season, and the animals are most conspicuous during the mating season (mid-May to early July). On a daily cycle, individuals seem to be inactive until midmorning, then bask briefly in the canopy before foraging. They are active through midday and return to shelter sites 1.5 to 2 hours before sunset (Sweet, unpublished observations). There are several firsthand accounts of crepuscular activity and foraging in the Kimberley population (e.g., Wilson and Knowles 1988; Rokylle 1989; Ehmann 1992). However, radiotracked animals in Arnhem Land did not change locations between sunset and early morning (Sweet 1999).

Thermoregulation

There are no published reports of field body temperatures for *V. glauerti*. Arnhem Land animals were observed to bask in the canopy before fully emerging from overnight retreats, and they often paused to bask in sunflecks on limbs and trunks as they foraged throughout the day (Sweet 1999).

Foraging Behavior

Varanus glauerti is an active forager. In Arnhem Land, individuals were observed to pursue small skinks (*Carlia* and *Glaphyromorphus* spp.) flushed from leaf litter in vine thickets and to catch and eat small invertebrates encountered while traveling on the ground. However, most foraging appeared to occur in rock crevices and (especially) in tree hollows. Radiotracked animals located in hollow trunks and limbs often spent considerable time excavating rotted wood. On one occasion, a monitor was seen to excavate a limb stub and capture a gecko (*Gehyra* sp.), and another was observed carefully investigating crevices in upright dead branches emerging from the canopy, from which it flushed and captured a large katydid (*Nicsara* sp.). Individuals were also seen pursuing small arboreal skinks (*Cryptoblepharus* sp.) on tree trunks and ledges. Feces of wild-caught animals contained remains of geckos, *Cryptoblepharus, Nicsara,* and large roaches that inhabit rotted wood in standing trees (Sweet 1999). No prey items that are usually found on the forest floor were recovered from feces, suggesting that most successful foraging occurs in tree hollows and rock crevices. Unlike the sympatric arboreal monitors *Varanus scalaris* and *V. tristis, V. glauerti* were never observed to employ tree trunks as foraging perches while seeking terrestrial prey.

Diet

James et al. (1992) reported stomach contents from 47 *V. glauerti* from the Kimberley as containing 38.3% orthopterans, 29.8% spiders, 12.8% roaches, 8.5% lizards (*Heteronotia* and *Cryptoblepharus* sp.), and minor items. These records and observations noted above indicate that *V. glauerti* finds most of its prey by searching concealment sites.

Reproduction

Varanus glauerti is not known to display ritualized male combat. In multiple observations of courting and mating individuals, I observed a single pursuit involving two males. In all other cases, two or three males closely attended a female, sometimes being in direct contact with a mating pair, with no evidence of aggressive behavior among them. Field observations in Arnhem Land place the breeding season from mid-May to early or mid-July (most matings occur in late May and early June), with a single record of eggs being laid inside a large hollow tree in mid-July. Male *V. glauerti* were observed to "trapline" females during the breeding season; a single male visited four females on five, three, two, and two occasions for totals of 2 to 9 days each over a 21-day period in May and June, and each of four other radiotracked males (plus one unmarked male) also visited some of the same females during this interval. Males appeared to retain information about the last known location for each female, as they returned directly to that spot, then rapidly tracked the female through her subsequent locations to her current site (Sweet 1999). James et al. (1992) report that both sexes mature at about 150 mm SVL (Kimberley specimens), and that males have enlarged testes from June into August. Two females were in reproductive condition in August and November, with clutch sizes of three eggs. Storr (1980) reports a SVL of 70 mm for a presumed hatchling, which agrees with a mean of 72.6 mm predicted by the allometric equation presented by Thompson and Pianka (2001). Sweet (1999) observed an apparent recent hatchling on November 1.

Movement

Sweet (1999) presents the only detailed account of the spatial ecology of *Varanus glauerti,* based on the Arnhem Land population. Because animals there are chiefly arboreal in a mesic setting, information on their spatial ecology may not be applicable to the Kimberley population, which is saxicolous in a much drier environment. In Arnhem Land, 75.2% of 513 observations were in monsoon forest dominated by *Allosyncarpia,* 20.7% in *Allosyncarpia* woodland, and 4.1% in dry eucalypt woodland. Animals had very stable home ranges of 1.25 to 7.36 ha. Only a single female was radiotracked, but observations of other females with distinctive markings support the idea that females have relatively small home ranges in comparison to males. Males moved an average of 41 m per day, versus 21.7 m per day for the single female, and both sexes were often sedentary for 1 to 2 days at a time. Animals were observed to move between adjacent trees in the canopy or via nearby rock outcrops, but most long-distance travel was on the ground. In these cases, the monitors briefly climbed or entered nearly every tree encountered. They tended to travel in straight lines, passing through or over outcrops rather than diverting around them, and they frequently crossed their entire home range in a single bout. Home range margins were sharply defined, and animals repeatedly returned to the same peripheral sites over a period of months. These monitors appear to have remarkable "map knowledge" of their home ranges.

Physiology

No data have been reported on physiological variables in *V. glauerti.*

Fat Bodies, Testicular Cycles

There are no data on fat body cycles per se, but recaptures of radiotracked animals showed consistent weight gains throughout the wet and early dry season, followed by a dramatic loss of condition by the end of the mating season in July. Males recaptured in late July had lost 23% to 36% of their premating masses and appeared emaciated (Sweet 1999). James et al. (1992) reported that Kimberley population males had enlarged testes from June to October (but not in May), suggesting a later and perhaps longer breeding season than in the Arnhem Land population.

Parasites

Varanus glauerti hosts ticks of the genera *Amblyomma* and *Aponomma,* with *Amblyomma glaueri* and *Aponomma glebopalma* recorded only from this species and *V. glebopalma* (Kierans et al. 1994; see also King and Green 1999b). These ticks attach to the lateral neck, axilla, and groin and are seldom present in large numbers (Sweet, unpublished observations). Gastric nematodes of the genus *Abbreviata,* which feed on ingested prey items, are commonly found in *V. glauerti.* Jones (1991) records three species in animals from the Kimberley population. Malarial and microfilarial parasites infecting the blood are very likely to occur in *V. glauerti* but remain to be studied.

7.9. *Varanus glebopalma*

Samuel S. Sweet

Figure 7.20. Varanus glebopalma *(photo by Samuel S. Sweet).*

Nomenclature

Varanus glebopalma was described by Mitchell (1955), who provided good illustrations of its diagnostic features. The holotype is South Australian Museum (SAM) R3222; the type locality is the south end of Lake Hubert (Arnhem Land), Northern Territory (Fig. 7.20).

Geographic Distribution

Varanus glebopalma occupies rocky habitats across monsoonal northern Australia, from the western Kimberley through the Top End of the Northern Territory, and follows the escarpments and rocky outliers of the Gulf of Carpentaria drainages southeast into northwest Queensland. It occurs on a number of islands off the Kimberley and Arnhem Land coasts but is absent from islands that lack significant rocky terrain (Fig. 7.21).

AM R14057, R14058: Forrest River Mission—15.18 127.85; AM R18859, R66264, R107883: Mt. Isa—20.73 139.483; AM R38317, R38327, R38332: Koongarra, Brockman Range, Arnhem Land—12.78 132.65; AM R39969 Nourlangie Rock, Mt. Brockman Range—12.88 132.833; AM R48798 Macarthur River Base Camp—16.42 136.15; AM R48807 Caranbirini Waterhole, Bluff near waterhole—16.27 136.083; AM R51900, R51901: Mt. Carr, Adelaide River Township—13.25 131.1; AM R53465 37 km N Macarthur River Camp on Borroloola Rd.—16.1 136.117; AM R60189 Mt. Carr, Adelaide River—13.23 131.083; AM R76802 Bullo River Stn. Rd. ~40 km NW Victoria Hwy.—15.67 129.65; AM R135926 Groote Eylandt, Gemco Mining Lease Area—13.95 136.467; AM R136112 Surveyors Pool on Tributary of Mitchell River—14.67 125.717; ANWC R0366 Nourlangie Rd.—12.85 132.783; ANWC R0457 Nourlangie Rd.—12.75 132.767; ANWC R0545 Tin Camp Creek, ~27 km SE of Oenpelli—12.45 133.267; ANWC R0572 E End of Nourlangie Rock—12.83 132.817; ANWC R1113

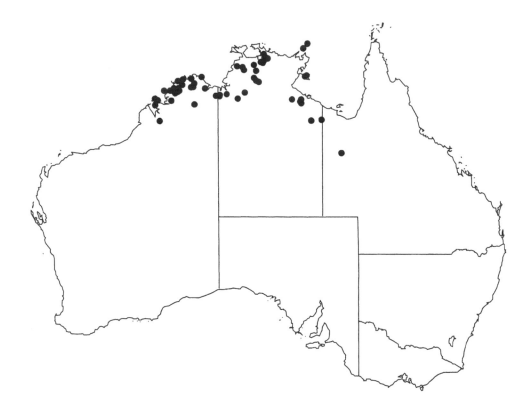

Figure 7.21. Known localities for
Varanus glebopalma.

Caranbirini Water Hole—16.27 136.083; NTM R25763, R25765: Katherine Caves—14.58 132.467; NTM R25764 Mt. Brockman—12.75 132.933; NTM R25766 Edith River—14.17 132.067; NTM R25767 Jasper Gorge, 79.2 km W Victoria River Downs—16.03 130.683; NTM R25768 Katherine—14.47 132.267; NTM R25769 Oenpelli—12.33 133.033; NTM R25770 North Island; NTM R25771 Jensen Bay Marchinbar Is.—11.17 136.683; NTM R25772 25 km E Victoria River Crossing—15.55 131.267; NTM R25773 Stuart Hwy.—13.48 131.183; NTM R25774, R25776, R25777: Cutta Cutta Caves Katherine—14.58 132.467; NTM R25775 Bamboo Creek, Litchfield National Park—13.15 130.633; NTM R25778 The Rock Hole 68 km E Pine Creek—13.55 132.25; NTM R25779 Yadikba—14.08 136.45; NTM R25780 Kununurra—15.78 128.733; NTM R25781 Ayakamindadina Groote Eylandt—13.97 136.683; NTM R25782 Jabiru Area—12.67 132.9; NTM R25783 Raragala Is. (North)—11.57 136.336; NTM R25784, R25785: Mt. Borradaile—12.05 132.917; NTM R25786 Nicholson River Reserve—17.9 137.883; NTM R25787 Nicholson River—Block, West—17.93 137.017; NTM R25788 Keep River—15.75 129.083; NTM R25789 Bauhinia Downs Stn.-Alligator Yard—16.08 135.367; NTM R25790 Keep River—15.85 129.05; NTM R25791 Deaf Adder Creek—13.05 132.032; WAM R11840, R11841: Wotjulum—16.18 123.617; WAM R19905 Katherine—14.47 132.267; WAM R28036, R28037: Kalumburu—14.3 126.633; WAM R29586 Koolan Is.— 16.15 123.75; WAM R32086 Mt. Anderson—18.03 123.933; WAM R32214, R32215: Oenpelli Mission—12.32 133.05; WAM R32242 Manning Creek. Mt. Barnett HS—16.57 126.933; WAM R41487 Katers Is., Bonaparte Archipelago— 14.47 125.533; WAM R43121 Surveyor's Pool. Mitchell Plateau—14.67 125.733; WAM R44106 Boongaree Is., Prince Frederick Harbour—15.08 125.2; WAM R44121 SW Osborne Is., Bonaparte Archipelago—14.35 125.95; WAM R44134 Uwins Is. Entrance to St. George Basin—15.28 124.833; WAM R44135 Uwins Is. Entrance to St. George Basin—15.25 124.833; WAM R44148 Byam Martin Is., Bonaparte Archipelago—15.37 124.35; WAM R45740 Mt. Anderson—18.03

123.933; WAM R46843, R46844, R46845: Prince Regent River Reserve—15.53 125.233; WAM R46874 Prince Regent River Reserve—15.12 125.55; WAM R47032 Prince Regent River Reserve—15.32 125.583; WAM R47034 Prince Regent River Reserve—15.53 125.317; WAM R47036, R47037: Prince Regent River Reserve—15.47 125.483; WAM R50450, R50451: Drysdale River National Park—15.02 126.65; WAM R50545, R50546, R50547: Drysdale River National Park—15.05 126.75; WAM R50599 Drysdale River National Park—14.73 126.933; WAM R60668 Camp Creek Mitchell Plateau—14.83 125.833; WAM R80045 Lachlan Is., Buccaneer Archipelago—16.62 123.517; WAM R80063, R80064: Hidden Is., Buccaneer Archipelago—16.25 123.517; WAM R80072 Gibbings Is., Buccaneer Archipelago—16.17 123.517; WAM R80091 Bathurst Is., Buccaneer Archipelago—16.05 123.533; WAM R97575, R97580: Kalumburu—14.3 126.633; WAM R97898 3.7 km NW Mt. Daglish—16.25 124.95; WAM R99052 4 km W King Cascade—15.63 125.265; WAM R99109 5.6 km W Evelyn Is.—14.12 127.519; WAM R99227 13.5 km NE Crystal Head on SW Osborne Is.—14.38 125.95; SAM R15546 Cannon Hill—12.38 132.95.

Fossil Record

No fossils of *Varanus glebopalma* have been reported.

Diagnostic Characteristics

- Moderate size, to just over 1 m TL.
- Tail approximately 1.7 times the body length, last third uniformly pale yellow.
- Dorsum gray, often with faint network of light or dark scales.
- Throat and chest pale with faded reticulum of darker scales.
- Palms and soles with enlarged, rubbery, shiny black scales.

Description

Varanus glebopalma is a slender, dark monitor with an extremely long, pale-tipped tail, restricted to rugged rocky habitats. Recent hatchlings are dark gray with numerous white scales forming ocelli or diffuse crossbands on the back, and narrow light rings on the base of the tail; the pale tail tip has faint dark crossbands (photograph in Horn and Schurer 1978). Subadults and adult females often retain a lacy reticulum of white scales dorsally; mature females have a rusty wash on the ventral surfaces of the thighs and tail base. Adult males become uniform gray dorsally, with scattered darker scales sometimes forming an irregular reticulum. The throat and venter are pale gray or dirty white, usually with diffuse purplish-gray netlike markings. Limbs are darker than the dorsum, with numerous small pale spots in all but large males. Iris is red-brown with gold inner ring, and the tongue is pink. The tail is rounded in cross section, depressed near the base, and becomes slightly compressed distally. Nostrils are lateral, about halfway from tip of snout to eye. Dorsal scales are small, uniform, smooth, in 132 to 170 (mean 150) rows at midbody; lateral and ventral scales of the tailbase are very slightly keeled (Storr 1980). Scales of the undersides of the digits, palms, and soles are enlarged, black, and rubbery in life;

these provide traction on rock surfaces and are the basis of the species name, which translates as "dirty palms." Claws are strongly curved and sharp. Males are larger and proportionally more robust than females; in addition to color differences noted above, males have five to nine slightly enlarged, domed postcloacal scales in paired ventrolateral patches on the tail base (illustrated by Sweet 1999).

Size

SVL ranges from about 90 mm at hatching to 397 mm (TLs ~190–1100 mm, though complete tails are rare in adults). Both sexes mature at ~170 mm SVL. Average SVL and mass are 290 mm, 240 g for adult males, and 245 mm, 130 g for adult females (data from James et al. 1992; and Sweet 1999). Tail length averages 1.67 times SVL (Thompson and Withers 1997c).

Habitat and Natural History

Varanus glebopalma is closely associated with heavily eroded rock outcrops throughout its range (Christian 1977; Horn and Schurer 1978; Sweet 1999). Surrounding habitats range from open spinifex grassland through a variety of open woodland types to closed-canopy monsoon forest but are typically very open woodlands. Appropriate rocky habitats feature numerous deep, horizontal crevices or boulder accumulations that are not infilled by soil, affording plenty of secure refuges. Although individuals will cross open terrain in traveling between outcrops, *Varanus glebopalma* is otherwise purely saxicolous and will not climb trees even if closely pursued. It is remarkably agile on rocks, able to ascend vertical surfaces or cross rugged boulder fields at full speed. An exceptionally alert and wary lizard, *V. glebopalma* is usually seen only when surprised at close range. Individuals will freeze if an observer appears within about 6 to 8 m, but their usual flight distance is well in excess of 50 m. These monitors have well-defined home ranges of 3.5 to 7.8 ha that are stable over time. Uniquely among *Varanus*, male *V. glebopalma* are strongly territorial (Sweet 1999). Although areas used by adjoining males overlap, sharp boundaries within each overlap zone separate regularly used regions from areas of brief, minor incursions into the territory of an adjoining male. Home ranges of females show about twice the areal overlap (~40%) of males, and each male territory includes parts of the home ranges of two to four females (Sweet 1999). Actual contact between males seems to be rare, and no visual displays are known; instead, territorial integrity appears to be maintained by scent marking and perhaps by long-range visual contact. Territories of missing males are rapidly usurped by neighbors.

Terrestrial, Arboreal, Aquatic

Varanus glebopalma is exclusively terrestrial and saxicolous; 97% of over 1400 observations were on rocks, and none were on trees, logs, or stumps (Sweet 1999).

Time of Activity (Daily and Seasonal)

Like most monitors in the Australian monsoonal tropics, *V. glebopalma* displays greatly reduced activity in the latter half of the dry season (July–September). Activity increases dramatically with the advent of the first thunderstorms of the approaching wet season, and breeding apparently occurs at this time (late September to November) (James et al. 1992; Sweet 1999). Individuals are active from first light to dusk, but select shaded perches during the hottest part of the day. There are several accounts of *V. glebopalma* being active after dusk (e.g., Swanson 1979; Ehmann 1992), but I was unable to document crepuscular or nocturnal activity in radiotracked animals in Arnhem Land (Sweet 1999). I noted that individuals were easily flushed from crevices around dusk and suspect that this may account for observations of crepuscular activity or foraging.

Thermoregulation

There are no published reports of field body temperatures for *V. glebopalma*. Throughout most of its range, the principal thermal challenge would appear to be remaining cool enough: exposed rock surfaces quickly reach surface temperatures of 50 to 60°C and often remain above 35°C until well after dark.

Foraging Behavior

Varanus glebopalma is a classic visually oriented ambush predator. Individuals rarely tongue-flick or pause to investigate crevices as they travel (but will pursue insects and lizards that they flush). Instead, this monitor seeks low ledges and boulders at the margins of an outcrop and remains motionless at an edge affording a broad field of view, with the head raised high and held horizontally. Insects (chiefly large grasshoppers), skinks, and frogs detected up to 15 m away are rushed and seized, then carried back to a crevice, where they are beaten on the substrate and rapidly swallowed. Large individuals have been seen to jump over 3 m from a 1-m-high ledge to drive their heads into grass clumps and leaf litter concealing a frog or skink. These predatory attacks and retreats to shelter are conducted at full speed and are completed in only 3 to 5 seconds. Foraging lizards will remain at an ambush station for 20 to 40 minutes if unsuccessful, then swiftly move to a new site that is often 30 to 90 m away (Sweet 1999 and unpublished observations). Direct observations and analyses of scats show that lizards are foraging for skinks at the onset of the wet season, then add frogs as the wet season progresses, but shift markedly toward large grasshoppers in the early dry season, and consume little else in the early winter months.

Diet

James et al. (1992) reported stomach contents for 70 *V. glebopalma* as 54.9% lizards, 27.7% orthopterans, 12.3% frogs, and minor items. Lizard prey included 10 species of terrestrial skinks, a pygopodid, and two *Varanus* spp. All frogs eaten were *Litoria* spp.; orthopterans were not identified further. Sweet (1999) documented predation on the skink *Ctenotus coggeri* and a frog (*Litoria* sp.) and collected several hundred fecal pellets. These show that *V. glebopalma* feeds heavily on skinks (*Carlia, Ctenotus,* and *Glaphyromorphus* spp.) from October to April, but thereafter shifts almost exclusively to grasshoppers (~30 spp. recorded; Sweet, unpublished observations) until dry-season fires eliminate grasshopper populations in July.

Reproduction

Male combat has not been reported in *V. glebopalma.* James et al. (1992) report that all males examined had enlarged testes from August to the end of October, and Sweet (1999) saw no evidence of reproductive behavior in radiotracked animals between November and July. Thus, mating probably occurs coincident with the resumption of activity following the first rains in September and October. Both males and females reach sexual maturity at about 170 mm SVL (James et al. 1992). There are no reports of egg deposition in nature; Barnett (1977) reported a clutch of seven eggs laid by a captive female in mid-November, and James et al. (1992) reported oviductal eggs in December and January, with one clutch count of five. The smallest juveniles in collections are 85 to 90 mm SVL, collected in July, and Sweet (1999) observed an apparent recent hatchling in late July. However, Christian (1977) reports finding three hatchlings on March 9 in Queensland.

Movement

Sweet (1999) presents the only detailed account of the spatial ecology of *V. glebopalma.* This work was conducted on a western outlier of the Arnhem Land Plateau and represents a suite of relatively mesic habitat types. Overall, animals used dry woodland (21.7%), *Allosyncarpia* woodland (50.4%), and monsoon forest (27.6%) habitats, although the frequency of use of monsoon forest declined from 48.6% of observations in the early wet season to 14% by early in the dry season, perhaps correlated with seasonal shifts in the distribution and accessibility of prey items. Home ranges varied from 3.5 to 7.7 ha; the rank order of home range sizes by sex does not support the idea that home ranges of males are larger, but larger individuals of either sex had larger home ranges. Male *V. glebopalma* moved farther per day than did females (84.7 vs. 59.8 m). Animals of both sexes moved daily until mid-June, but thereafter became increasingly sedentary.

Physiology

There are no published reports on physiological variables in *V. glebopalma.*

Fat Bodies, Testicular Cycles

Fat body cycles have not been reported. Testes and epididymides of male *V. glebopalma* begin to enlarge in June (one of eight males) and

July (two of four males) and are enlarged in all animals from August through October, with regression commencing in November (James et al. 1992). This is consistent with mating taking place at the dry season–wet season transition, as suggested by Sweet (1999).

Parasites

Varanus glebopalma hosts specialized ticks of the genera *Amblyomma* and *Aponomma*. *Amblyomma glauerti* and *Aponomma glebopalma* are known only from *V. glebopalma* and *V. glauerti* (Kierans et al. 1994; see also King and Green 1999b). These small ticks attach to the sides of the body and the tail base but are seldom present in large numbers. Three species of gastric nematodes of the genus *Abbreviata* are commonly found in *V. glebopalma* (Jones 1991; King and Green 1999b); these consume the lizards' prey, but are not otherwise known to affect their hosts. Malarial and other protozoans infecting the blood are very likely to occur in *V. glebopalma*, but remain to be studied.

7.10. *Varanus gouldii*

GRAHAM THOMPSON

Figure 7.22. (top) V. gouldii *(photo by Jeff Lemm). (bottom)* V. g. flavirufus *(photo by Eric Pianka).*

Nomenclature

Initially named *Hydrosaurus gouldii* (Gray 1838, p. 394), this species included *Varanus rosenbergi* and *Varanus panoptes* until 1980. Gray's (1838) description is at best vague: "*Hydrosaurus Gouldii*, with two yellow streaks on the sides of the neck; scales over the orbits small, flat." It was placed in the suborder Leptoglossa, family Monitoridae. Gray (1845) also referred to this species as *Monitor gouldii* and provided an almost identical description, but added that it inhabited the western and northwestern coast of Australia. Storr (1980) raised *Varanus rosenbergi* to full species status from a subspecies *Varanus gouldii rosenbergi* (Mertens 1942, 1957a, p. 17). Before this time, there were two recognised subspecies, *V. g. flavirufus* and *V. g. rosenbergi*. At the same time, Storr (1980) separated *Varanus panoptes* from *V. gouldii*, and created *Varanus panoptes rubidus* and *Varanus p. panoptes* (Storr 1980). *Varanus gouldii* is commonly referred to as the sand goanna, bungarra, or Gould's goanna (Fig. 7.22).

Böhme (1991, p. 38), in respect of the nomenclature of *V. gouldii*, reported that Storr (1980) did not examine "the type material of the traditionally known species before diagnosing and naming the new one." The lectotype of *V. gouldii* was a stuffed and dry mounted specimen in the British Museum (BM 1946.9.7.61), which has the color and pattern of *Varanus p. panoptes.* The outcome of this confusion was that the sand monitor, which had long been recognized as *V. gouldii,* should have henceforward been known as *Varanus flavirufus* and the species known as *V. p. panoptes* (Storr 1980) should have been known as *V. g. gouldii* (Gray 1838); the species known as *V. p. rubidus* (Storr 1980) should have been known as *V. g. rubidus* (Storr 1980); and the New Guinea species known as *V. p. horni* (Böhme 1991) should have been known as *V. g. horni* (Böhme 1991). This change caused considerable confusion. Subsequently, Sprackland et al. (1997) argued to the International Commission on Zoological Nomenclature that on the basis of near-universal usage of the name *Varanus gouldii* (Gray 1838) for the sand goanna or Gould's goanna, the lectotype of *gouldii* should be set aside and a neotype designated in accord with common usage. After considering a number of submissions (King 1997; Böhme and Ziegler 1998; Shea and Cogger 1998; Hoser 1999), the International Commission on Zoological Nomenclature (Anonymous 2000) set aside all previous fixations for the nominal species *Hydrosaurus gouldii* Gray 1838 and designated the specimen collected by G. Thompson at Karrakatta, Perth, Western Australia (BMNH 1997.1, in the Natural History Museum, London), as the neotype. Nomenclatures as proposed by Storr (1980) were placed on the Official List of Specific Names in Zoology. Some books and articles between 1991 and 2000 have used the nomenclature proposed by Böhme (1991) (e.g., Eidenmüller 1997; Bennett 1998), which will lead to some confusion.

Varanus gouldii belongs with the subgenus *Varanus* and is closely related to *V. panoptes, V. rosenbergi, V. giganteus,* and *V. mertensi* (Fuller et al. 1997; Ast 2001).

Geographic Distribution

Varanus gouldii is found in most parts of Australia with the exception of southern Victoria, central Queensland, southern Western Australia, and Tasmania (Fig. 7.23). Its preferred habitat is sandy soils, but it is found in a range of soil types. It is easily confused with *V. panoptes,* and examples in the literature record *V. panoptes* as *V. gouldii.* Before 1980, *V. rosenbergi* was known as *V. gouldii,* and the subspecies name (e.g., *V. g. rosenbergi*) was not always provided. This leads to confusion because Dennis King and Brian Green published a considerable amount of information about *V. rosenbergi* under the name of *V. gouldii* before Storr (1980) changed its taxonomic status. Geographic distribution of *V. gouldii* overlaps that of *V. rosenbergi,* although the two species are easily distinguished. Geographic distributions of *V. gouldii* and *V. panoptes* also overlap. In some localities, these species are sympatric; at others, they are not. Partitioning of habitat seems to be based on soil type. The confusion as to the extent of sympatry is enhanced by *V. gouldii* being more than a single species.

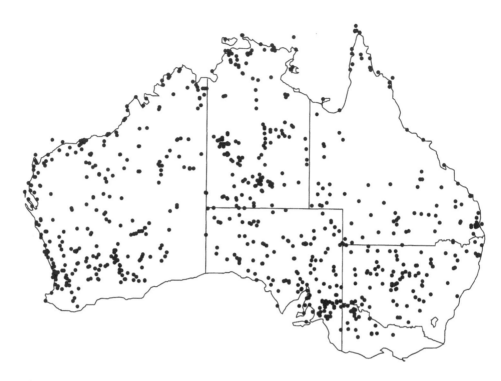

Figure 7.23. Geographical distribution of Varanus gouldii *in Australia, based on voucher specimens in Australian museums.*

AM R81, R82 Derby—17.3 123.62; AM R846 Branlin—34.73 148.05; AM R1981 Narromine—32.23 148.25; AM R2687 Perth—31.95 115.85; AM R3093, R3140 Boulder—30.78 121.48; AM R3888 Beecroft—33.75 151.07; AM R5219 Windsor—33.62 150.82; AM R7125 Mt. Lyndhurst, 20 mi E of Farina—30.2 138.57; AM R7636, R7637 407 mi EW Line—30.5 132.15; AM R7638 Fisher—30.55 130.97; AM R9510 Kootingal—31.05 151.07; AM R9907, R9908, R9909, R9910 Roper River N.T.—14.73 134.73; AM R10567 Barolin Stn. Bundaberg—24.88 152.48; AM R10851, R10852, R10858 Hermannsberg—23.95 132.77; AM R11042 Groote Eylandt, Gulf of Carpentaria—13.98 136.47; AM R11076 Flinders Is.; AM R11335 Alexandra Stn.—19.05 136.7; AM R11686 Kaimkillenbun, Moola —27.1 151.53; AM R11901 The Granites—20.58 130.33; AM R11902 Alice Springs—23.7 133.87; AM R12046, R12047 The Plains, Nyngan—31.77 147.12; AM R12427 Yirrkala Mission—12.25 136.88; AM R12875, R13240 Katherine— 14.47 132.27; AM R14059 3 mi W of Ayers Rock—25.35 130.98; AM R14060 Tanami—19.98 129.72; AM R14061, R14062 Forrest River Mission—15.18 127.85; AM R14170 825 km S of Alice Springs, on Old South Rd.—23.95 133.87; AM R14284 92 km N of Hamilton Hotel—22.77 140.6; AM R14391 Sandy Camp, Coonamble—30.87 147.73; AM R17862, R18624 Mt. Isa—20.73 139.48; AM R18554 Nymagee—32.07 146.32; AM R19273 Benagarie Stn. ~80 mi NW Broken Hill—31.42 140.42; AM R20190 Silver Plains via Coen Cape York Peninsula— 13.98 143.55; AM R20208 Darwin—12.45 130.83; AM R20257 Mataranka— 14.93 138.07; AM R20835 1 mi W of Windorah—25.42 142.63; AM R21408, R21418 Todd Rd. Stn., Alice Springs—23.85 134.5; AM R21439 Silver Plains, Port Stewart, Cape York—13.98 143.55; AM R26272 Smith Point, Port Essington— 11.13 132.15; AM R26378, R29621 Gilgandra—31.72 148.65; AM R28493 Gilgandra—31 153.02; AM R28496 Bald Hill, Canobolas State Forest; AM R28540 Macquarie University, North Ryde—33.77 151.1; AM R31698 Bulloo Downs— 28.53 142.95; AM R31735 Kinchega National Park—32.48 142.35; AM R31736 28 mi N Innamincka Bore No. 2—28 140.93; AM R32754 56 mi E Broken Hill— 31.75 142.37; AM R32769 Ruins of Yalpunga, between Woman and Warri Warri

Gaps—29.03 142.03; AM R33113 between Mt. Hope and Nymagee—32.42 146.03; AM R33173 42 mi S Gilgunnia on Euabalong Rd.—32.98 145.9; AM R33236 10 mi Ex. Warren on Carinda Rd.—31.57 147.88; AM R33296 55 mi N Katherine—13.85 131.85; AM R37758 4 mi S Rocky Creek Cape York—12.97 142.75; AM R38328, R38329 Koongarra, Brockman Range, Arnhem Land—12.78 132.65; AM R38728 Atula Stn., Plenty River—23.13 135.8; AM R40351 Wilcannia—31.57 143.37; AM R40358 Roto—33.05 145.47; AM R40520 Vicinity of Behn River—16.33 128.83; AM R40647 Cannon Hill—12.38 132.95; AM R41870 Liverpool River, 15 mi S Maningrida—12.33 134.08; AM R45995 NE Arnhem Land, M.R. Arnhem Bay S.D. 53-3 303391; AM R47284 Belaringar (between Nyngan and Nevertire on Mitchell Hwy.)—31.78 147.6; AM R47286 75 km W of Nyngan on Barrier Hwy.—31.57 146.5; AM R48077 Moa Is., Torres Strait—10.18 142.27; AM R48110 Prince of Wales Is., Torres Strait—10.68 142.15; AM R48949 Friday Is., Torres Strait.—10.6 142.17; AM R49230 100 km S Mt. Isa—21.63 139.5; AM R49399 20 km NW Mt. Doreen—21.97 131.25; AM R49561 20 km N Kulgera—25.68 133.3; AM R49665 98 km N Windorah on road to Winton—24.03 143.45; AM R49770 45 mi SW Alice Springs—24.22 133.37; AM R49771 3 mi N Orange Creek on Stuart Hwy.—24.07 133.72; AM R51311, R51312 Karumba—17.48 140.83; AM R51920 Mandorah, Darwin Harbour—12.45 130.75; AM R53400 5 km W Macarthur River Camp—16.43 136.07; AM R53464 Caranbirini Water Hole, 21 km N Macarthur River Base Camp—16.22 136.15; AM R53731 38.1 mi N Neale Junction Great Victoria Desert—28.12 125.92; AM R54256 10 mi NW of Birdsville—25.83 139.2; AM R54915 Goomadeer River—11.87 133.8; AM R57762 Somerset Cape York—10.75 142.58; AM R59735 110 mi. from Julia Creek on Normanton Rd.—19.53 140.83; AM R59736 12 mi N of Bundarra—30.02 151.1; AM R59737 1 mi S Diehard Creek, Old Gwydir Hwy.—29.63 152.1; AM R59738 58 mi S Monto—25.83 151.12; AM R59893 10 mi SW of Euroa—36.85 145.7; AM R59894 Near Gladstone—23.85 151.27; AM R60203, R60204, R60205 Mandorah, Cox Peninsula—12.47 130.67; AM R60206 1.5 km SE of Glenariff—30.85 146.57; AM R60207 9.6 km S of Wyandra on Mitchell Hwy.—27.4 145.93; AM R60208 48 km N of Blackall on Landsborough Hwy.—24 145.35; AM R60209 32 km S of Larrimah on Stuart Hwy.—15.82 133.37; AM R60210 77 km S of Dunmarra on Stuart Hwy.—17.45 133.47; AM R60211 89.6 km S Renner Springs on Stuart Hwy.—19.12 134.17; AM R60212 101 km S Renner Springs on Stuart Hwy.—19.25 134.17; AM R60213 88 km S Tennant Creek on Stuart Hwy.—20.43 134.23; AM R60214 64 km SW of Alice Springs on Stuart Hwy.—24.23 133.45; AM R60215 94 km E Broken Hill on Barrier Hwy.—31.75 145.53; AM R60216 41.6 km E Wilcannia on Barrier Hwy.—31.65 143.7; AM R60217 86.4 km E of Wilcannia on Barrier Hwy.—31.23 144.07; AM R62437 Horn Is. Torres Strait—10.62 142.28; AM R64787 Port Keats—14.23 129.52; AM R64788 Ootha, 33.12 147.45; AM R64803 Townsville Common—19.27 146.82; AM R67345, R67346, R67347, R67348 Kinchela—30.42 146; AM R68812, R86869 Townsville District—19.27 146.82; AM R68814 Baradine—30.95 149.07; AM R68815 Wondabyne—33.5 151.27; AM R68816 Alice Springs Area—23.7 133.87; AM R68817 Tamworth District—31.08 150.93; AM R71620 7 mi E of Yetman—28.75 150.88; AM R73744 5 km NW Griffith—34.27 146; AM R73774 4 mi on Kingston Rd. from Yarrowyck—30.45 151.28; AM R73938 Reedy Creek George Gill Range—24.3 131.58; AM R75084 55 mi N of Julia Creek—20.08 141.2; AM R75332, R75569 Upper Reaches of the Mckinlay River—13.27 131.72; AM R75333 Upper Reaches of the Mckinlay River—13.25 131.7; AM R75571 Lower Reaches of Mckinlay River—12.95 131.65; AM R77360 Angurugu Mission, Groote Eylandt—13.97 136.45; AM R81384 Mallina—20.88 118.03; AM R81387 Wooramel Rd., Continue 1 km S—25.6 114.28; AM R82557 Batavia Outstation Landing, N of Weipa—12.18 141.88; AM R82558 Powerhouse at Lorim Point, Weipa—12.62 141.9; AM R86285 just S of Maroon Dam on the Burnett Creek Rd.—28.17 152.67; AM R86870 Baradine Area, on Coonamble Rd.—30.95 149.07; AM R87690 19 km ENE of Laverton on Whitecliffs Rd.—28.57 122.57; AM R87691 2.5 km SE of Condun Well—28.43 123.08; AM R87692 9 km E of Karragullen Darling Ranges—32.5 116; AM R90339, R90340 Terrara, Gilgandra —31.72 148.65; AM R90869 4.3 km N Devil's Marbles along Stuart Hwy.—20.53

134.27; AM R90871 S of Mt. Doreen—22.13 131.42; AM R90874 Vicinity of The Granites—20.58 130.68; AM R90877 12.6 km E Nyngan—31.58 147.08; AM R90941 23.0 km SE The Granites via Rabbit Flat-Yuendumu Rd.—20.73 130.5; AM R91900 9 mi N of Mataranka—14.82 133.08; AM R91901 40 mi W Quilpie—26.6 143.67; AM R91902 50 mi W of Frewena—19.33 134.68; AM R91903 20 mi W Frewena—19.33 135.15; AM R91937 Windorah—25.42 142.65; AM R91962 Karrumba; AM R92744 8 km N Sandringham Stn. HS, 60 km NW Bedourie—23.98 139.03; AM R93175 Darnick—32.85 143.62; AM R93193 7 mi E of Yetman—28.9 150.9; AM R94601 10.7 km SE of Mt. Hope on the Euabalong West Rd.—32.9 145.98; AM R94602 Yathong Nature Reserve, near Yathong Homestead—32.63 145.57; AM R94629 9.7 km NE of TO to Mt. Hope via Roto-Gigunnia Rd.—32.58 145.75; AM R94660 25.9 km NE TO to Mt. Hope via Roto-Gilgunnia Rd.—32.48 145.87; AM R94661 27.4 km NE of TO to Mt. Hope via Roto-Gilgunnia Rd.—32.48 145.88; AM R94672 33.3 km NE of Nymagee PO via road to Hermidale—31.88 146.48; AM R94673 8.0 km S of Gwabegar by road—30.67 148.98; AM R94674 ~9.6 km E of Pilliga by road—30.37 148.98; AM R96718 Flaggy Creek, N of Glenreagh—30.05 152.98; AM R97508 Georgetown Billabong, Magela Creek, Jabiru Project Area—12.68 132.93; AM R98616 Gulungul Creek—12.65 132.88; AM R100030 11.5 km N of Wilcannia on Wanaaring Rd.—31.45 143.45; AM R100345 85 km S of Menzies—30.38 121.3; AM R100495 ~79 km W of Wiluna (by road)—26.47 119.52; AM R100496 3.7 km S of the 26th parallel via the Great Northern Hwy.—26.03 118.68; AM R100497 2.1 km S of the TO to Neds Creek and Wiluna on the Great Northern Hwy.—25.4 119.3; AM R100498 18.9 km S of the middle branch of the Gascoyne River via the Great Northern Hwy.—25.37 119.3; AM R100499 28.4 km E of the Port Hedland Rd. via the Wittenoom-Newman Rd.—22.45 118.88; AM R100564 1.8 km SW of the "Broome 400 km" marker via the Northern Hwy.—19.95 120.38; AM R100877, R100878, R100879 Rubbish Tip at Sandfire Flat Roadhouse—19.73 121.22; AM R101414 Rubbish Tip at Kalgoorlie—30.75 121.45; AM R101415 2.3 km SW of The "Broome 340 km" marker via the Great Northern Hwy.; AM R101537 ~9.6 NE of the Yule Rd. via the West Coastal Hwy.—20.67 118.4; AM R101538 Ca 35.8 km N of the road to "Bullara" via the road to Exmouth—22.32 114.08; AM R102596 87.0 km S of Coolgardie via Great Eastern Hwy.—31.57 121.58; AM R104922 Black Point Swamp, Port Essington—11.15 132.17; AM R105416 8.8 km SW of Bumbaldry on Mid-Western Hwy.—33.93 148.37; AM R105418 22.5 km ENE of Cargabal. Grid reference 602818 Forbes 1:250000 Map—33.82 147.98; AM R105796 48.8 km N of Norseman Rail Crossing on Coolgardie Rd.—31.82 121.63; AM R110554 Palparara Stn., NW of Windorah (11 km SW of Homestead)—24.83 141.37; AM R112442 Jabiru—12.65 132.88; AM R112855 27.8 km S Gwydir Hwy. via Terrie Hie Hie-Gravesend Rd.—29.8 150.2; AM R113136 Simpson Desert 90 km W of Birdsville—25.88 138.43; AM R114127 30 km W of Birdsville—24.87 138.97; AM R114161 18.8 km E of Chibnalwood on on Turlee Rd.—33.92 143.13; AM R114181 29.3 km SW of Bellnar Hmstd. on Mildura Rd.—34.02 142.4; AM R114280 15.5 km N of Balranald on Ivanhoe Rd.—34.5 143.6; AM R114309, R114310 0.2 km N of Prungle on Arumpo Rd.—34.23 142.98; AM R114311 1.6 km E Top Hut on Zanci Rd.—33.68 142.95; AM R114472 1.6 km E Top Hut Hmstd. on Zanci Rd.—33.67 142.95; AM R114491 72 km NE of Ivanhoe on Cobar Rd.—32.47 144.78; AM R114551 8.1 km from Boronga on road to Mungo N.P.—34.12 142.2; AM R116991 Hillston to Griffith Rd.; AM R118102 "Jersey," 12 km NW Cobar—31.42 145.7; AM R118897 Scenic Hill, Griffith; AM R122939 5 km from Goolgowi on Hilston Rd.—33.97 145.7; AM R123634 57.9 km from Bourke on Mitchell Hwy.—29.68 145.87; AM R123635 45 km from Tilpa on Wanaaring Rd.—30.65 144.22; AM R123636 44.9 km along Tilpa Rd. from TO on Wilcannia—Cobar Rd.—31.27 144.73; AM R123637 34.1 km N of Walgett on Castlereagh Hwy.—29.73 148.12; AM R123696 14.3 km from Wanaaring on Bourke Rd.—29.72 144.27; AM R125074 Narrogin—32.93 117.18; AM R125084, R125085, R125086, R125087, R125088 Mt. Liebig—23.3 131.37; AM R130597 25 km E of Wanaaring on Bourke Rd.—29.72 145.38; AM R130732 3.5 km W Mt. Dangar Rd. at Baerami via Bylong—32.38 150.43; AM R130815 67.9 km W Wanaaring on Tibooburra Rd.—29.7 143.42;

384 • Australian Varanid Species

AM R131190 Port Denison—20.05 148.25; AM R132984 ~10 km N of Hermidale—31.45 146.75; AM R134035 Rubbish Tip at Boulder—30.78 121.45; AM R134472 3.2 km W of Kununoppin by road—31.12 117.88; AM R138112 Ellerslie Stn. (from within a 12- km radius)—29.13 146.4; AM R140584 20 km SW Millstream Hmstd. TO on Pannawonnica Rd.—21.68 116.93; AM R142826 Thurloo Downs—29.28 143.48; AM R142834 12 km S of Alice Springs on Old South Rd.—23.83 133.87; AM R142852 88 km N of Lake Nash TO, on Main North Rd.; AM R142858 Macquarie Marshes, near Glencoe TO—30.88 147.62; AM R142893 5 km N of "Kulgera"—25.95 133.3; AM R143562 Putty Rd., 8.9 km S Garland Valley; AM R143848 46 km N Wyandra on Mitchell Hwy.—26.85 146.07; AM R143922 Bundjalung National Park (North), Bombing Range Rd.—29.13 153.42; AM R146174 Mullumgum, Bruxner Hwy.—28.85 152.8; AM R146185 Evans Head, Broadwater Rd.—29.12 153.43; AM R146256 Mootwingee National Park—31.28 142.3; AM R146639 Newfoundland Forest Way (Burns Forest Rd.) 100 M SW of Barcoongere Way-29.93 153.17; AM R147202 Newfoundland State Forest: Browns Knob Forest Rd.—29.9 153.12; AM R147203 Barcoongere Way, Barcoongere State Forest—29.93 153.2; ANWC R0259 Bourke—30.08 145.93; ANWC R0298 Urana—35.33 146.27; ANWC R0544 Bodallin—31.37 118.85; ANWC R1139 165 mi E of Threeways on Barkly Hwy.—19.92 136.37; ANWC R1278 Mt. Magnet—28.07 117.85; ANWC R1444 10 mi S of Alice Springs—23.87 133.8; ANWC R1777 Albion Downs, S of Wiluna—27.28 120.38; ANWC R1794, R1804 Lake Way, Millbillillie Stn., SE of Wiluna—26.7 120.35; ANWC R2715, R2716 Western Side of Lake Urana—35.33 146.08; ANWC R3018, R3020 Moolawatana Stn.—29.9 139.73; ANWC R3021 Yathong—32.47 145.57; ANWC R3070 Black Range, 10 km SW of Stawell—37.08 142.63; ANWC R3118 Yunta—32.58 139.57; ANWC R3122 Moolawatana Stn.—29.9 139.73; ANWC R3412 5 km S of Evans Head-29.17 153.42; ANWC R3435 Moolawatana Stn.?; ANWC R4623 Ouyen Water Reserve; ANWC R4672, R4745 Walpeup; ANWC R5243 Eastern Mcillwraith Range Lowlands, Cape York Peninsula—13.78 143.5; ANWC R5286 Griffith? District—34.28 146.03; ANWC R5287 Griffith—34.28 146.03; ANWC R5316 Coutt's Crossing, N of Coffs Harbour—29.84 152.9; ANWC R5881 Bundjalung National Park, Evans Head—29.17 153.42; NMV D38699 Ardlethan—34.37 146.9; NMV D15597 Chalka Creek—34.72 142.38; NMV D10154 Balladonia—32.47 123.87; NMV R912 Perth—31.95 115.85; NMV D14219 Kangaroo Is., 3.2 km N of Hawkes Nest—35.83 138.05; NMV D16436 46.7 km NE of Wyalong—33.63 147.6; NMV D41573 22.5 km NNE of Billeroo Waterhole—31.07 140.47; NMV D16435 16 km N of Peak Hill—32.58 148.18; NMV D16433 Renmark—34.17 140.75; NMV D5893 Adelaide—34.93 138.6; NMV R4856 Purnong—34.85 139.63; NMV D67720 Larapinta Drive, W of Alice Springs—23.77 133.6; NMV D67800 Gosse Bluff—23.82 132.31; NMV D1096, D1097, D1098 Lake Eyre, Killalpaninna—28.57 132.53; NMV D40136 12.8 km S of Culcairn—35.77 147.03; NMV D67961 22 km W, 6.1 km S of Erldunda Roadhouse—25.27 133; NMV D65855 10 km NE of Buningonia Springs—31.38 123.58; NMV D67907 Watarrka, Dune Site—24.31 131.54; NMV D68670 White Hills, Bendigo—36.73 144.27; NMV D57478 Murray Valley Hwy., 0.1 km W of Entrance to Hattah Kulkyne N.P.—34.75 142.33; NMV D4984 Tennant Creek—19.65 134.18; NMV D51962 1.6 km E of Poochera—32.72 134.85; NMV D68133 Lake Boga—35.45 143.63; NMV T1277, T1298 Roper River—14.68 134.73; NMV D3070, D3071, R4553 Purnong Landing—34.85 139.63; NMV D358 Ooldea—30.45 131.83; NMV T1282, T1256, T1257, T1258, T1259, T1260, T1261, T1279, T1285, T1290 Katji Lagoon—12.35 134.78; NMV T1292 Roper River—14.68 134.73; NMV T1296 Groote Eylandt—13.82 136.63; NMV T1274 Glyde River—12.37 135.02; NMV T1300 Cape Stewart—11.95 134.75; NMV T1283 between Derby Creek and The Glyde River—12.35 134.78; NMV T1081 Upper Glyde River—12.37 135.02; NMV T1278 Groote Eylandt—13.82 136.63; NMV T1275, T1276, T1280, T1281, T1286, T1288 Roper River—14.68 134.73; NMV T1087, T1088 Labbi Labbi—21.52 128.63; NMV T1284, T1291, T1299 Milingimbi, Crocodile Is.—12.12 134.92; NMV T1086 12.8 km W of Singleton—20.72 134.13; NMV T1085 Cape Stewart—11.95 134.75; NMV T1262 Yonko River, 40 km N of The Kendall River—14 141.77; NMV T1301 Sefton Creek, near

Lockhart River—12.97 143.52; NMV T1289 Lower Archer River—13.5 142; NMV D59643 16.4 km NE of Patchewollock—35.27 142.32; NMV D651 Kewell—36.5 142.42; NMV D5649 Karawinna—34.38 141.7; NMV D57485 15.5 km WSW of Hattah—34.88 142.22; NMV R4853, R4854 Ouyen—35.07 142.32; NMV D59660 15.4 km NE of Patchewollock—35.3 142.32; NMV D57498 7.2 km ESE of Meringur—34.42 141.4; NMV D59657 16 km SSW of Hattah—34.83 142.13; NMV D16288 1.6 km S of Nandaly—35.32 142.7; NMV D9815 Karawinna—34.38 141.7; NMV D59659 12.6 km SE of Walpeup—34.18 142.15; NMV D16434 8 km N of Hattah—34.68 142.25; NMV D59642 14.9 km NW of Lascelles—35.52 142.46; NMV D58417 30 km WNW of Kiamil—34.87 141.98; NMV D47389 30 km S of Millewa South Bore—34.03 141.07; NMV D53804 6.6 km ENE of Chinaman Well—35.86 141.73; NMV D58488 4 km ENE of Chinaman Well—35.86 141.7; NMV D59658 23.5 km WSW of Merringur—34.45 141.08; NMV D56418 Pyramid Hill—36.05 144.13; NMV D66491 4 km SW of Tarnagulla—36.78 143.8; NMV D16437 Mildura Airport—34.18 142.17; NMV D51790 8 km SSW of Kiata—36.42 141.75; NMV D52182 7 km SSW of Lillimur S—36.48 141.13; NMV D62021 0.7 km E of Llanelly—36.73 143.85; NTM R25792 4 mi N Alice Springs—23.63 133.87; NTM R25793 El Sharana—13.52 132.52; NTM R25794 Anzac Hill, Alice Springs—23.7 133.87; NTM R25795 Yanumbabore—25.95 131.78; NTM R25796 Anzac Hill, Alice Springs—23.7 133.87; NTM R25797 Katherine Farms Rd. Katherine—14.47 132.27; NTM R25798 2.25 mi E Rabbit Flat—20.28 130.03; NTM R25799 8 mi N Katherine—14.4 132.22; NTM R25800 Angas Downs—25.05 132.28; NTM R25801 2.75 mi E Yuendum—22.3 131.83; NTM R25802 Brunette Downs Racecourse—18.65 135.95; NTM R25803 13 mi S Darwin—12.57 131.05; NTM R25804 12 km N Pine Creek—13.75 131.83; NTM R25805 Alice Springs—23.7 133.87; NTM R25806 4 mi N Alice Springs—23.6 133.87; NTM R25807 Deep Well Alice Springs—23.53 134.02; NTM R25808 63.7 km W Yuendumu TO—23.42 133.18; NTM R25809 Alice Springs—23.7 133.87; NTM R25810 137.9 km W Yuendumu—21.45 130.92; NTM R25811 59 km W Ayres Rock—25.2 130.5; NTM R25812 Maryvale—24.67 134.07; NTM R25814 Jensen Bay Marchinbar Is.—11.17 136.68; NTM R25815 Berrimah, Darwin N.T.—12.43 130.83; NTM R25816 Tanami—20.01 131; NTM R25817 10 km S Alice Springs—23.78 133.92; NTM R25818 Marchinbar Is.—11.17 139.7; NTM R25822, R25825 Tanami—21.37 130.9; NTM R25823 Barrow Creek—21.53 133.88; NTM R25824 Sangsters Bore—12 km SW—20.87 130.27; NTM R25827 Frewena—19.92 135.42; NTM R25828 65.2 km W Ayres Rock—25.15 130.67; NTM R25829, R25830 Curtin Springs Stn.—25.32 131.75; NTM R25831 218.2 km W Stuart Hwy. on Yuendumu Rd.—22.45 132.07; NTM R25832 8 mi N Humpert Crossing, Victoria River Downs—16.42 130.55; NTM R25833 Darwin River Dam—12.83 130.97; NTM R25834 5.6 km E S Alligator River Arnhem Hwy.—12.63 132.55; NTM R25835 31.5 km W E Alligator TO Arnhem Hwy.—12.58 132.88; NTM R25836 11.5 km N Noonamah—12.58 131.15; NTM R25837 13.7 km N Noonamah—12.53 131.05; NTM R25838 49 km SW Borroloola—16.3 136.08; NTM R25839 Borroloola Area—16.2 136; NTM R25840 Barrow Creek—21.53 133.88; NTM R25841 6 km E Kurandi Stn. Homestead—20.47 134.68; NTM R25842 Connell's Bore, 30 km NW Alexandria Stn.—18.88 136.55; NTM R25843 Fannie Bay, Darwin—12.43 130.83; NTM R25844 75 km W Frewena—19.48 134.72; NTM R25845 Yadikba—14.08 136.45; NTM R25846 15 m S Wauchope—20.9 134.23; NTM R25847 82 km W of Frewena—19.37 135.72; NTM R25848 Frewena, 40 km E—19.62 135.75; NTM R25849 Avon Downs Police Stn.—20.03 137.83; NTM R25850 Mundabollangana Stn., Cape Thouin WA—20.52 118.07; NTM R25851 Tanami Sanctuary—20.87 130.58; NTM R25852 Jabiru Area—12.67 132.9; NTM R25853 Arnhem Hwy., Gulungul Creek Crossing—12.63 132.83; NTM R25854, R25855, R25856 Jabiru E—12.67 132.9; NTM R25857 Mt. Barkly—21.8 132.25; NTM R25858 Sangsters Bore, 17 km W—20.78 130.23; NTM R25859 Tanami Sanctuary—19.01 130.03; NTM R25860 Dalmore Downs, 5 km S No. 5 Bore—19.01 135.03; NTM R25861 Tanami Sanctuary—20.01 131.68; NTM R25862 Simpson Desert, Illogwa Creek—24.13 135.97; NTM R25863 Tanami Sanctuary—20.87 130.58; NTM R25864 Tanami—20.87 130.58; NTM R25865 Todd River Stn.—24.15 134.7; NTM R25866 Tempe Downs Stn.—24.28 131.58; NTM R25867 Nymagee Area—31.98

146.27; NTM R25868 Darwin, 16 mi S—12.53 131.05; NTM R25869 Undoolya Stn.—24.08 133.03; NTM R25870 Tanami—20.01 130.92; NTM R25871 Tanami —20.01 130.95; NTM R25872 Tanami—20.01 131.68; NTM R25873 Keep River —15.78 129; NTM R25874 Tanami—20.53 131.03; NTM R25875 Alice Springs— 23.77 133.88; NTM R25876 Livingstone Pass—24.9 129.1; NTM R25877 Todd River Stn.—24.03 134.88; NTM R25878 Monkira HS, 8 mi SE—24.92 140.07; NTM R25879 Stuart Hwy.—20.08 134.22; NTM R25880 Stirling Creek—21.75 133.68; NTM R25881 Tanami—21.77 131.92; NTM R25882 Tanami—20.07 130.45; NTM R25883 Tanami, Mt. Ptilotus—20.27 130.23; NTM R25884 Todd River Stn.—23.97 134.7; NTM R25885 Todd River Stn.—23.92 134.68; NTM R25886 Burt Plain—23.27 133.77; NTM R25887 Todd River Stn.—24.3 134.88; NTM R25888 Yoolperlunna Creek; NTM R25889 Brunchilly—18.87 134.05; NTM R25890 Corandirk Stn.—21.55 130.03; NTM R25891 Tanami—20.55 130.37; NTM R25892 Tanami—19.01 130.03; NTM R25893 Tanami—20.25 130.01; NTM R25894 Petermann Ranges—24.8 129.05; NTM R25895 Rapid Creek, Darwin—12.57 131.05; NTM R25896 Stuart Hwy.—13.48 131.18; NTM R25897 Woodgreen Stn.—22.4 134.23; NTM R25898 Keep River National Park— 15.75 129.08; NTM R25899 Hay River—23.88 137.27; NTM R25900 5 km S Milikapiti—11.46 130.67; NTM R25901 12 Mile Stock Yards—14.95 133.22; QM R432 Noosa Heads—26.38 153.1; QM R653 Brisbane—27.47 153.02; QM R2112 Burpengary—27.17 152.95; QM R2146 Eidsvold—25.37 151.12; QM R2168 Kimberley, NW Aust—32.82 141.1; QM R2406 Darling Downs—27.25 151.5; QM R2711 Bundaberg—24.87 152.35; QM R2830 Eudlo—26.73 152.95; QM R3238 Brisbane, Morningside—27.47 153.05; QM R4829 Cooktown—15.47 145.25; QM R6990 Brisbane, Hemmant—27.45 153.13; QM R7071 Brisbane, St. Lucia—27.5 153.02; QM R7301, R7302 Brisbane—27.47 153.02; QM R7350 Brisbane, Mitchelton—27.42 152.97; QM R7846, R7847 Brisbane, Nudgee— 27.37 153.08; QM R8374 Camp Mt—27.4 152.87; QM R10225 Birdsville—25.9 139.35; QM R10791 Doomadgee Mission Stn.—17.93 138.82; QM R14218 Grey-mare—28.17 151.77; QM R14362 Cunnamulla, Gilruth Plains, National Field Stn.—28.07 145.68; QM R14397 Wacol, near Wacol Migrant Hostel—27.58 152.92; QM R15362 Brisbane—27.47 153.02; QM R15696 Highland Plains Stn., 48 km E Clermont—22.83 148.07; QM R16135 Rockwood Stn., Chinchilla— 26.93 150.37; QM R18009 Roma, 67.2 km W—26.5 148.12; QM R18011 Charle-ville, 112 km E—26.45 147.47; QM R18361 Kunwarara–Pt. Arthur Rd., 70.4 km from Kunwarara—22.27 150.03; QM R20467 Marburg, 13 km W—27.57 152.48; QM R20520, R20521 Baralaba—24.18 149.82; QM R20600 McIvor Rd. mouth, 1.6 km N—15.1 145.22; QM R20824 Lizard Is.—14.67 145.47; QM R21301 Brisbane, Stafford—27.42 153.02; QM R21657 Urandangie, 48 km SW—21.92 137.97; QM R21672 Tailem Bend, 32 km E—35.32 139.77; QM R24017 Emerald, 16–24 km W—23.53 147.97; QM R24845 Wakes Lgn, 89 km NW Adavale—25.58 144.83; QM R29843 Emerald, Outskirts of Town—23.53 148.17; QM R29844, R29845, R29847, R29848 Dynevor Downs—28.1 144.35; QM R29846 Bulloo Downs—28.53 142.95; QM R30305 Birdsville Track, Etadunna Airstrip—28.72 138.63; QM R31913 Bollon, 80 km W—28 146.73; QM R33146 L Naryilco, SW of Naryilico Stn.—28.72 141.43; QM R33147 Wyandra, 24 km N, Charleville to Cunnamulla, on rd—27.02 145.98; QM R33149, R33151 L Naryilco—28.72 141.43; QM R33150 Naryilco, Omicran Rd., 10–15 km NW L Naryilco—28.72 141.42; QM R34782 Moombah Stn., ~64 km W Westmar—27.98 149.3; QM R38470 Injune to Taroom Rd., Hornet Bank TO—25.75 149.4; QM R38981 Mt. Isa, 24.2 km W, on Barkly Hwy.—20.53 139.45; QM R40759 Charleville, 45 km W—26.4 145.8; QM R40762 Cunnamulla, 100 km N—27.18 145.68; QM R40763 Brisbane Valley?—27.4 152.48; QM R41493 Glenmorgan—27.25 149.68; QM R41579 Benditoota Waterhole, Durrie Stn.—25.62 139.8; QM R41857 Coora-bulka Stn., Terriboah Waterhole—24.07 140.07; QM R43380 Townsville area?— 19.27 146.82; QM R46768 Roma, 20 km E—26.58 148.78; QM R46769 Gayndah/ Hivesville Rd.—25.62 151.62; QM R47306 Highvale, via Samford—27.38 152.8; QM R47351 Two Mile Creek, Pallaeida Rd.; QM R48501 Bollon, 43 km E—27.98 147.03; QM R48502 Karmona Stn., 40 km W Jackson—27.38 141.95; QM R49849 Townsville Common—19.27 146.82; QM R49945 Mt. Crosby, via Bris-bane—27.53 152.8; QM R51643 Stuart Hwy., 25 km N of TO to Kings Canyon;

QM R51687 Townsville, Rowes Bay, Pallarenda Rd.—19.2 146.77; QM R51754 Moreton Is., Ocean Beach—27.32 153.43; SAM R00134, R00135, R00137 Eidsvold—25.37 151.12; SAM R00550 Purnong River Murray—34.85 139.63; SAM R00611 Murray Bridge—35.12 139.27; SAM R00750 Lake Crossing—29.55 139.88; SAM R00751, R00752, R00753 Kanowana—27.85 139.63; SAM R01031 Ooldea—30.45 131.83; SAM R01678 Cockatoo Creek—22.05 132.03; SAM R01678 Cockatoo Creek—22.05 132.03; SAM R01695 Koolunga—33.58 138.33; SAM R01936 Flinders Is.—14.18 144.25; SAM R01946 Flinders Is.—14.18 144.25; SAM R01947 Flinders Is.—14.18 144.25; SAM R01948 Flinders Is.— 14.18 144.25; SAM R01949 Flinders Is.—14.18 144.25; SAM R02515 Dareton— 34.08 142.07; SAM R03087 Snowtown—33.78 138.22; SAM R03176 Mern Merna —31.65 138.35; SAM R03297 Muloorina HS—29.23 137.9; SAM R03299 5 mi NE of Muloorina HS—29.18 137.95; SAM R03505 Yirrkala Arnhem Land—12.25 136.88; SAM R03801 Muloorina HS—29.23 137.9; SAM R03898 100 mi W of Oodnadatta—27.28 133.7; SAM R04324 Dunmarra—16.68 133.42; SAM R04325 Donalds Lagoon—12.73 131.23; SAM R04964, R05123 Mornington Is.—16.6 139.35; SAM R05335, R05336 Mt. Davies—26.22 129.27; SAM R05338, R05361 Angepena HS—30.57 138.85; SAM R06085 Mt. Burrell—24.62 133.97; SAM R08115 Lake Nash Stn.—20.97 137.92; SAM R08490 Waikerie—34.18 139.98; SAM R11764 4 mi SW Maynards Bore Everard Rgs—27.33 132.33; SAM R12052 81 mi E Wilkannia—31.03 144.38; SAM R12053 16 mi E Terowie—33.2 139.17; SAM R13011 Ardrossan Yorke Penin.—34.43 137.92; SAM R13364 Etadunna HS—28.72 138.63; SAM R13369 Kulgera—25.83 133.3; SAM R13818, R13819, R13820, R13821, R13822 Muloorina Stn. (Lake Eyre S Shore)—29.17 137.72; SAM R13823 20 mi W of Chilla Well—21.53 130.68; SAM R13850 Muloorina Stn. (S edge of Lake Eyre)—29.07 137.9; SAM R14395 Maralinga "Bomb Site"—30.17 131.68; SAM R14473 William Creek—28.92 136.33; SAM R13785 Edithburgh— 35.08 137.75; SAM R13824 Copley—30.55 138.42; SAM R13826 Mt. Mary— 34.12 139.47; SAM R13827 Renmark—34.17 140.75; SAM R13828 Batavia Downs—12.67 142.67; SAM R13848 Witchitie—32.15 138.92; SAM R13849 25 mi S of Narromine—32.6 148.22; SAM R13851 Chilla Wells—21.52 130.98; SAM R13854 10 mi S of Hayes Creek—13.63 131.57; SAM R13896 4 mi N Oraparinna—31.32 138.7; SAM R13988 Balcanoona Stn.—30.53 139.3; SAM R15000 11 km SE of Ooldea—30.57 131.98; SAM R15030 near Mt. Gunson TO— 31.55 137.12; SAM R15335 Whyalla (Outskirts)—33.03 137.58; SAM R15575 Carnes Bore—30.22 134.5; SAM R16332 halfway between Wangary and Coulta— 34.47 135.47; SAM R16334 15 km W of Waikerie—34.28 139.77; SAM R16336 Mt. Elizabeth Stn.—16.3 126.18; SAM R17858 between Hesso and Port Augusta— 32.13 137.45; SAM R17892 5 mi NE of Blanchetown—34.32 139.65; SAM R17893 Western Approach to Kingston Bridge River Murray—34.23 140.35; SAM R17894 6.6 mi W of Waikerie—34.25 139.85; SAM R18042 Honey Moon Creek Uranium Site Koongarra Stn.—31.75 140.67; SAM R18114 7 km N of Bute on Mundoora Rd.—33.8 138.03; SAM R19073 Moolawatana Stn. NE Flinders Rgs— 29.9 139.73; SAM R20104 1 mi S of Brinkley Hall—35.25 139.23; SAM R21353, R21354 Olympic Dam Area Roxby Downs—30.43 136.85; SAM R21462 15 km S Whyalla—33.08 137.5; SAM R21945 St. Peter Is.—32.28 133.58; SAM R22501 Moolawatana HS—29.9 139.73; SAM R23798 12 km N Port Augusta/Woomera TO—32.4 137.65; SAM R24119 between Wudinna and Warramboo—33.25 135.5; SAM R25333 near Morgan—34.08 139.65; SAM R25567 Muckera Campsite No1—30.07 130.07; SAM R26208 30 km SW Mabel Creek HS—29.16 134.11; SAM R26558 SW of Granite Downs HS—27.42 132.88; SAM R26863 12 km N Lake Phillipson—29.37 134.47; SAM R26865 7 km S Mabel Creek Camp—29.22 134.25; SAM R26871 Danggali Cons Pk—33.5 140.7; SAM R27030, R27031, R27032 Mt. Liebig—23.3 131.37; SAM R27309 151 km N of Kingoonya—29.9 135.12; SAM R27782 10 km NW Renmark—34.13 140.68; SAM R28522 125.8 km ENE Minnipa—32.47 136.43; SAM R28545 115 km NE Minnipa—32.27 136.18; SAM R28598 130.8 km ENE Minnipa—32.35 136.45; SAM R28642 92.8 km NE Minnipa—32.38 135.98; SAM R29247 18 km S Kalgoorlie—30.85 121.3; SAM R29585 Monarto Zoo Area—35.05 139.13; SAM R30253 Olympic Dam— 30.45 136.88; SAM R30380, R30381, R30382. R30383 Wardang Is.—34.5 137.37; SAM R31821 Mintabie Opalfields—27.32 133.3; SAM R31850 17 km ESE Mt.

Christie Siding—30.63 133.28; SAM R31856 15 km SSE Mt. Christie Siding—
30.63 133.26; SAM R31880 6.3 km NE Mt. Finke—30.91 134.08; SAM R31891
9.2 km NE Mt. Finke—30.89 134.1; SAM R31892 11 km NE Mt. Finke—30.87
134.11; SAM R31912, R31913 12 km W Pinjarra Dam—32.13 134.66; SAM
R31982 1 km W Yumbarra Rock Hole—31.77 133.47; SAM R31991 7 km S
Mitcherie Rockhole—31.5 132.85; SAM R32075 5.5 km S Immarna Siding—30.55
132.15; SAM R32190 47 km N Muckera RH—29.6 130.14; SAM R32492 Narlaby
Stn.—32.28 135.03; SAM R32625 Ca 20 KM E Carnadinna Native Well-29.25
132.75; SAM R32626 Commonwealth Hill Stn.—27.48 133.14; SAM R32627 4
km SE Cleanskin Swamp Bore Mabel Creek Stn.—29.17 134.12; SAM R37535 14
km E Morgan on H/Way 64—34.02 139.8; SAM R37835 3 km W of Zoo Agistment
Area—35.15 139.2; SAM R38380 E Margin Pooginook Cons Pk—34.1 140.13;
SAM R38430 5 km NE of Pilcherra Bore—35.02 140.19; SAM R39222 10 km N
of Iron Knob—32.68 137.23; SAM R39284 1.5 km N of Veitch—34.65 140.51;
SAM R39287 10 km ENE of Swan Reach—34.54 139.7; SAM R39298 3 km W of
Jabuk—35.37 140.04; SAM R39314 1 km N of Cambrai—34.64 139.28; SAM
R39417 5 km NW of Tauragat Hill—35.55 139.95; SAM R39721 4 km E of
Geranium PO—35.38 140.2; SAM R39738 Yookamurra HS—34.0 139.47; SAM
R39743 1 km NE of Blanchetown—34.33 139.63; SAM R39749 6 km N of
Wildhorse Plains—34.33 138.28; SAM R40822 100 km N of Port Augusta—31.73
137.25; SAM R40955 near Mannum—34.92 139.3; SAM R41414 5.3 km NW
of Round Dam—33.53 140.65; SAM R41674 York Dam—32.72 140.74; SAM
R41760 2.5 km (Air) NE of Tobacco Bush Dam—32.86 140.09; SAM R41802
Horrocks Pass—32.63 138.03; SAM R41964 Yellabina RR—30.85 132.07; SAM
R42495 Sam Campsite—29.02 133.27; SAM R42515 2 km SE of Tallaringa Well—
29.05 133.32; SAM R42661 Bedourie Dump—24.33 139.45; SAM R43385 10 km
S of William Creek—28.97 136.4; SAM R43914 1.5 km E of Clifton Hills Outstn—
26.54 139.46; SAM R44831 15 km WSW of Womikata Bore—26.15 132.02; SAM
R44906 6.3 km SW of Pile Hill—29.14 138.4; SAM R45204 5 km W of Loxton on
Waikerie Rd.—34.46 140.51; SAM R45389 10.2 km E of Ampeinna Hills—27.08
131.23; SAM R46567 4 km SSE of Patsy Dam—28.64 135.94; SAM R46758 near
Calperum HS—34.03 140.72; SAM R46946 1.1 km SSW of Top Camp Well—
26.47 134.95; SAM R48621 7.7 km E of Mt. Cheesman—27.41 130.41; SAM
R48763 3.1 km WNW of Mt. Lindsay—27.02 129.85; SAM R48803 13.2 km NW
of Mt. Cheesman—27.34 130.23; SAM R50146 10.5 km S of Sentinel Hill—26.18
132.46; SAM R14633 20 km E of Ammaroodinna Hill Granite Downs Stn.—27.6
133.6; SAM R15001 7 km W Immarna RS—30.48 132.08; SAM R15526 14 km
NW of Emu—28.57 132.07; SAM R15540 Muloorina Stn.—29.23 137.9; SAM
R15573 94 mi W of Vokes Hill Cnr—28.52 129.13; SAM R15620 Mungeranie HS
Birdsville Track—28.02 138.67; SAM R21009 Olympic Dam Roxby Downs—
30.45 136.88; SAM R24554 Etadunna Stn. 7.2 mi N of HS on Cooper Creek
Track—28.63 138.7; SAM R24716 2 km N Toolache W/H Strzelecki Creek—28.35
140.42; SAM R25948 36 km S Moomba on Strzelecki Trk—28.38 140.43; SAM
R25949 53 km S Kudriemitchie Outstation on Innamincka Rd.—27.7 140.52;
SAM R29916 24 km along Mulga Pk Rd. SSE Curtin Springs HS—25.52 131.82;
SAM R29930 9 km along Mulga Pk Rd. SSE Curtin Springs HS—25.4 131.77; SAM
R30969 Coongie—27.18 140.15; SAM R30988 10 km SSE "Coongie"—27.27
140.17; SAM R31759 Fregon—26.77 132.03; SAM R31789 64.3 km E
Pipilyatjara—26.13 129.77; SAM R37145 Curtin Springs Stn.—25.32 131.75;
SAM R38113 Purni Bore—26.28 136.1; SAM R44212 Ruin of Charlotte Waters
Telegraph Stn.—25.92 134.9; SAM R44236 2 km N of Moonabie Gap—33.27
137.23; SAM R22941 Balcatta—31.88 115.82; WAM R1869 Woodlands. Tam-
bellup—34.05 117.65; WAM R2851 Marvel Loch—31.47 119.48; WAM R3770
Seabrook. Northam—31.67 116.73; WAM R3996, R3997 Wells 49 and 50, Canning Stock Route—20.2 126.83; WAM R3997 Wells 49 and 50, Canning Stock
Route—20.2 126.83; WAM R3999 Well 49 or 50. C.S.R.—20.17 124.68; WAM
R4949 Yalgoo—28.35 116.68; WAM R5519 North Perth—31.93 115.95; WAM
R5815 Mosman Park—32.02 115.77; WAM R7191, R7192 Kellerberrin—31.63
117.72; WAM R7975 Rivervale—31.97 115.9; WAM R8930 Irwin River—29.2
115.43; WAM R9732 Swanbourne—31.97 115.77; WAM R10017 Clackline—
31.72 116.52; WAM R10239 Kings Park—31.95 115.83; WAM R10437 Mt.

Lawley—15.73 127.42; WAM R11249 Kimberley Research Stn., N of Kununurra—15.65 128.7; WAM R11425 Balladonia—32.45 123.87; WAM R11436 Yellanup. Via Narrikup—34.73 117.92; WAM R12406 Kathleen Valley. Near Leonora—27.4 120.65; WAM R12636 Koordarrie Stn.—22.3 115.03; WAM R12686 Jurien Bay—30.25 115.02; WAM R12701 Nollamara—31.88 115.85; WAM R12722 Woolgangie—31.17 120.55; WAM R12997 Cundeelee Mission—30.73 123.42; WAM R13078, R13079, R13081, R13379 Woodstock Stn.—21.62 118.95; WAM R13260 Bernier Is.—24.93 113.13; WAM R13425 E Cannington—32.02 115.97; WAM R13437 10 km E of Pingelly—32.53 117.18; WAM R13449 Cliff Head. Dongara—29.53 114.98; WAM R13527 Bernier Is.—24.93 113.13; WAM R13792 Kulgera—25.83 133.3; WAM R13965 Mellenbye Stn. Ederga River—28.83 116.3; WAM R13975 New Forrest—27.37 115.65; WAM R14134 Cloverdale—31.95 115.93; WAM R14661 Warburton Range—26.13 126.58; WAM R14666 32 km ESE of Wittenoom—22.35 118.62; WAM R14882 Mouth of Turner River—20.35 118.48; WAM R14883 25 km E of Mundabullangana—20.52 118.3; WAM R14885 Yanrey—22.5 114.78; WAM R14886 New Mundiwindi—23.8 120.25; WAM R14887 28 km SW of Doolgunna [543mp Great Northern Hwy.]—25.85 119.03; WAM R14888 Mileura—26.5 117.42; WAM R14892 10 km N of Goomalling—31.22 116.82; WAM R14893 Kings Park. Perth. on Park Rd.—31.95 115.85; WAM R14894 Kings Park (Perth)—31.97 115.83; WAM R14895 Samson Dam. Waroona—32.88 116.03; WAM R14899 45 km S of Norseman—32.6 121.77; WAM R14900 Seemore Downs—30.73 125.3; WAM R14907 Lesmurdie—32 116.03; WAM R14908, R14909, R14910–R14912. R14913 Lesmurdie—32.02 116.05; WAM R15782 Mileura. Bulleen Mill—26.37 117.33; WAM R15783, R15784 Nookawarra—26.3 116.87; WAM R16894 Woodanilling, between Katanning and Wagin. ~258 km from Perth—33.57 117.43; WAM R16919 18 mi NW Tenindewa—28.37 115.12; WAM R17079 20 km ENE of Mundabullangana HS—20.45 118.23; WAM R17650 13 km W of Moorine Rock—31.3 118.98; WAM R17695 Mt. Vernon HS—21.23 118.23; WAM R17854 Kalamunda—31.98 116.07; WAM R18480 Byford. 2 km E of Town—32.22 116; WAM R18582 5 km S of Pithara—30.45 116.67; WAM R18583 Lesmurdie—32.02 116.05; WAM R18602 18 km S of Nerren Nerren—27.32 114.63; WAM R18603 27 km S of Nerren Nerren—27.42 114.63; WAM R19147 Coolgardie—30.95 121.17; WAM R19230 Yarloop—32.97 115.9; WAM R19751 Yornaning—32.75 117.15; WAM R19864 Kalamunda—31.97 116.05; WAM R21206 Warburton Range Mission—26.13 126.58; WAM R21850 North Lake (N of Bibra Lake)—32.08 115.82; WAM R21889 Brookton—32.37 117; WAM R21942 Presumably Carnarvon—24.88 113.67; WAM R22023 Presumably Warburton Range—26.13 126.58; WAM R23834 Cloverdale. Perth—31.95 115.93; WAM R24475 32 km W of Victoria Downs—16.4 130.72; WAM R24735 Dianella. Perth—31.9 115.88; WAM R24744 Monte Bello Is.—20.43 115.55; WAM R24811 Presumably Wyndham—16.75 123.95; WAM R25148 3 km SE of Turee Creek—23.65 118.7; WAM R25824 Hyden—32.45 118.87; WAM R25984 Gnangara. Sand plain E of Plantation—31.8 115.92; WAM R26918 110 km N of Charles Knob. Gibson Desert—24.05 124.98; WAM R26939 7 km SE of Kidson Camp. Gibson Desert—22.58 125.05; WAM R26995 7 km S of Gary Junction—22.57 125.25; WAM R27051 Well 24. Canning Stock Route—23.12 123.33; WAM R28007 Parry Creek—15.67 128.25; WAM R28133 Imbin Rock-Hole, 23 km NE of Earaheedy—25.45 121.75; WAM R28281 Windjana Gorge—17.43 125; WAM R28329 Kings Park. Narrows Bridge Approaches. Perth—31.95 115.87; WAM R29149 129 km ESE of Wittenoom—22.63 119.4; WAM R29397 City Beach. Perth—31.93 115.77; WAM R30322 32 km NNW of Gingin—31.08 115.75; WAM R30506 12 km E of Jurien Bay—30.33 115.17; WAM R30689 61 km E of Southern Cross—31.23 119.95; WAM R30918 Newman—23.35 119.73; WAM R30970, R30971 Albion Downs, within 13 km of HS—27.28 120.38; WAM R31071 Kings Park—31.95 115.83; WAM R31146 northern part of Chichester Range—21.52 117.7; WAM R31148 64 km N of Norseman—31.62 121.78; WAM R31207 24 km N of Broome—17.73 122.23; WAM R31963 70 mi NNE of Rawlinna—30.0 125.75; WAM R33421 Bamboo Creek—20.87 120.17; WAM R33512 7 km E of Kalbarri. Meanarra Hill—27.7 114.22; WAM R33537 7 km E of Kalbarri. SE Shoulder of Meanarra Hill—27.7 114.22; WAM R33849 Hawks Head Lookout. Kalbarri National Park—27.8 114.48; WAM R34013 Tennant

Creek—19.65 134.18; WAM R34246 19 km E of Mt. Olga—25.3 130.93; WAM R34569 112 km N of Rawlinna—29.42 125.33; WAM R37639 Kalbarri Township (Red Bluff Rd.)—27.7 114.2; WAM R37725 Kalbarri. Red Bluff Rd.—27.75 114.2; WAM R37813 Split Rocks, 29 km N of Mt. Holland, 381 m high—31.95 119.62; WAM R39084 Keysbrook—32.43 115.98; WAM R39141 48 km W of Talawana. 48 km W of Well 26 C.S.R.—22.95 120.57; WAM R39142 Well 23. C.S.R.—23.08 123.22; WAM R39769 Callagiddy—25.05 114.03; WAM R40282 Coulomb Point —17.37 122.15; WAM R41191 Walyahoming Rock—30.63 118.75; WAM R41227 80 km NNE of Rawlinna—30.33 125.6; WAM R42297 19 km SE of Jiggalong. Nianihya Roadhouse—23.48 120.92; WAM R44138 Boongaree Is., Prince Frederick Harbour—15.08 125.2; WAM R45231 50 km N of Windy Corner—22.98 125.2; WAM R45232 22 mi N of Windy Corner—23.37 125.2; WAM R45696 Lochada Stn. Perenjori. 11 km SW of Old HS—29.2 116.55; WAM R45741 Nedlands. Perth—31.98 115.8; WAM R45777 10 km NE of Port Pirie. 81 km from Port Augusta—33.13 138.1; WAM R46317, R46318 Trayning Reserve—31.05 118; WAM R46517 42 km ENE of Mclarty Hills—19.32 123.95; WAM R46864 Prince Regent River Reserve—15.53 125.23; WAM R47800, 47801 Green Head— 30.07 114.97; WAM R48383 43 km N Beacon—30.07 117.83; WAM R48384 56 km N of Beacon—29.93 117.87; WAM R48730 Winduldarra Rock Hole—26.52 126.02; WAM R51616 Possibly Bernier Is. (Or Dorre Is.)—25 113.13; WAM R51924 Carnarvon Range—25.28 120.7; WAM R52125 2 km N of Yamarna HS— 27.97 123.77; WAM R52484 Badjaling Nature Reserve 11 km E of Quairading— 31.98 117.5; WAM R52632 8 km W of Kukerin—33.2 118; WAM R53584 10 km E of Yamarna—27.98 123.88; WAM R53587 19 km N of Warburton Range Mission—25.95 126.58; WAM R56149 Wallal. 69 km WSW of Sandfire Roadhouse—19.83 120.45; WAM R57049 48 km NW of Wilson Cliffs—21.75 126.75; WAM R58531 8 km NE of Lagrange TO—18.67 121.97; WAM R58541 Martins Well—16.57 122.95; WAM R58701 1.5 km S of Queen Victoria Spring—30.45 123.58; WAM R60137, R60138 Presumably Caron—29.58 116.32; WAM R60145 Kewson Mound Springs—near Coward Springs-Maree-Oodnatta Rd.—29.37 136.78; WAM R60260 Cranbrook—34.3 117.55; WAM R60281, R60282 Possibly Bernier Is.—24.93 113.13; WAM R60671 Mitchell Plateau—14.58 125.75; WAM R61006 13 km S Bullara HS TO on Main Exmouth Rd.—22.8 113.97; WAM R61494 Vlaming Head—21.8 114.1; WAM R62223, R62224 5 km NE of Jindabinbin Rock Hole—32.37 122.12; WAM R62225 4 km N of Jindabinbin Rock Hole—32.38 122.08; WAM R62230 Dryandra State Forest—32.32 116.9; WAM R63215 33 km 45° Bulga Mulgardy Soak—19.98 121.27; WAM R63233 30 km 56° La Grange—18.53 122.02; WAM R63378 2 km 21° Lens Bore—20.23 127.52; WAM R63635 9 km 026° Guli Tank Well No 42 Canning Stock Route—21.23 125.92; WAM R63958 12 km 21° Well No 29 Canning Stock Route—22.47 124.92; WAM R63961, R63962 25 km S Nooloo Soak—22.85 121.93; WAM R64089 0.5 km W Lens Bore—20.27 127.52; WAM R64112, R64127 Breaden Valley—20.28 126.53; WAM R64138 Breaden Pool—20.25 126.57; WAM R64191 Murguga Well No 39 Canning Stock Route—21.78 125.63; WAM R64192, R64193, R64224, R64225 Minjoo Well No 35 Canning Stock Route—22.22 125.05; WAM R64270 38 km 209° Mctavish Claypan—20.93 123.22; WAM R64295 38 km 210° Mctavish Claypan—20.93 123.25; WAM R65254 40 km W Lake Cronin—32.33 119.17; WAM R65434 30 km NW Heartbreak Ridge—31.8 122.32; WAM R65816 12 km NE Comet Vale—29.91 121.24; WAM R67382 11 km 134° Capel—33.62 115.62; WAM R68886 15 mile N of Beegull Waterholes— 27.58 124.2; WAM R68985 Wanneroo—31.75 115.8; WAM R69319 10.5 km SSE Banjiwarn HS—27.79 121.66; WAM R69320 12.5 km SE Banjiwarn HS—27.8 121.68; WAM R70026 12.5 km 309° New Lissadell HS—16.6 128.46; WAM R70345 5.3 km 167° Mt. Percy—17.67 124.95; WAM R71512 Gooseberry Hill— 31.93 116.05; WAM R71517 11 km NNW Minilya Bridge Roadhouse—23.7 114; WAM R71518 11.5 km W Mia Mia HS—23.38 114.32; WAM R71599 Mardathuna HS—24.48 114.55; WAM R71600 12 km NNE Minilya Bridge Roadhouse— 23.72 114.05; WAM R71745 4.5 km 070° Forrestonia Crossroads—32.4 119.8; WAM R71827 16.5 km 080° Toomey Hills—31.59 119.82; WAM R71878 12 km SE Woodstock HS—21.7 119.03; WAM R72063 18 km S Mt. Jackson—30.41 119.25; WAM R72608 6.75 km NE Comet Vale—29.9 121.17; WAM R72609 12.0

km ENE Comet Vale—29.92 121.25; WAM R72651 6.75 km NE Comet Vale—29.9 121.17; WAM R72758 9.75 km NE Comet Vale—29.89 121.2; WAM R73009 16 km SSE Newman—23.45 119.77; WAM R73084 18 km N Barradale Roadhouse—22.7 114.97; WAM R73159 23.2 km 231° Marillana HS—22.76 119.23; WAM R73288 12.5 km 248° Black Flag—30.6 121.12; WAM R74219 1.1 km WSW Lake Cronin—32.39 119.75; WAM R74455 24 km S Boorabbin—31.27 120.33; WAM R74714 7.5 km ENE Yuinmery—28.54 119.09; WAM R74826 Shay Gap Western Outskirts of Town—20.5 120.15; WAM R75524 12 km NW of New Lissadell Homestead—16.62 128.47; WAM R75809 about 30 km WNW of Mclarty Hills—19.38 123.25; WAM R76206 14 km NE Bungalbin Hill—30.3 119.72; WAM R76432 10 km SSW Cooya Pooya HS—21.12 117.12; WAM R77367 Mitchell Plateau—14.87 125.8; WAM R77598 Mitchell Plateau—14.73 125.73; WAM R77630 Camp Creek—14.89 125.75; WAM R78148 Kookynie—29.35 121.5; WAM R78489 6 km N of Yowie Rockhole—30.42 122.35; WAM R78516 7 km WSW of Black Flag—30.58 121.17; WAM R78634 22 km S of Mt. Elvire Homestead—29.55 119.6; WAM R78970 Capricorn Range—23.35 116.8; WAM R78996 15 km E of Toomey Hills—31.6 119.82; WAM R80427 17 km SE of Onslow—21.77 115.2; WAM R80439 11 km SW of Cane River Homestead—22.15 115.53; WAM R80942 Murdoch—32.07 115.82; WAM R81959 Chiddarcooping Nature Reserve—30.9 118.67; WAM R82625 Eagle Hawk Is., Dampier Archipelago—20.65 116.45; WAM R82639 Mt. Helena—31.87 116.2; WAM R82774 13 km N of Collie—33.25 116.15; WAM R82852 Montebello Is.—Hermite Is.—20.43 115.55; WAM R82932 Hermite Is.—Montebello Is.—20.43 115.55; WAM R83354 15 km SE of Wyndham—15.53 128.22; WAM R83357 12 km ESE of Mt. Rob—15.9 128.35; WAM R84183 10 km NW of Tallanalla—33.03 116.2; WAM R84184 10 km SSE of Contine—32.95 116.92; WAM R84410 7 km SE of Old Beyondie (Marymia)—25.07 120.15; WAM R84430 11 km SSW of Anketell Homestead—28.12 118.78; WAM R87195 Kununurra—15.78 128.73; WAM R89368 30 km NW of Balladonia—32.17 123.15; WAM R89385 Kalbarri—27.72 114.17; WAM R90617, R90623, R90712 Woodstock—21.61 118.96; WAM R90992 Perth Airport—31.92 115.98; WAM R93204 147 km N Rawlinna Railway Stn.—29.69 125.2; WAM R93389 12 km E Boingaring Rocks—32.47 123.35; WAM R93407 27 km S Mt. Malcolm—32.42 122.85; WAM R94769 Ca 80 km S Telfer—22.13 122.19; WAM R94823 10 km SSW Wallal Downs HS [53 km S Sandfire]—19.85 120.62; WAM R96143 Kambalda E—31.2 121.67; WAM R96207 Mandogalup—32.05 115.83; WAM R96246 Middle Swan—31.88 116.02; WAM R96247 Kalamunda—31.98 116.05; WAM R96577 10 km S Bimbijy HS—29.78 118.05; WAM R96607 12 km NNE Remlap HS—30 117.65; WAM R96614 15 km NNE Geranium Rock—29.85 117.35; WAM R96644 Pimlicarra Well—28.82 116.28; WAM R97572 Karridale—34.25 115.08; WAM R97573 Contine—32.88 117; WAM R97574 Maya—29.88 116.5; WAM R97579 Kalumburu—14.3 126.63; WAM R97581 Bayswater—31.92 115.9; WAM R97889 Murdoch—32.1 115.82; WAM R98046 Warwick—31.82 115.82; WAM R99160 Woodstock Stn.—21.61 118.97; WAM R100238 Rowles Lagoon—30.43 120.85; WAM R100370 Karrakata, Perth—31.97 115.8; WAM R100405, R100407 39 km E Laverton—28.47 122.83; WAM R100406, R100408, R100409 7–8 km WNW Point Salvation—28.2 123.6; WAM R100804 Gidgegannup—31.8 116.18; WAM R100844 Gnangara—31.77 115.88; WAM R101282 42 km N Wanneroo (16 km NNW Yanchep Park Entrance)—31.42 115.65; WAM R103493, R103499 Bungle Bungle National Park—17.53 128.35; WAM R104305 Mt. Lawrence Wells Area—26.8 120.2; WAM R104328 13 km E of Cardabia HS.—23.1 113.95; WAM R104361 41 km NE Whim Creek—20.78 118.17; WAM 08529, R108530, R108535, R108536, R108537, R108538, R108539, R108540, R108541, R108542, R1083, R108544, R108569, R108572, 7–8 km WNW Point Salvation—28.25 123.6; WAM R108789 Murdoch University Campus—32.07 115.83; WAM R113328 Safety Bay—32.32 115.73; WAM R114196 Girrawheen Senior High School—31.83 115.85; WAM R114778, R114779 Karrakatta Cemetery—31.97 115.8; WAM R115042 Bold Park—31.95 115.75; WAM R115105, R115277 Spalding Park, Geraldton—28.65 114.63; WAM R115246 Kings Park—31.98 115.83; WAM R115302 Gingin—31.35 115.9; WAM R115303 Geraldton—28.75 114.65; WAM R115305 Buller River—28.87 114.62; WAM R115306 Spalding Park, Geraldton—28.65 114.65;

WAM R117202 R117204 7–8 km NW Point Salvation—28.2 123.6; WAM R117205 R117208, R1172089, R117210 39 km E Laverton—28.47 122.83; WAM R117288 Doole Is.—22.45 114.18; WAM R119661 Glenfield, Geraldton—28.77 114.62; WAM R120036 De La Poer Nature Reserve—27.37 122.75; WAM R123771 Parkerville—31.88 116.13; WAM R125268 New Forrest—27.37 115.65; WAM R126373 Bungalbin sand plain ~0.5 km W of grid D—30.27 119.78; WAM R126523 Kambalda E—31.17 121.62; WAM R127021 Mardathuna Stn.—24.44 114.51; WAM R127546 Narrogin—Wandering Rd. (Dryandra)—32.75 117; WAM R100780 Woodstock, Site 0—21.66 119; WAM R104032 Woodstock, Site 0—21.66 119; WAM R104138 Woodstock, Site 4—21.59 119.08; WAM R122360 Mr5—24.41 114.44; WAM R125844 Gj2—25.18 115.49; WAM R126712 Na—26.25 113.8; WAM R126919 Bo5—24.41 113.76.

Diagnostic Characteristics

After further taxonomic examination, *V. gouldii* will probably be divided into a number of species. Color, dorsal pattern, and adult size vary considerably across its range, and no doubt when the taxonomy and morphology of the current *gouldii* complex is further examined (e.g., limb dimensions, scale counts), differences will be evident.

Generally, *V. gouldii* can be distinguished from *V. rosenbergi* and *V. p. panoptes* by the yellowish distal third of its tail; and from *V. p. rubidus*, *V. p. panoptes*, and *Varanus giganteus* by the lack of large black spots on the dorsal surface of the abdomen. However, not all *V. gouldii* have a yellow end to their tail (e.g., figs. 228 and 292 in Hoser 1989). *Varanus gouldii* can also be distinguished from *Varanus giganteus* by the black stripe through the eye and the lack of a reticulated pattern under the throat.

Coloration

For most specimens, the distal end of the tail is yellow without any dark bands. There are numerous color morphs.

The dorsal surface of the inland desert red sands morph is reddish yellow with black transverse blotches, or broad bands of small black dots. Alternating tail bands are yellowish (circular spots) on a black-brown background; the yellowish markings can be as wide as the black-brown background. Head and dorsal surface of the southwest Australian morph (neotype) is predominantly dark green-black with transverse yellow spots, often with a black center mark. Tail is dark green to black with narrow transverse yellow bands. The Kimberley morph is darker and less patterned than the desert morph and is heavily flecked with black. Some eastern Australian morphs have a green-gray dorsal surface with little markings on the dorsal surface of the abdomen and tail. Lighter bands on this morph's tail are much thinner. Central desert and Kimberley morphs (and perhaps others) can develop a very thick tail (fat stores) toward the base when well fed.

Hatchlings always have brighter and more distinct markings. The southwest Australian morph has a series of transverse yellow spots on the dorsal surface of the abdomen and alternating yellow and brown bars on the tail. In contrast, inland NSW hatchlings have yellow bands across their back and tail, but no yellow tail tip (see fig. 292 in Hoser 1989).

Hemipenes

Branch (1982) describes the hemipenes of *V. gouldii* being long and shallowly forked, with the sulcus rising from the base and running diagonal to the crotch, where it forms a shallow groove. A symmetrical V-shaped zone of stiffened frills occurs on the distal, asulcal surface. There are 10 to 11 frills, and the most distal 4 to 5 frills are bisected by a triangular extension of the smooth apical zone. The two horns are of equal size, strap-shaped with split ends that bear denticulate edges.

Size

Hatchling size varies across its geographical range. Neonates that I have measured have hatched at about 12 g, with a SVL of 108 mm and a tail length of 150 mm. Shine (1986) reports two hatchlings from the Northern Territory with SVL of 94 and 96 mm. I measured an adult female that grew to a SVL of 361 mm with a tail of 570 mm, and a male to a SVL of 590 mm with a tail of 820 mm; the Kimberley form is larger. *Varanus gouldii* caught in the Gibson Desert were generally the smallest I have seen. Shine (1986) reports adult male *V. gouldii* in the Northern Territory to be generally three times the mass of females. Maximum adult size varies across its geographical distribution.

For males, mean ± SE forelimb length (% SVL) is 26.8 ± 0.238% (n = 49), hind limb length is 38.0 ± 0.34% (n = 49), head length (snout to rear of tympanum) is 17.8 ± 0.34% (n = 48), head width is 8.3 ± 0.14% (n = 48), and head depth is 6.2 ± 0.11% (n = 48). For females, forelimb length is 26.5 ± 0.19% (n = 33), hind limb length is 39.0 ± 0.32% (n = 33), head length (snout to rear of tympanum) is 18.3 ± 0.35% (n = 33), head width is 8.5 ± 0.16% (n = 33), and head depth is 6.2 ± 0.16 (n = 33).

The relationship between SVL and body mass on a log–log plot is linear (Fig. 7.24) and can be represented by the equation

$$\text{SVL (mm)} = 51.25 m^{0.301} \ (n = 148, \ r^2 = 0.99),$$

where m = body mass in g.

Case and Schwaner (1993) report the Wandang Island population of *V. gouldii* differs in size from that of the mainland. However, their island sample was small, and there is considerable variability in the adult size of *V. gouldii* on the mainland of Western Australia, which would suggest further investigation was required before it could be concluded that island populations differed in size to those on the mainland.

Habitat and Natural History

Seasonal and Daily Patterns of Activity

Varanus gouldii are seasonally active. They emerge in September in the Great Victoria Desert and in October in Perth and Ora Banda, Western Australia, and are active on the surface until late February or March (Pianka 1970b, 1994a; S. Thompson, personal observation).

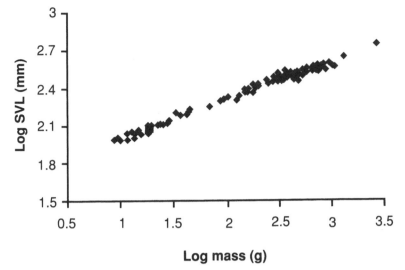

Figure 7.24. Relationship between body mass (g) and SVL for V. gouldii.

Although predominantly diurnal (Shine 1986), *V. gouldii* have been observed active at night (Valentic 1995).

Habitat

Varanus gouldii typically inhabits spinifex-covered sand plains (Shine 1986) but are found in a variety of habitats (e.g., eucalypt woodlands, *Acacia* thickets, chenopod shrublands), most of which have a sandy substrate.

Diet

Varanus gouldii has a catholic diet, eating almost anything they can catch and devour (Pianka 1994a). Great Victoria Desert specimens feed predominantly on lizards and lizard eggs (Pianka 1994a). In Karrakatta Cemetery, Perth, Western Australia, where there are few lizards, their diet consists predominantly of grasshoppers, and to a lesser extent spiders and invertebrate larvae (Thompson 1996). Shine (1986) reports *V. gouldii* in the Northern Territory eat predominantly invertebrate larvae and a lesser number of grasshoppers, beetles, and reptile eggs. Losos and Greene (1988) report lizards and orthopterans to be the most important component of their diet.

When digging out burrowing spiders, they leave a characteristic inverse V-shaped hole with the apex abutting the spider's hole (personal observation). In grazed mulga woodland of northeastern New South Wales, *V. gouldii* excavate many holes in search of spiders and beetle larvae. Whitford (1998) reports the mean number of holes ranging from 12 holes/ha in paddocks to 229 holes/ha in water "run-on" sections of heavily grazed areas. Many holes are associated with log mounds.

Varanus gouldii will scavenge off carrion on roadside verges and will learn to scavenge in human rubbish (personal observation). If given regular access to quantities of rubbish or food, they become obese (personal observation). I can confirm Berney's (1936) observation that these monitors eat fowl eggs and in captivity will readily eat a range of meats and small vertebrates. Berney (1936) also indicates that *V. gouldii* fall prey to black-headed pythons (*Aspidites melanocephalus*).

Activity Area

Varanus gouldii appear to have a defined activity area that may shift as resources within the area are depleted (Limpus 1995; Thompson 1992, 1994, 1995). When tracked with a spool and line in Karrakatta Cemetery, Western Australia, the mean daily distance traveled was 112 m, with a positive correlation between maximum daily temperatures, hours of sunshine, and daily distance traveled (Thompson 1992). Activity areas at the same location was 8.91 ha, with no evidence of territoriality for males or females. Activity area size was positively correlated with body size. Animals often forage in dense leaf litter toward the periphery of their activity areas and retreat to burrows that were more centrally located (Thompson 1994). Like most other large terrestrial goannas, they will emerge from their overnight retreat, which is generally a burrow in the ground, and bask to increase their body temperature before moving off to forage. On hot days, activity may be bimodal because these monitors will retreat underground during the middle part of the day, when ambient and soil temperatures are too hot for them to regulate their body temperature or as ambient temperature restricts foraging to shaded areas (Thompson 1994).

Pianka (1994a) reports mean ± SD field active body temperature of 37.7 ± 2.86°C (n = 78) for *V. g. flavirufus* specimens in the Great Victoria Desert. Licht et al. (1966) report the mean body temperature of *V. gouldii* in southern Western Australia as 37.1 ± 1.87°C (note that the latter data could have included *V. panoptes* because it was considered a subspecies of *V. gouldii* when these data were collected). Christian and Weavers (1996) provide a much more comprehensive analysis of the body temperature of *V. gouldii* in the Kakadu National Park, indicating a mean of 35.1°C, with significant variation among seasons (wet 35.9 > late dry 32.9 > dry 28.2°C), with a set point range of 34.0 to 36.2°C.

Burrows

Varanus gouldii generally dig their burrow, but they will use and enlarge an existing burrow (e.g., rabbit warren, *Oryctolagus cuniculus*). Burrow entrances are normally located next to an obstacle, such as a log, tree trunk, or spinifex bush. Digging is done with the strong forefeet to remove substrate from the hole, and then the accumulated pile of sand will be moved further away with the hind feet (Glazebrook 1977). Thomson and Hosmer (1963) suggest that *V. gouldii* is not arboreal, but Pianka (1970b) reports they will climb trees to escape capture. I have also seen *V. gouldii* climb trees to avoid capture and have seen them climb into shrubs and small trees, presumably in search of prey (Thompson 1995).

Burrows are normally quite shallow, less than 1 m deep, with a couple of branches and a chamber at the end of one or both branches. Burrows can range in length from 2 to 5 m. The chamber roof is often within 30 cm of the surface, and most often these lizards construct a pop hole from the chamber to the surface. When digging *V. gouldii* from their burrows, they will often rapidly enlarge the pop hole, and emerge running to escape capture. Aboriginals use their knowledge of this chamber to capture *V. gouldii*. On one field trip to the Kiwirrkurra Community (22°49'S, 127°46'E) in the Gibson Desert during July, when *V. gouldii* would have been inactive in their burrows since at least March, Aboriginal women showed us how they located and dug *V. gouldii* from their burrows. These women knew which burrows did and did not contain *V. gouldii* (although they all looked much the same to us), and using either a digging stick or a heel of their foot, they would bang the ground in a circle about 1 or 2 m from the burrow entrance until they found the section of ground that collapsed (the chamber roof) or until the sound was different. They would immediately dig the *V. gouldii* from the ground and would kill it by banging its head on the ground. Once dead, they would break both femurs with their fingers, which ensured that if it was merely stunned, it could not run off. After catching a number of goannas, they would light a small fire and allow time for hot coals to accumulate and the surrounding soil to get hot before cooking the goannas. Dead goannas would be buried in hot ground and covered with coals and hot sand. Within minutes, they would turn the goannas over to cook on the other side. Aboriginals considered goannas cooked when intestines protruded out the cloaca. By Western standards, the meat was raw; it was eaten by pulling pieces of flesh off the bone. After the cat (*Felis catus*) and the Australian bustard (*Otis australis*), the sand goanna was considered to be a prize catch as a food source.

Movement and Posture

Mean speed of movement for *V. gouldii* between foraging sites is about 27.6 m·min⁻¹ and while foraging is 2.6 m·min⁻¹ (Thompson 1994). *Varanus gouldii* have a number of characteristic body postures (Fig. 7.25). Standing erect on the hind feet and tail has been observed on occasions when a goanna wanted a better view of its surroundings (Glazebrook 1977). The vigilant posture is the most frequently seen and is characterized by the body remaining motionless, the abdomen held in either the prone or erect position, and the head and neck held high (Thompson 1995). The tail is rested on the substrate, and the head is often slowly rotated to give a clear view of the surroundings.

When cornered or unable to escape, *V. gouldii* displays a threat posture with back arched, erect on all four limbs, gular pouch inflated, and lungs inflated to increase its apparent size. It will hiss loudly and sometimes flick the tail (Johnson 1976; Pianka 1970b). Delean (1981) also reports this aggressive behavior among captive specimens. Johnson (1976) reports *V. gouldii* adopting a bipedal stance with extreme provocation, a behavior I have often seen in *V. rosenbergi*.

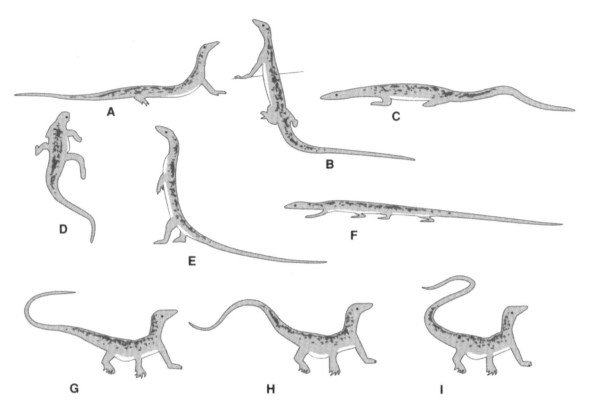

Figure 7.25. Body postures adopted by V. gouldii *at Karrakatta Cemetery.*
(A) Lying on a grave cover with the posterior abdomen flattened onto the gray concrete surface and the head and neck raised. (B) On the side of a grave cover with the abdomen dorsoventrally flattened and directed toward the sun. (C) Cloaca and anterior proportion of the tail being wiped on the ground after extruding a scat. (D) Lateral sigmoidal trotting action shown while walking. (E) Standing erect by balancing on the hind feet and tail. (F) Walking with only the distant end of the tail dragging on the ground. (G–I) Tail swipes with nonaggressive posture (redrawn from Thompson 1995).

Senses

A well-developed olfactory sense enables *V. gouldii* to locate and dig up lizard eggs laid in the ground (Garrett and Card 1993; Pianka 1970b, 1982; Thompson 1995). Garrett and Card (1993) report naive neonate *V. gouldii* being able to distinguish the odor of crickets from experimental controls. Visual and auditory senses are also very acute, being able to detect small movements of potential prey meters away, and being able to distinguish particular bird calls and to move to avoid being swooped, or being able to locate small reptiles in leaf litter from the sound of moving litter (Garrett and Card 1993; Pianka 1970b, 1982; Thompson 1995). Like *Varanus tristis*, some *V. gouldii* have a curious habit of curling their tail over their abdomen and head as if to provide shade, although its width is hardly likely to provide much protection from solar radiation.

Reproduction

Limpus (1995) provides the following description of mating between *V. gouldii* at Mos Repos Beach near Bundaberg in southern Queensland. The male courted the female by biting her on the neck and intermittently distending the throat. The male then held the female on the neck with his mouth and mounted her, after pinning her shoulders to the ground with his forelimbs and rotating the pelvis by holding his tail under hers and inserting his hemipene.

Great Victoria Desert *V. g. flavirufus* specimens are sexually mature at SVL of about 250 mm (Pianka 1994). *Varanus gouldii* mate in September–November and lay eggs around December, and hatchlings emerge in the following summer during January and February in the Great Victoria Desert (Pianka 1970b, 1982). In Karrakatta Cemetery, Perth, Western Australia, eggs are laid mid-December to mid-January, and hatchlings emerge about late October the following year. Around Ora Banda, Western Australia, hatchlings are first seen in January. Pianka (1994a) reports mean clutch size for *V. gouldii flavirufus* in the Great Victoria Desert as 6.2 eggs. For *V. gouldii* in the Northern Territory, Shine (1986) reports oviductal eggs in January (n = 1) and February (n = 2) and inactive ovaries in April, June, and October.

Combining the data that are available from a number of sources (Barnett 1979; Brooker and Wombey 1978; Doles and Card 1995; Horn and Visser 1989; Irwin 1986; Mitchell 1989; Shine 1986; S. Irwin, unpublished data; G. Thompson, unpublished data), mean clutch size is 5.9 eggs for wild-caught and 7.3 eggs for captive specimens. As for most varanid eggs, hatching time is influenced by incubation temperature (Thompson and Pianka 2001). At 30 to 32°C, eggs take about 170 days to hatch (Barnett 1979), and at 30°C eggs take 223 to 270 days (S. Irwin, personal communication). S. Irwin provided unpublished data on clutches laid in north Queensland on November 4, 8, 26, and 27, December 7, and December 25. For a clutch incubated in the same container, the incubation period was spread over a span of 223 to 245 days, suggesting either variability in incubation period or a nonuniform temperature in the incubator. Card (1995) reports *V. gouldii* double clutching without access to a male between the two clutches, which indicates that they have the capacity to store sperm.

Berney (1936) provides an interesting commentary on the behavior of a *V. gouldii* in western Queensland (it could have been a *V. panoptes*, but there is no way of knowing this). This particular lizard was observed digging up varanid eggs (presumably its own) and depositing them in a new hole that it dug for the purpose. Berney (1936) indicates that it actually swallowed the eggs and then regurgitated them into the new hole.

Combat Ritual

Thompson et al. (1992) report observing the combat ritual between male *V. gouldii* in December 1990. On this occasion, both goannas were erect on their hind limbs, with their ventral surfaces in contact and forelimbs wrapped around each other. These two animals fell to the ground, embracing each other with fore- and hind limbs, writhing and rolling on the graded road. Marks on the road indicated that the battle had gone on for some time before they were observed. On another occasion, two *V. gouldii* were observed fighting in a similar fashion, and this time they bit each other; one lizard sustained two lacerations about 60 mm in length on its dorsal surface of the abdomen. Healed scars on the same lizard suggested that it had been in fights on earlier occasions.

Physiology

Bennett (1973) reports ventilation frequency increases with body temperature at rest and when active for *V. gouldii*. The relationship between ventilation frequency and body temperature at rest is

$$\log f \text{ (breaths min}^{-1}) = 0.0506\, T_b - 0.635,$$

where f is ventilation frequency and T_b is body temperature; and when active, the relationship is

$$\log f = 0.0281 T_b - 0.062.$$

Tidal volume increases with body temperature at rest:

$$\log V_t \text{ (cm}^3 \text{ breath}^{-1}) = 0.240 + 0.0127 T_b;$$

and when active,

$$\log V_t \text{ (cm}^3 \text{ breath}^{-1}) = 1.156 + 0.009 T_b.$$

Thus, the minute volume for *V. gouldii* at rest and when active also increases with temperature at rest:

$$\log V_E \text{ (cm}^3 \text{ min}^{-1}) = 0.0636\, T_b - 0.395,$$

and when active,

$$\log V_E \text{ (cm}^3 \text{ min}^{-1}) = 0.0371 T_b + 1.218.$$

Standard metabolic rate for *V. gouldii* is within the range expected for other reptiles:

$$\log_{10} \text{mL } \dot{V}o_2 \text{ h}^{-1} = -2.78 + 1.09 \log \text{mass (g)} + 0.45\, T_b \text{ (°C)}$$

(Thompson and Withers 1994, 1997b).

Maximal metabolic rate for *V. gouldii* is within the range of other similar size goannas is

$$\dot{V}o_2 \text{ mL h}^{-1} = 10.5 m^{0.62},$$

where m = mass in g at 35°C, providing a factorial aerobic scope (maximal/standard metabolic rate) of 9.6 at 35°C. Evaporative water loss at 35°C is

$$H_2O \text{ (mg h}^{-1}) = 0.97 m^{0.82},$$

where m = mass (in g) (Thompson and Withers 1997a).

Exercise after long periods of inactivity (e.g., remaining underground for most of winter) and captivity have no affect on maximal metabolic rate of *V. gouldii* (Thompson 1997).

Parasites

Jones (1983a) reports six species of nematode in the genus *Abbreviata* in stomachs of *V. gouldii* (*A. hastaspicula, A. barrowi, A. antarctica, A. levicauda* sp. nov., *A. tumidocapitis* sp. nov., and an unnamed species). Jones (1991) later adds a seventh species (*A. confusa*) to the list.

7.11. *Varanus keithhornei*

STEVE IRWIN

Figure 7.26. Varanus keithhornei *(photo by Jeff Lemm).*

Nomenclature

Varanus keithhornei was first described as *Odatria keithhornei* (Wells and Wellington 1895) on the basis of three specimens held at the Queensland Museum (Fig. 7.26). The first specimen was collected by Gregory Czechura of the Queensland Museum in August 1978. The type specimen (holotype) is QM J31566, type locality Buthen Buthen, Cape York Peninsula, Queensland. Two additional specimens were then collected by officers in the Queensland National Parks and Wildlife Service, in close proximity to the first specimen. They are paratypes marked QM J35450 and QM J35451 (Czechura 1980). Sprackland (1991) described *Varanus teriae* on the basis of the same material without any reference to Wells and Wellington's earlier description. The two descriptions have been synonymized as *Varanus keithhornei* (Covacevich and Couper 1994).

Geographic Distribution

Varanus keithhornei, the canopy goanna, has the most restricted range of all currently recognized species of Australian monitors occurring in the McIlwraith and Iron ranges, Cape York Peninsula, Queensland. This species is listed as rare in *An Atlas of Queensland's Frogs, Reptiles, Birds and Mammals,* Queensland Museum, by Glen J. Ingram and Robert J. Raven (1991), and rare or insufficiently known in the *Action Plan for Australian Reptiles,* Australian Nature Conservation

Figure 7.27. Known localities for
Varanus keithhornei.

Agency Endangered Species Program Project 124, by Cogger et al. (1999). (Fig. 7.27).

QM R31566 Buthen Buthen, Nesbit Range, Cape York—13.4 143.45; QM R35450 Cape York, Leo Creek Rd., 17 km NE Mt. Kroll—13.7 143.283; QM R35451 Lankelly Creek, 10 km NE Coen—13.9 143.25; QM R45860 Iron Range—12.7 143.283; QM R48677 McIlwraith Range, 12 km E Peach Rd.—13.8 143.25.

Fossil Record

No fossils of *Varanus keithhornei* are known.

Diagnostic Characteristics

• Medium-sized monitor.

• Uniformly dark with keeled body scales.

• Slender, with long digits and long, prehensile tail.

• Palms and bottom of toes that tend to stick to surfaces.

Description

Uniformly dark in color, scales along back are keeled. Dorsum mainly black in color with some silver gray. Belly scales silver gray, as are scales under the tail and the neck. Snout blue-gray in color. Most specimens have an orange and white stripe across the top of and in contact with the top of the ear opening, which may in some specimens extend down in front of the ear. Midbody scales 90 to 99 rows (Czechura 1980). Digits long with sharp claws, palms black, and sticky or adhesive to surfaces. Tail very prehensile. Hatchlings and juveniles are more vibrantly marked with silver chevrons across the back. Subadults are more brightly colored than adults, although much of the color has merged. An arboreal monitor lizard, which is a part of the *prasinus* group. Distinct from other Australian species of monitor in appearance and very similar in its slender build to *Varanus prasinus* and *V. beccarri.*

Size

Adult SVL ranges from 210 to 260 mm. Tail length 1.7 times SVL. TL of adults ranges between 450 and 650 mm. Adult body weight ranges from around 190 to 270 g.

Habitat and Natural History

Occurring in a variety of forest types in the McIlwraith and Iron Ranges: semideciduous mesophyll vine forest, deciduous vine thicket, and eucalyptus-dominated open forest close to deciduous vine thicket (Czechura 1980), at altitudes from as low as 60 to 540 m (Czechura 1980). Upland and lowland rain forest (altitude 10–500 m) and *Melaleuca,* bulgaroo swamp (altitude 0–10 m) (Irwin 1996a).

Canopy goannas are an arboreal monitor lizard capable of moving through foliage, trees, canopy, and vines very quickly and unobtrusively (Irwin 1996a). Their black coloration possibly aids in heat absorption in the dappled light of the canopy, and along with their slender build, aids in their camouflage. When disturbed, they will take to a tree, or if already in a tree, they will move to the opposite side of the branch. They live, hide, and sleep in hollows averaging 7 m in height from ground level (Irwin 1996a).

I have seen three wild Iron Range lizards, two adults and one subadult in the same home tree, which was riddled with hollows, during winter. I presume they are gregarious to some extent. However, in captivity (Australia Zoo 2000), we have witnessed ritualized combat between males and aggression between females (Irwin, personal observation).

Terrestrial, Arboreal, Aquatic

Varanus keithhornei are arboreal, adept for life in the canopy, although they do move around the forest floor to forage.

Time of Activity (Daily and Seasonal)

These lizards are diurnal, with the peak of activity possibly occurring when foraging. Midmorning and midafternoon are preferred foraging times (Irwin 1996a). Although they are active all year round, lizards do not move far from home trees during the winter months except close to the middle of the day on warmer days with no wind. Activity peaks during the summer monsoon season (Irwin, personal observation).

Thermoregulation

No data exist on the thermal biology of *V. keithhornei*. Being very melanistic may be a result of the need to absorb heat in the dense forest. Canopy goannas are masters at basking. Within their territory, they have designated and well-used sunning branches. These are situated close to their home hollows, and their early morning basking location, where they heat up and watch for threatening predators before moving on, usually within 2 m of the hollow. Throughout their territories, they are often seen on exactly the same branch or vine at the same time of day, generally close to a good foraging area.

Foraging Behavior

Canopy goannas forage for insects and prey on the ground, among leaf litter and rotting timber. They use their forelimbs to scratch out and locate their prey, particularly insects (Irwin 1996a). Foraging occurs at every level of the rain forest. Katydids, leaf mimics, and grasshoppers in the high-level canopy to lower-level bushes and shrubs to the ground, where prey are usually located visually, stalked, then rushed, seized, and swallowed. *V. keithhornei* are frequently observed foraging in leaf litter where they locate insects with their tongue. Perhaps the most interesting foraging technique is scratching and tearing apart large dead trees with their forelimbs. They easily dismember the soft, rotting wood where they locate crickets embedded deep in particularly soft timber, like figs (*Ficus* sp.).

When *V. keithhornei* capture live or dead food items, they demonstrate unusual behavior. They rake their head and food with their forelimbs. This raking technique appears to remove green ants and also aligns sticky insect legs for swallowing.

Diet

Fecal samples taken in the field have contained beetles, crickets, cockroaches, and insect mandibles (Irwin 1996a). *V. keithhornei* have also been observed eating katydids, leaf mimics, stick insects, and grasshoppers. In captivity (Australia Zoo 2000) they relish rats, mice, chickens, fish, crickets, and grasshoppers, so it is assumed they are mainly insectivorous but also opportunistically feed on rodents and birds in the wild.

Reproduction

Nothing is known of their reproduction in the wild. In captivity, copulations have been observed in September to May. Clutches have been laid from December to June, with females sometimes laying two clutches within this period. Clutches contain two to four eggs. Eggs are elongate, measuring 40 to 55 mm in length and 13 to 21 mm in width, and they weigh 4 to 13 g. Eggs in captivity have taken between 170 to

182 days to hatch. Measurements from four hatchlings SVL 97 to 101 mm (mean, 98.75 mm), TL 232 to 253 mm (mean 243.75 mm), and weight 10 to 12 g (mean, 10.5 g).

Movement

V. keithhornei appear to have distinct communal territories where more than one lizard can be observed within 20 m of each other, returning to the same home hollows for refuge. When threatened, they scurry straight up the nearest tree to extreme branch tips, from which they will traverse from tree to tree across the canopy. If the threat continues, they launch themselves out with all four legs splayed—any contact with any foliage will result in the lizard clinging to it. Jumping from tree to tree to avoid capture or predators is a fluid motion aided by their prehensile tail, which has keeled scales for additional traction.

Physiology

No work has been done on their physiology to date.

Fat Bodies, Testicular Cycles

No work has been done to date.

Parasites

Nothing is known about internal parasites of this species. Ticks, ectoparasites in the family Ixodidae, are often seen embedded in the skin around the neck, legs, and vent.

7.12. *Varanus kingorum*

Max King

Figure 7.28. Varanus kingorum *(photo by Robert Browne-Cooper).*

Nomenclature

This small, long-tailed, rock-dwelling varanid was described by Storr in 1980 (Fig. 7.28). The holotype (R.60374, Western Australian Museum) was discovered by Dr. Max King in June 1978 at 10 km west-northwest of Timber Creek in the Northern Territory at 15°37'S, 130°23'E. The holotype was collected from between rock exfoliations. This species was named after Drs. Max King and Dennis King for their work on varanids.

Geographic Distribution

Specimens of *V. kingorum* are found in the eastern Kimberley area in the far northeast of Western Australia. Their known distribution extends from Wyndham in the north to Turkey Creek in the south, and east from there to Timber Creek in the Northern Territory, where appropriate rocky habitat is available (Fig. 7.29).

NTM R25903 Turkey Creek—16.93 128.25; NTM R25904 Kununurra—15.78 128.73; NTM R25905 Turkey Creek—16.93 128.25; NTM R25906 Turkey Creek —17.03 128.22; NTM R25907, R25908, R25909 Turkey Creek 170 km N Halls Creek—16.9 128.32; WAM R60374 10 km WNW of Timber Creek. 0.2 km W of Junction of Big Horse Creek and Victoria River—15.62 130.38; WAM R63340 Turkey Creek—17.03 128.22; WAM R63341 Kununurra—15.77 128.73; WAM R87199, R87200 7 km SSE of Turkey Creek—17.08 128.18; WAM R103184 Bungle Bungle National Park—17.73 128.15.

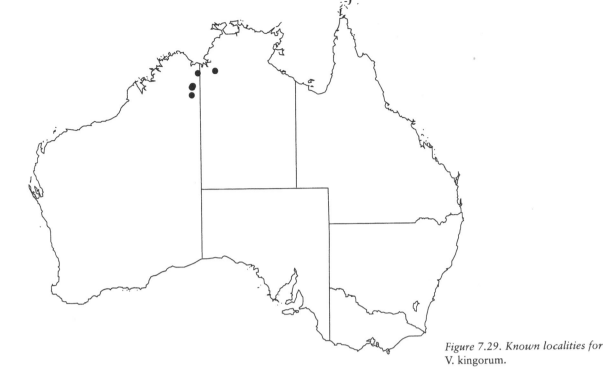

Figure 7.29. Known localities for V. kingorum.

Fossil Record

No fossils of *V. kingorum* are known.

Diagnostic Characteristics

V. kingorum is a small red-brown monitor with occasional dark brown to black spotting on the dorsal surface. This rock-dwelling varanid, of the subgenus *Odatria* Gray 1838, is distinguished from all other species by its size and coloration and by its extraordinarily long tail (*V. caudolineatus* Boulenger 1885, *V. gilleni* Lucas and Frost 1895, *V. eremius* Lucas and Frost 1895); its tail coloration (*V. eremius*); and head scalation (*V. eremius*).

Description

Storr (1980) described *Varanus kingorum* as being dull reddish brown dorsally and laterally with small black spots on the dorsal surface. In young specimens, the black spots sometimes form a reticulate pattern. The ventral surface is buff to white with occasional spotting, particularly on the throat.

The long, prehensile tail is proximally compressed, subtriangular in section at midlength, and elliptical distally. The tail is not striped or banded. The scales on the tail are sharply keeled with a small mucron (Bedford and Christian 1996). Claws are short, thick, and curved.

Size

SVLs of the type specimens (n = 5) ranged from 67 to 114 mm (mean, 98.4 mm). Tail lengths ranged from 200 to 270 mm (mean, 228 mm). The tail of adults was approximately 2.3 times the SVL. The forelegs ranged from 25 to 27 mm long (mean, 25.8 mm). Hind legs ranged from 33 to 37 mm in length (mean, 35.4 mm) (Storr 1980).

A series of 21 hatchlings bred in captivity had SVLs ranging from 44 to 55 mm (mean, 50 mm). Tail lengths of these specimens ranged from 80 to 92 mm (mean, 85 mm). Body mass of hatchlings ranged from 1.5 to 2.1 g (Eidenmüller 2001). These appear to be particularly large hatchlings, approaching half of the adult SVL. The tails of juveniles are typically shorter than in adult, being only 1.7 times the SVL.

Habitat and Natural History

These small goannas are found in the tropical north of Australia in areas of suitable habitat. Specimens of *V. kingorum* live in crevices between rocks and under rock exfoliations. They have been collected in a variety of rock types including sandstone, granite, and metamorphic conglomerate, and they live on outcrops in either escarpment, hilly or flat country. Such outcrops are generally surrounded by red, often sandy soil, and the vegetation is open *Eucalyptus* woodland or tropical savanna grassland, although this vegetation becomes sparser in the south of the distribution of *V. kingorum*.

Specimens of *V. kingorum* are relatively common in the western section of their distribution. Indeed, on a field trip in 1979, I collected *V. kingorum* from seven new localities between Wyndham and Turkey Creek in a single day. Specimens are far less common in the east of the distribution of *V. kingorum*.

Terrestrial, Arboreal, Aquatic

Varanus kingorum is a specialized terrestrial rock-dwelling varanid.

Foraging Behavior

In captivity, adult specimens of *V. kingorum* have been observed actively hunting in sandy areas adjacent to their rocky lairs. In the wild, these varanids probably shelter and hunt on the rock outcrops on which they live, and also use the surrounding terrain for hunting.

Diet

An analysis of stomach contents of three wild-caught *V. kingorum* revealed that orthopterans made up 50% of the total number of prey items; the rest of the diet comprises blattids (25%), isopterans (12.5%), and insect eggs (12.5%) (James et al. 1992). Small though this sample size is, the predominance of grasshoppers suggests foraging in areas adjacent to the rock outcrops in which these small monitors live. These data also suggest that *V. kingorum* is a general insectivore only limited by prey size.

Reproduction

An examination of preserved specimens of *V. kingorum* for their reproductive condition (James et al. 1992) found reproductive activity in males and females collected toward the end of the wet season in February. Because the sample size was small, this suggests either a prolonged reproductive season or wet season reproduction.

In captivity, specimens of *V. kingorum* have been observed digging burrows in soil and beneath rocks and laying eggs in these burrows. Eidenmüller (2001) reported successfully breeding *V. kingorum* in captivity for two generations. Clutch sizes ranged from 3 to 6 eggs with a mean of 4.5 eggs in five clutches. Eidenmüller reported that the adult female laid multiple clutches 2 to 3 months apart, and on the first occasion, there were six eggs in each. This finding suggests a particularly high reproductive rate for *V. kingorum*. Eggs were 15 mm long and 7 mm in diameter, and at incubation temperatures of from 27° to 29°C, hatching occurred between 97 and 119 days.

Unfortunately, to my knowledge, nothing has been published on the time of activity, thermoregulation, physiology, water balance, energetics, fat bodies, or the parasites found on *Varanus kingorum*.

7.13. *Varanus mertensi*

KEITH CHRISTIAN

Figure 7.30. V. mertensi *(photo by Robert Browne-Cooper).*

Nomenclature

Described by Glauert in 1951 from the type locality of Moola Bulla, Western Australia (18°12'S, 127°30'E) (Fig. 7.30).

Geographic Distribution

V. mertensi is found near permanent water across northern Australia from the Kimberley region to western Cape York Peninsula (Fig. 7.31).

QM R10551 Normanton, Walkers Creek—17.5 141; QM R14321 Dismal Channels, Julia Creek to Normanton—20.1 141.1; QM R20654 Wakooka, 50 m from Wakooka Creek—14.6 144.6; QM R25985 Leichhardt Rd.—20.6 139.7; QM R37577 Coen Airport, 23 km NNW Coen—13.8 143.1; QM R44993 Natal Downs Stn.—21.1 146.2; QM R46280 Lower Leichhardt Rd.—20.6 139.7; QM R47276 E Leichhardt Dam; QM R48640, R48678 E Leichhardt Dam, E of Mt. Isa—20.8 139.8; QM R58822 Wenlock River—12.3 142.1; QM R58823 E Leichhardt River —20.9 139.8; AM R14056 Forrest River Mission—15.2 127.9; AM R14810 Bulliwallah Stn. via Clermont—21.6 146.6; AM R18655 Mt. Isa—20.7 139.5; AM R26379 Port Bremer Cobourg Peninsula—11.2 132.3; AM R38051, R38331 Koongarra, Brockman Range, Arnhem Land—12.8 132.7; AM R40224 Cannon Hill. 12.4 132.4; AM R41646 Cannon Hill—12.4 133; AM R53624 Caranbirini Water Hole, 21 km N Macarthur River Base Camp—16.2 136.2; AM R54912 48.5 km Upsteam Tomkinson River—12.3 134.3; AM R54913, AM R54914 Andranangoo

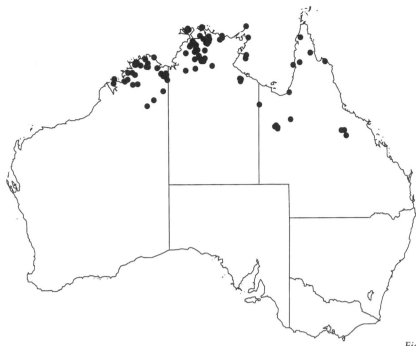

Figure 7.31. Known localities for V. mertensi.

Creek Melville Is.—11.4 130.9; AM R57124 Macarthur River Region—16.4 136.1; AM R64801 Blowhard Creek, ~70 mi S Charters Towers—21.1 146.5; AM R64802 Mt. Isa—20.7 139.5; AM R75334 Upper Reaches of the Mckinlay River—13.2 131.7; AM R75362 Mid Reaches of the Mckinlay River—13 131.7; AM R75568 Mid Reaches of Liverpool River—12.5 134; AM R76963 Emerald River Groote Eylandt—14.3 136.6; AM R79264 Arnhem Hwy. 30 km W Mary River Bridge—12.8 131.5; AM R97540 3.5 km along the Oenpelli Rd. from its intersection with the Arnhem Hwy.—12.6 132.8; AM R112433 Gulungul Creek At Arnhem Hwy., vicinity of Jabiru; AM R112434 Jabiru Area, TO to E Jabiru from Hwy.—12.6 132.9; AM R112441 Gulungul Creek Crossing of Arnhem Hwy., vicinity of Jabiru—12.7 132.9; AM R123876, R123877 Mitchell Plateau, Upstream of Mitchell Falls—14.8 125.7; AM R126199 Mitchell Plateau, Upstream of Little Mertens Falls—14.8 125.7; AM R126200 Mitchell Plateau Upstream of Mitchell Falls; AM R136106 Surveyors Pool on Tributary of Mitchell River, Mitchell Plateau—14.7 125.7; AM R140170 The Grotto, 69.7 km W of Kununurra on Wyndham Rd.—15.7 128.3; AM R140171 King Edward River, Mitchell Plateau—14.9 126.2; AM R140278 Manning Gorge, Mt. Barnett Stn.—16.7 125.9; AM R142902 Mataranka Homestead; AM R147263 Cullens Creek (~65 km N Katherine) crossing of Main Darwin—14 131.9; ANWC R0258 Tindal, Katherine—14.5 132.4; ANWC R0400 Nourlangie Rock—12.9 132.8; ANWC R1115 7 km W of Ryan Bend on S side of Track following—16.2 136; ANWC R1146 Mt. Bundy Rd., Humpty Doo—12.6 131.3; NMV D9148 St. George; NTM R25910 Daly River—13.8 130.7; NTM R25911 Edward River—14.7 142.1; NTM R25912 UDP Falls—13.5 132.5; NTM R25913 Katherine, CSIRO Research Stn.—14.5 132.3; NTM R25914 190 km S Darwin on Stuart Hwy.—13.6 131.5; NTM R25915 Maningrida—12.1 134.2; NTM R25916 13 km N Adelaide River on Stuart Hwy.—13.1 131.1; NTM R25917 Roper River Crossing, Mataranka HS—15 133.2; NTM R25918 Alyangula, Groote Eylandt; NTM R25919 Katherine Gorge Rd.—14.4 132.3; NTM R25920 Katherine River—14.5 132.3; NTM R25921 Katherine, 11 km W—14.5 132.2; NTM R25922 Valentine Springs via Kununurra—15.4 130.7; NTM R25923 Daly River—

13.8 130.7; NTM R25924 Claravale—14.4 131.6; NTM R25925 Katherine Gorge
National Park—14.3 132.5; NTM R25926, R25928 Katherine River, Katherine—
14.5 132.3; NTM R25927 Umbrawarra Gorge—13.4 131.8; NTM R25929 46.5
km N Katherine—14.2 132; NTM R25930 Katherine River, Katherine—14.5
132.3; NTM R25931, R25932 Ebirramurrumanja Groote Eylandt—14 136.7;
NTM R25933 Ivanhoe Crossing, Ord River—15.7 128.7; NTM R25934 Batchelor
TO Stuart Hwy.—13 131.1; NTM R25935 Katherine—14.5 132.3; NTM R25936
Marchinbar Is.—11.2 136.7; NTM R25937 Magela Creek—12.6 132.9; NTM
R25938 Arnhem Hwy., Boralil Creek—12.6 132.8; NTM R25939 Jabiru area—
12.7 132.9; NTM R25940 Ja Ja Billabong—12.5 132.9; NTM R25941, R25942
Jabiru area—12.7 132.9; NTM R25943, R25944, R25945, R25946 Daly River—
13.8 130.7; NTM R25947 Cooper Creek—12.1 132; NTM R25948 Cooper
Creek—12.1 132.9; NTM R25949 Daly River—13.8 130.7; NTM R25950 Mussel-
brook Res, Border Waterhole—18.6 138; NTM R25951 Katherine River; NTM
R25952 Sandy Creek, near Taracumbie—11.6 130.7; NTM R25953 Goose Creek—
11.5 130.9; WAM R5819 Moola Bulla—18.2 127.5; WAM R11813, R11814,
R11815, R11816, R11817, R11818, R11819, R11820, R11821, R11822, R11823,
R12333 Wotjulum—16.2 123.6; WAM R13518, R13526 Yirrkala—12.3 136.9;
WAM R21933, R21934 Katherine, in a Small Billabong—14.5 132.3; WAM
R24938 Katherine—14.5 132.3; WAM R26781 Soda Creek near Wyndham—15.7
128.3; WAM R28030 May River, Near Derby—14.3 126.6; WAM R28031 Kalum-
buru—14.3 126.6; WAM R43120 Surveyor's Pool, Mitchell Plateau—14.7 125.7;
WAM R46518 Mary River At Gt Northern Hwy.—18.7 126.9; WAM R47018,
R47019 Prince Regent River Reserve—15.8 125.6; WAM R47035 Prince Regent
River Reserve—15.6 125; WAM R47039 Prince Regent River Reserve—15.6 125.4;
WAM R50335 Drysdale River National Park—15 126.9; WAM R50548 Drysdale
River National Park—15.1 126.8; WAM R50602 Drysdale River National Park
—14.7 126.9; WAM R50974 Drysdale River National Park—15 126.8; WAM
R51615 Beverley Springs—16.7 125.5; WAM R51818 Stewart River Kimbolton—
16.6 123.5; WAM R60112, R60113, R60114 Ord Dam Lake Argyle—16.1 128.7;
WAM R60126 Katherine—14.5 132.3; WAM R60131 Saw Ranges 35 km WSW of
Kununurra—15.9 128.4; WAM R61731 5 km S Walsh Point Port Warrender—14.6
125.9; WAM R83358 Lake Argyle—16.3 128.8; WAM R83835 Walcott Inlet—
16.4 124.7; WAM R87182 Kununurra—15.8 128.7; WAM R94822 Piccaninny
Gorge—17.4 128.4; WAM R97576, R97577 Kalumburu—14.3 126.6; WAM
R97899 3.7 km NW Mount Daglish—16.3 125; WAM R1001915.6 km W Evelyn
Is.—14.1 127.5; SAM R03517 Yirrkala Arnhem Land—12.3 136.9; SAM R15543
Tribitary of Alligator River—12.4 133; SAM R19227 15 mi W Edward River—14.9
141.3; SAM R25531 115 km SW Katherine on Victoria Hwy.—15.2 131.6.

Fossil Record

Unknown.

Diagnostic Characteristics

- Medium size, with typical adult size being slightly over 100 cm TL.
- Dorsal surface black or dark brown with scattered small yellow or cream spots.
- Ventral color yellow or creamy with some gray mottling on neck and chest.
- Tail very compressed with distinct dorsal keel, and about 1.5 times SVL (Bedford and Christian 1996).
- Nostrils on dorsal surface of snout.

Description

V. mertensi is the most aquatic of the Australian varanids, and several morphological characteristics are related to this lifestyle. Its tail is extremely compressed to facilitate swimming (Bedford and Christian 1996). The position of the nostrils allow the animals to breathe with only a small part of the snout above the water, and they can be closed during dives. The drab coloration makes the animals very difficult to see when they are in the water. Scales are very smooth, giving this long-necked species a glossy appearance, particularly when wet.

Size

V. mertensi grows to a maximum SVL of about 48 cm and reaches TLs up to 130 cm.

Habitat and Natural History

These monitors are always found in or near permanent water. They may explore more widely (and presumably colonize new billabongs) during the wet season when surface waters are expansive, but during other times, they are found very near bodies of water including rivers, streams, lakes, and more permanent billabongs. These lizards sometimes sleep in shallow water, but also sleep in burrows hollowed out under rocks near the water's edge, in hollow logs or snags, or on branches of trees (either living or dead) that are either surrounded by or that hang over water, allowing a quick escape into the water. When pursued, these goannas usually submerge, where they can remain for considerable periods, before briefly rising between dives to take a breath with only their nostrils in the air.

Terrestrial, Arboreal, Aquatic

Aquatic, but also use trees in and around water.

Time of Activity (Daily and Seasonal)

V. mertensi are active year round (Christian et al. 1996c), unlike some terrestrial species of varanids living in the same area in the wet-dry tropics of Australia (Christian et al. 1995, 1996b, 1996c). This is presumably related to availability of water and food in their habitat compared with woodland habitats in the dry season (Christian et al. 1996c, 1999). Activity within a day is spread fairly evenly across the daylight hours (Christian and Weavers 1996), and they spend about 1 hour or less a day in locomotory activity (Christian and Weavers 1996). Sleeping sites change seasonally with many individuals sleeping in trees during the build-up season (hot months of October and November), but in the wet and dry seasons, they tend to sleep in more sheltered sites like hollow trees or burrows. However, these animals can occasionally be found sleeping in water throughout the year.

Thermoregulation

V. mertensi thermoregulates around a mean midday body temperature of 34°C both in the field and in a laboratory thermal gradient, which is significantly lower than that of terrestrial species in the same area (Christian and Weavers 1996; Christian and Bedford 1996). Indices of thermoregulation indicate that, compared with sympatric terrestrial goannas, *V. mertensi* is a moderately "careful" thermoregulator (Christian and Weavers 1996). The solar absorptance of this species is about 89% (Christian et al. 1996a), which is at the high end of the range of values measured in 12 species of varanids. Thus, when basking, these lizards can gain heat quickly.

Foraging Behavior

Varanus mertensi forages along the edge of water bodies as well as in the water. This monitor can sit in shallow water and use its tail to sweep small fish toward its head (Hermes 1981). It also forages underwater by walking on the bottom, completely submerged (sometimes in water several meters deep), stopping to scratch the muddy bottom for food in the same way it does on land (personal observation).

Diet

This species eats aquatic prey, as well as terrestrial prey that can be found near the water's edge. Freshwater crabs are an important component of the diet (Shine 1986), as well as crayfish, beetles, fish, spiders, various insects (both aquatic and terrestrial), amphipods, frogs, reptile eggs, snakes, birds, and small mammals (Shine 1986; Losos and Greene 1988). These water monitors are also known to find, excavate, and consume eggs from underwater nests of the turtle *Chelodina rugosa* (Kennett et al. 1993).

Reproduction

Males engage in ritual combat in which the animals face each other on their hind legs and attempt to wrestle each other to the ground (Greer 1989). *V. mertensi* females have been found gravid between February and July (Shine 1986; Vincent and Wilson 1999). A gravid female with eight eggs was killed by a dog at Berry Spring, Northern Territory on February 22, 2001, and a female collected at Berry Springs laid eight eggs on February 9, 1998, with masses ranging from 45.8 to 57.7 g (G. Husband, unpublished data). Clutch sizes range from 3 to 14 eggs (Vincent and Wilson 1999).

Movement

Some individuals move between bodies of water in the wet season, but generally they remain close to water. Those living around billabongs or lakes have large overlapping home ranges, and some individuals will eventually use the entire area of the lake (personal observation).

Physiology

The field metabolic rate (FMR) is slightly lower in the dry season than in the wet season, largely as a result of cooler nighttime body temperatures at this time of year (Christian et al. 1996c). This seasonal pattern differs from that of sympatric terrestrial varanids, which have markedly decreased FMRs in the dry season as a result of decreased activity, decreased daytime body temperatures, and metabolic depression.

The standard metabolic rate of *V. mertensi* is similar to that of other varanids (which do not differ from other lizards in this respect) (Christian and Conley 1994; Schultz 2002). The standard metabolic rate of this species does not change between seasons (Christian et al. 1996c), unlike that of the sympatric *V. scalaris* (Christian et al. 1996b).

During exercise, *V. mertensi* does not have a high aerobic capacity as is found in some varanids, but rather, its activity metabolic rate is similar to that of a similar-sized iguana (Christian and Conley 1994; Schultz 2002). Oxygen transport during exercise in *V. mertensi* is not limited by ventilation of the lungs (Frappell et al. 2002a). The ventilation rate in *V. mertensi* increases in proportion to metabolism, indicating adequate pulmonary ventilation, even without the gular pumping displayed by *V. exanthematicus* (Owerkowicz et al. 1999). Available evidence suggests that the limitation to oxygen transport during exercise is related to the circulatory system. Although oxygen transport of the circulatory system does increase with exercise, the agents of change (increase in heart rate, and oxygen extraction at the tissues) appear to reach their maximum rates before maximal oxygen consumption is reached (Frappell et al. 2002a, 2002b).

Interestingly, although this species spends much of its time in the water, water flux in the field is no different from that of sympatric terrestrial varanids during their active seasons (Christian et al. 1996a). Thompson and Withers (1998) report on evaporative water loss in juvenile *V. mertensi*.

Fat Bodies, Testicular Cycles

Unknown, but this species is more stable with respect to seasonal body condition than sympatric terrestrial varanid species in the wet dry tropics, which generally decrease in mass during the dry season (Christian et al. 1995, 1996c).

Parasites

Gastric nematodes of at least two species (*Abbreviata hastaspicula* and *Tanqua tiara*) have been found in *V. mertensi* (Jones 1988).

7.14. *Varanus mitchelli*

TIM SCHULTZ AND SEAN DOODY

Figure 7.32. Varanus mitchelli *(photo by Robert Browne-Cooper).*

Nomenclature

Varanus mitchelli was named after Francis J. Mitchell, the former curator of reptiles at the South Australian museum. It is commonly known as Mitchell's water monitor and was first described by Mertens (1958). The holotype (SAM R. 3230) is kept in the South Australian Museum and was collected 5 mi west of Oenpelli in the Northern Territory (Fig. 7.32).

Geographic Distribution

V. mitchelli occur in the wet-dry tropics of northern Australia. Their range stretches north of 18.5° latitude and west of the Northern Territory–Queensland border through to the Kimberley region of Western Australia (Fig. 7.33).

AM R9913 Roper River N.A.—14.7 134.7; AM R39538 Deaf Adder Gorge—13.1 132.9; AM R41645 Nourlangie—12.8 132.7; AM R48659 Batten Creek Crossing between Borroloola and Bing-Bong—15.9 136.2; AM R51915, AM R51916 Adelaide River Township—13.3 131.1; AM R54911 ~70 km upstream Tomkinson River—12.4 134.3; AM R55052 Batten Creek—16.3 135.7; AM R64806, AM R64807, AM R64808 Katherine River—14.4 132.4; AM R75342, R75343 upper reaches of the Mckinlay River—13.2 131.7; AM R75360, R75363 mid reaches of the Mckinlay River—13 131.7; AM R76803, R76804 Bullo Rd. Stn. Rd. 23 km NW of Victoria Hwy.—15.8 129.7; AM R112435 Jabiru—12.7 132.9; AM R112436

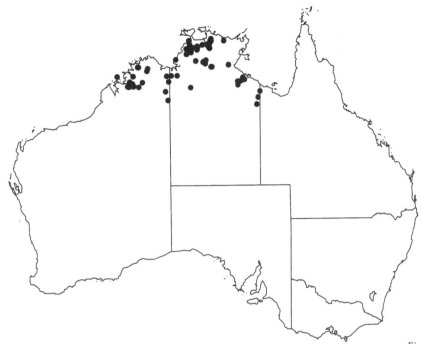

Figure 7.33. Known localities for
Varanus mitchelli.

Magela Creek 500 m upstream from Mudginberri Billabong; AM R128129 Mataranka—14.9 133.1; AM R141738 Little Roper Creek 6 km E of Mataranka—14.9 133.1; AM R142915 Roper River, Mataranka Homestead; ANWC R0933 Darwin Area ?; ANWC R0944 5.1 km N of Borroloola Rd. on road to Bauhinia Downs—16.6 135.7; ANWC R0945 Batten Creek 13 km W of Ryan Bend—16.2 136; ANWC R1655 Burketown Crossing, Borroloola—16.1 136.3; ANWC R4577 Deaf Adder Gorge, Kakadu National Park—13.1 132.9; NMV D52116 3.2 km E of Mataranka; NTM R25954 Oenpelli—12.3 133; NTM R25955 Ludmilla School, Darwin—12.4 130.9; NTM R25956 Claravale—14.4 131.6; NTM R25957 Rapid Creek, Darwin—12.4 130.9; NTM R25958 Tumbling Waters—12.8 131; NTM R25959 Oenpelli—12.3 133; NTM R25960 Katherine—14.5 132.3; NTM R25961 Coomarlie Creek—13.1 131.2; NTM R25962 Nicholson River—Spring Valley—18.5 137.6; NTM R25963 Nicholson River Gorge—18.1 128.8; NTM R25964 Annaburroo Homestead, 7 km E—13 131.7; NTM R25965 8 km SE Oenpelli—12.3 133.1; NTM R25966 17 km SW Oenpelli, E Alligator River—12.4 133; NTM R25967 Katherine Low Level—14.6 132.5; NTM R25968 W Alligator River Arnhem Hwy.—12.8 132.2; NTM R25969 Katherine Gorge National Park—14.3 132.5; NTM R25970 Daly River, ~10 km downstream Wooliana—13.7 130.6; NTM R25971 Adelaide River, Daly River Rd. Bridge—13.5 131.1; NTM R25972 Gold Creek, Wollogorang Stn.—17.2 138; NTM R25973 Litchfield National Park, Tjaynera Falls Area—13.3 130.7; NTM R25974 Bamboo Creek, Annaburroo Stn.—12.9 131.7; NTM R25975 Nicholson River—17.8 137.7; NTM R25976 Dee Creek—14.2 129.6; NTM R25977 Florence Falls, Litchfield National Park—13.2 130.6; NTM R25978, R25979 Keep River National Park—15.8 129.1; WAM R24060 Mt. Hart within 6 km of HS—16.8 124.9; WAM R26344 Katherine near a spring—14.5 132.3; WAM R32250 Oenpelli Mission—12.2 133.1; WAM R32309 Hann River. 14 km SW of Gibb River HS—16.4 126.3; WAM R46959 Prince Regent River Reserve—15.8 125.3; WAM R50544 Drysdale River National Park—15.1 126.8; WAM R50678 Drysdale River National Park—15.3 126.7; WAM

R56459, R56462 Plain Creek, Beverley Springs Stn.—16.7 125.4; WAM R56468 15 km NW of Mt. Hart HS—16.7 124.8; WAM R58848 Kununurra—15.8 123.7; WAM R60110 Kununurra at Diversion Dam—15.8 128.7; WAM R60111 Behn River 3 km from its confluence with the Ord—16.3 128.8; WAM R60132, R60139 ~50 km S of Victoria River Downs—16.9 131; WAM R77001, R77144, R77409, R77605 Mitchell Plateau—14.9 125.8; WAM R83355, R83356 Dunham River-Near Kununurra—15.8 128.7; WAM R83687 Galvans Gorge—16.8 125.9; WAM R83920 4 km SSE of Mt. Daglish-Calder River—16.3 125; WAM R99060 6 km N Lake Gilbert Near Beverley Springs—16.5 125.3; WAM R103081 Bungle Bungle National Park—17.3 128.5; SAM R03230 5 mi W of Oenpelli—12.4 133; SAM R05920 Katherine Gorge—14.3 132.5; SAM R34266 Pine Creek/Jabiru Rd. crossing at Nourlangie Creek—12.8 132.7.

Fossil Record

No fossils of *V. mitchelli* are known.

Diagnostic Characteristics

- A small- to medium-sized varanid with slender build.
- A long black tail, circular in section proximally, strongly laterally compressed distally with a median double keeled crest.
- Medium-sized "polished" head scales, 90 to 130 scales around the midbody and 27 to 31 lamellae under the fourth toe (Cogger 1986; Storr 1980).
- Throat and mouth with black vertical bars.
- Nostrils positioned dorsolaterally, slightly closer to the snout tip than the eye.

Description

Adult *V. mitchelli* are generally dark colored above with light colored ventral surfaces. The patterning may vary geographically (Storr 1980), but dorsal surface is blackish gray to olive gray, flecked with numerous cream to yellow spots or pale black-centered ocelli. The dorsal surface of the upper back, neck, and head may also be flecked with orange, and the throat and side of the neck is bright yellow; these characteristics are more distinct in Northern Territory specimens. The limbs are dark with white-yellow spots and long, curved nails. The tail is black above and below. Aside from the tail, limbs, and throat, the ventral surface is unblemished with creamy yellow to white coloration.

Size

There is no sexual dimorphism in size of adult *V. mitchelli* (Shine 1986). However, there may be some difference in body size between populations. In Kakadu National Park (Northern Territory), adult specimens ranged in SVL from 22 to 32 cm (Shine 1986), whereas maximum SVL of specimens examined by Storr (1980) from the Kimberley was only 25 cm. Tail length ranges from 1.7 to 2.1 times SVL

(Storr 1980); therefore, maximum TL is almost 100 cm. On the basis of Shine's (1986) data, adult mass ranges from 110 to 550 g. Hatchlings emerge at about 8 cm SVL (Shine 1986).

Habitat and Natural History

V. mitchelli is restricted to riparian habitats, primarily freshwater, although it does venture occasionally into brackish water. During the wet season (November–April) *V. mitchelli* exploit small creeks and inundated paperbark forests. When these habitats dry out in the dry season, *V. mitchelli* retreat to areas of permanent water (e.g., billabongs, large rivers) (Shine 1986). Like many varanids, this species will also make use of man-made habitats and are found along concrete drainage channels.

V. mitchelli use many different types of vegetation, for example pandanus (*Pandanus aquaticus*), paperbarks (*Melaleuca* sp.), freshwater mangroves (*Barringtonia acutangulum*), and bamboo (*Bambusa arnhemica*). Vegetation (dead or alive) may be close to, in, or overhanging water. Rocky habitats are also used extensively, when adjacent to watercourses.

This species is particularly alert, wary, and difficult to approach. Shine (1986) reports that escape is typically attempted via an arboreal, rather than an aquatic, pathway. However, Shine's (1986) findings may reflect his (aquatic) approach, as our observations indicate that when capture is attempted from the land, animals attempt escape by swimming and diving. Otherwise, tall tree hollows, and the roots of *Pandanus* are highly favored refuges, as are rock crevices in a saxicoline environment.

This species is common within suitable habitat. Doody (unpublished observations) counted an average of 0.73 lizards per kilometer along 35 km of one side of the Daly River (Northern Territory). Although this is only a crude estimate of density and underestimates the true population size because inactive or hidden lizards could not be counted, *V. mitchelli* was the most commonly encountered varanid.

Doody's survey work has been initiated to measure any impacts caused by the imminent arrival of the exotic cane toad (*Bufo marinus*), which is expected to become totally sympatric with *V. mitchelli* (Sutherst et al. 1996). On the basis of the diet of *V. mitchelli* and habitat use of both species, Burnett (1997) categorized *V. mitchelli* as a species likely to be severely affected if it ate poisonous toads.

Mitchell's water monitors are preyed upon by other varanids and by raptors. Doody (unpublished observations) has witnessed two successful predation events by *V. panoptes* and one success (from three attempts) by collared sparrow hawks (*Accipiter cirrhocephalus*).

Terrestrial, Arboreal, Aquatic

Although *V. mitchelli* is typically regarded as being semiaquatic, it is much more commonly encountered on vegetation, and may also be found foraging on the riverbank. In some areas, *V. mitchelli* are probably more reliant on rock crevices than trees.

Time of Activity (Daily and Seasonal)

Shine (1986) reported a peak in activity corresponding with late morning, but animals are active throughout the day. They mainly sleep in hollow logs or rock crevices, but may occasionally sleep on exposed branches overhanging water. During the early dry season, *V. mitchelli* are often seen basking in early morning (0830–0930 hours) near their overnight retreats (Doody, unpublished data).

Shine (1986) noted a reduction in activity of *V. mitchelli* in the dry season in Kakadu National Park. Additionally, Doody (unpublished observations) counted 40% fewer *V. mitchelli* in July than in May, indicating that these monitors become less active as the dry season progresses. This behavior closely matches that of its sister species *V. scalaris* (Christian et al. 1996a), but contrasts to that of sympatric *V. mertensi,* which is active year round (Christian et al. 1996b).

Thermoregulation

Little is known about the operating temperatures of this species. *V. mitchelli* generally start basking at around 0830, and bask most frequently on horizontal branches or rocks. Their peak in activity between 1000 and 1100 hours (Shine 1986) indicates that this species is easily able to achieve adequate body temperatures, aided by the high year-round temperatures within its range.

Foraging Behavior

Dietary studies (Shine 1986; Losos and Greene 1988) suggest that *V. mitchelli* forage on trees, on the ground, and in water. Only one account on foraging in this species has been published. Vincent and Wilson (1999) report several individuals diving into the Daly River from overhanging branches to catch passing fish.

Diet

Mitchell's water monitor's diet is wide ranging, undoubtedly as a result of the different environments in which they forage. Spiders, crabs, snails, insects (dragonflies, cockroaches, grasshoppers, hemipterans, beetles, ants, caterpillars, cicadas), fish, frogs, reptiles and their eggs, birds, and mammals have all been found in this species' guts (Shine 1986; Losos and Greene 1988). In quantitative terms, spiders, grasshoppers, and fish are the most important prey items, reiterating the range of habitats in which *V. mitchelli* forage.

In captivity, *V. mitchelli* can be maintained on live insects, small mice, fish, and minced meat.

Reproduction

Sexual maturity in *V. mitchelli* from Kakadu is attained at a SVL of about 22 cm (Shine 1986). As with many small- to medium-size monitors, there is no sexual dimorphism in size at sexual maturity. Oviposition (clutch size 7 to 12) occurs between the end of the wet season (April) and the middle of the dry season (June) (Shine 1986), with

TABLE 7.12.
Summary of breeding data for *V. mitchelli* at Perth Zoo.

Characteristic	Value
Copulation date	Apr
Laying date	May–Aug
Hatching date	Feb
Egg incubation period (d)	190
Egg incubation temperature (°C)	30–31.5
Clutch size	16–20
Egg weight (g)	4.21–6.08
Egg length (mm)	31.4–27.0
Egg diameter (mm)	17.0–19.3
Hatchling SVL (mm)	68
Hatchling weight (g)	3.94

mating presumably occurring late in the wet season. Hatchlings have been observed late in the dry season (September), indicating that incubation lasts between 3 and 5 months in the wild.

In captivity, clutch sizes of up to 20 eggs (each of 5 g) have been observed (Gavin Bedford, personal communication). One female (SVL 26 cm) laid in late April, with copulation occurring some 6 weeks previously. The lizard did not excavate a burrow, and eggs were laid on the bare substrate. Table 7.12 summarizes a breeding event at Perth Zoo, Western Australia.

Movement

Although virtually nothing is known about movement patterns in *V. mitchelli*, given its small to medium size and strong association with water, it probably has a fairly limited home range. Shine (1986) reports movement between different habitats in different seasons.

Physiology

The standard metabolic rate of *V. mitchelli* has been measured in six specimens at 6°C intervals between 18 and 36°C. At a mean mass of 116 g, the standard metabolic rate (measured in m g^{-1} h^{-1}) averaged: 0.021 ± 0.001 at 18°C, 0.036 ± 0.003 at 24°C, 0.070 ± 0.004 at 30°C, and 0.122 ± 0.005 at 36°C (Schultz 2002). These values, and the rate of increase in metabolism with temperature, are typical of varanids, and indeed all types of lizards (Thompson and Withers 1997b; Schultz 2002).

Maximal metabolic rate was also measured in the same specimens during treadmill exercise at 36°C. In general, *V. mitchelli* could sustain low to moderate treadmill speeds (0.8 km/h) for less than 10 minutes. This poor performance was reflected in a low maximal metabolic rate of 0.97 ± 0.07 mL g^{-1} h^{-1}. This is about half the expected value for a lizard of its size and is significantly lower than most other varanids studied (Thompson and Withers 1997b; Schultz 2002).

Fat Body Cycles

No information is available on fat bodies in this species.

Parasites

Jones (1988) reports infection by four species of gastric nematode in *V. mitchelli*: two species of *Abbreviata* and one each of *Tanqua* and *Hastospiculum*. More than 50% of individuals were infected, although the intensity of infection was generally low, with an average of 2.8 worms per lizard. Larval *Abbreviata* were found in 35% of stomach walls of *V. mitchelli*.

7.15. *Varanus panoptes*

Keith Christian

Figure 7.34. Varanus panoptes. *(upper)* horni. *(lower)* rubidus. *(Both photos by Jeff Lemm.)*

Nomenclature

Varanus panoptes is described by Storr (1980) as consisting of two subspecies: *V. panoptes panoptes* (holotype from Lake Argyle, Western Australia, 16°03'S, 128°47'E) and *V. panoptes rubidus* (holotype from 60 km north-northwest of Cue, Western Australia, 26°06'S, 117°42'E) (Fig. 7.34). Both specimens are held in the Western Australian Museum. A third subspecies from New Guinea (*V. panoptes horni*) was subsequently described by Böhme (1988a). The status of this nomenclature was thrown into doubt when Böhme (1991) discovered that the lectotype of *V. gouldii* was, in fact, conspecific with the taxon described by Storr (1980). However, the nomenclature was subsequently restored to the commonly accepted distinction between these two species (Shea and Cogger 1998), with the associated ruling in 2000 (Anonymous 2000).

Geographic Distribution

V. panoptes is found across northern parts of Western Australia, Northern Territory (as far south as the Barkley Tablelands), and Queens-

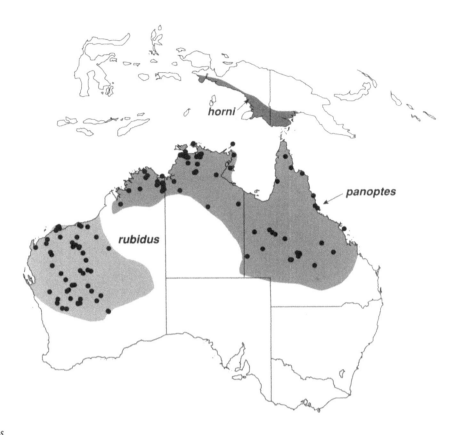

Figure 7.35. Known localities where V. panoptes *has been collected.*

land (as far south as the channel country). It is also found in arid western regions of central Western Australia and in southern New Guinea (Fig. 7.35).

AM R100438 The Middle Branch of the Gascoyne River at the Great Northern Hwy.—25.2 119.3; AM R100500 25.4 km E of the Port Hedland Rd. via the Wittenoom-Newman Rd.—22.5 118.9; AM R100563 ~3 km SW of the De Gray River via The Northern Hwy.—20.3 119.2; AM R111350 Smith Point, Cobourg Peninsula—11.1 132.1; AM R111351 Black Point, Cobourg Peninsula—11.2 132.1; AM R112438 1 km E of Sandy Crossing, vicinity of Jabiru; AM R112439 1 km W of Jabiru on Arnhem Hwy.—12.7 132.9; AM R112440 Arnhem Hwy., vicinity of Jabiru—12.7 132.9; AM R128311 Wenlock Goldfields, Batavia River (Batavia Downs, Wenlock Rvr)—12.7 142.7; AM R131853 Nankeen Billabong, Magela Creek—12.4 132.9; AM R141735 Katherine—14.5 132.3; AM R143871 74 km NW of Boulia TO on Lansborough Hwy.—21.9 142.6; AM R143872 26 km SE Mckinlay on Lansborough Hwy.—21.4 141.5; AM R143873 17 km NW Mckinlay (Gidyea Creek) on Landsborough Hwy.—21.2 141.2; AM R143874 39 km SE Cloncurry on Landsborough Hwy.—20.9 140.8; ANWC R0549 Koolpin Creek—13.5 132.6; ANWC R0550 Cannon Hill—12.4 133; ANWC R1140 Humpty Doo Rice Fields—12.6 131.4; ANWC R1443 Katherine Gorge—14.3 132.5; ANWC R4077 Arnhem Hwy., Kakadu National Park—12.7 132.5; NMV D4912 Oenpelli—12.3 133.1; NMV D5155 Roper River—14.7 134.7; NMV D34184 Behn River, Ord River Dam Area—16.3 128.8; NTM R25980 Stuart Park, Darwin—12.5 130.9; NTM R25981 Ivanhoe Crossing, Ord River, 5 km W—15.6 128.7; NTM R25982 Arnhem Hwy., toward Alligator Head, 2 km N—12.8 131.5; NTM R25983 9.7 km S Winnellie PO—12.5 131; NTM R25984 Nightcliff High School Darwin—12.4 130.8; NTM R25985 Darwin—12.4 130.9; NTM R25986 Malgala, Groote Eylandt—14 136.4; NTM R25987 Dinah Beach Rd., Darwin—12.5 130.8; NTM R25988 Dubbo; NTM R25989 Mandorah, Cox Peninsula—12.5 130.7; NTM R25990 De Gray River—20.4 119.4; NTM R25991 Beatrice Hill—12.7 131.3;

NTM R25992 Humpty Doo—12.6 131.3; NTM R25993 Humpty Doo (Arnhem Hwy.)—12.6 131.3; NTM R25994 3.5 mi S Katherine—14.5 132.3; NTM R25995 222.9 km from Stuart Hwy. Arnhem Hwy.—12.6 132.7; NTM R25996 Humbert River Crossing, Victoria River Downs—16.5 130.6; NTM R25997 Fogg Dam—12.6 131.3; NTM R25998 E Marchinbar Is. opposite Jensen Bay—11.2 136.7; NTM R25999 Ban Ban Springs—13.4 131.5; NTM R26000 Humpty Doo—12.6 131.3; NTM R26002 Junction of Ord/Behn Rivers—16.4 128.8; NTM R26003 Cape Hotham Is.—12.1 131.3; NTM R26004 Renner Springs—18.3 133.8; NTM R26005 Wyndham—15.5 128.1; NTM R26006 Fannie Bay, Darwin—12.5 130.8; NTM R26007 Arnhem Hwy., 0.8 km W Jim Jim TO—12.6 132.8; NTM R26008, R26009, R26010 Jabiru Area—12.7 132.9; NTM R26011 Adelaide River, on Arnhem Hwy., 2 km W—12.7 131.4; NTM R26012 Elizabeth River, 3.8 km N Noonamah—12.6 131.1; NTM R26013 Victoria River—15.5 130; NTM R26014 Beatrice Hill—12.7 131.3; NTM R26015 Nicholson River—18 137.3; NTM R26016 Darwin, Fannie Bay Beach—12.4 130.8; NTM R26017 Darwin, Fannie Bay—12.5 130.8; NTM R26018 Katherine River; QM R2981 Port Darwin—12.5 130.7; QM R4721 Barcarolle, Longreach—23.5 144.3; QM R5131 Aloomba, near Cairns—17.1 145.8; QM R12934 Cardwell—18.3 146; QM R14752 Mitchell Rd. Mission—15.5 141.8; QM R20641 Wakooka Outstation, Granite Creek, 3.2 km E—14.6 144.6; QM R24016 Capella, Emerald Rd.—23.3 148.1; QM R32798 Silver Plains, via Coen—14 143.4; QM R38106 Epping Forest Stn., 136 km NW Clermont—22.3 146.8; QM R45864 Hinchinbrook Is., Georges Point—18.5 146.3; QM R45865 Hinchinbrook Is., Georges Point—18.5 146.3; QM R46292 Carlisle Is.—20.8 149.3; QM R46770 Blackall, 63 km S—24.7 146; QM R48291 Boulia, 16 km E, on road to Winton—23 140.1; QM R48943 Mt. Isa, 40 km S, on road to Dajarra—21.1 139.1; QM R52589 Longreach-Jundah Rd., N side of Ernestina Creek—23.8 144.1; QM R54281 Noonbah Stn., near Cottage—24.1 143.2; QM R54365, R54367, R55347 Noonbah HS—24.1 143.2; QM R54577 Silsoe Rd., 25 km W of Longreach—23.4 144; QM R55304 Simpson Desert National Park—23.7 138.3; QM R57213 Longreach, 10 km S, Longreach-Jundah Rd.—23.5 144.2; QM R57214 near Thomson River on Tonkoro Rd.—24.1 143.4; WAM R72891 3 km NE Mt. Linden—29.3 122.4; WAM R78236 2 km NE of Goolthan Goolthan Hill—28.1 116.7; WAM R84016 20 km NNE of Karalundi—26.6 118.7; WAM R125765 Dale Gorge Hamersley Ranges—22.6 118.6; WAM R22368 Kimberley Research Stn. N of Kununurra—15.7 128.7; WAM R23450 E Point Darwin—12.4 130.8; WAM R26782 Grotto Creek Bridge—15.7 128.3; WAM R28009, R28017 Mt. Anderson—18 123.9; WAM R28033, R28035 Kalumburu—14.3 126.6; WAM R32173 Mt. Anderson—18 123.9; WAM R44791 vicinity of Main Dam Lake Argyle—16.1 128.7; WAM R44792, R44793 Lake Argyle, vicinity of Main Dam—16.1 128.8; WAM R44794, R44795 Lake Argyle—16.3 128.8; WAM R44796 ~12 mi S of Main Ord Dam Site Lake Argyle—16.3 128.7; WAM R47717 3 km S of Old Lissadell—16.6 128.6; WAM R50492 Drysdale River National Park—14.8 127; WAM R50549 Drysdale River National Park. Carson River—15.1 126.8; WAM R60217, R60218 Lake Argyle Main Ord River Dam Site—16.1 128.7; WAM R74057 Gove—12.5 136.8; WAM R75547 3 km WSW of Kununurra—15.8 128.8; WAM R86967 7 km N Mount Elizabeth HS—16.3 126.1; WAM R99258 6 km E Mount Talbot—16.5 124.8; WAM R12907 S of Fields Find. Salt River—29.1 117.3; WAM R13307 Ullawarra Stn. Barlee Ranges—23.5 116.1; WAM R14331 Dolphin Is., Dampier Archipelago—20.5 116.8; WAM R14884 52 km W of Tambrey—21.6 117.1; WAM R14889 37 km E of Mt. Magnet—28.1 118.2; WAM R14890 Billabalong. 5 km N of Homestead—27.4 115.8; WAM R19132 Wilgie Mia. 60 km NNW of Cue near Glen Stn Rd.—26.1 117.7; WAM R26261 11 km E of Wittenoom—22.2 118.5; WAM R27233 Wanjarri Stn. Kathleen Valley—27.3 120.6; WAM R30912 11 km S of Yalgoo—28.5 116.7; WAM R51524 35 km N of Ilgararri Creek—24.1 120; WAM R51614 Peedamulla. 80 km NNE of Onslow—21.9 115.6; WAM R53058 Yinnietharra HS—24.7 116.2; WAM R53507 16 km S of Yalgoo—28.5 116.7; WAM R53689 Errabiddy Stn.—25.5 117.1; WAM R61509 Myaree Pool Maitland River—20.9 116.6; WAM R62868 21 km SSE of Mt. Keith—27.3 120.6; WAM R63087 16 km 210° Meekatharra—26.7 118.4; WAM R63807 6 km W Giralia HS—22.7 114.3; WAM R65958 2.5 km N of Mt. Linden.

Camp—29.3 122.4; WAM R65959 7.75 km SSE Mt. Linden—29.4 122.5; WAM R65960 2.5 km N Mt. Linden—29.3 122.4; WAM R69071 8 km ENE Yuinmery HS—28.5 119.1; WAM R69072 7.5 km E Yuinmery HS—28.6 119.1; WAM R69129 18 km NE Yuinmery HS—28.5 119.2; WAM R69198 Vickers Creek 7.5 km SE Banjiwarn HS—27.8 121.7; WAM R69928 15 km E of Big Bell—27.3 117.8; WAM R70802 25.7 km 235° Marillana HS—22.8 119.2; WAM R70964, R70965 Dandaraga Stn.—28.1 119.3; WAM R73006 13 km SW Tuckanarra—27.2 118; WAM R73165 22 km SW Marillana HS—22.6 119.2; WAM R74095 Tooncoonaragee Pool Oakover River—21.6 121.2; WAM R74698 24 km ENE Yuinmery—28.5 119.2; WAM R74825 Shay Gap—20.5 120.2; WAM R82626 Hauy Is., Dampier Archipelago—20.4 117; WAM R84070 30 km S of Glenflorrie Homestead—23.3 116; WAM R84408 25 km SW of Mt. Methwin—25.2 120.5; WAM R84409 13 km SE of Old Beyondie (Marymia)—25.1 120.2; WAM R87071 23 km N of Mt. Murray—22.3 115.5; WAM R97353 Neds Creek HS—25.5 119.7; WAM R98075 Bundarra Stn.—28.3 121.2; WAM R100214, R100215 Arafura Dune—14.8 128.7; WAM R100351 Mount Lawrence Wells—26.8 120.2; WAM R102099 13 km NW Mt. Cotten—22.8 122.6; WAM R103950 20 km N Paynes Find—29.1 117.7; WAM R104128 Woodstock, Site 4—21.6 119; WAM R114759 Cleaverville—20.7 117; WAM R119905 Emma Gorge, Cockburn Range—15.8 128; WAM R125765 Dale Gorge Hamersley Ranges—22.6 118.6; WAM R135025 Mt. Whaleback—23.3 119.7.

Fossil Record

Unknown.

Diagnostic Characteristics

- Large, males up to 160 cm TL; females rarely more than 100 cm TL.
- Background color ranging from dark reddish brown (in arid Western Australia) to dark gray or black with small yellow spots and flecks and dark spots forming bands.
- Bright yellow horizontal stripes on face.
- Tail banded to the tip, compressed, and about 1.4 times SVL (Bedford and Christian 1996).

Description

This robust species is characteristically aggressive and wary. Apart from the diagnostic color patterns, this species differs from sympatric populations of *V. gouldii* in several respects. Although *V. gouldii* is generally restricted to sites with sandy soils, *V. panoptes* live in a very wide range of habitats. When approached, the former species either hides or runs a short distance before hiding, whereas the latter species either displays aggressively or runs to shelter, which may be some considerable distance away. Indeed, the two species can be distinguished while being held in cloth collecting bags: if the bag is prodded, *V. gouldii* typically do not respond, but *V. panoptes* almost invariably respond with a jerk or a hiss.

Size

Varanus panoptes are strongly sexually dimorphic in size (Shine 1986), with the largest males having a mass of 7 kg; adult females are typically about 1.5 kg, rarely above 2 kg (Christian et al. 1995). Maximum SVL recorded is 740 mm.

Habitat and Natural History

V. panoptes is sometimes called the floodplain goanna in the far northern parts of the Northern Territory, and they can be found in fairly high densities along edges of floodplains and in riparian habitats. However, this species lives in a wide range of habitats, including coastal regions, savanna woodlands (where it is often sympatric with *V. gouldii*), mangrove fringes, and urban environments.

Terrestrial, Arboreal, Aquatic

Although generally terrestrial, these goannas can climb, although more clumsily than other species. Along margins of floodplains, they seem to prefer milkwood trees (*Alstonia actinophylla*), presumably because the soft, corky bark makes climbing easier. These lizards typically sleep in burrows dug in the ground, but occasionally they roost in hollow trees or logs. They sometimes hunt in shallow water of floodplains and are capable swimmers.

Time of Activity (Daily and Seasonal)

In the wet-dry tropics of northern Australia, many species, including some varanids, substantially reduce their activity in the dry season (Christian et al. 1995, 1996a, 1996b, 1999). The seasonal activity pattern in *V. panoptes* is dependent on the habitat an individual inhabits (Christian et al. 1995). In woodlands, this species exhibits the same pattern as the sympatric *V. gouldii*, which burrow underground early in the dry season and remain largely inactive until rains begin about 4 or 5 months later. However, individuals living along floodplains and near billabongs or creeks remain active as long as water and food resources are available. Along floodplains, individuals are most active in the middle of the dry season because many prey species congregate in or near the remaining water as the floodplain dries. These individuals typically retreat underground late in the dry season after the floodplain has dried and are inactive for about a month before the season changes. Individuals living near permanent water are apparently active throughout the year.

Daily activity of this species has been quantified in detail (Christian et al. 1995; Christian and Weavers 1996; see below). During active seasons, these goannas are active throughout the daylight period (Shine 1986; Christian and Weavers 1996).

Thermoregulation

Thermoregulatory indices reveal that, compared with *V. gouldii* and *V. mertensi* from the same area, *V. panoptes* are less careful thermoregulators (Christian and Weavers 1996). Nevertheless, *V. panoptes* exhibit typical basking behavior and attain high daytime body temperatures (mean, 36.4°C) when they are active. The solar absorptance of this species is about 87% (Christian et al. 1996a), which is at the high end of the range of values measured in 12 species of varanids.

Foraging Behavior

V. panoptes is an active forager, and this is particularly evident in those individuals living near floodplains during the dry season. On a typical day for these animals, they walk from their burrows at the edge of the floodplain to the water's edge (as far as 2 km), then forage along the edge of the water for most of the day before returning to their burrow late in the afternoon. The mean amount of time spent walking is 3.5 hours per day, with a maximum of 6.6 hours per day (Christian et al. 1995; Christian and Weavers 1996). Although comparable data are scarce, no other lizard has been measured walking for this length of time in a day (Christian et al. 1997). During the wet season, *V. panoptes* typically spend less than 1 hour per day in locomotory activity, but given rapid gains in mass during this highly productive season, they presumably can obtain abundant prey in short forays (Christian et al. 1995; Christian and Weavers 1996). A *V. gouldii* was eaten after a *V. panoptes* dug it out of a burrow (in which it had been for at least 5 days) not via the plugged entrance, but rather straight down to the animal (the *V. gouldii* was being tracked by radio, so its exact location was known to me—and apparently also to the *V. panoptes*). This raises an interesting question about how the *V. panoptes* knew the exact location of the *V. gouldii* (~1.5 m from the plugged burrow entrance) through about 40 cm of soil (Christian 1995). Similarly, *V. panoptes* extract small grubs from deep holes and they regularly raid turtles nests, suggesting that they have an extraordinary acute sense of smell.

Diet

These monitors will eat virtually any animal material, including carrion, eggs, invertebrates, and vertebrates (Shine 1986; Losos and Greene 1988, personal observation). Some noteworthy prey items of note include frillneck lizards (*Chlamydosaurus kingii*), death adders (*Acanthophis praelongus*), file snakes (*Acrochordus arafurae*), and a *V. gouldii* that had a mass equivalent to 11% of the body mass of the predator (Christian 1995).

Reproduction

Few direct records pertain to reproduction in *V. panoptes,* but available information indicates that oviposition occurs around April (Blamires and Nobbs 2000; Shea and Sadlier 2001; Christian, personal observation), with hatching occurring around October (Shea and Sadlier 2001; G. Bedford, personal communication). However, contractors uncovered a clutch of 12 eggs at Casuarina Coastal Reserve, Darwin, Northern Territory, on December 26, 1998. These eggs (mass range, 54–61 g) began to hatch July 14, 1999, after being incubated at an unspecified temperature. Hatchlings ranged in mass from 27.8 to 32.2 g (G. Husband, unpublished data), and SVL was 120 to 129 mm. Clutch sizes range between 6 and 14 eggs.

One interesting, but poorly understood, aspect of the reproductive biology of *V. panoptes* relates to the use of aggregated burrows by females. I have seen five such "warrens," which consist of 10 to 20 burrows in a small area (approximately ranging between 25 and 60 m²). Although observations of behavior at these sites have proved difficult, I have observed unusually high densities of individuals (mostly females) in the vicinity. I have been told by Aboriginal women in Kakadu National Park that the females deposit eggs in these burrows. If this proves to be correct, it raises additional questions related to the function of communal nesting, the degree of interconnectedness of the burrows, and behavioral interactions among individuals that used the site. Communal nesting in iguanas has generally been explained by availability of appropriate nesting conditions, including soil type (Christian and Tracy 1982). A similar explanation may be appropriate for aggregated burrows of *V. panoptes*. All sites I have seen were in well-drained soils, and four of five sites were on slightly elevated ground near the edge of floodplains, and similar observations have been made on the Maningrida floodplains in western Arnhem Land (T. Schultz, personal observation).

Movement

In the dry season, individuals living near floodplains move substantial distances while foraging (as much as 6 km), but during other seasons, they move less (Christian et al. 1995; Christian and Weavers 1996). During the wet season, these goannas probably obtain all the food they need without moving great distances, as evidenced by rapid increases in mass associated with only moderate amounts of activity (Christian et al. 1995).

Physiology

Seasonal field metabolic rates (FMRs), water flux rates, and body temperatures generally follow the seasonal activity patterns described above (Christian et al. 1995; Christian and Weavers 1996). *V. panoptes* has a high aerobic capacity compared with lizards generally (Thompson and Withers 1992; Christian and Conley 1994), but resting metabolism is typical for reptiles of their size (Christian and Conley 1994; Schultz 2002). Resistance to water loss is relatively low compared with arboreal varanids and with some smaller terrestrial species of varanids (Thompson and Withers 1997a).

Fat Bodies, Testicular Cycles

Testes of one male were enlarged in July (Blamires 1999), corresponding with observations of courtship (Shine 1986) and copulation (personal observation) in July.

Parasites

Gastric nematodes of at least two species (*Abbreviata hastaspicula* and *Tanqua tiara*) have been found in large numbers in *V. panoptes* (Jones 1988).

7.16. *Varanus pilbarensis*

DENNIS R. KING

Figure 7.36. Varanus pilbarensis
(photo by Brad Maryan).

Nomenclature

The holotype was collected in the Chichester Range in 1971, and the Pilbara monitor was described by Storr in 1980. The type specimen is in the Western Australian Museum (WAM R39782) (Figs. 7.36, 7.37).

Geographic Distribution

The distribution of this species is restricted to the Pilbara region of Western Australia.

WAM R10811 Abydos—21.4 118.92; WAM R13082, R60432: Woodstock—21.6 118.95; WAM R14901 Rim of Dales Gorge. Hamersley Range—22.5 118.6; WAM R20017 Black Hill Pool—21.3 117.25; WAM R28011, R28012 Woodstock—21.6 118.95; WAM R39158 Nullagine. 2 km W of Town—21.9 120.08; WAM R39782 Chichester Range. Cockeraga River—22.1 118.8; WAM R66282 Carawine Gorge —21.5 121.03; WAM R69695 Knox Gorge—22.4 118.3; WAM R90856 Wood-stock—21.6 119.02; WAM R102103 20 km SSW Cooya Pooya. Millstream-Chichester National Park—21.2 117.07; WAM R113094 42 km NNE Auski Road-house [20 km S White Springs]—22 118.83; WAM R125100 15 km E of Newman —23.4 119.9; WAM R125456 30 km E Newman—23.3 120.03; WAM R125766 Circular Pool, Dale Gorge Hamersley Ranges—22.6 118.56; WAM R100766 Wood-stock, Site 17—21.6 118.98.

Fossil Record

No fossils are known for this species.

Figure 7.37. Known localities for Varanus pilbarensis.

Diagnostic Characteristics

- A slender, small, red rock monitor (up to 47 cm in TL).
- Tail long (about 1.7 to 2 times the SVL), banded and circular, with weakly keeled scales, and clusters of pointed scales on each side of the base of the tail.
- Superior position of nostril, head scales small and smooth.
- No canthus rostralis.
- Head dotted or flecked with dark reddish brown (Storr 1980).

Description

A small, reddish rock monitor with a long circular tail with black and cream banding distally. Legs also mottled with black and cream. Throat and venter grayish white, irregularly spotted and reticulated with gray. Clusters of enlarged spiny scales at the base of the tail on either side of the vent. Nostril laterodorsal, about midway between tip of snout and orbit. Claws very short and thick and strongly curved (Cogger 1992; De Lisle 1996; Storr 1980; Storr et al. 1983). *V. pilbarensis* can be distinguished from *V. kingorum* both by locality (*V. kingorum* occurs in the Kimberley rather than in the Pilbara) and tail (the tail of *V. kingorum* is more strongly keeled and not banded). Also, *V. kingorum* has a weak canthus rostralis, whereas *V. pilbarensis* does not have one.

TABLE 7.13.
Summary of data on breeding of *V. pilbarensis* at the Perth Zoo.

Characteristic	Value
Laying date	Dec–Feb
Hatching date	Mar–Jun
Egg incubation period (days)	102, 103, 90
Egg incubation temperature (°C)	30–31.5
Maturation age	2 years, 5 months (male)
Clutch size	2–4 eggs
Egg weight (g)	2.48–3.5
Egg length (mm)	26–29
Egg diameter (mm)	12.8–14.2
Hatchling snout–vent length (mm)	55–56
Hatchling weight (g)	2.24–2.36

Size

SVL up to 169 mm, tail length 175% to 205% of SVL.

Habitat and Natural History

Rock dwelling, sheltering in crevices or in cavities among piles of boulders (Cogger 1992; Wilson and Knowles 1988). Also found in exfoliating granite outcrops and ironstone rock faces and gorges often associated with watercourses (Brad Maryan, personal communication).

Terrestrial, Arboreal, Aquatic
Terrestrial, mainly on rocky hillsides.

Time of Activity (Daily and Seasonal)
Diurnal, active throughout the year.

Foraging Behavior
Unknown; mostly seen on rocky slopes. Bennett (1995) reports that Johnstone (1983) records that *V. pilbarensis* forages for orthopterans.

Diet

Stomachs of 6 of 17 specimens contained food, which consisted of 3 with orthopterans, 2 with unidentified invertebrates, and 1 with a lizard (*Ctenotus* sp.) (James et al. 1992; Losos and Greene 1988).

Reproduction

All adult females (six) collected between September and March were reproductively active, whereas three collected at other times were not. All eight males collected between February and June had enlarged testes. One female collected on September 25 had five shelled eggs in her oviducts.

Eidenmüller and Langner (1998) reported that a pair of captive *V. pilbarensis* laid six clutches in less than 11 months, with a total of 24 eggs. Incubation time at $27 \pm 1°C$ was 99 to 136 days. The average SVL of the hatchlings was 13.9 cm, with an average mass of 2.6 g. Of the 24 eggs, eight hatched, and of these, five hatchlings died without taking any food. The other 16 were fully developed but died shortly before hatching. The three living hatchlings developed normally.

V. pilbarensis hatchlings, as in some other monitors, are more brightly colored than adults. They have a black background overall, an orange head, and spots on the dorsal surface (Eidenmüller and Langner 1998; King and Green 1999).

Movement

There is no information regarding movements of *V. pilbarensis*.

Physiology

No physiological data have been collected for this species.

Fat Bodies, Testicular Cycles

Large fat bodies were found in two animals captured in September. Those found in two animals in December and February were medium sized.

Parasites

An unidentified nematode was collected. *Amblyomma* sp. ticks were recorded on *V. pilbarensis* (King and Green 1999).

7.17. *Varanus primordius*
GRANT HUSBAND AND KEITH CHRISTIAN

Figure 7.38. Varanus primordius
(photo by Brad Maryan).

Nomenclature

Varanus primordius was originally described as a subspecies of *V. acanthurus* in 1942b by Mertens from a specimen collected some 80 years earlier (Husband 2001). This small varanid was elevated to the status of a separate species by Storr (1966) (Fig. 7.38). Holotype ZMB 6204 is from an undesignated locality in Northern Australia (Cogger et al. 1983).

Geographic Distribution

Varanus primordius is found in tropical parts of the Northern Territory, Australia (Fig. 7.39), and appears to be restricted to an area extending from Darwin in the north to Katherine in the south, and from Kakadu in the east to the Daly River region in the west (Husband 2001).

QM R25947, R25949 Berry Springs—12.7 130.967; AM R41650, R46045, R47508 Berry Springs—12.7 130.967; AM R49087 Pine Creek—15.85 135.933; NTM R26019, R26020, R26021, R26022 Berry Springs—12.72 131.05; NTM R26023 Berry Springs Reserve—12.7 130.967; NTM R26024 17 mi SE Darwin on

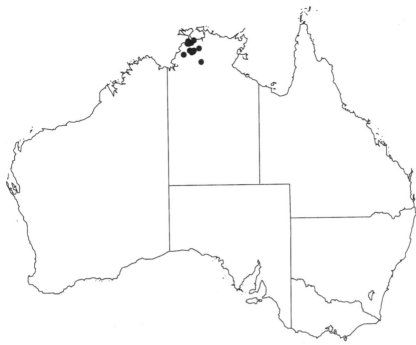

Figure 7.39. *Known localities for* Varanus primordius.

Stuart Hwy.—12.6 131.05; NTM R26025 Kakadu National Park, Stage Three—
13.2 132.05; NTM R26026 Berry Springs Reserve—12.7 130.967; NTM R26027
Katherine, Low Level Reserve—14.47 132.267; NTM R26028 Ban Ban Springs—
12.38 131.5; NTM R26029 Berry Springs Reserve—12.7 130.967; NTM R26030
Ban Ban Springs—13.38 131.5; NTM R26031 Darwin, 17 mi S—12.45 130.983;
NTM R26032 Adelaide River Town, 7 mi S—13.35 131.15; NTM R26033 Foun-
tain Head Rail Siding—13.47 131.483; NTM R26034, R26035, R26037 17 mi S
Darwin—12.62 131.05; NTM R26039 Darwin, 17 mi S—12.45 130.983; NTM
R26040 Elizabeth Downs Stn.—13.73 130.533; WAM R23779, R23780 58 km S of
Adelaide River Town—13.53 131.35; WAM R48820, R48821 Berry Springs—12.7
130.967; SAM R20588 17 mi S of Darwin on Stuart Hwy.—12.58 131.067; SAM
R30265 Humpty Doo—12.63 131.25.

Fossil Record

Unknown.

Diagnostic Characteristics

- Small, usually less than 30 cm TL.
- Dorsal color ranging from gray to reddish-brown interspersed with darker scales.
- Ventral color creamy.
- Tail round to slightly flat, about 1.4 times SVL (Bedford and Christian 1996).

Description

The only varanid smaller than *V. primordius* is *V. brevicauda*. Scales of the tail are very coarse, consistent with other species that live under rocks (Bedford and Christian 1996), but not as spinose as the tails of *V. acanthurus* and other spiny-tailed monitors (Vincent and Wilson 1999). *V. primordius* is secretive, and most observations of this species result from active searches under rocks or discarded building material (Husband 2001).

Size

Adult SVL is about 12 cm. Of 14 hatchlings from the Territory Wildlife Park near Darwin, mean SVL was 4.4 cm and mean TL was 10 cm (Husband 2001).

Habitat and Natural History

V. primordius is associated with rocky outcrops in tropical savanna woodland, living in shallow burrows under rocks on soil (Bedford and Christian 1996), but during the dry season, they can be found on black soil floodplains (Husband 2001). These black soil habitats are seasonally inundated, and given that these small monitors are poor climbers, they presumably move from this habitat to rocky outcrops during the wet season (Husband 2001). Although they are distributed patchily within their limited geographic range, they are common in some places (Husband 2001).

Terrestrial, Arboreal, Aquatic

Terrestrial.

Time of Activity (Daily and Seasonal)

Few records exist of activity, but presumably activity is restricted to daylight hours. In at least some populations, these lizards apparently occupy rocky outcrops in the wet season and nearby floodplains in the dry season (Husband 2001).

Thermoregulation

Unknown. Solar absorptance of this species is about 87% (at a body temperature of 35°C) (Christian et al. 1996a), which is at the high end of the range of values measured in 12 species of varanids. This fact, in conjunction with their small body size, implies that these animals would warm quickly when basking.

Diet

Diet consists of a range of arthropods (ants, crickets, cockroaches, neuropteran larvae, and centipedes), skinks, geckos, and reptile eggs (Losos and Greene 1988; Husband 2001).

Reproduction

The following information is from captive animals at the Territory Wildlife Park (Husband 2001). Courtship involves males following females, flicking their tongues and displaying with small jerky head movements. Females lay from two to five eggs (mean, 3.3, n = 9) about 2 to 3 weeks after copulation. Incubation periods (at temperatures ranging from 28.5 to 30.5°C) ranged from 87 to 120 days. Sexual maturity in captivity is reached in 9 to 12 months.

Movement

As mentioned above, in some populations, *V. primordius* move between rocky outcrops in the wet season and black soil floodplains in the dry season. Although there is no direct evidence for these movements (which could be more than 1 km), it is difficult to imagine how they could remain in inundated floodplains.

Nothing is known about the foraging behavior, physiology, fat bodies, testicular cycles, or parasites of *V. primordius*.

7.18. *Varanus rosenbergi*

Dennis R. King and Ruth Allen King

Figure 7.40. Varanus rosenbergi
(photo by Dennis King).

Nomenclature

Varanus rosenbergi was considered to be a subspecies of *Varanus gouldii* (Mertens 1957a) until 1980, when it was elevated to the status of a full species (Storr 1980). Type specimen is held at the Western Australian Museum (WAM R822); the type locality is Moingup Pass, Stirling Range, Western Australia. *Varanus rosenbergi* belongs with the subgenus *Varanus,* along with other members of the *gouldii* group (Fuller et al. 1998; Ast 2001) (Fig. 7.40).

Geographic Distribution

Geographic distribution is in the south coastal areas of Western Australia (to Mussel Pool, slightly north of Perth), South Australia and Victoria, Australian Capital Territory, and southern New South Wales (near Sydney, Ehmann et al. 1991). Eastern populations differ from those in the west (Schmida 1985, p. 166) and appear to have disjunct distributions—they may ultimately be split off as a new species (Fig. 7.41).

AM R3694 Loftus—34.0 151.05; AM R7126 Birchmore Lagoon, 15 mi from Kingscote, Kangaroo Is.—35.7 137.483; AM R7634 Cranbrook, SW Aust.—34.3 117.55; AM R11109 Woodlands, Tambellup—34.0 117.633; AM R13077 Collaroy Plateau—33.7 151.3; AM R15965 Waterfall—34.1 151; AM R19029, R19030, R19031, R19136, R19723, Penneshaw Kangaroo Is.—35.7 137.933; AM R20323 Mt. Keira, Wollongong—34.4 150.85; AM R21614 2 mi E Goulburn—34.7 149.75; AM R49193 Yalwal Plateau, SW of Nowra—34.9 150.383; AM R68811 Royal National Park, Sydney—34.1 151.067; AM R68813 Helensburgh, Sydney—34.1 150.983; AM R81556 Kangaroo Is.—35.8 138.05; AM R86871 Royal National

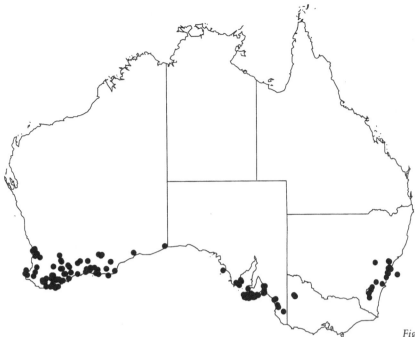

Figure 7.41. Map showing known localities for V. rosenbergi.

Park, Sydney—34.1 151.067; AM R86872 Helensburgh—34.1 151.983; AM R95810 Cooma Area, "Cundumbul" property of Mr. N. B. Carter, N of Cooma—36.0 149.1; AM R102312 approx. middle section of Mason Rd., Albany—35 117.867; AM R103531 Belrose—33.7 151.217; AM R105595 12.4 km W of Pallinup River Crossing on Eyre Hwy.—34.5 118.667; AM R105600 Hanraun Rubbish Tip, Albany—35 117.683; AM R105608 near Sleeman River Crossing on South Western Hwy.—34.9 117.467; AM R107267 The Forge, 20 mi N Bathurst—33.1 149.683; AM R111030 Terry Hills, Sydney—33.6 151.217; AM R116992 Bobbin Head, Ku-Ring-Gai National Park—33.6 151.15; AM R119494 ~30 km E of Moingup Springs Campsite, Stirling Ranges N.P. N—34.4 118.417; AM R120944 Waterfall, Royal National Park—34.1 151; AM R123331 19 km N of Kulnura on Wollombi Rd.—33.0 151.117; AM R123333 Wakehurst Parkway; AM R127397, R127398 Lucas Heights, Sydney—34.0 150.983; AM R131191 Port Lincoln—34.7 135.867; AM R132056 Menai—34.0 151.017; AM R133750 10.5 km E of Caiguna—32.2 125.567; AM R143983 Mangrove Mountain—Wollombi Rd., 200 m N of entrance to BU-33.1 151.183; AM R144656; AM R146188 Cowan, 5 km N on Pacific Hwy.—33.5 151.183; ANWC R0260 Mt. Thisby (=Prospect Hill) Kangaroo Is.—35.8 137.75; ANWC R5110 3 km S of Queanbeyan—35.3 149.233; ANWC R5149 7 km S of Billy Goat Gully, Kangaroo Is.—35.7 136.917; ANWC R5151 4 km S of Mt. Donnell, Kangaroo Is.—35.6 137.317; ANWC R5152 Ravine Des Casoars West Bay Rd., Kangaroo Is.—35.8 136.617; ANWC R5155 Denly Drive Off Macs Reef Rd, NE of Canberra—35.2 149.317; ANWC R5574 Boboyan Rd., S of Canberra—35.6 149.017; NMV D55455 17 km WNW of Chinaman Well—35.8 141.633; NMV D55628 0.6 km NW of Chinaman Well—35.8 141.665; NMV D53500 1.3 km N of Chinaman Well—35.8 141.665; WAM R419 Keysbrook—32.4 115.967; WAM R821 Moir Pass—34.4 118.117; WAM R822 Moingup Pass, Stirling Range—34.4 118.1; WAM R1870 Woodlands. Tambellup—34.0 117.65; WAM R3835, R3836, R3837, R3838 Balingup—33.7 115.983; WAM R5381, R5382, R5383, R5384, R5385, R5386, R5387, R5388 N

Fremantle—32.0 115.733; WAM R5950 Witchcliffe—34.0 115.1; WAM R9581 Scaddan—33.4 121.733; WAM R12298 Dumbleyung—33.3 117.733; WAM R13043 near Jerramungup—33.9 118.917; WAM R13080 Bremer River—34.3 119.25; WAM R14891 Geraldton [Editors' note: probably misidentified]; WAM R14896 11 km SW of Albany—35.0 117.883; WAM R14897 6 mi E Jacup—33.8 119.433; WAM R14898 Young River—33.5 120.983; WAM R17626 13 km SW of Israelite Bay—33.7 123.767; WAM R18481 Eucla—31.6 128.717; WAM R18483, R18484 Hopetoun—33.9 120.117; WAM R19251 6 km NNW of Bannister. 90 km S of Perth on Albany Hwy.—32.5 116.45; WAM R21732 23 km SE of Newdegate—33.2 119.2; WAM R21733 44 km SE of Newdegate—33.3 119.35; WAM R22534 Needilup—33.9 118.767; WAM R28148 Kuender Siding. Lake Grace—32.9 118.533; WAM R34543 32 km N of Lake Grace—32.8 118.467; WAM R36349 Two People Bay Reserve. In Gully, Opposite Coffin Is.—34.9 118.183; WAM R36350 Two Peoples Bay Reserve—35 118.183; WAM R36455 21 km Nnw of Bannister—32.2 116.183; WAM R41072, R41073 Chinocup Reserve. 6 km NW of Pingrup—33.5 118.567; WAM R42528 Hellfire Bay. Cape Le Grand National Park—34.0 122.167; WAM R42964 20 km W of Rocky Gully—34.5 116.783; WAM R42968 Lake Chinocup—33.4 118.433; WAM R44662 Kent River Area. between Denmark and Nornalup—35 117.017; WAM R51102 8 km NE of Bendering—32.3 118.35; WAM R51559 Mussel Pool. 14.5 km NNE of Perth—31.8 115.9; WAM R52112, R52113 Esperance—33.8 121.883; WAM R52115, R52116, R52117 Kangaroo Is.—35.8 137.25; WAM R58086 6 km NE of Clear Streak Well (i.e., 68 km ESE of Norseman)—32.4 122.433; WAM R58818 Bluff Knoll Stirling Ranges—34.3 118.25; WAM R58871 15 km S of Mt. Barker—34.7 117.7; WAM R60129 11 km S of Nannup—34.0 115.767; WAM R60348 20 km N of Rocky Gully—34.3 117.017; WAM R60375 ~70 km SSW of Balladonia Hotel—32.9 123.333; WAM R61790 Stirling Range—34.4 118; WAM R62222 35 km E of Jindabinbin Rock Hole—32.4 122.467; WAM R67817 1 km E Willanup Spring Cape Naturaliste—33.8 115.033; WAM R68096 2 km NE Yallingup Leeuwin Naturaliste National Park—33.6 115.05; WAM R73736 Kangaroo Is.—35.7 137.333; WAM R78157 Lort River Stn.—33.4 121.35; WAM R78404 Frank Hann National Park—32.8 120.317; WAM R78468 Peak Charles National Park—32.9 121.083; WAM R82935 10 km E of Ravensthorpe—33.0 120.133; WAM R83864 Albany-Narrikup—35.0 117.883; WAM R83867 Albany Town—35.0 117.883; WAM R84495 Crusoe Beach—34.9 117.426; WAM R84661 Middle Is., Lake Hillier—34.1 123.183; WAM R86965 Albany—35.0 117.883; WAM R86976 7 km SSE Munglinup-33.7 120.917; WAM R86977 10 km S Lake Muir—34.6 116.7; WAM R93552 Mount Clarence; WAM R93581 26 km ENE Gibson—33.6 122.1; WAM R93638 9 km SW Clyde Hill—33.5 122.933; WAM R93665 15 km ENE Jyndabinbin Rocks—32.3 122.217; WAM R93670 Frank Hann National Park—32.9 120.367; WAM R94510 NE side of Lake Muir—34.4 116.7; WAM R94792 Albany—35.0 117.883; WAM R95408 E side of Lake Muir—34.4 116.7; WAM R96245 Esperance Area—33.8 121.883; WAM R98119 16 km W Rocky Gully—34.6 116.85; WAM R98748 Hopetoun Townsite—33.9 120.125; WAM R100325 Cannington Botany Reserve—32.0 115.983; WAM R105983 Frenchmans Bay—35.0 117.95; WAM R106094 halfway between Collie and Arthur Rivers—33.4 116.083; WAM R120298 Hassell Hwy.—34.5 118.5; SAM R01032 Blackwood—35.0 138.617; SAM R01326, R01327, R01328, R01329, R01330, R01331, Harriet River—35.9 137.15; SAM R02185 Aldgate—35.0 138.733; SAM R02548 Thistle Is.—35 136.15; SAM R03277 Ravine De Casoars Kangaroo Is.—35.8 136.6; SAM R06016 Galston—33.6 151.05; SAM R06088 Flinders Is.—33.7 134.517; SAM R11075 Big Heath Cons Park—37.1 140.567; SAM R13809 Cowaramup Bay—33.8 114.983; SAM R13837 Pt. Lincoln—34.7 135.867; SAM R16578 Crows Nest/Cut Hill Victor Harbour—35.5 138.633; SAM R16694 ~2 km S Penneshaw Kangaroo Is.—35.7 137.933; SAM R17895 near Salt Creek Coorong—36.1 139.65; SAM R18624 Reevesby Is.—34.5 136.283; SAM R18983, R18984 Flinders Chase National Park—35.9 136.667; SAM R21183 Tulka Rd. Pt. Lincoln—34.7 135.817; SAM R21436 D'Estrees Bay—35.9 137.633; SAM R21473 near Blackford on Main Rd. Kingston to Keith—36.7 140.033; SAM R21506 Bald Hill Rd. near Kingston SE—36.6 139.967; SAM R22943, R22944 Lort River—33.6 121.3; SAM R23426

3 km E Cape Borda—35.7 136.617; SAM R23427 1 km SW Cygnet River and Main E-W Rd. Junct—35.7 137.517; SAM R23471 1 km W Prospect Hill Kangaroo Is.—35.8 137.75; SAM R23472 Just E of Corramer Junct—35.8 137.383; SAM R23473 "Ketley Post Box"—35.7 136.85; SAM R23531 3 km SW Salt Creek Coorong National Park—36.1 139.633; SAM R23600 W of D'Estrees Bay Kangaroo Is.—35.8 137.483; SAM R23833 near Rocky River—35.9 136.733; SAM R24890, R24891, R24892 Pelican Lagoon Cons Park Kangaroo Is.—35.8 137.783; SAM R27189 Islet 475 Pelican Lagoon Cons Park Kangaroo Is.—35.8 137.775; SAM R27269, R27270, R28189 Reevesby Is.—34.5 136.283; SAM R27482, R27483, R27484, R27485, R27486, R27487, R27488, R27489, R27490, R27491, R27492, Thistle Is.—35 136.15 SAM R30487, R30488, R30489, R30490 Taylors Is.—34.8 136.017; SAM R30491, R30492, R30493, R30494 Spilsby Is.—34.6 136.333; SAM R30495, R30496 Thistle Is.—35 136.15; SAM R31157 Kangaroo Is. Flinders Chase—35.9 136.767; SAM R35059 6 km E Marion Bay on South Coast Rd.—35.2 137.033; SAM R36889 2 km E Flinders Chase Headquarters—35.9 136.767; SAM R37324 7.2 km NE Cape Younghusband—35.9 136.864; SAM R37488 5.5 km W Kelly Hill Caves—35.9 136.848; SAM R37489 4 km E Rocky River—35.9 136.789; SAM R42981 15 km N of Rocky River Camping Ground—35.8 136.767; SAM R45139 12.5 km ENE of Salt Creek—36.0 139.785; SAM R46628 near Nangkita—35.3 138.683; SAM R47041 6 km S of Barip HS—36.7 140.055; SAM R49242 0.5 km N Hindmarsh Valley Reservoir—35.6 138.607.

Fossil Record

Possibly from Pleistocene deposits, Naracoorte Caves, South Australia (Smith 1976). Smith identified a single humerus as being that of a *V. gouldii*. At this time, *V. rosenbergi* was considered to be a subspecies of *V. gouldii*.

Diagnostic Characteristics

• Medium size.

• Dark dorsal color with dark tail tip.

• Pale ventral surface often with dark reticulation or banding.

• Distal tail compression.

Description

A medium-sized monitor up to slightly over 1 m. Head mostly black with speckling of white, black eye stripe edged with white, gray, or yellow, neck and body with narrow black bands. Tail proximally black with dull yellowish bands, distally narrowly banded with yellowish brown, and tip often dark. Lower surfaces yellowish or white often with a black or gray reticulum from chin to abdomen. Tail proximally circular, distally strongly compressed. Claws long, slender, and curved.

There is some regional variation in coloration and body size. Specimens from Kangaroo Island, South Australia, can be larger and darker than those from Western Australia. Mertens (1958, p. 253; original text in German) described *Varanus gouldii rosenbergi* as a "distinctly melanistic varanid . . . basically different from the nominate form as well as from *flavirufus*," and he alluded specifically to the "black Gould's varanids from Kangaroo Island."

Size

A medium-sized monitor with a SVL ranging 67 to 395 mm in Western Australia (Storr et al. 1983) and 140 to 470 mm on Kangaroo Island (King, unpublished data). The use of tail length/SVL ratios (Storr et al. 1983), which are 137% to 179%, would give a maximum TL of 1.03 m from Western Australia and 1.31 m from Kangaroo Island. The heaviest of 75 adult specimens weighed on Kangaroo Island was 1935 g, and average weight was 1130 g. Case and Schwaner (1993) found that island populations of goannas, including *V. rosenbergi,* tended to be larger than their mainland counterparts.

There is apparently no sexual dimorphism between male and female Rosenberg's goannas, and the only reliable way of sexing the animals is to examine their gonads (King and Green 1979).

Hatchlings on Kangaroo Island weigh about 18 g and are 140 to 170 mm in TL. Like other species of *Varanus,* young grow rapidly during their first year and increase their weight more rapidly than their TL.

Sex Ratio

The sex ratio of *V. rosenbergi* is apparently biased toward males. In a study on Kangaroo Island, of 45 specimens, 37 were found to be males (King and Green 1979). This disparity in sex ratio is found in other monitor species and may well be the result of behavioral differences between the sexes. Males move longer distances than females, and they are more likely to be captured. Other factors may also affect the apparent sex ratio.

Habitat and Natural History

Varanus rosenbergi occur mainly in open woodlands, sclerophyll forest, and heathlands. Home ranges vary in size (1.7 to 43.7 ha on Kangaroo Island, mean, 19.44 ha, n = 13), and activity areas vary seasonally from mean daily values of 0.18 ha during winter to 1.37 ha during summer (King and Green 1993b). Males have larger home ranges and activity areas than females.

Kangaroo Island *V. rosenbergi* use different areas in summer and winter. Animals are found in flat areas during summer but move to more elevated, better drained areas during winter, as the flatter areas become waterlogged. Sandridges may be used throughout the year (King 1977, 1980).

Burrows are used extensively. They provide refuge from predators and protection against adverse weather. Burrow temperatures are more stable than soil surface or air temperatures. Burrows investigated on Kangaroo Island were fairly shallow, slightly curved, and had enlarged terminal chambers. The maximum temperature variation found in these burrows was 8.4°C over 3 days in December, compared with variation in air temperature of 15.1°C. The minimum variation was 1.2°C in early August compared with a 10.3°C variation in air temperature (King 1980). The depth of the burrows varied according to the season: during summer, they were approximately 14 cm deep, whereas those used during winter were about 18 cm deep (King 1980). The difference between burrow temperatures in winter and summer is about 12°C (King 1980). Burrows were generally about 100 cm long.

Several burrows within the home range of individual animals are used, but it is unusual for more than one animal to inhabit a burrow at the same time, although this has been observed (Green and King 1978). Because home ranges may overlap considerably, some burrows may be used by different animals on different nights. Most burrows appear to be dug by the monitors that occupy them, but burrows dug by other species of animals are also used.

Terrestrial, Arboreal, Aquatic

Varanus rosenbergi are almost entirely terrestrial, but they, and the young especially, will climb trees to escape from potential predators. They will forage along beaches or watercourses but don't appear to actually forage in the water.

Time of Activity (Daily and Seasonal)

On Kangaroo Island, *Varanus rosenbergi* are active throughout the year. Their activity areas are smaller during winter, and body temperatures in winter are lower than those during other seasons. They are active on fewer days during winter, because there are more days on which weather conditions are unfavorable (Green and King 1978). The length of time lizards spend out of their burrows depends on ambient temperature and time of emergence. During summer, they remain out of their burrows for most of the daylight hours, returning to them about 1 to 2 hours before sunset as the ambient temperature decreases. In winter, animals emerge later and may return to the burrow if conditions are unfavorable for basking (King 1980).

Christian and Weavers (1994) classified three types of behavior: resting in burrows, sitting out of burrows, and activity. In winter, a smaller proportion of time was spent in activity than in spring and summer. They found that the time spent in actual activity is not great at any time of year—in summer, a mean of 47.6 minutes per day is spent in activity, in spring, 22.2 minutes per day, and in winter, only 10.5 minutes per day. Daily activity is "significantly and positively related to daily solar radiation" and body temperature (Christian and Weavers 1994, p. 292). Therefore, activity is limited in cooler weather.

Christian and Weavers (1994) compared the activity pattern of *V. rosenbergi* in winter with the inactive periods of two goannas in tropical areas. These tropical goannas may spend weeks or months in their burrows during their inactive periods. The advantage to *V. rosenbergi* of being active in winter (its low activity period) is unknown, but there is an indication that some food is eaten during this season. These goannas will emerge from their burrows even in overcast or rainy conditions.

Morning emergence times between and within individuals on Kangaroo Island vary over about 3 hours at all times of year and mean emergence times were approximately 1 hour earlier during summer than in winter, and appear to occur when temperature outside the burrow has risen to approximately the same temperature as inside the burrow (King 1980). At this stage, it would be possible to increase body temperature by basking. Animals generally remained outside their burrows for most daylight hours during summer unless conditions became unfavorable, largely because of heavy rain. Activity periods are much shorter during winter because of the longer basking periods required to achieve normal activity temperatures and the frequent inclement weather.

Individuals within a population show different activity and body temperature patterns, with females tending to be active after dark, especially late in the breeding season. The body temperature of these females may be up to 20°C higher than air or soil temperature (Rismiller et al. 1997).

Thermoregulation

Varanus rosenbergi thermoregulates very carefully during spring and summer when it is physically possible (Christian and Weavers 1996) but poorly during winter when optimal conditions occur rarely, and when they generally stay in burrows for most of the day. When possible, they gain heat by basking, but basking periods are longer and heat gains are slower during winter than in warmer periods. In winter, animals sometimes bask on piles of dead branches, which may prevent heat loss to the soil (King 1980). Skin color also affects thermoregulation. *V. rosenbergi* is a melanistic species, and its dark color absorbs more solar energy, enabling it to heat up more rapidly and remain active longer in cooler seasons (King and Green 1999a).

Body temperatures of free-ranging animals fitted with transmitters varied seasonally. Mean ± SE body temperatures during November–December were 33.4 ± 1.5°C, 32.0 ± 1.0°C in March, 29.3 ± 1.1°C in May–June, and 24.5 ± 0.9°C in August. Mean daily maximum temperature of animals active on clear days in summer was 36.7 ± 0.2°C (King 1980).

Mean cloacal temperature at time of capture on Kangaroo Island during summer was 35.1 ± 0.4°C (n = 14). Mean temperature of seven specimens captured during summer in southwestern Western Australia was 36.4 ± 0.3°C. Mean cloacal temperature of *Varanus rosenbergi* in temperature gradients was similar at 35.43 ± 0.16°C (n = 42), and mean values for individuals taken in gradients more than once, at intervals of at least 4 weeks, varied by as much as 3°C on different occasions.

Cloacal temperatures are approximately at their lowest just after emergence of the animal. Lizards may stay with their head at the burrow mouth or just outside it for up to about an hour before moving onto the mound of earth outside the burrow to bask. This may warm the head of the animal and increase the function of the central nervous system, or the lizard may just be watching for predators outside the burrow (King 1980).

Thermistors were implanted inside the skulls of four *Varanus rosenbergi*. Their head and body temperatures were measured simultaneously in thermal gradients. Before basking, head temperatures were slightly lower than body temperature, but during basking, they increased more rapidly than body temperatures and remained higher during the remainder of their activity periods. Head and body then cooled at similar rates. If only the body was heated, both head and body temperature increased, whereas if only the head was heated, there was no increase in body temperature (King 1977; King and Green 1993b).

Gular fluttering is used to reduce body temperature but is very costly in terms of water loss. Captive *Varanus rosenbergi* began gular fluttering when body temperature rose above head temperature, at about 38°C. Gular fluttering resulted in a reduction in head temperature and can continue for up to 75 minutes.

Foraging Behavior

Varanus rosenbergi has a wide-ranging foraging pattern utilizing a wide range of vegetation types from open beaches to dense forests, investigating disturbed sites, and frequently digging through litter and soil with their snout or claws in search of prey. They consume a variety of organisms that are often located by scent, even when prey is hiding in rubble, under stones, or in burrows. They walk with a swinging gait while foraging, and their head is usually held close to the ground. Their constantly flickering tongues swing widely through the air and collect scent particles from their prey, which are transferred into the Jacobson's organs. This is obviously important in the foraging strategy of these monitors (King and Green 1979).

Diet

Like most monitors, *Varanus rosenbergi* are carnivorous. They have a high aerobic metabolic rate and will actively chase prey. Green and King (1978) state that these animals appear play a similar role to that of small carnivorous mammals

V. rosenbergi forages over wide areas and digs in loose soil and decaying vegetation for prey. They are opportunistic feeders, as are most goannas, eating anything that is an appropriate size, and they would certainly uncover invertebrates in the course of this foraging activity (King and Green 1979). A dietary study on Kangaroo Island showed high levels of invertebrates (33.3% volume), mammals (29.8% volume), and reptiles (21.5% volume) were consumed by adult *V. rosenbergi* there (King and Green 1979). A small percentage of the diet consisted of birds (3.7%) and frogs (*Lymnodynastes* sp., 8.9%). The greatest numbers of invertebrates eaten were Blattoidea, Acrididae, Arenidae, and Coleoptera.

The most commonly eaten mammals were *Rattus fuscipes*, *Macropus fuliginosus,* and *Trichosurus vulpecula* (the latter two species were probably eaten when feeding on road kills, as a number of goannas were observed scavenging on carcasses). Overton (1987) reported seeing a goanna apparently preying on a juvenile echidna (*Tachyglossus aculeatus*), which showed injuries consistent with goanna bites. Nine species of reptiles were eaten, with *Aprasia striolata,* unidentified eggs, and *Lerista bougainvillea* being eaten most frequently. One lizard contained the remains of a small *V. rosenbergi,* but it was unclear whether this was carrion or a case of cannibalism. Cannibalism has been reported in other species of *Varanus* (King and Green 1979).

Adult *V. rosenbergi* from Western Australia had far fewer vertebrates in their stomachs than those from Kangaroo Island. Only 1 of 33 adult specimens examined contained a mammal (unidentified), 7 contained reptiles or their eggs, and 22 specimens contained invertebrates (mostly Coleoptera or Acrididae) (King, unpublished data).

Green et al. (1999) examined scats from active juveniles on Kangaroo Island and found they consumed mainly invertebrates (insects and spiders) and also some vertebrates (possibly small lizards). Scats from hatchlings before emergence from the termitarium indicates that they eat termites. Only 5 of 15 juveniles from Western Australia (ranging 9.8–17.3 cm SVL) contained prey. Of these, four contained invertebrates (mostly Acrididae), and one contained a small mouse (*Mus musculus*).

Reproduction

On Kangaroo Island, maximum testis and ovary size is reached between November and February, and mating occurs during midsummer (King and Green 1979). Animals from the southwest of Western Australia in November and January showed testis development at a similar stage to Kangaroo Island animals at about the same time of year.

Fewer data are available for females than for males. Five animals captured in late November and December had enlarged ovarian eggs with a mean diameter of 6.5 mm, and a female captured in February had oviductal eggs with a mean length of 30.4 mm (King and Green 1979). This timing coincides with that of peak testes development in males.

Mating occurs near a female's burrow and continues over a period of several days. The male limits access to his mate during this time (King and Green 1993b). The same pair of individuals commonly mate successively.

For egg laying, a female digs a burrow in an active termite (*Nasutitermes exitiosus*) mound and deposits her leathery-shelled eggs in a chamber 45 to 70 cm deep. Digging takes 2 to 3 days, continuing into the night. The outer wall of the termitarium is fairly soft, but about 10 cm inside is a very hard layer that is difficult to dig through. Inside this hard wall are the nursery galleries, consisting of flaky material. At the end of the tunnel, an enlarged egg chamber is hollowed out (Ehmann et al. 1991; Green et al. 1999). Rismiller et al. (1997) observed that females may be active after sunset, especially in the late breeding season, and oviposition often occurs at night. The usual clutch size is 10 to 17 eggs, and eggs average 26 g (Green et al. 1991). Eggs are extremely rich in lipids.

After egg laying, the female begins to refill the excavation, with very limited assistance by the male. Both animals remain in the vicinity of the mound for several days, possibly to limit the possibility of predation on the eggs or disturbance of them by, for example, echidnas (*Tachyglossus aculeatus*) using the excavations as access to the inside of the termitarium (Ehmann et al. 1991).

The termites reconstruct the mound and completely cover the nesting holes within a short time (Ehmann et al. 1991; Green et al. 1999). The mound provides an excellent incubator because internal temperature is fairly constant at about 30°C throughout the year, and humidity in the mound is also near saturation (Ehmann et al. 1991; King and Green 1993b; Green et al. 1999). Eggs are protected from predators by the tough outer shell of the termitarium.

Ehmann et al. (1991) found that the eggs absorbed moisture in the humid mound, and increased in size by up to 75% within the first few months. Temperature monitors showed that the egg temperatures remained relatively constant (range, 33.3 to 38.6°C; mean, 36.2°C; n = 95; all months). CO_2 levels near the eggs were also high (range, 9670–14,000 $\mu L \cdot mL^{-1}$, mean 11,650 $\mu L \cdot mL^{-1}$, n = 5), as was the relative humidity, which remained 92% or above. Packard et al. (1997, pp. 87, 94, cited in Ehmann et al. 1991) found that the high CO_2/low O_2 concentrations in monitor eggs in sealed nesting sites such as termitaria increase the incubation time.

Ehmann et al. (1991) found that eggs in the termite mounds in their study near Sydney became completely cemented into the termite mound within about 5 months after laying. They suggest that initial uptake of water by the eggs/embryos and increase in egg size must be rapid enough to give the embryo room to develop once the eggs are encased by termite construction. Examining the mounds in their study, they found remains of eggs from previously laid clutches, and X-ray comparisons of the skeletal remains matched those of hatchling *V. rosenbergi*, in some cases fully developed. Ehmann et al. (1991) concluded that hatchlings may be too weak to dig their own way out and may be dependent on adult monitors to release them from the mound. However, a study of animals on Kangaroo Island between 1991 and 1996 (Green et al. 1999) determined from the small size of the exit burrow that the hatchlings dig a tunnel to exit from the mound without assistance from their parents. Green et al. (1999) did not observe adult monitors digging in the termite mounds around the time the hatchlings emerged.

Eggs incubate for about 8 months, and young usually hatch in September. In common with other species of monitors, hatchlings of *V. rosenbergi* are brightly colored. The reason for this bright coloration is not known, and it fades within months of hatching. It may contribute to camouflage of the hatchlings as they forage, or it may deter predators, as many toxic species of animals use bright colors as a warning (King and Green 1999a).

Hatchlings may not emerge from the termite mound for 2 to 3 weeks after hatching, and they may continue to use it as a refuge for some months after hatching; Green et al. (1999) captured hatchlings at termite mounds 4 months after hatching. The young monitors wait for warm, sunny conditions before emerging to bask (Green et al. 1999; King and Green 1999a). However, most hatchlings are taken by birds (mostly ravens) or snakes within weeks of their first emergence (King and Green 1999a). Adult *V. rosenbergi* are also known to kill juveniles (Green et al. 1999). Green et al. (1999) found that the short-beaked echidna (*Tachyglossus aculeatus multiaculeatus*) will enlarge the hatchling access tunnel to feed on termites, and hatchlings may be killed by punctures from the claws and spines of the monotremes.

Young *V. rosenbergi* forage actively once they leave the mound and quickly gain weight.

Movement

Young *Varanus rosenbergi* are very active and are adept at climbing.

Adults move over large areas on Kangaroo Island. The mean value for home ranges of 13 animals radiotracked for periods from 5 to 74 months was 19.44 ha, with considerable overlap between individuals. Daily activity areas ranged from 0.16 ha during winter to 2.43 ha in summer. Most lizards studied inhabited cleared flat areas during summer but moved to elevated locations in sand hills during the winter when the flat areas became wet and cold (Green and King 1978).

Physiology

As with other species of varanids, *Varanus rosenbergi* has many modifications to its lungs, heart, and blood parameters that result in a high aerobic capacity, enabling it to forage widely (King and Green 1993b).

Water Balance

The rate of water loss through the skin of *Varanus rosenbergi* is similar to that of most other varanids but very low compared with most other animals. *V. rosenbergi* loses 0.12 mg of water per square centimeter at 30°C, and double that at 38°C (King and Green 1999a). Evaporative water loss is much lower in adults than in juveniles because of the difference in surface area to mass ratios: that of a hatchling is about five times that of an adult. Because the eyes of hatchlings are relatively much larger than those of adults, evaporative water loss from the eyes is also greater.

Respiratory evaporative water loss also occurs, the rate being relative to the breathing rate. Gular fluttering, used to reduce body temperature, greatly increases the rate of respiratory water loss.

A study by Thompson and Withers (1997a) measured standard evaporative water loss near preferred body temperatures for a range of species of *Varanus*. They found that *V. rosenbergi,* which lives in a mesic environment, has a higher standard evaporative water loss than species from arid and semiarid regions.

Water is resorbed from feces and urine in the cloaca and hind gut. Further water is conserved by the action of salt-secreting glands located in the nasal capsules.

Water use is generally quite low, with marked seasonal differences in water use of *Varanus rosenbergi*. Summer turnover values are more than four times those during winter, mostly as a result of differing activity levels over the different seasons (Green 1972b; Green et al. 1991). In summer, water influx comes totally from food. Less free water is available, and the lizards are more active and consume more food in this season. As carnivores, their food contains a high percentage of water. In winter, the humidity is high, and lizards spend the majority of time in their burrows. During this time of the year, and in spring and autumn, pulmocutaneous water exchange and/or drinking account for more water influx (Green et al. 1991). Although the climate in spring and autumn is similar, water turnover in spring is higher than that in autumn because of the increased activity related to reproduction (Green 1972b).

Monitor lizards possess nasal salt-secreting glands that assist in the removal of this excess. Saint Girons et al. (1981) found that the nasal glands of *V. rosenbergi* showed moderate development of striated cells, indicating an increased ability to excrete salt over animals with less development of these cells. This would be advantageous for a predominantly coast-dwelling lizard species. The secretion of the glands is predominantly sodium chloride, with some potassium present (Green 1972a)

Energetics

Energy acquisition and use by *Varanus rosenbergi* in the field have been measured (Green et al. 1991) during different seasons and an annual energy budget has been calculated. FMRs and water turnover were measured by means of doubly labeled water $^{3}H_{2}^{18}O$. Results showed that FMR and water turnover was highest in summer and lowest in winter, and intermediate in spring. Green et al. (1991) estimated that a 1-kg *V. rosenbergi* requires 4.7 kilograms of prey per year to stay in energy balance. If an animal is building fat reserves or is breeding, more is needed.

The energetic cost of producing a clutch (an extra 40%) has also been calculated (Green et al. 1991). Female *V. rosenbergi* may not breed every year unless prey or carrion are particularly abundant, and they may not be able to accumulate the extra reserves needed. However, if a female does not breed, she will come into winter with extra fat reserves. This may increase her chances of being able to breed in the following season (Green et al. 1991).

Juveniles that still used termitaria as refuges during the night were demonstrated to show a higher FMR than those not able to use termitaria. This is probably the result of the high temperatures within the termite mound compared with typical spring conditions—cloudy and cool. Young with no access to mounds have a similar FMR to adult rates. They may not be able to bask for long enough to become active, whereas those kept warm overnight in a termite mound will have a distinct advantage. Not only are they warmer before basking, the reduced basking time needed to become active means more time for foraging and less time when they are exposed to predation (Green et al. 1991). FMRs of hatchling *V. rosenbergi* that overnight in termitaria are about five times that of adults in spring, whereas those that do not have rates similar to adults.

Varanus rosenbergi use far less energy than similar-sized birds or mammals.

Fat Bodies

Fat bodies in the abdomen are laid down as energy reserves by *Varanus rosenbergi* and constitute up to 7.6% of their body weight. Fat is also stored in the tail.

Parasites

A high percentage (85%) of *Varanus rosenbergi* from Kangaroo Island were infested with the tick *Aponomma fimbriatum* (Sharrad and King 1981), whereas only 53% of adults and no juveniles from Western Australia had ticks (the majority *Aponomma fimbriatum*, only three animals with *Ammblyomma limbatum*) (King, unpublished data). Male ticks from both localities are generally concentrated near the tail base and cloaca, and may congregate around wounds. Female ticks are found in heavily vascularized areas on the head and upper back (King and Green 1999).

Specimens from Western Australia contained endoparasites; 74% were infested with nematodes, and 7.5% had cestodes (King, unpublished data).

Jones (1983b) found three species of nematodes (*Abbreviata antarctica, A. levicauda, A. tumidocapitis*) in seven Western Australian *V. rosenbergi*. Almost all *V. rosenbergi* in the southwest of Western Australia were infested with *A. antarctica*, both adult and larval forms. In 47 *V. rosenbergi* from the Western Australian Museum collection, nematodes of seven species of the genus *Abbreviata* (*A. antarctica, A. levicauda, A. hastaspicula, A. tumidocapitis, A. anomala, A. perenticola, A. confusa*) and one *Maxvachonia* sp. were found in 41 animals (the other six were eviscerated) (King, unpublished data). In one *V. rosenbergi* from Kangaroo Island, a specimen of *Abbreviata antarctica* was found. Less investigation of nematodes has been carried out in South Australian lizards.

Varanus rosenbergi, like other large species of *Varanus*, may harbor hundreds of adult worms and even larger numbers of larvae (Jones 1991).

Two genera of cestodes are found to occasionally infest Australian varanids (King and Green 1999b). Unidentified tapeworms were found in three adult *V. rosenbergi* in the sample of 47 from the Western Australia Museum (King, unpublished data).

Because *Varanus rosenbergi* was not elevated to full species status until 1980, in some early literature (Green 1969, 1972a, 1972b; Green and King 1978; King 1977, 1980), this species was misidentified as *Varanus gouldii*.

7.19. *Varanus scalaris*

LAWRENCE A. SMITH, SAMUEL S. SWEET, AND DENNIS R. KING

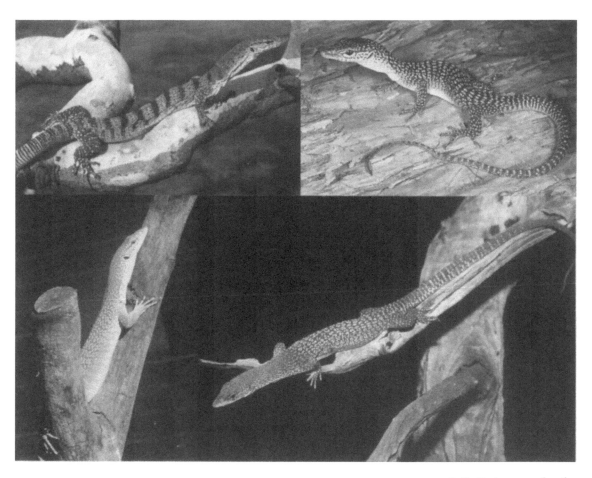

Figure 7.42. Various morphs of "Varanus scalaris," *clockwise from upper left:* V. scalaris *(sensu stricto) from Beagle Bay (photo by David Knowles), from near Kununurra (photo by Robert Browne-Cooper), from Darwin (photo by Jeff Lemm), and the "V. similis" morph from Kakadu (photo by Eric Pianka).*

Nomenclature

V. scalaris was described by Mertens in 1941, originally as a subspecies of *V. timorensis*. In 1957b, Mertens informally elevated *V. scalaris* to species status but later, in 1958, relegated it to a subspecies of *V. timorensis* again (Fig. 7.42). The holotype is from Beagle Bay Mission, Dampier Land, northwest Australia, in the Senckenberg Museum (specimen SMF 32806). It is now considered to be a species complex (Bennett 1995; L. A. Smith, personal communication). In 1958, Mertens described *V. similis* as another subspecies of *timorensis*, thus promulgating the concept that *V. timorensis* ranged across northern Australia. *V. similis* is now considered to be a subspecies of *V. scalaris* but will eventually be elevated to full species status, along with several other as yet undescribed new species found in northern Australia (L. A. Smith, personal communication).

Taxonomy

Mertens had great difficulty understanding the relationships between *Varanus tristis, V. timorensis*, and *V. scalaris* (Mertens 1957b, 1958). He noted whether enlarged cloacal spines were present or absent but did not make much of their structure. Although he noted that *V. timorensis* lacked cloacal spines, he did not differentiate between the nature of the spines in his forms of *V. tristis* and *V. t. scalaris*. Once this distinction is made (see Storr et al. 1983), the enormous variety in color and pattern forms with *scalaris*-like cloacal spurs becomes apparent. Mertens' figure of the holotype of *V. t. scalaris* (plate 679, p. 321; Wilson and Knowles 1988) and Cogger's plate of a *V. "timorensis"* from Cape York Peninsula (Cogger 1992, p. 377) illustrate just some of the variety in the group. Wells and Wellington (1985) described the morph shown by Wilson and Knowles as *V. pengilleyi* and the morph shown by Cogger as *V. kuranda*, but it is equivocal whether these names will be able to be used.

An uncompleted morphological study by L. A. Smith at the Western Australian Museum identified seven distinctive morphs, including the three illustrated by Mertens, Wilson and Knowles, and Cogger.

Recent molecular studies have confirmed that *V. scalaris* is polytypic. Ast (2001) identifies two *"scalaris,"* one from the Northern Territory, Australia, and another from New Guinea (morphologically, the New Guinean and Cape York population appear to be the same taxon). Furthermore, unpublished electrophoresis data obtained by R. A. How of the Western Australian Museum and colleagues at the University of Western Australia indicate that in Western Australia alone, *V. similis* (the most widespread of the seven morphs identified by Smith) is three species. K. Grams (Louisiana State University, Baton Rouge) is currently sequencing tissue samples that will give a more precise indication of how many taxa are in the *scalaris* group—perhaps as many as 10 species.

Geographic Distribution of *V. scalaris* (sensu stricto)

West Kimberley region of Western Australia from Dampier Land and Kimbolton Peninsula east to Fitzroy Crossing (Fig. 7.43).

Geographic Distribution of the Species Group

Northern Kimberley region of Western Australia south to Beagle Bay, Fitzroy Crossing, Lake Argyle. In the Northern Territory south to about Mataranka and the Roper River. In Queensland, the extreme northwest and Cape York, almost exclusively east of the Great Dividing Range and south to about Townsville. Also the Torres Strait Islands and southern New Guinea. This corresponds roughly to the distribution of stringybark, *Eucalyptus tetrodonta,* and thus to a minimum annual rainfall isohyet of around 600 mm (Fig 7.43).

Figure 7.43. Localities of collection of members of the Varanus scalaris *species group. The rectangle shows the range of scalaris (sensu stricto).*

QM R455 Herbert Gorge—18.2 145.53; QM R1790 Port Darwin—12.4 130.7; QM R2138 Stannary Hills, near Herberton—17.3 145.22; QM R2169, R2171 Cardwell—18.2 146.02; QM R4054 Mt. Surprise, Georgetown District—18.1 144.32; QM R8161 Cape York Peninsula, Portland Rds.—12.6 143.4; QM R15141 Cardwell—18.2 146.02; QM R15561 Marlborough, near 22.8 149.88; QM R21013 Wakooka Outstn. 182 m along Wakooka–C. Melville Rd.—14.5 144.55; QM R21789 Humpty Doo district, ~64 km SE Darwin—12.6 131.25; QM R25131 Home Rule, 30 km S Cooktown—15.7 145.28; QM R27081 Spear Creek, via Mt. Molloy—16.7 145.4; QM R27275, R27276, R27277 Gap Creek, 55 km S Cooktown—15.8 145.25; QM R28842 Hopevale Mission, 4 km S McIvor River crossing—15.3 145.12; QM R28862 Gulf of Carpentaria, Mornington Is.—16.6 139.35; QM R28864 Ravenshoe area, Tall Timbers Caravan Park—17.6 145.48; QM R33161 Mt. Spec, Echo Creek—18.9 146.13; QM R33162 Paluma Rd.—19.0 146.22; QM R34456 Buthen Buthen, Nesbit R., E of Coen—13.3 143.45; QM R34534 Coen, 13 km NE of Steens Hut—13.4 143.23; QM R34826 Beatrice Hill Reserve Stn.—12.6 131.33; QM R37575 Silver Plains HS, 14 km SW—14.0 143.45; QM R37579 Mt. Croll, 10 km NNE—13.7 143.22; QM R37623 Silver Plains—13.9 143.53; QM R39824 Red Beach, 8 km S Cullen Point—12.0 141.88; QM R39825 Mapoon, 5 km S Cullen Point 11.9 141.9; QM R39839 Mapoon—11.9 141.87; QM R39849 Port Musgrave, Namaleto Creek, N bank, 5 km SE Cullen Point—11.9 141.93; QM R42000 El Arish, Tree Farm—17.8 146; QM R44507 Helenslee Stn.—20.5 145.7; QM R48105 Ingham, Longpocket—18.6 146.17; QM R48956 Carron Creek, on Koombooloomba Rd.—17.8 145.58; QM R54013 Heathlands Junction, 10.7 km S, on Cape rd—11.8 142.65; QM R58958 Noah Valley, Cape Tribulation National Park; QM R58994 3 km N-Y jnct Capt. Billy Landing Peninsula Development Rd.—11.7 142.68; QM R58995 ½ way between Lake Eacham car park and National Park boundary—17.2 145.5; NTM R26043 Bing Bong Stn.—15.0 135.37; NTM R26044 Winnellie—12.4 130.9; NTM R26045, R26047, R26051 Darwin—12.4 130.83; NTM R26046 Reynolds River, Litchfield National Park—13.2 130.68; NTM R26048 Darwin Showground—12.4 130.83; NTM R26049 Angurugu, Groote Eylandt—13.9 136.43; NTM R26050 Ban Ban Springs—13.3 131.5; NTM R26052 Pine Creek—13.8 131.83; NTM R26053, R26054 Howard Springs—12.4 131.05; NTM R26055 East Veron Is.—12.0 131.1; NTM R26056 Katherine—14.4 132.27; NTM R26057 Keep River

National Park—15.7 129.08; NTM R26059 Balbirini Stn., 3 km W O.T. Bore—16.7 135.73; NTM R26060 Litchfield National Park, Tjaynera Falls Area—13.2 130.73; NTM R26061 Victoria Hwy., 7 km E WA Border—15.8 129.12; NTM R26062, R26063, R26064 Murgenella—11.5 132.87; NTM R26065 Annaburroo Homestead—12.9 131.67; NTM R26066 Berry Springs, Northern Territory—12.7131.02; NTM R26067 Katherine Gorge National Park—14.2 132.5; NTM R26068 8 km E Adelaide River Bridge—12.7 131.38; NTM R26069 Deaf Adder Creek—13 132.92; NTM R26070 Katherine—14.4 132.27; NTM R26071 Gregory River National Park; NTM R26072 Keep River National Park—15.7 129.08; NTM R26073 Katherine—2 km N—14.4 132.25; NTM R26074 East Veron Is. —12.0 131.1; NTM R26075 Gimbat Stn., Coronation Hill Area—13.5 132.58; NTM R26076 Katherine Farms Rd.—14.4 132.27; NTM R26077 Melville Is.—11.0 130.38; NTM R26078 Humpty Doo District—12.6 131.25; NTM R26079 Bathurst Is.—11.1 130.38; NTM R26080 20 km S Nabarlek Mine—12.5 133.35; NTM R26081, R26082 Bullita Stn.; NTM R26083 Katherine District—14.4 132.27; NTM R26084 Donydji (Arnhem Land)—12.9 135.32; NTM R26085 Berry Springs Reserve—12.7 130.98; NTM R26086 Keep River National Park—15.7 129.08; NTM R26087, R26088 Winnellie, Darwin—12.4 130.83; NTM R26089 Stuart Hwy.—22.6 132.03; NTM R26090 Keep River National Park—15.7 129.08; NTM R26091 Katherine Gorge National Park—14.2 132.5; NTM R26092, R26093, R26094, R26095, R26096, R26097, R26098, R26099, R26100, R26107 Kakadu National Park—Stage 3—13.7 132.57; NTM R26101 Long Billabong, Wulunurrayi Creek—15.3 135.35; NTM R26102 Keep River National Park —15.7 129.08; NTM R26103 Cape Fourcroy, Bathurst Is.—11.7 130.02; NTM R26104 Cape Hotham Is.—12.0 131.28; NTM R26105 Murgenella—Talbot Rd.—11.5 132.87; NTM R26106 Kakadu National Park—Stage 3—13.6 132.27; NTM R26108 Milingimbi Is., east end of Airstrip—12.1 134.89; NTM R26109 Yimendumanja Groote Eylandt—14.1 136.47; NTM R26110 Cape Fourcroy, Bathurst Is.—11.7 130.02; NTM R26111 Casuarina, N University Campus—12.3 130.87; NTM R26112 Wangi Stn Rd., 10 km S Mandorah Rd.—12.8 130.82; NTM R26113 Jabiru E—12.6 132.9; NTM R26114 Jabiru Area—12.6 132.9; NTM R26115 Arnhem Hwy., on Woolner Stn Rd., 3 km NE—12.7 131.47; NTM R26116 Cape Arnhem—12.3 136.95; NTM R26117 Cape Arnhem—12.4 136.85; NTM R26118 Litchfield National Park, Tjaynera Falls Area—13.2 130.73; NTM R26119 Shoal Bay, Military Reserve—12.3 130.97; NTM R26120, R26121 Keep River National Park—15.7 129.08; NTM R26122 Taracumbie Falls—11.6 130.71; NTM R26123 Maria Is.—14.8 135.72; WAM R10278 Ivanhoe Research Stn. Ord River—15.6 128.7; WAM R11205, R11833, R11834, R11835 Wotjulum—16.1 123.62; WAM R11839 Beverley Springs—16.7 125.47; WAM R13520, R13521, R13522 Yirrkala—12.2 136.88; WAM R13566, R13580, R13663 Kalumburu—14.3 126.63; WAM R13664 Beverley Springs, Kimberley—16.7 125.47; WAM R13953, R13955, R16511, R16512, WAM R19904, R23890 Katherine—14.4 132.27; WAM R24062, R24063, R24064, R24072, R24083 Mt. Hart, within 6 km of HS—16.8 124.92; WAM R26227 Darwin—12.4 130.83; WAM R26677 Cape Don—11.3 131.77; WAM R26783 ~40 km from Wyndham—15.7 128.38; WAM R28032 Kalumburu—14.3 126.63; WAM R29758 Katherine—14.5 132.22; WAM R32252, R32253 Oenpelli Mission—12.2 133.05; WAM R34115 Darwin—12.4 130.83; WAM R40999 Kildurk Stn.—16.4 129.62; WAM R42830, R42831 Lake Argyle. Banana Springs, 11 km S of Main Ord Dam—16.3 128.8; WAM R42852, R42853 Lake Argyle. Hicks Creek, tributary of Ord River—16.3 128.8; WAM R43180 Mitchell Plateau—14.8 125.83; WAM R46931 Prince Regent River Reserve—15.7 125.33; WAM R47597 Edith River—14.1 132.07; WAM R47670 Fitzroy Crossing—18.1 125.58; WAM R50570 Drysdale River National Park—15.2 127.2; WAM R50697, R50698 Drysdale River National Park—15.2 126.72; WAM R51203 Kimbolton Spring—16.6 123.72; WAM R51833 Stewart River. Kimbolton—16.6 123.52; WAM R58315 Mitchell Plateau Surveyors Pool—14.6 125.73; WAM R58528 Martins Well—16.5 122.85; WAM R60115 Lake Argyle at Dam Site—16.1 128.73; WAM R60117 Lake Argyle at Spillway of Dam—16.0 128.25; WAM R60118 Lake Argyle—16.0 128.78; WAM R60119, R60120, R60121 Lake Argyle—16.1 128.73; WAM R60143, R60474 Kalumburu—14.3 126.63; WAM

R60409 Lake Argyle, presumably Banana Spring, Ord River—16.3 128.8; WAM R60487 Napier Broome Bay—13.8 126.72; WAM R60581 Camp Creek Mitchell Plateau—14.9 125.93; WAM R60905 16 km N Beagle Bay—16.8 122.82; WAM R70112 12.2 km 280° (New) Lissadell HS—16.6 128.44; WAM R70363 17.5 km 273° (New) Lissadell HS—16.6 128.39; WAM R75535, R75550 12 km NW of New Lissadell Homestead—16.6 128.47; WAM R75551 18 km W of New Lissadell Homestead—16.7 128.4; WAM R77028 Mitchell Plateau—14.6 125.8; WAM R77145, R77198, R77222, R77223, R77410, R77533, R77600 Mitchell Plateau—14.8 125.83; WAM R77604, R77606 Mitchell Plateau—14.8 125.8; WAM R77610 Mitchell Plateau—14.7 125.82; WAM R81224 Bow River Kimberley—16.8 128.45; WAM R83688 19 km NW of Gibb River HS—16.3 126.32; WAM R87196 20 km NNE of Kununurra—15.6 128.77; WAM R94413 8 km SE Beagle Bay Mission—17.0 122.72; WAM R96094, R96095, R96096, R96312, R96313, R96764, R96765, R96766, R96767 Yirrkala—12.2 136.88; WAM R96357 Beverley Springs—16.7 125.47; WAM R97090 4 km SW Mount Hart—16.8 124.9; WAM R99542 17 km NE Beagle Bay—16.9 122.8; WAM R108725 Geike Gorge National Park—18.0 125.72; WAM R127329 Walcott Inlet—16.4 124.77; AM R6976 Napier Downs Kimberley District—17.2 124.73; AM R9700 Groote Eylandt, N.T.—14.0 136.58; AM R10207, R10208, R10208, R10209, R10210, R10211, R10212, R10213 Groote Eylandt, N.A.—13.9 136.47; AM R10537, R10814, R10818, R11318 Almaden—17.3 144.68; AM R11039, R11040, R11041 Groote Eylandt, Gulf of Carpentaria—13.9 136.47; AM R12371, R12372, R12373, R12374, R12375, R12376, R12378 Yirrkala via Darwin—12.2 136.88; AM R12874, R13080 Katherine—14.4 132.27; AM R13550 Cape Arnhem—12.3 136.98; AM R14066, R14068, R14069, R14070, R14099 Port Keats Mission—14.2 129.52; AM R16540 Coen, Cape York Peninsula—13.9 143.2; AM R16886, R19615; Silver Plains via Coen, Cape York—13.9 143.55; AM R18760 Groote Eylandt; AM R29980 Port Essington, Cobourg Peninsula—11.2 132.13; AM R31649 Humpty Doo District—12.6 131.25; AM R32621, R32623, R32625, R32626 Black Point, Port Essington, Cobourg Peninsula—11.1 132.15; AM R32624 Port Essington—11.2 132.15; AM R33498 Gove Peninsula—12.2 136.83; AM R38000, R38323 Koongarra Brockman Range, Arnhem Land—12.7 132.65; AM R38503 13 mi E Iron Range—12.6 143.35; AM R39792, R39793 Woolwonga Ranger Stn.—12.7 132.65; AM R41871 Environs of Maningrida—12.0 134.22; AM R41910 Maningrida—12.0 134.22; AM R42383 Sue on Warraber Is., Torres Strait—10.2 142.83; AM R43892 Hammond Is., Torres Straits; AM R45492, R45493 Naulunbuy-Gove—12.3 136.93; AM R48076 Moa Is., Torres Strait—10.1 142.27; AM R48425, R48439 Horn Is., Torres St.—10.6 142.28; AM R48581 Mabuiag Is. Torres Strait—9.96 142.17; AM R48660 6 km E MacArthur River Base Camp—16.4 136.2; AM R48799 6 km E MacArthur River Base Camp—16.4 136.18; AM R51910 Near Empire Springs, Reynolds River Area—13.3 130.67; AM R55053 6 km E Macarthur River Camp—16.4 136.15; AM R55058 3 km S Batten Point—15.8 136.33; AM R55681 West Is. Sir Edward Pellews—10.3 142.07; AM R59000, R59059, R59065 Badu Is., Torres Strait—10.1 142.12; AM R59137 Horn Is.—10.6 142.28; AM R60193 Maranboy—14.5 132.78; AM R60194 61 km E of Stuart Hwy. on Roper Hwy.—14.8 133.67; AM R60195 Borroloola—16.0 136.3; AM R61223 Near Batten Landing, N of Boorrooloola—15.8 136.52; AM R61488 Darwin—12.4 130.83; AM R61662, R61834, R61848, R61849, R62425, R62426, R62427 Yam Is. Torres Strait—9.9 142.1; AM R64225 5 km SW St. Pauls on Kubin Rd., Moa Is. Torres St—10.1 142.33; AM R68818, R68819 Atherton Tableland, W of Cairns; AM R68820 Kuranda—16.8 145.63; AM R68821 Cairns District—16.9 145.77; AM R73760 Yorke Is., Torres Strait—9.73 143.42; AM R77362 Angurugu Mission, Groote Eylandt—13.9 136.45; AM R82601 Batavia Outstation Landing, 70 km N Weipa; AM R86867, R86881 Atherton Tablelands—17.2 145.48; AM R86868 Kuranda, Atherton Tablelands—16.8 145.63; AM R86882 Cairns District—16.9 145.77; AM R88941 Jabiluka Project Area—12.4 132.87; AM R93183 9 km N of Daly River—13.7 130.72; AM R93625 Batavia Outstation Landing, N of Weipa—12.1 141.87; AM R94228 18 km E of Pascoe River Crossing on Iron Range Rd.—12.7 143.07; AM R94494 40 km E of Pascoe River Crossing on road to Iron Range—12.7 143.17; AM R95478 41 km N of

Mataranka—14.7 132.8; AM R95479 9 km S Daly River TO—13.5 131.2; AM R96644 Jabiluka Project Area. 12.5 123.93; AM R97086 Regenerated Bauxite Mines, Weipa—12.6 141.88; AM R97295 Jabiru E Township—12.6 132.9; AM R97384, R97385, R97386 Georgetown Billabong, Magela Creek, Jabiru Project Area—12.6 132.93; AM R97454 Magela Creek, 1.5 km W of 009 Gauge Stn. by road—12.6 132.88; AM R98501 Magela Creek, vicinity of 009 Gauge Stn.—12.6 132.9; AM R99787, R99836 Vrilya Point, Cape York Peninsula—11.2 142.12; AM R104135 8 km S of Pascoe River Crossing on road to Iron Range, Cape York—12.4 140.52; AM R104921 Cobourg Peninsula; AM R105153 Weipa Regeneration Site—12.6 141.88; AM R107240 Magela Creek, 009 Gauge Stn.—12.6 132.9; AM R112430 Jabiru E Township—12.6 132.9; AM R112431 Arnhem Hwy., vicinity of Jabiru—12.6 132.88; AM R112432 Jabiru Rd. near Retention Pond no. 1, E Jabiru—12.6 132.9; AM R113930 ~18.6 km S of Lappa RR Stn. (on Chillagoe-Mareeba line, via Mt. Garnet Rd.)—17.5 144.92; AM R128130 Wenlock Downs, Batavia River (Batavia Downs, Wenlock Rvr)—12.6 142.67; AM R128163 Melville Is.—11.5 130.67; AM R128230 Malaya Bay—11.4 132.87; AM R128242 Cooktown—15.4 145.25; AM R128555 Silkwood—17.7 146.02; AM R128713 Darwin—12.4 130.83; AM R128789, R128793, R128794 Mandalee, Innot Hot Springs—17.7 145.27; AM R128797 Mt. Carbine—16.5 145.13; AM R128872 Silkwood—17.7 146.02; AM R129069, R129070 Fanning Downs, Burdekin River, Macrossan—20.0 146.45; AM R133265, R133287 Doyndji [Donydji], Eastern Arnhem Land—12.4 135.47; AM R135301 Groote Eylandt, Gemco Mining Lease —13.9 136.43; AM R138687 Groote Eylandt, Gemco Mining Lease Area—14.0 136.43; AM R138712 Groote Eylandt, Gemco Mining Lease Area—14.0 136.42; AM R142916, R142930, R142931, R142932 51 km S of Katherine—15 132.27; ANWC R0447 Nourlangie Rd.—12.8 132.77; ANWC R0640 5.5 mi S of Kapalga Lagoon—12.3 132.42; ANWC R0642 4.1 mi S of Kapalga Lagoon—12.6 132.42; ANWC R0643 Kapalga Lagoon—12.6 132.42; ANWC R0886 14 mi N of main road at Bmr Hut, Kapalga—12.5 132.3; ANWC R0923 3 mi E of Shady Camp (Mary River)—12.5 131.77; ANWC R0924 Berrimah Darwin—12.4 130.92; ANWC R0925 R0926, R0927 ~1 mi N of Fogg Dam, Humpty Doo—12.5 131.29; ANWC R0926 R0927 1 mi N of Fogg Dam, Humpty Doo—12.5 131.29; ANWC R0942 2.2 km N of Borroloola Rd. on road to Bauhinia Downs—16.6 135.67; ANWC R0943 4.8 km N of Borroloola Rd. on road to Bauhinia Downs—16.6 135.68; ANWC R2078 Pascoe River, Cape York Peninsula—12.9 143; ANWC R3305 2 km along Flying Fox Rd., Kapalga, Kakadu National Park—12.7 132.35; ANWC R3306 R3307 N of Airstrip, Kapalga, Kakadu National Park—12.5 132.32; ANWC R3309 R3310, R3311 1.5 km S of Appletree Point TO, Kapalga, Kakadu National Park—12.5 132.33; ANWC R3312 Obiri Rock, Kakadu National Park— 12.7 132.9; ANWC R3522 R3523 South Point Rd., Kapalga, Kakadu National Park—12.5 132.38; ANWC R3665 R3762 Munmarlary Stn. Area, Kakadu National Park-12.5 132.7; ANWC R3666 Redlily, Kapalga, Kakadu National Park— 12.6 132.31; ANWC R3760 North Rd., Kapalga, Kakadu National Park—12.6 132.38; ANWC R3664, R3761, R3820 Kapalga, Kakadu National Park—12.5 132.33; ANWC R5280 Lake Eacham, Atherton Tableland—17.2 145.63; NMV D1088 Kuranda—16.8 145.63; NMV D5202, D5203, 5204, D5206 Oenpelli, E Alligator River—13.1 132.37; NMV D1017, D1018, D1020 Port Darwin—12.4 130.83; NMV T1095 Upper Glyde River—12.3 135.02; NMV T1094 Glyde River —12.3 135.02; WAM R13581 Kalumburu—14.3 126.63; SAM R04905, R04908, R04965, R04966, R04967 Mornington Is.—16.6 139.35; SAM R05856, R05858 Doomadgee Mission—17.9 138.82; SAM R13661 Phillips Range—16.8 125.8; SAM R00351, R00352, R00353 Stewart River—14.1 143.53; SAM R00362, R00363 Melville Is.—11.5 131.17; SAM R03227 Milingimbi Is.—12.0 134.88; SAM R03817 Woodstock—19.6 146.83; SAM R05018, R05019 Doomadgee Mission—17.9 138.82; SAM R05024 Mornington Is. Mission—16.6 139.17; SAM R05136, R05137 Doomadgee Mission—17.9 138.82; SAM R05332 Mornington Is.—16.6 139.35; SAM R06011 Pt. Essington—11.2 132.15; SAM R09947 Strathgordon HS—14.6 142.17; SAM R13542 Bing Bong Stn.—15.6 135.35.

The following description and notes on the diet, reproduction, thermoregulation, physiology, etc., should only be taken as generalizations, because it is often uncertain to which taxa within the *scalaris* group the data refer.

Fossil Record

None known.

Diagnostic Characteristics

- Medium-sized, brownish-gray arboreal monitor with moderately broad dark bands or rows of ocelli on the back.
- Tail round in cross section, with strongly keeled mucronate scales.
- Males with several rows of small spinose white scales on lateral surface of tail behind vent that are of a different structure to those of *V. tristis*.
- Males are only slightly larger than females, but have broader and deeper heads.
- *V. scalaris* has a shorter snout and deeper head than *V. tristis*.
- Sides of head and neck (and often the top of the head) are spotted (throat often spotted).

Description

A medium-sized climbing pygmy monitor with a ground color of brownish gray to black. Its long prehensile tail (about 1.5 times SVL) is round in cross section, with strongly keeled scales and pale banding on the proximal half; the distal half is black. The sides and top of its head and neck are spotted. Prominent, transversely aligned series of pale dark centered ocelli (occasionally spots) on dorsal surfaces; ocelli often incomplete. Limbs spotted to ocellated with pale pigment. Ventral surfaces whitish, sometimes with yellow on throat and flanks, and sometimes with transverse rows of darker spots (Wilson and Knowles 1988).

Size

SVL 71 to 253 mm (average adult, 250 mm), tail 132% to 166% of SVL, maximum TL of about 60 cm. At Kakadu (Sweet, personal observation), SVL of adult animals is 220 to 250 mm, weight is 120 to 150 g.

Habitat and Natural History

Occurring in areas with trees throughout northern Australia, these monitors inhabit hollows in trees and logs. At Kakadu (Sweet, personal observation), *V. scalaris* occupies savanna woodlands where *Eucalyptus miniata* and *E. tetrodonta* are codominant trees. They are apparently absent from monsoon forest, *Allosyncarpia* forest, and paperbark- and *Acacia*-dominated woodlands. Preferred residence sites are small hollow horizontal limbs, 4 to 8 cm in diameter, located at the lower margin of the canopy (usually 6–9 m above ground level). Animals spend the night in such limbs, and they often spend several hours each day with only their head exposed. Hollow tree trunks, ascending hollow limbs, larger horizontal hollow limbs, and logs are investigated during foraging and used as temporary refugia.

Varanus scalaris is an alert and wary lizard (Sweet, personal observation). Individuals approached on the ground either move quietly into dense vegetation, or, more commonly, run to the nearest tree and rapidly ascend on the far side, and almost invariably enter a hollow. Animals already in a tree shift quite carefully to maintain concealment. Radiotracked lizards located daily for many months never habituated (Sweet, personal observation).

Juvenile and subadult *V. scalaris* are subject to heavy predation by blue-winged kookaburras, but adult lizards are not taken. Of 40 adults radiotracked in Kakadu, one was taken by a hawk, one killed but not eaten by a dingo, two each by *Varanus gouldii* and a python *Antaresia childreni*, and nine were killed by a single feral cat (Sweet, personal observation).

All of 16 adult lizards survived a moderately intense bushfire in mid-June but ceased any terrestrial activity for about 6 weeks, perhaps owing to the loss of ground cover and terrestrial prey (Sweet, personal observation). Loss of foraging opportunities at this time, when fat bodies were depleted at the end of the breeding season, may have led to significant mortality from starvation in the late dry season (Sweet, personal observation).

Terrestrial, Arboreal, Aquatic

V. scalaris is a strongly arboreal monitor (Shine 1986) that nonetheless does much of its foraging, and typically travels between trees, on the ground. Short-limbed and rather stocky in comparison to other arboreal goannas, *V. scalaris* usually avoids branches of less than its body diameter, where it climbs only with evident difficulty. Unlike *V. glauerti* or *V. tristis*, *V. scalaris* has not been observed to jump between limbs or between trees, and it cannot travel on the undersides of horizontal limbs. Individuals sometimes fall off vertical trunks they are attempting to climb. Sweet's observations do not match with Shine's. Sweet observed that "*V. scalaris* are really quite clumsy in trees, but that's where people usually see them on chance encounters. Radio-tracking shows that most of their 'business' is conducted on the ground, and that trees are mostly used for shelter when not foraging, or for escape from predators. Compared to a *V. tristis* or (especially) a *V. glauerti*, a *V. scalaris* is most inelegant in a tree!" (Sweet, personal observation).

Time of Activity (Daily and Seasonal)

Active around the year, daily from early morning to late afternoon, *V. scalaris* is wholly diurnal. At Kakadu (Sweet, personal observation), lizards are relatively inactive during the later half of the dry season, July to October, and may spend 10 to 25 days in the same tree. Activity rises

TABLE 7.14.
Body temperature of *V. scalaris* at different times of day
during the wet and dry seasons.[a]

Time	Wet Season (Mean °C)	Dry Season (Mean°C)
Predawn	21.7	18.5
Morning	33.7	24.7
Midday	38.9	35.4
Late afternoon	33.7	34.1
Number of lizards	86	69

[a] Data from Christian and Bedford (1996).

with the first rains and peaks on dry days during the wet season. From late December through May, radiotracked animals used different trees each day and were often observed to visit four to eight trees per hour while foraging (Sweet, personal observation).

Thermoregulation

Body temperatures of free-living animals during the wet season were compared with those during the dry season by Christian and Bedford (1996). Table 7.14 summarizes results of this study.

In captivity, mean body temperatures selected by captive lizards in gradients in the wet season (38.1°C) were also higher than those (35.2°C) in the dry season. In the field, lizards seemed to stay in fence posts during the night as they retain higher temperatures, and they are often seen basking on the outside of the posts in the mornings. In trees at Kakadu (Sweet, personal observation), these lizards usually bask with only their head exposed. They often emerge from such shelters without pausing to bask further and quickly descend to the ground to begin foraging.

These varanids exploit thermal refugia to maintain high body temperatures for most of the day while minimizing exposure to predators (Christian and Bedford 1996; Sweet, personal observation).

Foraging Behavior

V. scalaris is a very active forager and expends a large proportion of its energy in hunting for insects and small vertebrates in trees. At Kakadu (Sweet, personal observation), foraging in trees is comparatively rare. These monitors also sometimes forage away from trees, to which they retreat when threatened. In captivity, both juveniles and adults sometimes hang from branches by their prehensile tails (Ruegg 1974). Although they can do this, Sweet never saw one engage in it unless it slipped off a limb. At Kakadu (Sweet, personal observation), *V. scalaris* does not dig and rarely searches into leaf litter. They do enter tiny apertures, force under loose bark, and wedge under the edges of downed logs. They seem to catch much of their prey on the ground surface or from the lower few centimeters of grass stems and coppiced saplings, and spend a lot of time examining dense vegetation above them as they forage. Most prey are snatched at short range (rarely pursued, and then not often successfully).

Diet

Stomachs of 40 of 80 specimens, captured in all months from Western Australia, contained mainly invertebrates. Most vertebrates eaten were lizards. Most items eaten were grasshoppers (n = 12), cockroaches (n = 8), spiders (n = 5), beetles (n = 5), and lizards (n = 6). Stomachs of four specimens examined by Losos and Greene (1988) contained a skink, a scorpion, and birds. Sweet (personal observation) found more or less the same at Kakadu: the lizards snap at and often catch flying insects (such as cicadas and dragonflies) that are attracted to the ends of the small hollow limbs used as refuges. On the ground, grasshoppers are the dominant prey, but the lizards appear to be searching for large stick insects much of the time. They attempt to catch small skinks such as *Carlia* spp., but Sweet never saw one actually catch any lizard. Feces contain quite a few roaches, which are probably captured inside logs and tree hollows.

Reproduction

Peak testis size was found from March to June, and the smallest SVL with enlarged testis size was 125 mm. Sex ratio of animals collected was 71 males to 31 females. Sex ratio at Kakadu was close to 1:1 (Sweet, personal observation). Enlarged follicles were found in females from May to September, and one female was gravid in late July. At Kakadu, the mating season extends from early May to mid-June. Males travel to trees where females reside and may remain in close attendance for 3 to 5 days. A series of two to four males usually attend the female in her home tree. Eggs are laid in July, either in deeper recesses of disused termite mounds or in old root channels accessed via hollow tree trunks (Sweet, personal observation). Mean clutch size was 7.7 eggs (range, 3–12 eggs). Incubation time in captivity ranges from 115 to 140 days. Hatchlings are from 156 to 175 mm in TL, weigh 3.4 to 5 g, and are similar to adults in pattern and coloration (Horn and Visser 1989; Bennett 1998).

Movement

V. scalaris runs for an estimated 1.1 hours per day in the dry season and 1.6 hours per day during the wet season (Christian et al. 1996b). No studies of home range have yet been published. At Kakadu (Sweet, personal observation), home ranges of 40 animals were surprisingly small, less than 1 ha for females and less than 1.5 ha for males, and are quite stable throughout the year. Males have strictly nonoverlapping home ranges that are probably defended as territories (although male home ranges that are vacated by the death of the resident are not absorbed by adjoining males but instead remain vacant). Female home ranges overlap to some extent, but not extensively, and are generally along the borders of the home ranges of two or three adult males. Subadult animals also have stable home ranges and seem to be tolerated by adults of both sexes.

Varanus scalaris attains high densities in suitable habitat, with as many as two or three adults and several subadults per hectare (Sweet, personal observation).

Physiology

Physiological studies on *V. scalaris* and other small varanids have been limited. There are two extremes of the tropical climate in the region they inhabit: one is very wet and the other is dry. Body temperatures differ during the wet and dry seasons. Temperatures are higher during the morning in the wet season and are similar in the latter part of the day in the dry season. These differences are not due to thermal limitations of the environment (Christian and Bedford 1996).

Water flux rates are much higher in the wet season than in the dry season, reducing the FMR by 47%. This also occurs in *V. gouldii* and *V. panoptes* (Christian et al. 1996b). Significant seasonal differences in energy and water fluxes occur between seasons. *V. scalaris* remain active during the dry season, unlike most other species of varanids inhabiting the same region (Christian et al. 1996b). At Kakadu (Sweet, personal observation), they are microsympatric with the yellow-headed form of *V. tristis*, which is a lot more active than *V. scalaris* once the dry season sets in.

Fat Bodies, Testicular Cycles

Peak testicular size occurs from March to June, when mating presumably occurs. At Kakadu (Sweet, personal observation), mating was observed from early May to mid-June, none earlier, although males begin "visiting" females in mid- to late April. Fat bodies increase in size from November to May, most rapidly in March–April. Animals examined in October were emaciated, with no macroscopic fat bodies (Sweet, personal observation).

Parasites

Adult nematodes *Abbreviata confusa* have been found in *V. scalaris,* and small numbers of the nematodes *Dioctowittus denisoniae, Trichostrongyloidea* sp., *Hastaspiculum* sp., and *Oswaldofilaria* sp. have also been found (Jones 1988). At Kakadu (Sweet, personal observation), most individuals have one to three small ticks and can be heavily infested with tiny orange mites, perhaps trombiculids, during the wet season. The mites are not evident after March. Also, an elongate flattened "worm," about 100 mm long, resides in the lateral trunk musculature and often in and around the lungs (Sweet, personal observation). Lizards with more than a couple of these parasites appear to be in pretty poor condition.

7.20. *Varanus semiremex*
Eric R. Pianka

Figure 7.44. Varanus semiremex
(photo by Jeff Lemm).

Nomenclature

Peters described *Varanus semiremex* in 1869. The type specimen (holotype), collected at Cape York, Queensland, is in the Zoologisches Museum, Universitat Humboldt, Berlin, as specimen ZMB 5776 (Cogger et al. 1983) (Fig. 7.44).

Geographic Distribution

Varanus semiremex occurs along the northeast coast of Australia in Queensland from Brisbane to the tip of Cape York as well as on the western side of upper Cape York. Its distribution extends inland about 50–100 km in vegetation bordering freshwater streams and swamps (Cogger 1992) (Fig. 7.45).

Figure 7.45. Map showing known localities for Varanus semiremex.

QM R454 Cardwell—18.2 146.02; QM R2151, R2152 Stannary Hills?, near Herberton—17.3 145.22; QM R2172 Burketown (not certain); QM R24867 Coen, 3 km N—13.9 143.18; QM R26696 Gregory, 11.2 km NW?; QM R35334 Jeannie Rd., ~112 km N Cooktown, Starcke Stn.—14.7 144.85; QM R37574 Bourne Creek, 11 km NW Mt. Croll—13.7 143.07; QM R37592 Coen, 12 km NW—13.8 143.15; QM R38268 Lloyd Is., near Lockhart Rd. Community—12.7 143.4; QM R46793 Gladstone, Boyne Is. T'off—23.9 151.3; QM R51184 Wathanhiin—13.4 141.62; AM R6144, R6735 Townsville—19.2 146.82; AM R17745, R17746, R17938 Yepoon—23.1 150.73; AM R17965 Gladstone—23.8 151.27; AM R21410 Cliff Is., Princess Charlotte Bay—14.2 143.77; AM R31796 Townsville Common—19.2 146.82; AM R33288 Rockhampton District—23.3 150.53; AM R86945 Morris Is.—14.5 144.92; AM R86950 Ingram Is.—14.4 144.88; AM R93739 on freshwater lagoon between Andoom and Pine River, Weipa—12.5 141.78; AM R94491 Rocky Point, Weipa—12.6 141.87; AM R97997, R97998, R97999, R98000 Andoom Camp Swamp, NE of Botchitt Swamp, Weipa; AM R107537 Townsville—19.2 146.82; ANWC R5497 Shoalwater Bay Army Training Reserve, N of Rockhampton—22.6 150.73; ANWC R5536 Shoalwater Bay Army Training Reserve, N of Rockhampton—22.3 150.22; NMV D56406 Townsville—19.2 146.82; NTM R26124 Townsville Common—16.6 145.33; NTM R26125 Dawson River, via Rockhampton—23.9 149.83; WAM R115177 Andoom Swamp, NE of Botchitt Swamp, Weipa [presumably Botchet]—12.5 141.77; SAM R13812 Edward River—14.6 142.05.

Fossil Record

No fossils of *Varanus semiremex* are known.

Diagnostic Characteristics

- Medium size, less than 600 mm TL.
- Gray-brown, reticulated with numerous black flecks and spots.
- Tail round at base, last two-thirds laterally compressed.

Description

Gray-brown, with numerous scattered blackish flecks and spots forming a fine reticulum over the dorsal surface. White, cream, or pale yellow irregularly banded with brown below. Tail about 1.6 times longer than SVL, round at base but laterally compressed distally. In the northern part of its range in Cape York, lizards are darker and have a stronger pattern than those from farther south (Wilson and Knowles 1988). Northern lizards have reddish brown to white ocelli aligned transversely on their body and neck. Body proportions are similar to those of other pygmy monitors (Christian and Garland 1996).

Size

Varanus semiremex is a pygmy monitor, averaging about 250 mm SVL, reaching a maximal TL of about 600 mm. Greer (1989) reports a maximal SVL of 263 mm, but they may reach 270 mm SVL. Ten males averaged 214 mm SVL and seven females averaged 228 mm. Six specimens weighed from 150 to 294 g.

Habitat and Natural History

The prime habitat of *Varanus semiremex* is mangrove swamps, but these lizards follow freshwater streams inland and are also found in freshwater swamps, creeks, and lakes, especially along *Melaleuca*-dominated margins. They occur in woodland and open forests adjacent to such riverine areas. *Varanus semiremex* hide in hollows in tree branches overhanging water and in mangroves (Bustard 1970).

Terrestrial, Arboreal, Aquatic

Varanus semiremex is semiaquatic and arboreal. It is a capable swimmer, taking to water or ascending trees when approached.

Time of Activity (Daily and Seasonal)

This wary and uncommon diurnal monitor is seldom seen.

Thermoregulation

Nothing has been reported about the thermal biology of *V. semiremex*.

Foraging Behavior

This predatory lizard climbs trees and mangroves looking for arboreal gecko prey, but it also forages on the ground, on mudflats when the tide is out, along beaches, and probably in the water as well.

Diet

V. semiremex feeds on large insects, geckos (*Gehyra australis*), frogs, crustaceans including mud crabs, and fish (Bustard 1970; Cogger 1992; James et al. 1992). Small mammals may also be eaten occasionally (Dunson 1974).

Reproduction

Sexual maturity is reached at an SVL of about 150 mm, and breeding occurs in the late wet season (between February and April; James et al. 1992). Clutch size varies from 2 to 14 eggs (mean, 5.9 eggs). Ritual combat in males was described by Horn (1985).

Movement

No one has done any detailed studies of movement in this species.

Physiology

V. semiremex have nasal salt-excreting glands that enable them to consume marine prey (crabs) (Dunson 1974).

Testicular Cycles

Testes are enlarged in the late wet season during February through April (James et al. 1992).

Parasites

Nothing is known about parasites of this species.

Conservation

Steve Irwin (personal communication) has a study site for *V. semiremex* at the southern extent of their range between Gladstone and Yeppoon. Irwin and his colleagues have captured, studied, and released approximately 100 animals since the mid-1970s.

Introduction of toxic cane toads (*Bufo marinus*) from South America coupled with habitat (mangrove) destruction have decimated populations of *V. semiremex*. Irwin and colleagues placed some adults into captivity to work out the intricacies of breeding them, just in case the population crashes. They are already on the ragged edge of extinction in the southern half of their range. They will become critically endangered if habitat destruction is not stopped. Luckily, they are slowly recovering from the cane toads, but an occasional lizard is still found dead with a toad in its mouth.

Irwin and colleagues have worked out their breeding requirements and are currently releasing all adults and juveniles back where they were captured. In captivity, they lay two clutches of eggs. Irwin and colleagues are monitoring wild populations and trying to slow down mangrove destruction.

7.21. *Varanus spenceri*

JEFFREY M. LEMM AND GAVIN S. BEDFORD

Figure 7.46. Adult and juvenile Varanus spenceri *(photo by Gavin Bedford).*

Nomenclature

Spencer's monitor, or the plains goanna, was first described by Lucas and Frost in 1903 and was named after its collector, W. B. Spencer, former professor of biology at the University of Melbourne (Fig. 7.46). The holotype was found at Frewena on Rockhampton Downs Station in the Barkly Tablelands, 50 mi northeast of Tennant Creek in the Northern Territory. Boulenger (1906) erroneously described the same species as *Varanus ingrami* from a specimen from Alexandria Station in the Northern Territory. Mertens (1942a), realizing Boulenger's mistake, chose the primary name over *V. ingrami* and placed *V. spenceri* in the subgenus *Varanus*. Studies of chromosomal groupings (King and King 1975; King 1990), mitochondrial DNA (Ast 2001), and microcomplement fixation (King et al. 1991) support Mertens' conclusion. On the basis of most of these studies, *V. spenceri* is most closely related to *mertensi, varius, giganteus, rosenbergi,* and other *gouldii*-type monitors.

Geographic Distribution

Possibly the least known of the large Australian goannas, *V. spenceri* is found on the black soil plains of western Queensland, through the Barkly Tableland to the eastern parts of the Northern Territory (Fig. 7.47; Cogger 1992).

QM R15694 Hughenden, 129 km S—22.02 144.33; QM R15695 Hughenden, 129 km S—22.02 144.47; QM R21656 Boulia, 22.4 km E—22.97 140.13; QM R24535 Boulia, Marion Downs—23.37 139.65; QM R37169 Winton—22.38 143.03; QM

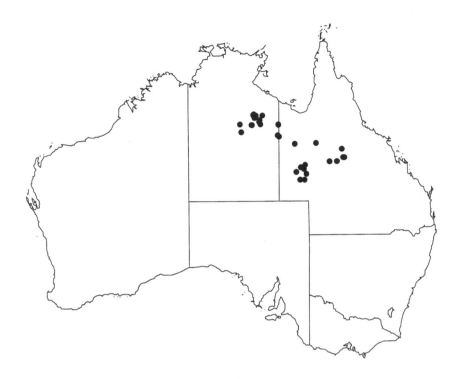

Figure 7.47. Known localities for
V. *spenceri.*

R41654 Coorabulka Stn., Terriboah Waterhole—24.07 140.07; QM R42022
Hughenden, ~43 km S—21.23 144.2; QM R47753 Springvale Stn; QM R47915
Boulia, 62 km S, Boulia to Springvale Rd.—23.1 140.45; QM R48641 Springvale
Stn., via Boulia—23.55 140.7; QM R52796 Brunette Downs—18.65 135.95; QM
R58096 Marita Stn.—22.42 143.73; AM R11900 Brunette Downs—18.65 135.95;
AM R21594 5 mi W Julia Creek—20.67 141.67; AM R43539 Kennedys Well, S Mt.
Isa—20.73 139.48; AM R64793 Mt. Isa Area P—20.73 139.48; AM R110514,
R110610 No. 2 Bore, Davenport Downs, NW of Windorah—24.12 140.53; AM
R142850 Pollygammon HS Tip—22.73 140.58; AM R142851 30 km W of
Camooweal—19.92 137.78; ANWC R1141 Brunette Downs, Barkly Tablelands—
18.42 136; NMV D 5891 Tennant Creek—19.65 134.18; NTM R26128 No. 8
Bluebush Bore—20.05 137.95; NTM R26129 Rockhampton Downs—18.95
135.18; NTM R26130 2k S Anthony Lagoon Homestead—18 135.52; NTM
R26131, R26133, R26135 Barkly Stock Route—17.98 135.33; NTM R26132
Rockhampton—18.95 135.2; NTM R26134 8 km W No 1 Bore Barkly Stock
Route—17.9 135.43; NTM R26136 Rockhampton Downs—18.95 135.2; NTM
R26137 Barkly Stock Route—17.95 135.37; NTM R26138 Rockhampton Downs
—18.95 135.2; NTM R26139 37 km S Brunette Downs—18.88 136.07; NTM
R26140 Rockhampton Dwns/Brunette Downs Rd.—18.25 135.5; NTM R26141
Tablelands Hwy.—18 135.62; NTM R26142 Anthony Lagoon—17.98 135.53;
NTM R26143, R26144 Eva Downs/Anthony Lagoon Rd.—17.98 136.25; NTM
R26145 Brunchilly Stn.—18.87 134.05; NTM R26146 Rockhampton Downs—
18.95 135.18; NTM R26147 Gallipoli Stn.—18.9 137.87; NTM R26148 Gallipoli
Stn.—18.93 137.88; NTM R26149 Brunchilly Stn.—18.87 134.05.

Habitat

A stout burrowing species, *V. spenceri* is mainly confined to the
arid plains and Mitchell grass–dominated (*Astrebla* spp.) grasslands,

where it uses large soil cracks and burrows as refugia (De Lisle 1996; Vincent and Wilson 1999). These cracks may be up to 1.5 m deep and very complex, and in some cases, they may extend for hundreds of meters. In the Barkly Tableland of the Northern Territory, the species may also be found on bare red soil plains, where it is sympatric with *V. panoptes*. Because of the openness of the habitats where it is found, one of its main defenses is to lie close to the ground, motionless, when approached. Greg Fyfe (personal communication) reports that *V. spenceri* is easily approached and rarely flees. When they do run, they are easily overtaken if safety is not found in a deep soil crack. Once inside a black soil crack, they are nearly impossible to extricate as they wedge the body as deeply as possible into the crack then fill their body cavity with air. Some animals have been known to flare their throats, stand bipedally, and whip their tails when disturbed (R. Jackson, personal communication), and in one case the tail of a large male was known to draw the blood of an adult human (G. Bedford, personal observation). The bipedal stance and body puffing has also been known to thwart the attack of a brown falcon (*Falco berigora*) near Mt. Isa, Queensland (Vincent and Wilson 1999).

Fossil Record

No fossils of *V. spenceri* are known.

Diagnostic Characteristics

- Head scales irregular and smooth.
- Nostrils lateral, oval, closer to the tip of the snout than the eye.
- Total of 150 to 175 scales around the middle of the body.
- Tail more or less round in section at the base with the last half laterally compressed with a double median dorsal keel.
- Tail in adults roughly the same or slightly longer than head and body (103%–107% of SVL).

Description

Adult *V. spenceri* range from gray to cream or brownish-gray in color, with darker brown or rusty-brown specimens known from some locales. Most specimens are irregularly banded with yellowish or gray crossbands on the neck, body, and tail, while some may also be flecked in dark brown and cream spots. Lower surfaces are light-colored with gray or brown flecking, especially on the throat. The lips are distinctly barred. Hatchling and juvenile *V. spenceri* are more colorful than the adults with bright yellow or cream bands on a brownish-black dorsum. Some are also known to have a bright pink coloration in the light bands upon hatching, but this fades with age and is no longer present by 1 year of age.

Size

V. spenceri attain a total adult length of 1.0 to 1.25 m (Cogger 1992; Schmida 1985), with a SVL measuring up to 50 to 55 cm (Vincent and Wilson 1999). Adult males may reach 6 kg in mass (G. Fyfe, personal communication), and males grow larger than females. The largest male recorded was 1.35 m long and was missing the tip of its tail. Hatchlings measure 22 cm TL and 13 cm SVL (Peters 1968). This is one of the few varanids hatched with a SVL longer than its tail length, although at some stage within the first year, the tail grows faster than the SVL (Bedford and Christian 1996).

Habitat and Natural History

Confined mainly to the black soil plains, *V. spenceri* is a heavy-bodied monitor built for a terrestrial lifestyle. Bennett (1998) states that they can climb quite well in captivity and may use trees in the wild; however, few trees are present in this region, and Bedford (personal observation) observes that they can climb, but very poorly because of their short claws. Inhabiting an arid area on the fringe of a monsoonal climate, the species is subject to extreme conditions ranging from drought to flood. *V. spenceri* has adapted well to these harsh conditions, becoming most active in the months from August to October (Schmida 1985). *V. spenceri* may be very common where they occur, and Bedford (unpublished data) has found as many as 13 animals in 1 hour on a windy 36°C day. All animals were found before 1100 hours. In the spring (September), animals become active around 0930 hours, and basking animals may be found up to an hour earlier (Fyfe, personal communication). In late summer and early autumn, these lizards are active from 0800 to 1030 hours, and depending on how hot the day becomes, they may reappear late in the afternoon, from 1700 to 1800, and may sometimes stay active after dark. In September 1989, Fyfe (personal communication) reports finding an adult female foraging on the road an hour after dark. Valentic (1995) also reports one record of nocturnal activity in the species that occurred after a heavy rain. Temperature and rainfall data from the middle of the range (Windorah, Queensland) for August–October show rainfall between 8 to 18 mm and temperatures recorded between 6 to 15°C (Bennett 1998).

No data exist on thermoregulation or active body temperatures of wild animals, but Fyfe (unpublished data) kept and bred the species outdoors in Alice Springs, Northern Territory. Temperatures there range from the upper 40°C in summer down to −7°C in winter. Animals were provided a heat pad in winter so that temperatures never dipped below 20°C. However, one year, the heater failed, and the lizards went without supplemental heat for the entire winter. They still emerged at the same time of day to bask and produced a successful clutch that year as well. From that point on, no supplemental heat was provided and the lizards survived and bred annually with no ill effects, illustrating *V.*

spenceri's tolerance for cold temperatures. Habitat and temperature extremes suggest that *V. spenceri* are probably only active for part of the year; however, more data are needed to substantiate this.

From gut contents and observation, the foraging behavior of *V. spenceri* is that of a slow, methodical forager that may consume whatever it can subdue. Studies of gut contents revealed that animals had eaten mammals, large elapid snakes, agamids, and insects, including beetles and locusts (Stammer 1970; Pengilley 1981; Valentic 1997). Much of the diet consists of locusts when they are in plague proportions on the Barkly Tableland; however, *V. spenceri* also consumes vertebrate prey. After a substantial wet season, long-haired rats (*Rattus villosisimus*) are overabundant and are an obvious prey item. Rodent hair was found in eight stomachs of *V. spenceri* and was believed to be that of *R. villosisimus* (Pengilley 1981), and Fyfe (personal communication) observed several stout monitors after one such plague occurrence of rats.

There is a significant size difference between sexes in *V. spenceri*, with adult males growing to nearly twice the size of adult females. Large adult males are known to engage in bipedal combat during the breeding season (King and Green 1993c). This behavior was first recorded by Waite (1929), who mistook them for perenties (Horn 1981). Mating is believed to occur around August, with egg laying occurring around October, but can vary from September to November (Greer 1989). Pengilley (1981) found large numbers of females looking for nest sites near piles of red soil dumped near the road in late September and early October, while Peters (1971) recorded a wild-caught female laying eggs in the beginning of November. Furthermore, Fyfe (personal communication) has collected and observed wild gravid females in the field from mid-September to early October. Oviposition by captive animals in the Northern Territory corresponds to that of wild counterparts (G. Fyfe, personal communication: G. Bedford, personal observation), with clutches being laid in late September and early October. Wild gravid females construct nesting tunnels prior to oviposition and intermittently bask near the entrance throughout the day, with the remainder of the day spent within the burrow (G. Fyfe, personal communication). Burrow sites are situated on any areas of raised soil or gravel, and animals have been observed digging into the raised edges of the road base, some burrows actually extending under the asphalt edge of the road (G. Fyfe, personal communication). The deepest burrow found under the asphalt was nearly 1 m deep. In the absence of roads, animals have been found digging into the soil around or between lone rock piles to nest. Similarly, Fyfe observed nesting on slight rises of laterite "islands" scattered through parts of the black soil plains. According to Fyfe, however, the most favored sites seem to be around the margins of large emergent termite nests that are scattered among the plains. The lizards begin to dig just before the edge of the nest and continue to dig around the circumference of the nest while progressively getting deeper. Total burrow length may be over 2 m and descends to a depth of 700 to 800 mm. Eggs are laid at the end of this tunnel in a separate chamber dug into the floor of the tunnel. The egg chamber is then backfilled with soil and the earth tightly packed (G. Fyfe, personal communication).

Bedford (personal observation) has observed five wild nests that were roughly 750 mm deep and tended to dogleg either to the right or left, with a large chamber at the end where the eggs were deposited. Nests similar to these have been excavated for egg laying in captivity (G. Fyfe, personal communication, G. Bedford, personal observation). *V. spenceri* has the largest clutch on average for any Australian goanna, with up to 35 eggs (range, 11–35 eggs) laid in a single clutch, and clutch size is positively correlated with body size (Christian 1979; Peters 1969, 1971; Pengilley 1981). The smallest female observed to lay in captivity was a 300-mm SVL animal that laid 12 eggs, whereas the largest female had a SVL of 450 to 470 mm and laid 35 eggs (G. Fyfe, unpublished data). The largest males in both captivity and in the wild to show breeding behavior measured 500 mm SVL, whereas the smallest male to evert hemipenes in the field measured 280 to 300 mm SVL (G. Fyfe, unpublished data). Eggs average 50 mm long by 35 mm in diameter (Pengilley 1981; Peters 1971). Bedford (personal observation) reports incubation at roughly 90 days at 32 to 33°C, whereas Fyfe (personal communication) incubated eggs at 30 to 31°C, with a total incubation period between 118 to 124 days. Peters (1986) incubated eggs between 133 to 140 days at a temperature of 29°C. An incubation time of 123 to 129 days at 28 to 32°C was reported by Horn and Visser (1989). Hatchlings emerge from late January to early February. Mass varies greatly with records from 14.5 to 25 g, with most hatchlings measuring between 19 to 22 mm SVL (Horn 1997; Peters 1971). No data exist on neonate ecology; however, Vincent and Wilson (1999) mention that newly hatched animals were observed in Winton, Queensland, in mid-December, earlier than when hatchlings are expected in the Northern Territory.

No major field studies of the ecology or behavior of *V. spenceri* have yet been undertaken, and they remain one of the least known of Australian monitors. They are believed to be slow-moving, wide-ranging foragers; however, studies in the laboratory have shown that their respiratory physiology is that of low maximum oxygen consumption, similar to other Australian varanids such as *indicus* and *mertensi,* yet significantly lower than that of *V. gouldii.* The metabolic rate of *V. spenceri* is similar to that of *V. gouldii* and *V. panoptes.* Furthermore, no difference was found in the metabolic rate between adult and juvenile *V. spenceri* (Schultz 2002). Christian et al. (1996a) studied solar absorptance of the skin of a number of Australian lizards and found that skin of *V. spenceri* had a similar absorptance (~73%) to other lizards. Furthermore, solar absorptance did not change with temperature in *V. spenceri* as it did in other animals.

7.22. *Varanus storri*

BERND EIDENMÜLLER

Figure 7.48. Varanus storri
(photo by Richard Bartlett).

Nomenclature

Described as *Varanus storri* by Mertens in 1966. The type specimen is from Charters Towers, Queensland, SMF 59011. Storr (1980) described the subspecies *ocreatus,* holotype WAM R42717 from Argyle Downs airstrip, Western Australia (Cogger et al. 1983) (Fig. 7.48).

Geographic Distribution

Two distinct populations of this species (*V. s. storri* and *V. s. ocreatus*) occur in Australia: *V. s. storri* is distributed in the eastern parts of Australia from Charters Towers along the Black Soil area of Queensland through to the eastern parts of the adjacent parts of Northern Territory (e.g., Borolloola) (Cogger 2000), and *V. s. ocreatus* is restricted to the Kimberley Region of Western Australia to the adjacent parts of Northern Territory (e.g., Top Springs) (Storr 1980). Fig. 7.49 shows known localities.

QM R14240 Charters Towers—20.08 146.27; QM R24492 Archer Pt, 16 km S Cooktown—15.6 145.33; QM R32296, R37685 Mt. Isa—20.73 139.48; QM R37291 Ben Lomond Area, Keelbottom Creek, W of Townsville—19.72 146.15; QM R38907, R38908 Warrigal Ra, crest, 27.4 km E Torrens Creek—20.72 145.17; QM R44679, R44680 Mt. Leyshon Stn.—20.28 146.27; QM R44693 Battery Stn.—19.48 145.65; QM R52761 Lappa Junction—17.37 144.88; QM R52773 Borroloola, 210.8 km S—16.17 135.67; QM R52889 Lappa Junction, Chillagoe Line—17.37 144.88; QM R53890 Lappa Junction—17.37 144.88; AM R14346, R15377 Mt. Isa—20.73 139.48; AM R15454 Halls Creek—18.23 127.67; AM R16531, R16532 15 m S of Charters Towers—20.28 146.18; AM R21032 Charters

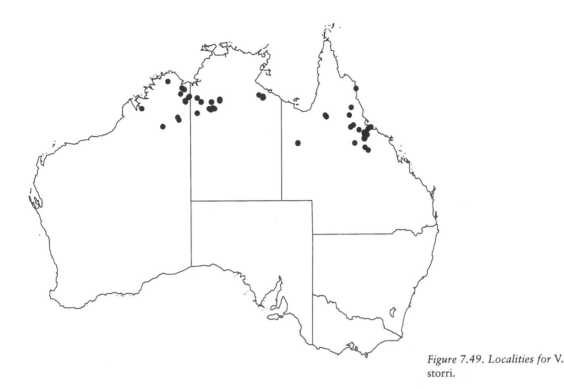

Figure 7.49. Localities for V. storri.

Towers—20.08 146.27; AM R25981 Mt. Isa—20.73 139.48; AM R27488, R27489, R27450 5 mi N Charters Towers—20.03 146.23; AM R29555 Charters Towers—20.08 146.27; AM R54645 Near Caranbirini Water Hole, ~21 km N Macarthur Rd.—16.27 136.08; AM R63108, R63109 ~35 km SE Charters Towers via Gregory Hwy.—20.38 146.18; AM R63177 19.7 km W Jcts. Kennedy and Gulf Hwy. S via Gulf Hwy.—18.13 144.67; AM R63672 8.1 km W Croydon PO—18.15 142.22; AM R64015 Mt. Isa—20.73 139.48; AM R64794, R64795, R64796, R64797, R64798 100 mi S Charters Towers—21.42 146.6; AM R68804, R68805, R68806, R68807, R68808, R68809, R68810 Charters Towers District—20.08 146.27; AM R86874, R86875, R86876, R86877, R86878, R86879, R86880 Charters Towers—20.08 146.27; AM R90500 6.6 km SE Greenvale by road—19.05 145.08; AM R95555 Charters Towers—20.08 146.27; AM R125956 between Kununurra and Turkey Creek; AM R129493, R129494, R129495, R129496 Macrossan, Burdekin River—20 146.45; AM R140476 El Questro Stn., Victoria Jackeroos Waterhole—16.02 128; AM R143912 Croydon, ~6 km E on Georgetown Rd.—18.25 142.33; ANWC R2044, R2045 Townsville—19.27 146.82; ANWC R3252, R3253 Laura/Mt. Carbine Region; NMV D56444 Charters Towers—20.08 146.27; NTM R26150 Charters Towers—20.08 146.27; NTM R26151, R26155, R26158, R26159, R26160, R26161, R26162 Mengala Range; NTM R26152, R26153, R26153 Wave Hill—17.48 130.95; NTM R26156, R26157 Blackwattle Creek—19.23 144.82; NTM R26163, R26164 Macarthur River Stn., Barney Hill—16.42 136.1; NTM R26165 50 km E Derby, on Gibb River Rd.—17.42 124.1; NTM R26166, R26167, R26168 Old Wave Hill Dump, 13 km E Police Stn.—17.48 130.94; NTM R26169 Gregory River National Park, Bullita; NTM R26170 Wave Hill Stn. Old Homestead Dump—17.48 130.95; NTM R26171 11 km S Top Springs—16.63 131.77; NTM R26172, R26173, R26174, R26175 Wave Hill Stn. Old Homestead Dump—17.48 130.95; NTM R26176 11 km S Top Springs—16.63 131.77; NTM R26177, R26183, R26184 Wave Hill Stn. Old Homestead Dump—17.48 130.95; NTM R26178 Old Wave Hill Stn Dump—17.48 130.95; NTM

R26179 11 km S Top Springs—16.63 131.77; NTM R26180 Victoria River By Wave Hill Settlement—17.48 130.95; NTM R26181, R26182 11 km S Top Springs —16.65 131.77; NTM R26185 Top Springs—16.53 131.8; NTM R26186 Wave Hill—17.48 130.95; NTM R26187 Wave Hill Stn., 3.9 km S, No. 2 Bore—17.43 131.25; NTM R26188 Wyndham, 9 km E—15.58 128.28; NTM R26189 Victoria River, By Wave Hill Settlement—17.48 130.95; NTM R26190 Near Borroloola; NTM R26191, R26192 Macarthur River Stn., Barney Hill—16.42 136.1; NTM R26193 Wave Hill—17.38 131.12; SAM R17692 Charters Towers—20.08 146.27; WAM R13720 Inverway—17.85 129.63; WAM R40998 Kildurk Stn.—16.43 129.62; WAM R42717 Argyle Downs Airstrip—16.33 128.77; WAM R42764 28 km SW of Argyle Downs HS—16.3 128.8; WAM R51266 Bull Flat Bore. Christmas Creek Stn.—19.1 126.12; WAM R53764, R53765 120 km S of Charters Towers. Blowhard Creek—21.18 146.27; WAM R55487 Charters Towers—20.08 146.27; WAM R57244 Old Theda Stn HS (Junction of Palmoondoora Creek and Morgan River)—14.82 126.67; WAM R60041 30 km SE of Halls Creek—18.48 127.78; WAM R60042 Wave Hill Police Stn. on creek bank at rubbish tip—17.45 130.83; WAM R60043 Gordon Creek 48 km S of Victoria River Downs 1 km From Creek— 16.83 131.02; WAM R60044 Gordon Creek 49 km S of Victoria River Downs on Banks of Creek—16.83 130; WAM R70171 13.5 km 242° (New) Lissadell HS— 16.72 128.45; WAM R70583 14.7 km 244° (New) Lissadell HS—16.73 128.43; WAM R75336, R75337, R75359, R75416, R75417, R75418, R75419, R75516, R75526, R75527 11 km W of New Lissadell Homestead—16.68 128.45; WAM R75362, R75420, R75421, R75508 15 km WSW of New Lissadell Homestead— 16.72 128.43; WAM R87357 4 km NW of Wyndham—15.45 128.12; WAM R57342 32 km ENE of Charters Towers—19.45 146.57.

Fossil Record

No fossils of *Varanus storri* are known.

Diagnostic Characteristics

• Small size up to 30 cm.

• Dorsally reddish brown, with numerous scattered dark brown or black scales, sometimes forming a reticulum, which is also seen sometimes on the upper surface of the limbs.

• Reticulum sometimes reduced to flecks.

• Head usually flecked with blackish brown.

• Lower surfaces and lower sides of neck white or cream.

• Tail strongly spinose.

Description

Head scales small, irregular, and smooth; nostrils dorsolateral, slightly closer to the tip of the snout than the eye. Midbody scale rows 70 to 84; transverse rows of ventrals 45 to 48. Tail round in section, basally slightly depressed with a two keeled crest. Tail between 1.4 (*V. s. storri*) and 1.6 (*V. s. ocreatus*) times longer than SVL.

V. storri differs from *V. primordius* in its more numerous midbody scale rows (70–91, vs. 60–66); *V. s. ocreatus* differs from *V. s. storri* mainly in the enlarged scales under distal part of hind leg. It also has a longer tail and limbs, fewer midbody scale rows, and transverse rows of ventrals (Cogger 2000; Storr 1980; Vincent and Wilson 1999).

V. storri is a small, spiny-tailed monitor species. Size, tail length, and proportion vary considerably between populations. At least two distinct populations of this species occur in Australia.

Size

SVL ranges from 49 mm to 132 mm (TL usually from about 150 mm to about 300 mm). Maximum size may exceed 40 cm (Peters 1973). No sexual size dimorphism in head size is discernible. Tail lengths average 1.4 to 1.6 times SVL.

Habitat and Natural History

Varanus storri occurs in rocky grassland with dead trees (Peters 1973). They appear to live in colonies: Peters (1973) found 22 lizards in 0.75 km² and suggested that the total population must have been about 50.

In Queensland, they are found in open woodland, sheltering under rocks (Stirnberg and Horn 1981) and in U-shaped burrows under rocks (Peters 1973).

Terrestrial, Arboreal, Aquatic

Varanus storri is terrestrial, with a preference for tussock and grassy areas with rocky outcrops. Both subspecies of *V. storri* live in large colonies.

Time of Activity (Daily and Seasonal)

Unlike many larger species of monitors, *Varanus storri* is generally active during the cooler parts of the day over most of the year, except for a few winter months. These very wary lizards are seldom seen.

Foraging Behavior

V. storri feeds mainly on insects, spiders, and sometimes on small lizards (James et al. 1992). The activity of *V. storri* is highly seasonal, and lizards rely on building up fat reserves during times of plenty to get them through lean periods. They avoid temperature extremes by retreating into burrows. They also overpower prey that also shelter in those burrows.

Diet

A survey of the stomach contents of 48 of these lizards (James et al. 1992) revealed that 47% of the diet of *V. storri* by volume consists of orthopterans. Formicidae (11%) and other lizards (9%) are also eaten. The 48 stomachs with food contained 53 prey taxa.

Reproduction

On the basis of enlarged testes and yolked ovarian eggs, both sexes appear to achieve sexual maturity at about 90 mm SVL (James et al. 1992). Reproductive males and females were found over many months of the year. This result may be in part due to variation in timing and different localities throughout the broad geographic range of this species. They are most active during February and March and between July and November. Clutch sizes are small, from one to six eggs (mean, 3.9 eggs). Eggs are laid in underground nests dug by females. Incubation time is short, around 100 to 129 days (mean, 113 ± 8.2 days).

Hatchlings are about 48 to 61 mm in SVL (weights about 3.0 g). Hatchlings have a spotted pattern, quite different from adults (Eidenmüller 1994, 1997; Eidenmüller and Horn 1985).

Movement

Home ranges are fairly small.

Reproductive Cycles

Animals with enlarged testes could be found at most times of the year, but were primarily collected in the late dry or early wet season (November). Females with yolking eggs are usually found during the period when testes of males were enlarged (James et al. 1992).

7.23. *Varanus tristis*

Eric R. Pianka

Figure 7.50. Varanus tristis (black-headed form from interior) (photo courtesy of Dr. Harold Cogger).

Nomenclature

Varanus tristis was described as *Monitor tristis* by Schlegel in 1839. The type specimen, which has been lost, was from the Swan River, Western Australia, exact whereabouts unknown (Cogger et al. 1983) (Fig. 7.50). *V. tristis* from various parts of their geographic range vary considerably in color and anatomy. Lizards from Arnhem Land are paler than those from the interior. They represent the northern and eastern "freckled form" sometimes called *V. t. orientalis*. The "black-headed" form *V. t. tristis* occurs in interior Australia. However, these subspecific names are not widely used and may be invalid because these forms intergrade with one another. Lizards from Queensland are brownish and morphometrically different from the western form (G. Thompson, personal communication). *V. "tristis"* will probably prove to be a species complex.

Geographic Distribution

Varanus tristis is widespread across most of Australia, except for cool temperate southern regions (Fig. 7.51).

AM R218 25 mi inland from Cairns—17 145.67; AM R229 Condoblin—33.08 147.15; AM R642 Cairns District—16.92 145.77; AM R2709 Gayndah—25.62 151.62; AM R3055, R3056 Perth—31.95 115.85; AM R3094, R3095, R3110 Boulder—30.78 121.48; AM R4094 Mt. Bartle Frere—17.4 145.82; AM R5313,

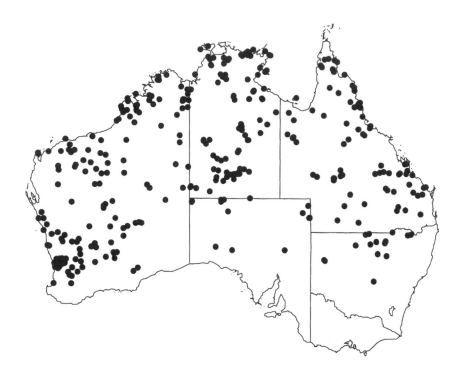

Figure 7.51. Known localities of collection of V. tristis *in Australia.*

R5328 Eidsvold, near Gayndah, Upper Burnett River—25.37 151.12; AM R5398 Dawson River; AM R5901, R5902, R5903, R6122, R6334, R6335, R63346 Eidsvold Upper Burnett River—25.37 151.12; AM R7045 Diamantina River; AM R7237 Kalgoorlie—30.75 121.47; AM R9320 Mt. Morgan—23.65 150.38; AM R9738 NW Queensland; AM R9914 Roper River N.A.—14.73 134.73; AM R10206 Groote Eylandt, N.A.—13.98 136.47; AM R10853 Hermannsberg—23.95 132.77; AM R11334. R11342 Alexandra Stn.—19.05 136.7; AM R12349 Northam—31.65 116.67; AM R12377 Yirrkala via Darwin—12.25 136.88; AM R13237 Katherine—14.47 132.27; AM R13637 E Alligator River; AM R13646 Cape Arnhem—12.35 136.98; AM R14064 Mt. Conway 28 mi W Alice Springs—23.75 133.42; AM R14065 Ayers Rock—25.35 131.03; AM R14067, R14072 Port Keats Mission—14.23 129.52; AM R14902 Maxland, Mungindi—28.92 149.02; AM R16530 Coen, Cape York—13.95 143.2; AM R17569 Mt. Davies, Tomkinson Ras.—26.22 129.23; AM R18322, R18607, R19007, R19135, R19199, R19200, R19201, R19664 Brewarrina—29.97 146.87; AM R19016 Mt. Isa—20.73 139.48; AM R29536 "The Brothers," North Star—29.17 150.57; AM R33499 50 mi NE Port Hedland—20.32 119.3; AM R39790 R39791 Woolwonga Ranger Stn.—12.75 132.65; AM R41873 20 mi S Maningrida—12.37 134.22; AM R41874 Liverpool River—12.25 134.12; AM R45479 Mootwingee Stn.—31.28 142.3; AM R48080 Prince of Wales Is., Torres Strait—10.68 142.15; AM R49374 25 km NW Refrigerator Bore—20.9 130.62; AM R49468 Todd St, Alice Springs—23.72 133.8; AM R49769 25 km NW Refrigerator Bore—20.9 130.62; AM R51902, R51903, R51904 4 mi NE of Alice Springs—23.67 133.92; AM R51905 Stuart Hwy. 18 mi SW of Barrow Creek—21.72 133.75; AM R51906, R51907, R51908 Stuart Hwy., 45 mi SW Alice Springs—24.22 133.47; AM R51909 Wiggleys Water Hole NE of Alice Springs—23.63 133.88; AM R55680 West Is. Sir Edward Pellews—10.35 142.07; AM R59740 Biggenden—25.52 152.05; AM R59761 Emerald Downs Stn.—23.48 148.13; AM R60190 155 km E of Daly Waters on Borroloola Rd.—

16.63 134.87; AM R60191 11 km N of Larrimah on Stuart Hwy.—30.47 133.22; AM R60192 99 km S Renner Springs on Stuart Hwy.—19.28 134.17; AM R64805 Charters Towers—20.08 146.27; AM R66256 1 mi E of Mt. Isa—20.72 139.5; AM R68822 Croydon District—18.2 142.25; AM R68823 25 mi SW of Collarenebri—29.83 148.28; AM R68824 40 km N Monto—24.65 150.75; AM R73937 Tempe Downs—24.38 132.42; AM R75884 Charters Towers—20.08 146.27; AM R77361 Angurugu Mission, Groote Eylandt—13.97 136.45; AM R86873 25 mi SW of Collarenebri—29.8 148.28; AM R106813 Widgee Downs, Jobs Gate Area W of Goodooga—29.07 146.72; AM R107516 Normanton—17.67 141.07; AM R110454 between King and Edith Rivers; AM R110536 Scott's Tank, Diamantina Lakes, NW of Windorah—23.97 141.53; AM R112437 Jabiru Drive, Jabiru—12.65 132.88; AM R125089, R125090, Mt. Liebig—23.3 131.37; AM R125339 Atholwood; AM R125937 23.2 km S Yetman-Goondiwindi Rd. via Wallangra Rd., along W side Macinty—29.12 150.88; AM R128367 Duaringa—23.72 149.67; AM R128430 separation, Duaringa—23.58 149.78; AM R12845 Northam—31.65 116.67; AM R128551—R128553 Duaringa—23.72 149.67; AM R128712 Wenlock Downs, Batavia River (Batavia Downs, Wenlock River)—12.67 142.67; AM R133751 ~20.3 km W of Balladonia by road—32.27 123.42; AM R13375 ~86.9 km W of Balladonia by road—32.02 123.8; AM R135424 Dunsandle Stn. (Near Ellerslie Stn.), NE of Engonnia—29.13 146.4; AM R136213 Mitchell Plateau, Victoria Mertens Creek, Mitchell River—14.82 125.72; AM R137610 Narran Lake Nature Reserve, ~70 km WNW of Walgett—29.73 147.4; AM R140473 Roebuck Bay, Broome Bird Observatory—17.97 122.33; AM R142903 51 km S of Katherine—14.68 132.67; AM R143919 1.8 km W Watsonville (W of Herberton) on Irvinebank Rd.—17.37 145.28; AM R146255 Mootwingee National Park: Mootwingee Gorge—31.32 142.33; AM R147168 "Calgory" (Collarenebri), Site 4—29.67 148.37; ANWC R0263 Gooseberry Hill, Perth—31.95 116.05; ANWC R0264 Tindal, Katherine—14.52 132.37; ANWC R0377 Nourlangie Rock—12.85 132.78; ANWC R0515 5 mi NW of Argyle Downs—16.2 128.74; ANWC R0641, R0887 Kapalga Lagoon Area, Just S of Lagoon—12.62 132.42; ANWC R0939 Amelia Creek (Loc 4)—16.6 136.18; ANWC R0940 between Castle Creek and Amelia Springs—16.58 136.17; ANWC R1805 Uramurdah Creek, Millbillillie Stn., SE of Wiluna—26.63 120.42; ANWC R1895 Macquarie Marshes, N of Warren—30.78 147.55; ANWC R3308 Berry Springs, S of Darwin—12.7 130.97; ANWC R5424, R5487, R5530 Shoalwater Bay Army Training Reserve, N of Rockhampton—22.78 150.29; NMV D1271, D1257 Alice Springs—23.7 133.87; NMV D5620, D5623 Barrow Creek—21.53 133.88; NMV D2930, D2945 Tennant Creek—19.65 134.18; NMV D5205 Oenpelli, E Alligator River—13.15 132.37; NMV D1019 Port Darwin—12.45 130.83; NMV D7621 Mingenew—29.2 115.43; NMV D6770 Pioneer Creek, Ormiston—23.7 132.71; NMV D8430 Mornington Is.—16.5 139.5; NMV D5196 2.4 km N of Alice Springs—23.67 133.87; NMV D1090 Ernabella Mission—26.3 132.13; NMV D2356 Port George 1V; NMV D1603 Mungeranie, Lake Eyre District—20.02 138.67; NMV D4918, D5232, D5233 Melville Is.—11.45 130.78; NMV D1021, D1022 Port Darwin—12.45 130.83; NMV D162, D163, D3566, R26263 Alice Springs—23.7 133.87; NMV D2920, D2938, D2941, D2950, D5636, D5637 Tennant Creek—19.65 134.18; NMV D1340 Ormiston Gorge—23.62 132.72; NMV D1098 Rose River Mission—14.28 135.73; NMV D77, D78 Tennant Creek; NMV D88, D4976 Barrow Creek—21.53 133.88; NMV D8431 Mornington Is.—16.5 139.5; NMV D5545 Claudie River—12.77 143.28; NMV D3564, D3565 Palmer River—15.98 143.65; NMV D1090, D1091 Ernabella Mission—26.3 132.13; NMV T42 Crocodile Is.—11.73 135.3; NMV T54 Groote Eylandt—13.82 136.63; NMV T53 Upper Glyde River—12.37 135.02; NMV T37, T38, T52 Roper River—14.68 134.73; NMV T46 Katji—12.35 134.78; NMV T55 Derby Creek—12.35 134.78; NMV T44 Crocodile Is.—11.73 135.3; NMV T36 Roper River—14.68 134.73; NMV T32 Blue Mud Bay—13.57 135.97; NMV T51 Cape Stewart—11.95 134.75; NMV T31 Millingimbi, Crocodile Is.—12.08 134.58; NMV T39, T41, T45, T47, T48, T49, T50 Milingimbi, Crocodile Is.—12.12 134.92; NMV T43 Crocodile Is.—11.73 135.3; NMV T30 Watson River—13.27 141.88; NMV T27 Bare Hill—12.97 143.52; NMV T26 near Lockhart River—

12.97 143.52; NMV T34 Groote Eylandt—13.82 136.63; NMV T40 Milingimbi, Crocodile Is.—12.12 134.92; NMV T28, T29, T33 Lower Archer River—13.5 142; NMV T35 Groote Eylandt—13.82 136.63; NTM R26195, R26196 Brunette Downs Stn.—18.65 135.95; NTM R26197 28 mi E Yuendumu—22.53 132.13; NTM R26198, R26199 Emily Gap, Alice Springs—23.75 133.95; NTM R26202 Yuendumu Settlement—22.33 131.75; NTM R26204 Brunette Downs Racecourse —18.65 135.95; NTM R26205 NT Doreen—22.03 131.33; NTM R26206 Katherine Caves—14.58 132.47; NTM R26210 Coonamble—30.95 148.4; NTM R26211 South Rd., Finke River—25.95 135.05; NTM R26212 Napperby Creek— 22.33 132.58; NTM R26213 Edith River—14.18 132.03; NTM R26214 Edith River, Katherine—14.17 132.07; NTM R26215 Horden Hill, Granites—20.65 130.32; NTM R26216 Helen Springs—19.43 133.88; NTM R26217 Dunmarra— 16.68 133.42; NTM R26218 20 km S Daly Waters, NT—16.48 133.4; NTM R26219 Petermann Ranges—24.87 129.08; NTM R26220 Finke Gorge National Park—24.03 132.72; NTM R26221 Rangers Residence, Simpsons Gap National Park—23.67 133.72; NTM R26225 Mt. Carr, Adelaide River—13.23 131.08; NTM R26227 Kakadu National Park—12.78 132.77; NTM R26228 Katherine Gorge National Park—14.27 132.5; NTM R26229 Rockhole Bore Camp, Loves Creek Stn.—23.53 134.87; NTM R26230 Melville Is., Snake Bay—11.47 130.7; NTM R26231 Ellery Creek, Site 17.—23.82 133.07; NTM R26232, R26233, R26234 Edith River—14.17 132.07; NTM R26235 Brunette Downs Homestead— 18.63 135.95; NTM R26236 Reedy Rockhole, George Gill Ranges—24.32 131.58; NTM R26237 Reedy Rockhole, George Gill Ranges—24.3 131.6; NTM R26238 6 km S Halls Creek—18.2 127.7; NTM R26239 Gosse River, 6 km NW Ooradidgee Rockhole—20.05 134.48; NTM R26240 Cape Fourcroy, Bathurst Is.—11.78 130.02; NTM R26241 Pinnacle Mines, Garden Stn.—23.28 134.42; NTM R26242 3 km W Racecourse Brunette Downs Stn. Rd.—18.63 135.95; NTM R26243 7 m S Adelaide River Town—13.35 131.15; NTM R26244 Katherine Low Level; NTM R26245 5 km W Ivanhoe Crossing Ord River—15.75 128.7; NTM R26246 Cape Fourcroy, Bathurst Is.—11.78 130.02; NTM R26247 6 km SW Ooradidgee Rock-hole, Davenport Rd.—20.15 134.47; NTM R26248, R26249 Bing Bong Stn.— 15.62 136.35; NTM R26250 Nicholson River—17.9 137.53; NTM R26251 4 km N on Fogg Dam Rd. From Arnhem Hwy.—12.57 131.3; NTM R26252 Jabiru— 12.67 132.9; NTM R26253 Alyawarre Desert Area, 24 km E Barkly Hst—19.85 136; NTM R26254, R26256 The Granites—20.58 130.35; NTM R26255 Bauhinia Downs Stn./Camp 3—15.92 135.32; NTM R26257 Attack Creek—19.03 134.13; NTM R26258 Kathleen Creek—24.03 131.67; NTM R26259 Hermannsburg— 23.97 132.77; NTM R26260, R26261 Kurundi Stn.—20.58 134.75; NTM R26262 Tanami—20.53 130.93; NTM R26264 The Garden Stn.—23.25 134.02; NTM R26265 Keep River—15.75 129.08; NTM R26266 Limmen Gate National Park— 15.78 135.34; NTM R26267 English Company Isles, Cotton Is.—11.89 136.48; NTM R26268 Petermann Ranges—24.97 129.01; NTM R26269 Palm Valley— 24.05 132.7; QM R129 Roper Rd.—14.73 134.73; QM R456 Bundaberg— 24.87 152.35; QM R457, R640, R2136, R2137, R2139, R2140, R2141, R2142, R2203, R3087 Eidsvold—25.37 151.12; QM R2153 Charleville—26.4 146.25; QM R2154 Rockhampton—23.37 150.53; QM R2170 Cardwell—18.27 146.02; QM R2558, R2559 Gyranda, Dawson Rd.; QM R6129, R6304, R6305, R6306, R6307 Capella, Retro Stn., via Clermont—22.87 147.9; QM R7481 Jackson— 26.65 149.62; QM R8101, R8102 Quilpie—26.62 144.27; QM R9066, R9067 Maneroo Stn., via Longreach—23.37 143.87; QM R9292 Wackon Stn., near Glenmorgan—27 149.5; QM R10861 Gregory Downs, via Burketown—18.65 139.25; QM R11153 Doomadgee Mission Stn.—17.93 138.82; QM R14361 Cun-namulla, Gilruth Plains, National Field Stn.—28.07 145.68; QM R14408 Warrego Rd., W of Bourke—30 145.38; QM R14742 Rockwood Stn., Chinchilla—26.93 150.37; QM R15625 Mt. Larcom, SF, foothills of mt—23.82 151.08; QM R15700 Rewan Stn., 80 km SW Rolleston—25.17 148.57; QM R17390 S Kolan—24.93 152.17; QM R17506 Hughenden area—20.85 144.2; QM R17566 Rewan Stn., 80 km SW Rolleston—24.97 148.37; QM R17802 Isabella Falls, 32 km NW Cook-town—15.28 145.03; QM R21207 Rockwood Stn., 32 km SW Chinchilla—26.93

150.37; QM R22479 Back Creek, Condamine Hwy.; QM R23618 Black Rock, near Lynd Rd.—18.93 144.48; QM R24004 Bogantungan—23.65 147.3; QM R24838, R24839 Wakes Lgn, 89 km NW Adavale—25.58 144.83; QM R27160 Shiptons Flat, Site 36—15.8 145.27; QM R27679 Magnetic Is., Picnic Bay-19.18 146.85; QM R29849 Bullawarra Stn., 35.2 km W Thargomindah—27.9 143.6; QM R30443 Robinson Gorge, via Taroom—25.28 149.15; QM R32299, R32300 Shute Harbour—20.3 148.78; QM R32324 Goombungee—27.3 151.85; QM R33354 Injune/Rolleston Rd.; QM R36994 Comet, E of—23.62 148.67; QM R36995 Anakie, near 23.53 147.75; QM R37594 Coen Airport, 11 km WNW—13.68 143.07; QM R38318 Peach Creek, 14 km NNW Mt. Croll—13.67 143.12; QM R38478, R38479 Robinson Gorge, 2 km from Shepherds Hut—25.28 149.15; QM R39068 Mt. Isa, 50 km W, on Barkly Hwy.—20.25 139.23; QM R39069 Mt. Isa, 56.9 km W, on Barkly Hwy.—20.37 139.2; QM R39852 York Downs, 50 km E Weipa—12.75 142.32; QM R41851 Oorida area, Diamantina Lakes—23.77 141.13; QM R43761 Isabella Falls, ~30 km NW Cooktown—15.28 145.03; QM R44347, R44625 Percy Springs Stn.—19.92 146.08; QM R44509 Spyglass Stn.—19.35 146.77; QM R44694 Battery Stn.—19.48 145.65; QM R45364 Mt. Mulligan, summit—16.85 144.85; QM R47024 Cooktown—15.47 145.25; QM R50007 The Hollow, Pitline trap 1—24.12 143.18; QM R50022 Longreach, 60 km S, on Jundah Rd.—23.92 143.9; QM R50047 Longreach, 40 km S on Jundah Rd.—23.77 144.07; QM R50726 Perth, 30 km N; QM R54252 Waterloo Stn., near gate, Filly Pd.–Bore Pd.—24.22 143.25; QM R57165 Noonbah Stn. HS—24.12 143.18; QM R57166 Valetta Stn.—24.25 143.18; QM R57651 Mosquito Waterhole—Oriners—15.48 142.8; QM R58090 Landsburough Hwy.; QM R58558 Ularang National Park, Hughenden; QM R58722 White Mtns—20.42 144.68; SAM R06119 Bunbury—33.33 115.63; SAM R14495 Maralinga Townsite—30.1 131.58; SAM R38779 5 km N of Tennant Creek—19.6 134.18; SAM R04643 Tambrey HS—21.63 117.6; SAM R25532 84 km SW Katherine on S side Victoria Hwy.—15.03 131.77; SAM R32491 Perm. Site 12e—27.27 140.17; SAM R41855 1 km W of Blue Hills Bore—27.12 132.85; SAM R44819 3.6 km SSW of Mount Gow—26.58 140.69; SAM R45307 2 km NW of New Bore Homeland—26.05 132.13; SAM R49067 7.2 km SW of Table Hill—27.63 140.83; SAM R50152, R50154 2.4 km NW of Sentinel Hill—26.06 132.44; SAM R00136 Eidsvold—25.37 151.12; SAM R01112 Groote Eylandt—13.98 136.47; SAM R01938 Flinders Is.—14.18 144.25; SAM R02186 Kairi North Queensland—17.22 145.55; SAM R03223 Innaminka—27.73 140.77; SAM R03224 Tennant Creek—19.65 134.18; SAM R04039 Haast Bluff—23.45 131.88; SAM R04487, R04488, R04489 Tambrey HS—21.63 117.6; SAM R05290 Giles—25.03 128.3; SAM R06014 Tennant Crk—19.65 134.18; SAM R06015 Leigh Creek—30.48 138.42; SAM R08317 Mt. Isa—20.73 139.48; SAM R42868 32 km S of Noonbah Stn.—24.25 143.18; SAM R42885 6 km E of Noonbah Stn.—24.1 143.25; SAM R42896 8 km E of Noonbah Stn.—24.1 143.27; WAM R74028 near Daly Waters—16.27 133.37; WAM R76135 Mt. Jackson Hill—30.25 119.25; WAM R81468 Wickham Well—22.05 120.63; WAM R83213 32 km SE of Karratha—21.12 117.12; WAM R86451 Laverton—28.63 122.4; WAM R87753 22 km SW of Yinnietharra H.S—24.8 116.02; WAM R91457 Mokine Nature Reserve—31.73 116.58; WAM R94372 Balgo Mission—20.15 127.97; WAM R10214 10 km NW Mt. Windell—22.6 118.46; WAM R102272 Barlee Range Nature Reserve—23.11 116.01; WAM R114669, R114671–R114672 Buller River—28.87 114.62; WAM R115773 Kings Park—31.97 115.83; WAM R117201, R127295, R127301–R127304, R127310–R127312 7–8 km NW Point Salvation—28.2 123.6; WAM R119536 Mirima National Park—15.78 128.78; WAM R121982 Cullacabardee—31.83 115.83; WAM R126374 Bungalbin Woodland Camp—30.31 119.72; WAM R127295, R127305, R127313 39 km E Laverton—28.47 122.83; WAM R127300 8 km NE Dunges Table Hill—28.13 123.92; WAM R104070, R106290 Woodstock, Site 12—21.62 118.95; WAM R408 Mt. Helena—31.87 116.2; WAM R742 Victoria Park—31.97 115.9; WAM R1352 Quairading—32.02 117.4; WAM R1736 Bunjil—29.65 116.35; WAM R2448 Cannington—32.02 115.95; WAM R2510, R2511 Kenwick—32.03 115.97; WAM R2778 Wadderin Hill—32 118.42; WAM R4170 28 km E of Kalgoorlie—30.73

121.75; WAM R4227 Northam—31.65 116.67; WAM R4358 Maya—29.88 116.5; WAM R4787 E Yuna—28.33 115; WAM R5749 Wialki—30.48 118.12; WAM R6168 N Fremantle—32.03 115.75; WAM R6820 Parkerville—31.87 116.12; WAM R6921 Midland—31.9 116; WAM R7577 Wagin—33.32 117.35; WAM R7901 E Perth—31.95 115.88; WAM R8085 Millendon E Swan—31.8 116.03; WAM R8564 Pingelly—32.53 117.08; WAM R8578 Merredin—31.48 118.27; WAM R8992 Bolgart—31.27 116.5; WAM R9659 Popanyinning—32.67 117.12; WAM R10417 Kalgoorlie—30.73 121.47; WAM R11298 Northam—31.65 116.67; WAM R11756 Bickley—32.02 116.1; WAM R11836, R11837, R11838 Wotjulum—16.18 123.62; WAM R12131 Bentley—32 115.9; WAM R12279 Mundiwindi—23.8 120.25; WAM R12308 Nullagine Area—21.88 120.12; WAM R12925 16 km N of Morawa—29.07 116; WAM R12926 Wyalunga Pool (near Gingin)—31.35 115.9; WAM R13308, R73484, R73531 Woodstock Stn.—21.62 118.95; WAM R13504, R13519 Yirrkala—12.25 136.88; WAM R13563 Wyndham—15.47 128.1; WAM R13567 Kalumburu—14.3 126.63; WAM R13666 Newman—23.35 119.73; WAM R13791 Kalumburu Mission—14.3 126.63; WAM R14914 Lesmurdie—32.02 116.05; WAM R16514 Katherine—14.47 132.27; WAM R17691 Turee Creek Stn.—23.62 118.65; WAM R18589 Witecarra Gully S of Kalbarri—27.8 114.2; WAM R19863 Gidgiegannup, 32 km NE of Midland Junction—31.8 116.18; WAM R20000 5 km E of Mt. Ulric—21.82 117.25; WAM R20001 Millstream—21.58 117.07; WAM R20785 Chirnside Creek. Peterman Ranges—25.03 129.62; WAM R20849 Owen Springs—23.88 133.47; WAM R21943 Upper Swan—31.75 116.02; WAM R22183 Warburton Range—26.13 126.58; WAM R22338 Gooseberry Hill—31.93 116.05; WAM R22750 Turee Creek—23.62 118.65; WAM R22922 Wembley Downs—31.93 115.75; WAM R22985 Goomalling—31.3 116.82; WAM R23919 Ullawarra—23.48 116.12; WAM R23984 Kumarina—24.75 119.58; WAM R24015 Neelaluna Claypan, 40 km S of Jiggalong—23.72 120.78; WAM R24061 Mt. Hart. Within 6 km of HS—16.82 124.92; WAM R24875 3 km E of Wialki (98 km E of Burakin)—30.48 118.17; WAM R24882 Floreat Park. Perth—31.93 115.78; WAM R24937 Katherine—14.47 132.27; WAM R26469 32 km N of Kellerberrin—31.35 117.72; WAM R26525 Mt. Newman—28.5 121.08; WAM R28010 13 km W of Bolgart—31.28 116.35; WAM R28013 Frazier Downs—18.8 121.72; WAM R28014 Injudinah Creek. La Grange—18.63 121.87; WAM R28016 Mt. Anderson—18.03 123.93; WAM R28018 Kalumburu—14.3 126.63; WAM R28154 Brookton Area—32.37 117; WAM R28393 Sorrento. Perth—31.83 115.75; WAM R31415 Exmouth Townsite Vicinity—21.93 114.12; WAM R31542 Murchison River—27.83 114.67; WAM R31543 Murchison River—27.83 114.67; WAM R31708 18 km S of Mt. Davies Camp—26.33 129.13; WAM R32270 Mt. Bell Leopold Range—17.17 125.32; WAM R32340 Manning Creek. Mt. Barnett—16.67 125.95; WAM R33400 Mukinbudin—30.92 118.2; WAM R33471 Lockwood Spring. 32 km ESE of Kalbarri. 8 m up from Gully Floor—27.32 114.47; WAM R34072 Gosnells—32.08 116; WAM R37939 29 km S of Yellowdine—31.57 119.65; WAM R39140 Balfour Downs Stn.—22.8 120.87; WAM R40235 Gidgiegannup—31.8 116.18; WAM R40702, R40703 Darlington—31.92 116.07; WAM R44959 Mileura—26.37 117.33; WAM R45054 Port Warrender—14.53 125.82; WAM R45078 2 mi N of Tambrey—21.6 117.6; WAM R45233 8 km W of Mt. Tietkens. Buck Hills—23.15 128.8; WAM R45234 Polllock Hills—22.87 121.98; WAM R45976 53 mi Peg Perth—Toodyay Rd.—31.67 116.33; WAM R46173 Derby—17.3 123.62; WAM R48394 51 km N of Beacon—29.85 117.87; WAM R48488 11 km E of Green Head. Eatha Spring—30.12 115.05; WAM R48801 Wanneroo—31.75 115.8; WAM R49121 Mt. Lesueur. On Slopes—30.18 115.2; WAM R51282 Wattle Creek Spring—19.22 126.12; WAM R51310 25 km E of Yornaning—32.75 117.37; WAM R51696 Kelmscott. Perth—32.12 116.02; WAM R51926 Durba Springs—23.75 122.52; WAM R53563 16 km ESE of Point Sunday. Gt Victoria Desert—28.18 124.23; WAM R53600 1 km S of Charles Knob—25.07 124.98; WAM R53935 Edgar Ranges Reserve—18.92 123.45; WAM R54005 Edgar Ranges Reserve—18.35 123.05; WAM R56140 Paraburdoo—23.2 117.67; WAM R58529 7 km NNE of Cape Borda—16.62 122.78; WAM R58530 7 km NNE of Cape

Borda—16.62 122.78; WAM R58615 21 km N of Turkey Creek Airstrip—16.85 128.22; WAM R58675 7 km SE of Willare Bridge Fitzroy River—17.78 123.7; WAM R59347 Neerabup National Park—31.65 115.72; WAM R60116 Lake Argyle at Dam Site-16.12 128.73; WAM R60133, R60134 32 km S of New Norcia—31.27 116.22; WAM R60140, R60141, R60142 Kalumburu—14.3 126.63; WAM R60431 13 km W of Bolgart—31.25 116.5; WAM R60674 Mitchell Plateau—14.58 125.75; WAM R60835 One Tree Mill NE of Derby—17.22 123.77; WAM R60912 10 km N of Beagle Bay Turn off Cape Leveque Rd.—16.92 122.8; WAM R60913 40 km S Coulomb Point—17.68 122.22; WAM R63164 Yarrie at De Gray River—20.68 120.2; WAM R63766 1 km N Talbot Soak—22.55 122.4; WAM R64000 Anketell Ridge—20.62 122.7; WAM R65801 11 km ENE Comet Vale—29.91 121.23; WAM R65842 9.75 km NE Comet Vale—29.89 121.2; WAM R65978 2.5 km Mt. Linden—29.31 122.42; WAM R68343 10 km SW of Pannawonica—21.72 116.27; WAM R68970 halfway between Paraburdoo and Tom Price—23 117.75; WAM R68981 5 km E Deep Creek—17.75 122.78; WAM R69342 9.0 km SSE Banjiwarn HS—27.78 121.66; WAM R69892 31 km 190° Gordon Hills—20.83 127.98; WAM R70136 11.5 km 265° (New) Lissadell HS—16.68 128.27; WAM R70384 11.5 km 260° (New) Lissadell HS—16.7 128.43; WAM R70591 17.5 km 273° (New) Lissadell HS—16.67 128.39; WAM R70796 25.7 km 235° Marillana HS—22.77 119.2; WAM R72757 9.75 km NE Comet Vale—29.89 121.2; WAM R73364 5 km 328° Mt. Manning Range SE Peak—29.96 119.62; WAM R74713 8.7 km ENE Yuinmery—28.53 119.09; WAM R75013 2 km NE of Willare Bridge Roadhouse—17.7 123.65; WAM R75534 11 km WNW of New Lissadell Homestead—16.65 128.47; WAM R77061 Mitchell Plateau—14.82 125.83; WAM R77270 Mitchell Plateau—14.59 125.76; WAM R77601 Mitchell Plateau—14.86 125.83; WAM R77603 Mitchell Plateau—14.74 125.78; WAM R77679 Walsh Point—14.57 125.83; WAM R81911 Moondyne Nature Reserve—31.58 116.22; WAM R83456 2 km S of Coulomb Pt.—17.38 122.15; WAM R84239 1 km SW of Mt. Daglish—16.28 124.95; WAM R84417 10 km N of Beyondie Homestead. (Marymia)—24.7 120.05; WAM R84772 Conzinc Is., Dampier Archipelago—20.53 116.77; WAM R85194 Julimar State Forest—31.4 116.23; WAM R87475 4 km N of Honeymoon Well.—26.87 120.42; WAM R94490 Balgo Mission—20.13 127.98; WAM R94716 Lorna Glen Stn.—26.25 121.33; WAM R95662 17 km S Onslow—21.79 115.09; WAM R96235, R96238 Mitchell Plateau—14.73 125.73; WAM R96244 Lesmurdie—32.03 116.05; WAM R96598 8 km NNE Bimbijy-29.63 118.07; WAM R96615 21 km NE Geranium Rock—29.87 117.43; WAM R96622 Damperwah State Farm—29.3 116.68; WAM R97382 Ord River—16.77 128.8; WAM R97618 Kalumburu—14.3 126.63; WAM R97897 ~10 km NNW Mitchell Plateau Mining Camp—14.77 125.78; WAM R98076 Lake Violet Stn.—26.53 120.67; WAM R99314 25.3 km WSW Mount Blythe on Charnley River—16.38 125.21; WAM R100381–R100383 7 km WSW Point Salvation 39 km E Laverton—28.47 122.83; WAM R100384–R100386, R100388–89 7–8 km WNW Point Salvation—28.2 123.6; WAM R100390–R100395 7–8 km WNW Point Salvation—28.2 123.6; WAM R100397–R100399 39 km E of Laverton—28.47 122.83; WAM R100400 39 km E Laverton—28.47 122.83; WAM R100401–R100404 7–8 km WNW Point Salvation—28.2 123.6; WAM R101374 6.5 km W Kununurra—15.77 128.67; WAM R103136 8 km W Brookton—32.37 117.92; WAM R103145, R103158–R103159, R10345 Bungle Bungle National Park—17.55 128.25; WAM R105944 47 km WSW Coolgardie—31.12 120.7; WAM R106054 Woodstock HS—21.62118.95; WAM R108547, R108559 7–8 km WSW Point Salvation—28.47 122.83; WAM R108562–R108563 7–8 km WSW Point Salvation—28.2 123.6; WAM R116096 Gibson Desert Nature Reserve 20 km to 40° to Lake Hancock—24.75 124.73; WAM R123942 10 km S Exmouth-22 114.08; WAM R126530–R126531 Kambalda West—31.22 121.58.

Fossil Record

No fossils of *Varanus tristis* are known.

Diagnostic Characteristics

- Medium size, up to 75 cm TL.
- Adults jet black above, especially head and shoulders and the posterior two-thirds of the tail (throughout most of the geographic range).
- Lateral and dorsal surfaces of body covered with numerous tiny rosettes, cream, blue gray, white, or pink with a dark center.
- Long round tail with strongly keeled scales, banded with narrow pale rings anteriorly in southern populations.
- Very sharp, strongly curved claws.
- Long snout, head unspotted in southern populations.

Description

Pale gray to brown to black above, with numerous pale, sometimes pinkish or blue gray, ocelli (often obscure in adults). Head and neck black in south, orange in Top End, tan from Gulf of Carpentaria west through Kimberley. Limbs black with small pale spots. Ventral surfaces whitish or gray. Yellow throat in Top End. Head scales small and smooth. Nostrils slightly closer to tip of snout than to eye. Tail round, slender, black, or gray and long, without a dorsal keel (Cogger 1992; Storr et al. 1983). Scales on tail sharply keeled. A widespread and variable arboreal monitor lizard.

Size

SVL ranges from 72 to 305 mm (TL ~55–80 cm). No sexual size dimorphism in head size is discernible (Pianka 1994a). Tail lengths average 1.5 to 2.3 times SVL. Average body weight of 41 adult lizards was 250 g.

Habitat and Natural History

Varanus tristis are found almost anywhere where trees with hollows occur. They also use rock crevices for shelter. They live in arid woodlands, subhumid tropical woodlands, and in rocky ranges and outcrops.

Terrestrial, Arboreal, Aquatic

Varanus tristis are strongly arboreal and have strongly curved and very sharp claws. They are seldom found far from trees, they but also use buildings and rock crevices for shelter.

Time of Activity (Daily and Seasonal)

Unlike many larger species of monitors, *Varanus tristis* are active to varying degrees over most of the year except for a few winter months (Pianka 1971, 1994a). Their dark color probably facilitates warming on cold days. These extremely wary lizards are seldom seen, but they

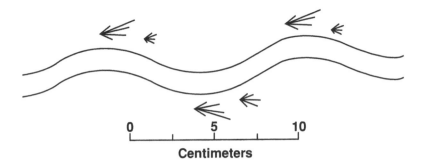

Figure 7.52. Trackway of
Varanus tristis.

can be studied and located by following their very conspicuous tracks (Farlow and Pianka 2000) because they drag the base of their tail, probably to leave a scent trail for conspecifics to follow (Fig. 7.52).

Thermoregulation

Mean ± SD body temperature of 46 active individuals was 34.8 ± 3.84°C. The slope of a least-squares linear regression of body temperature on ambient air temperature is 0.60, an indication that these lizards are, relative to most other monitors, thermoconformers. A radiotelemetric study revealed a daily cycle with lower body temperatures early and late in the day and higher ones during midday. Mean field active body temperature recorded in this study was 33.2°C (Thompson et al. 1999b).

Foraging Behavior

V. tristis consumes other lizards as well as baby birds (and probably bird eggs); its very distinctive track typically runs more or less directly from tree to tree (these monitors climb most trees looking for food). *V. tristis* activity is seasonal, and lizards rely on building up fat reserves during times of plenty to get them through lean periods. Once, camped at a study site where I had never seen a *V. tristis* track in many weeks of work over several months in the midst of a prolonged drought, I noticed a beady black eye peering out of a small black hole in a burned out *Eucalyptus* tree. Chopping open the hollow tree revealed an extremely emaciated *Varanus tristis,* literally skin and bones, waiting for the drought to break!

These large climbing predatory lizards must constitute a potent threat to hole-nesting parrots. I have found baby birds in their stomachs (Pianka 1971, 1982). Once, I heard a galah cockatoo screeching loudly as if in distress nearby. The bird was on the ground when first sighted, with its crest held high and wings partially outstretched. The galah flew up onto a fallen log under a marble gum tree, and then into the tree, which proved to be its nesting tree. A large *V. tristis* clambered over the same log toward the tree. The cockatoo continued to screech and then began to harass the lizard. When the lizard climbed about 3 m up the tree and went around out of sight on the other side, the galah attacked and actually drove the monitor back down the tree. The galah's mate was also present.

Samuel Sweet (personal communication) made the following observations at Kakadu. There, *V. tristis* (northern form) forage principally on the ground, actively investigating the edges and interiors of logs, grass clumps, and beds of dry leaves for skinks, which are pursued for up to 10 m and rarely escape. Shallow depressions filled with leaf litter are investigated intensively. The monitors quickly force their way through the leaves for two to four body lengths, then pop up and freeze for 10 to 20 seconds, heads held high. From this position, they often vigorously undulate the distal half of their tail, displacing many leaves. Any movement elsewhere in the leaf pile is pounced on immediately. Sweet saw a monitor capture up to four skinks (*Carlia* sp.) from a 1.5-m-diameter leaf deposit in about 10 minutes using this foraging strategy.

Diet

A survey of stomach contents of 100 of the southern form of these lizards (75 with food; Pianka 1994a) revealed that 70% of the diet of *V. tristis* by volume consists of other lizards. Large grasshoppers, cockroaches, and lepidopteran larvae are also eaten. A total of 75 stomachs with food contained 137 large prey items, including 35 individual lizard prey items representing some 11 other species. Prey species eaten by *V. tristis* include seven species of skinks (*Ctenotus brooksi*, *C. colletti*, *C. grandis*, *C. helenae*, *C. pantherinus*, *C. quattuordecimilineatus*, and *Lerista bipes*), two species of agamids (*Pogona minor* and a small *Moloch horridus*), and a gecko (*Gehyra variegata*), as well as the pygmy monitor *Varanus caudolineatus*. Very probably, nearly any other lizard species small enough to be subdued would be eaten.

One *V. tristis*, weighing 330 g, had eaten a *Pogona minor* with a body mass of 57 mL (17.3% of the *V. tristis*'s own body mass). Another, weighing 220 g, had eaten a *Pogona minor* with a body mass of 50 mL (22.7% of its own body mass) (Pianka 1994a).

Reproduction

On the basis of enlarged testes and yolked ovarian eggs, both sexes appear to achieve sexual maturity at about 200 mm SVL (Pianka 1994a). *V. tristis* are most active during the austral spring October–November when mating occurs. Tracks are often found on top of tracks, suggesting that males might follow female scent trails (alternatively, males could be following females visually). Clutch sizes are large, from 5 to 17 eggs, with a mean of 10.1 eggs (n = 20 clutches). Relative clutch mass of 11 females with eggs in their oviducts averaged 16.2% of female body weight (Pianka 1994a). Eggs are laid in underground nests dug by females in October–November, and hatchlings emerge in February (incubation time is short, probably around 114–117 days). Hatchling SVLs are about 72 to 73 mm (weight, ~4.3 g). Hatchlings have a tessellated checkered pattern, quite different from adults (Thompson and Pianka 1999a).

In Kakadu, mating occurs from early May to late June. Eggs are laid in late June and early July. One nest was in a root channel about 0.7 m deep inside a hollow stump. Another was in rotted wood about 3 m up in a large hollow tree (S. Sweet, personal communication).

Movement

Home ranges are quite large. A radiotelemetric study undertaken in the Great Victoria Desert during September and October 1995 (Thompson et al. 1999) showed that males move greater distances (187 m/d) than females (100 m/d). One male moved 890 m in a day. Another male traveled 723 m in a straight line into the wind in 1 day; it was found in a hollow marble gum tree with a female, suggesting that it may have followed an airborne scent trail to find her. Males have much larger activity areas than females (40 vs. 4 ha). This species is active all year long but exhibits distinct seasonal changes in activity, being most active during late winter and early spring.

Physiology

Thompson and Withers (1997a) report standard evaporative water loss rates (mg H_2O/g/h) at 35°C of 0.212 ± 0.035 (SE). Standard metabolic rates of *V. tristis* ($\dot{V}O_2$ and $\dot{V}CO_2$, both measured in mL/h) at 25°C are 5.3 ± 1.74 and 3.9 ± 1.31, respectively, and at 35°C are 15.7 ± 5.36 and 13.3 ± 5.12 (Thompson and Withers 1997b). Maximal metabolic rates at 35°C ($\dot{V}O_2$ and $\dot{V}CO_2$, both measured in mL/h) are 310 ± 104 and 663 ± 288 (Thompson and Withers 1997b). Factorial aerobic scope of *V. tristis* is 18.1. Metabolic rates are typical for small monitors but well above those measured for other small species of lizards (Thompson and Withers 1997b).

Fat Bodies, Testicular Cycles

Testes are large during September to November and regress in January–March. Fat bodies are larger during spring than during late summer, suggesting that these lizards forage and deposit fat during autumn and winter. When yolking up eggs, female fat bodies are small (Thompson and Pianka 1999).

Parasites

Varanus tristis from two study sites in the Great Victoria Desert examined by Jones (1995) contained six species of nematodes *Physalopteroides filicauda, Abbreviata hastaspicula A. levicauda, A. tumidocapitis, A. antarctica,* and *Skrjabinoptera goldmanae.* Prevalence of encysted physalopterid nematodes was high (33% at one site and 65.8% at the other). Jones suggests that prey lizard species, beetles, and cockroaches could be intermediate hosts.

7.24. *Varanus varius*

BRIAN WEAVERS

Figure 7.53. Varanus varius
(photo by Brian Weavers).

Nomenclature

Varanus varius is most commonly known as the lace monitor, but is also sometimes referred to as "tree goanna," "lacy," and various other names (Fig. 7.53). One particular color variation found west of the Australian Great Dividing Range is known as Bell's monitor.

The species was mentioned as *Lacerta varia* by White in his 1790 *Journal of a Voyage to New South Wales,* and White is given the naming priority by Cogger et al. (1983; White in Cogger et al. 1983).

Of interest, however, is the history of the scientific name described by Cogger et al. (1983). Apparently *Lacerta varia* was mentioned the previous year (1789) by Shaw in an unpublished manuscript. Hence, some earlier *V. varius* literature cites Shaw as the species' author. However, Shaw's use of the name was not actually the first description either, but a junior homonym. Wyttenbach had already used exactly the same name before Shaw (although just earlier in the same year), but Wyttenbach's name became forgotten and was not used for more than 50 years. Rules of nomenclature mean that Wyttenbach is not now credited because he was not recognized for more than 50 years, and Shaw is not recognized because he was apparently simply quoting Wyttenbach, leaving White as the recognized author because he "described" it afresh.

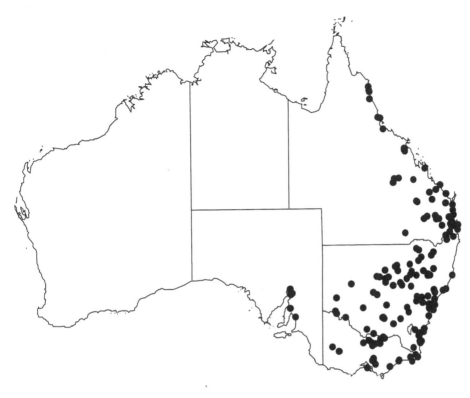

Figure 7.54. Known localities for
Varanus varius.

Geographic Distribution

Lacies are found in eastern Australia forests and woodlands from
Cape York Peninsula (Queensland) in the north, to Gippsland (Victoria)
in the south, and westward into western New South Wales as far as
Broken Hill and into South Australia (Fig. 7.54; Cogger 1992).

QM R932 Petrie—27.27 152.98; QM R2015 Woodford—26.95 152.78; QM
R2135 Eidsvold—25.37 151.12; QM R2288 Bell, via Dalby—26.93 151.45; QM
R2543 Bundaberg—24.87 152.35; QM R3062 Brisbane District—27.47 153.02;
QM R3595 Boonah—28 152.68; QM R3730, R3731, R3732: Stradbroke Is.,
opposite Southport—27.83 153.42; QM R4222, R4223: Brisbane District—27.47
153.02; QM R4464 Moreton Bay, Lamb Is.—27.63 153.38; QM R4828 Cook-
town—15.47 145.25; QM R5052 Narangba—27.2 152.97; QM R5598 Maroochy
Rd.—26.58 153.02; QM R6077 Brisbane, Everton Park—27.42 152.98; QM
R6730 Brisbane District—27.47 153.02; QM R7428 S Pine Rd., near Bald Hills—
27.33 153; QM R8140 Redbank Plains—27.65 152.85; QM R8412 Kin Kin, Como
SF—26.22 153; QM R8413 Kin Kin, Como SF—26.22 153; QM R8839 Condamine
Rd.—26.85 150.42; QM R9058 Emerald, 129 km E—23.65 149.35; QM R9059
Emerald, 129 km E—23.65 149.35; QM R10029 Eungella Ra.—20.92 148.5; QM
R10030 Aramara, via Maryborough—25.62 152.32; QM R10342, R10343: Dar-
ling Downs, SE of Chinchilla—26.73 150.63; QM R13263 Mt. Nebo, 24–32 km
NW Brisbane—27.38 152.78; QM R13578 Ironpot Creek, near Jandowae—26.67
151.42; QM R14393 Upper Brookfield, Brisbane—27.48 152.87; QM R14479
Brisbane, Indooroopilly—27.5 152.97; QM R14865, R14866 Nambour, near—
26.63 152.97; QM R15364 Cardwell—18.27 146.02; QM R15583 Coominglah

SF, Monto—24.75 150.83; QM R16156 Rockwood Stn., Chinchilla—26.93 150.37; QM R20325 N Stradbroke Is.—27.5 153.5; QM R21924 Brisbane, Mt. Coottha Reserve—27.48 152.95; QM R22572 N Stradbroke Is., Pt. Lookout—27.43 153.53; QM R24018 Anakie, Willows Rd.—23.57 147.58; QM R24019 Wilga Downs Rd., ~17.6 km WSW Emerald—23.53 147.98; QM R24144 Brisbane, New Farm Park—27.47 153.05; QM R24390 N Stradbroke Is., Flinders Beach—27.42 153.48; QM R24532 Toronto Zoo, via Newcastle; QM R25111 Mt. Molloy, Rifle Creek banks—16.6 145.32; QM R25984 Mt. Nebo—27.38 152.78; QM R26378 Hinchinbrook Is., Ramsay Bay—18.32 146.3; QM R27404 Tanny Morel, Farm Creek—28.27 152.3; QM R27564 Brisbane, The Gap—27.45 152.95; QM R29842 Capricorn Hwy. W of Anakie, near Bon Accord—23.68 147.5; QM R32739 Crediton, Site 7a—21.22 148.57; QM R33352 Mt. Glorious, 3 km Brisbane side of Maiala National Park—27.33 152.77; QM R33353 Conondale Ra., Sandy Creek—26.77 152.63; QM R33691 Bulburin SF, adjoining Site 1a—24.52 151.48; QM R36083 Kauri Creek Inlet, via Maryborough—25.82 152.95; QM R36992 Ducabrook—23.9 147.43; QM R38471 Glenhaughton Rd., about halfway from Taroom—25.48 149.63; QM R40917 Nanango, 43 km SW—26.97 151.73; QM R41391 Moreton Bay, Lamb Is.—27.63 153.38; QM R41481 N Stradbroke Is.—27.5 153.5; QM R41482 Samford, Brisbane—27.37 152.88; QM R41498 Taroom to Glenhaughton Rd., 15 km from Taroom—25.52 149.77; QM R43381 Shiptons Flat, via Cooktown—15.8 145.27; QM R45759, R47002, R48583, R48585: Highvale, via Samford—27.38 152.8; QM R47427 Flagstone Creek, via Helidon—27.45 152; QM R49893 Mt. Glorious, ~4 km below—27.4 152.77; QM R49905 Mt. Nebo, McAfees Lookout—27.42 152.87; QM R51467 Brisbane, Bunya Downs—27.38 153; QM R51759 Bundaberg, Cedar Creek Rd.—24.88 152.35; QM R54520 Mt. Nebo Rd., 7.5 km from The Gap—27.42 152.85; QM R54533 Mt. Nebo, Darcy Kelly Rd.—27.38 152.78; QM R55213 Sandy Creek, Conondale Ra.; QM R57327 Mt. Nebo area—27.4 152.78; QM R57359 Samford—27.4 152.85; AM R69 Greenwich—33.88 151.07; AM R949 Branlin—34.73 148.05; AM R962 Singleton—32.57 151.17; AM R1302 Hartley—33.55 150.18; AM R1463 Blacktown—33.78 150.88; AM R1469 Moree—29.47 149.85; AM R1536 Gladesville—33.83 151.13; AM R1538 Gosford—33.43 151.33; AM R1778 Narramine—32.23 148.25; AM R2058 Narellan—34.55 150.75; AM R2259 Bloomfield River, Cooktown—15.97 145.32; AM R3399 Narrandera—34.75 146.55; AM R3699 Campbelltown—34.07 150.82; AM R4911, R4912: Port Hacking—34.08 151.17; AM R5281 Werris Creek—31.35 150.65; AM R5395 Batemans Bay—35.73 150.25; AM R5396 Southern Queensland; AM R6503 Isis Scrub; AM R7904 Legges Camp, Bombah Pt., Myall Lakes—32.5 152.3; AM R8772 Narrabri—30.33 149.78; AM R8969 Fairfield, Sydney—33.88 150.95; AM R9739 NW Queensland; AM R9886 Sans Souci—34 151.12; AM R10035, R10036: Yarramalong Mt., Gosford—33.23 151.28; AM R10316 Sydney—33.88 151.22; AM R10610 Yanco, Blue Gate Swamp—34.6 146.42; AM R10611 Murrumbidgee R.Y.A. HS—34.63 146.37; AM R11048 CurlewIs.—31.12 150.27; AM R11049, R11050 Bullerana, Moree—29.33 149.62; AM R12247 Bombala—36.92 149.25; AM R12354 Wagga Wagga—35.12 147.37; AM R13462 Narrabeen Heights—33.72 151.3; AM R15061, R15062 Bugaldie, Near Coonabarabran—31.12 149.12; AM R16885, R16887 Tooraweenah—31.43 148.92; AM R17987 Nymagee—32.07 143.55; AM R17988 Cobar—31.5 145.83; AM R19181, R20511, R20512 Tooloom—28.62 152.42; AM R21616 Lindfield, Sydney—33.78 151.17; AM R21617 Richmond—33.6 150.75; AM R25937 Forestville, Sydney; AM R25951 20 mi N Bathurst; AM R25952 Bathurst District—33.42 149.58; AM R26271 20 mi E Cobar—31.53 146.17; AM R26384 between Eumungerie and Narromine; AM R26385 between Nymagee and Mount Hope; AM R27497 Deniliquin—35.53 144.95; AM R33122 Near Tottenham—32.23 147.35; AM R33206 Bathurst—33.42 149.58; AM R37226 Bringelly—33.95 150.73; AM R38022 Seven Hills, Sydney—33.78 150.93; AM R38195 30 mi S Hermidale—32 146.7; AM R40662 Gilgandra—31.72 148.65; AM R47285 8 km W of Hermidale on Mitchell Hwy.—31.07 146.67; AM R47899 Erimbula via Miriam Vale—24.18 151.83; AM R47900 Eungella—21.13 148.5; AM R55575 Copeton Dam—29.92 151.02; AM R60219 28 km E of Nevertire on Mitchell Hwy.—32.03 148; AM R64799, R64800: Sofala—33.08 149.68; AM R70496

Hawkesbury District Hospital; AM R70497 Richmond District? Hawkesbury Rd.; AM R71618 26 mi S of Uralla—30.95 151.17; AM R71619 11 mi N of Inverell on Ashford Rd.; AM R76739 ~10 km on Putty Rd. on grassy hill to Colo Track; AM R76833 22 km SSW Quambone—31.1 147.85; AM R81026 between Coonamble and Come By Chance; AM R84660 Quirindi—31.5 150.68; AM R86079 Earlwood—33.92 151.12; AM R90813 20 km S of Carinda on Warren-Carinda Rd.—30.63 147.75; AM R93191 11 mi N of Inverell on Ashford Rd.—29.65 151.17; AM R93192 26 mi S of Tamworth—31.45 150.87; AM R94626 8.8 km SE of Mount Hope on the Euabalong West Rd.—32.87 145.93; AM R94686 3.1 km N of Gwabegar by road—30.58 148.97; AM R94688 2.3 km SW of Miandetta by road—31.58 146.95; AM R94689 Yathong Nature Reserve, on SE side of Merrimerriwa Ra.—32.7 145.62; AM R94914 11.0 km N, 8.5 km E of Bega—36.57 149.93; AM R95874 12 km E of Capertee—33.13 150.08; AM R96478 1.5 km N of Bermagui TO, Princess Hwy.—36.4 149.9; AM R102891 5 km S of Tilba Tilba—36.42 150.07; AM R104940 100 m from the Intersecton of Becks Rd. on Cooranbong to Heaton LK; AM R106315 1.3 km ESE of High Wollemi Stn.—33.05 150.68; AM R107268 on Putty Rd. between Bulga and Howe's Valley, Singleton District; AM R111065 Blacktown—33.77 150.92; AM R112009 40 km W of West Wyalong on Mid Western Hwy.—33.7 146.77; AM R112853 9.5 km S junction of Bellata Rd. and Terry Hie Hie–Narrabri Rd.—30.02 150.1; AM R113956 just S of Finch Hatton National Park, Queensland—21.07 148.62; AM R113960 18.5 km W of Upper Ross River Rd. via Herveys Ra. Rd., Townsville Area, QlD-19.32 146.55; AM R118888, R118890: Near Menai—34.02 151.02; AM R120945 5 km NE of Trangie PO—32 148; AM R123332 Bathurst District—33.42 149.58; AM R123425 14.5 km S of Hermidale by road—31.65 146.63; AM R133160 13 km N Karuah on Pacific Hwy.—32.62 152.08; AM R133492 Border Ranges National Park, Lophostemon Falls on Brush Box Falls Track—28.4 153.07; AM R135356 Flat Rock—31.43 152.93; AM R135358 Garrawarra Farm (Victoria); AM R138464 Pokolbin on the Cessnock-Pokolbin-Broke Rd.—32.8 151.28; AM R138465 Mangrove Mountain, ~10 km S of PO on Wisemans Ferry Rd.—33.35 151.17; AM R142904 24 km E of Euston—34.58 143; AM R144676 Killarney State Forest, 2 km N—30.25 149.88; ANWC R0265 Black Mountain, Canberra—32.28 149.11; ANWC R0266 Captains Flat—35.58 149.45; ANWC R0267 Aranda, Canberra—32.26 149.08; ANWC R0916 15 km W of Booligal—33.9 144.62; ANWC R1806 Macquarie Marshes—30.78 147.55; ANWC R1992 Sandy Camp Stn., Macquarie Marshes—30.77 147.58; ANWC R2683, R2684 Sandy Camp Stn., Macquarie Marshes—30.77 147.6; ANWC R4103, R4104, R4105 Bendethera, Deua National Park, W of Moruya—35.97 149.73; ANWC R4106 Lake Yartla, Near Menindee—33.13 142.22; ANWC R4107 5 km from Mallacoota toward Genoa—37.55 149.7; ANWC R4108 Bendethera Fire Trail, 15 km From Moruya—35.93 149.92; ANWC R4109 Mallacoota—37.9 149.15; ANWC R5199, R5200, R5201, R5202, R5203, R5204, R5205, R5206, R5207, R5208, R5209, R5210, R5211, R5212, R5213, R5214, R5215, R5216, R5217, R5218, R5219 Deua River Valley, Deua National Park, SW of Moruya—35.77 149.93; ANWC R5477 Shoalwater Bay Army Training Reserve, N of Rockhampton—22.71 150.5; ANWC R5493 Shoalwater Bay Army Training Reserve, N of Rockhampton—22.68 150.39; ANWC R5594 7 km S of Cowra—33.9 148.68; ANWC R5832 3 km N of Frogmore, road between Boorowa/Nyangala Res—34.25 148.85; NMV D9211 Tocumwal—35.82 145.57; NMV D3344 Mt. Arapiles—36.77 141.85; NMV D300 Kyabram—36.32 145.05; NMV D9137 Tocumwal—35.82 145.57; NMV D301 Kyabram—36.32 145.05; NMV R1378 Mt. Martha—38.3 145; D1643 13.6 km NE of Gilgandra—31.67 148.7; NMV D1360 Shipwreck Creek, 10.4 km SW of Mallacoota Airport—37.63 149.67; NMV D1422 between Wangarabelle and Wroxham—37.35 149.47; NMV D1780 Craigs Swamp, 16 km S Sale—38.25 147.02; NMV D6741 3 km from Erinunda TO from Combienbar Rd.—37.6 148.92; NMV R1776 Metung—37.88 147.85; NMV D9119 Murray River, near Tocumwal—35.82 145.57; NMV D3568 Bunyip—38.1 145.72; NMV D8923 St. George—28.05 148.58; NMV D6173 Termeil—35.48 150.33; NMV D5263 Reevesby Is.; NMV D6835 Bruce Road, 7.4 km N of Princes Hwy.—37.78 147.9; NMV D6509 Red Cliffs—34.3 142.22; NMV D6866 Upper Lurg—36.58 146.18; NMV D1053 Wandin N—37.77 145.42; NMV

D5801 1 km NW of Violet Town, on Shepparton Rd.—36.62 145.72; NMV D4169 Kialla West—36.47 145.38; NMV D6869 Longwood area—36.75 145.47; NMV D6563 2.5 km S of Elmore—36.52 144.62; NMV D3572 Rutherglen—36.05 146.47; NMV D5741 Old Howlong Rd., Chiltern—36.15 146.6; NMV D1772 Chesney Vale, 16 km NW of Benalla—36.42 146.13; NMV D5716 0.3 km NW of Mt. Killawarra—36.23 146.17; NMV D9270 Healesville—37.65 145.52; NMV D1772 Reef Hills, 2.4 km S of Benalla—36.57 145.98; NMV D1764 8 km WNW of Benalla—36.52 145.9; NMV R1076 Launching Place—37.77 145.58; NTM R26270 Warrumbungle Mountains—31.43 149.6; NTM R26272 Cobar; SAM R00132, R00133 Eidsvold—25.37 151.12; SAM R00649 Lower Light—34.55 138.43; SAM R01704 Bute—33.87 138; SAM R02510 Dareton—34.08 142.07; SAM R02633 Fish Falls—37.12 142.4; SAM R13817 Narromine—32.23 148.25; SAM R13942, R13971 Mambray Creek National Park—32.82 138.08; SAM R23277 Mt. Remarkable National Park—32.83 138.08; SAM R33351 Mambray Creek—Alligator Gorge—32.75 138.05; SAM R33773 Stoney Creek—32.68 138.1; SAM R34444 Mambray Creek—32.85 137.98; SAM R35998 5 km NW of Melrose—32.75 138.17; SAM R36335 4 km N of Pichi Richi Village (Historic Site)—32.37 137.97; SAM R46759 Spring Creek—32.7 138.18.

Fossil Record

A number of Pleistocene (~10,000 years ago to 2 mya) fossil vertebrae, skull, bones, and teeth of *V. varius* have been found in Victoria Cave at Naracoorte (in South Australia, near the Victorian border). This area is within the current range of the lace monitor (Smith 1976).

Diagnostic Characteristics

- Large, up to nearly 2 m in TL.
- Tail rounded at the base, quickly assuming a triangular cross section with twin ridges of scales along the dorsal surface.
- Head scales moderate (Cogger 1992).
- Nostril lateral and about twice as far from the eye as from the tip of the snout (Cogger 1992).
- About 200 scales around the middle of the body (Cogger 1992).

Description

Clean adult animals give an overall appearance of having had strips of cream lace draped across a dark body—hence the common name of lace monitor and the specific name of *varius* (as in "variegated"). Large adult lace monitors are typically dirty, though, and from a distance may give the appearance being an overall dark brown or gray. Closer inspection, however, reveals lighter cream or off-white lacy or blotchy bands, with very broad cream solid bands toward the end of the tail. Clean dorsal skin between the scales is black, and scale color actually creates the lace patterning effect. There are often bands of blue scales around the snout and in patches on the side of the head. Banding elsewhere on the dorsal surface of the body is typically pale cream to yellow, creating the lace appearance. The ventral surface ranges from predominantly yellow or cream irregularly interspersed with a few dark bands (particularly in animals from west of the Great Dividing Range), to a much greater proportion of black in the irregular black and yellow banding of animals from southeastern coastal forests.

A banded phase of this species, known as Bell's monitor, is found mixed in with normally colored animals in western New South Wales and northern Victoria (inland from the Great Dividing Range). Bell's monitor has broad, fairly equal bands of solid cream and black coloring across the dorsal surface of the body, not looking at all like lace.

Size

Documented measurements of SVLs range from 105 mm for a small hatchling up to 765 mm for a very large adult. TLs range from 265 mm (again for a small hatchling) up to 1920 mm (Weavers 1988), although general accounts of the species often suggest a TL of about 2 m or more (e.g., Worrell 1963; Minton and Minton 1973; Houston 1978; Cogger 1992).

Mass at hatching is typically about 25 g (Carter 1999b), although it may be as little as 16 g (Boylan 1995). Adults may reach 14 kg (Weavers 1988).

Males mature at about 415 mm SVL (when they have fully differentiated seminiferous tubules and sperm present) (Carter 1999a). Lacies can reach this length in captivity in about 50 months, even when fed on a very modest diet (Weavers 1988), and probably achieve this size more quickly in the wild.

Females are mature at about 385 mm SVL (Carter 1999a). Carter (1999a) found maximum size of adult females to be much smaller than the biggest males (765 mm SVL compared with the longest female Carter found of 575 mm SVL in the Australian Museum collection).

Habitat and Natural History

Varanus varius is found in lowland open forest and woodland. In the northern lowland parts of its range, suitable habitat can extend into forests with denser canopy cover (probably because of the greater thermoregulatory opportunities in the more tropical climate). Inland from the Great Dividing Range, lacies are found in narrow fringes of woodland along margins of rivers and lakes (e.g., river red gums), otherwise surrounded by fairly arid and treeless countryside. Here lacies may be found almost side by side with other species such as *V. gouldii*, even though the latter species primarily occupies quite different low-scrub habitat outside the woodland fringe.

Termite mounds may become a social center for relatively large numbers of lacies, and lacies will sometimes take overnight refuge in termite mounds (personal observation).

Although fighting between males has been observed by many field workers and battle scars seen by most, Carter (1990) found no aggression exhibited by females either toward males or other females.

Captive ages of more than 15 years for lacies have been recorded by Flower (1937) and Kennerson (1979), and two lizards that I hatched in 1983 are still alive 21 years later (one in Melbourne Zoo and the other recently escaped; Birkett, pers. comm.).

Terrestrial, Arboreal, Aquatic

Varanus varius are very capable tree climbers, and they generally run and climb a tree if on the ground when disturbed by a human. As with other arboreal goannas, lacies spiral around the trunk to always keep out of sight of the observer. However, if the observer keeps them within sight and watches closely, the animal will often stop on the vertical trunk once it thinks it is out of reach—typically at about 5 m or more above the ground. There it will hang, and it seems to be capable of doing this for hours without tiring. Lace monitors have long and very sharp claws. Most of the digits seem to have a double-jointedness about them, and when handling animals, I have found it possible to rotate the longer digits through very wide angles and almost tie them like a piece of rope. Yet for climbing (and for scratching handlers!), there are obviously powerful muscles attached through the digits to the claws. It seems as though they are using their claws a bit like an ice climber uses an ice pick: they use their powerful muscles to throw the claw into the bark of the tree, and again to pull it out again for the next step. But once they stop, the animals seem to be able to relax the muscles and leave the extremely sharp claw embedded while the pads of the toes and feet rest on the trunk, enabling the animal to hang with very little apparent effort.

Despite their obvious skills at climbing trees and avoiding humans, and despite the fact that lace monitors seem to be confined to forests, woodlands, or the treed margins of inland lakes and rivers, I found many lace monitors that I radiotracked in southeastern Australia spent little or no time in trees (except when I actually disturbed them).

Water is generally nearby in most lace monitor habitat, although this species rarely seems to venture into it. However, on two or three occasions, I have observed lacies drinking, and once a lacy dived into a small pond and sat on the bottom for about 20 minutes to evade me.

Time of Activity (Daily and Seasonal)

If secluded overnight in a burrow or a hollow, a lace monitor will emerge to bask when it becomes warmer outside than inside—presumably the lizard senses a warm draft of air. Depending on how close already the animal's body temperature is to its operating temperature range (see "Thermoregulation" below), it will progressively emerge, headfirst, into better and better basking conditions. Experimentally, I used a small gas reflector-heater to entice a lacy out of its overnight burrow earlier than usual one morning. I placed the heater near the entrance to the burrow of a field-telemetered lacy very early one morning, and it emerged 30 to 60 minutes ahead of four other animals that morning, and about 40 minutes earlier than the same animal had been emerging on preceding days.

Animals typically emerge on sunny days throughout the year, even in cold southeastern Australian winters, where inland overnight air temperatures may fall well below 0°C. However, because winter days are considerably shorter than summer ones, emergence in winter (time of day, mean ± SD; 1012 hours ± 53 minutes) is much later than in summer (0736 hours ± 32 minutes) (Weavers 1983).

Despite being able to achieve suitable body temperatures on sunny days at any time of year (see below), lacies do not forage in southeastern Australia in winter (although conceivably they might do so in the northern Queensland extremes of their range). Even in southern Queensland, I found that captive lacies would not eat during winter despite being provided with suitable basking opportunities (heat lamps). Further south, under field conditions in a southeastern Australian winter, lacies have to spend all their time carefully orienting themselves toward the sun to achieve high body temperatures, and they would quickly cool if they dropped out of this position to go foraging. However, lacies do occasionally excrete during these winter periods of high body temperature, no doubt using the opportunity to metabolize some fat reserves.

Thermoregulation

As suggested by Magnuson et al. (1979) for ectothermic vertebrates generally, the thermal environment for lacies is an important ecological resource (a thermal niche) similar in importance to food availability.

The ability of the lace monitor to occupy the niche it does (see "Natural History" above) depends on its highly competent thermoregulation. Lace monitors are diurnal heliotherms. In other words, solar radiation is the most critical factor for their thermoregulation, although in extreme ambient conditions, air temperatures, and wind play a part too. Given suitable environmental conditions, free-ranging lace monitors maintain their body temperatures within an operating temperature range of 32.8 to 36.4°C (Weavers 1983).

Preferred body temperature (sometimes called "eccritic body temperature") is defined by Bligh and Johnson (1973, p. 37) as "the range of core temperature within which an ectothermic animal seeks to maintain itself by behavioral means." However, many workers apply the term to a single "specific body temperature maintained whenever possible" (e.g., Heatwole 1976). Consequently, "selected body temperature range" is proposed by Pough and Gans (1982) as a nonanthropomorphic alternate term, similar to the proposition to consider an upper and lower set-point temperature (e.g., Templeton 1970; Cloudsley-Thompson 1972; Firth and Turner 1982).

I prefer a term like "operating temperature range" rather than the more commonly used "preferred body temperature" for lace monitors, because these animals really do appear to cycle their body temperatures between clear and regular upper and lower limits (rather than randomly around a single temperature). Under ideal conditions (clear, sunny days), I found that about 60% of animal-days of thermoregulation plots showed a measurable periodicity, or cycling, when body temperatures were measured by radiotelemetry at about 10- to 20-minute intervals throughout the day (Weavers 1983). My alternative term also removes any temptation to credit lace monitors with some sort of conscious thought about their body temperature (smart though these animals sometimes seem to be). Lacies can achieve body temperatures of up to 38°C on a clear, sunny day in winter in southeastern Australia, despite air temperatures peaking at only 16°C and overnight frequently dropping below zero (recorded for a 6.1-kg animal; Weavers 1983).

Particularly in thermally unfavorable conditions, lacies may devote a significant amount of their time and behavioral activities attempting to maintain body temperatures within the operating temperature range. Indeed, when body temperatures are outside this range, human observers will generally not encounter lace monitors. Active lacies find observers, rather than observers finding lacies. In the field, most lizards I found had already observed me and were taking evasive action, which is what actually drew my attention to them.

I have only once seen a lacy panting, and this was in the manner of the "gular pumping" described by Heatwole et al. (1973), although the mouth was open for some time as well. In this case, the animal had been feeding at a carcass in full summer sun on a day of 35°C air temperature, and then went up a large dead tree (also in the sun). As I watched, in clear sight of the animal, it began panting after about 10 minutes, and continued intermittently to do so over the next 2 hours until it finally ran down the tree and evaded me. If the animal had reached a sufficiently high body temperature to induce the highly unusual panting activity within 10 minutes of being up the tree, the panting would seem to have been effective in keeping the animal below its critical temperature for the next 1 hour, 50 minutes.

Foraging Behavior

Lace monitors fit Regal's (1978) description of "intensive foragers," although much of their foraging success also seems to come from less active chance encounters, such as finding prey in tree hollows or burrows when the lacy is retiring to its overnight roost.

A keen sense of smell and/or the sensitivity of the vomeronasal organ are also thought to help these animals locate carrion without a great deal of physical search effort. However, I am not sure of the real effectiveness of this organ for that purpose. My own observations suggest that the forked tongue is actually used to touch surfaces to pick up molecules or particles that are larger than those normally involved in the conventional sense of smell. In the vicinity of carcasses, I have seen fresh lacy tracks that were well within range of my own sense of smell (!), but that showed that the lacies did not actually find the carcass until many days after I did.

Lacies seem to spend a relatively small proportion of their time actually moving around. Even on clear, sunny days during the animals' active season, lacies only spend an average ± SD of about 8.5 ± 5.4% of their time actually moving around during the operating period of the day (17 animals monitored for activity between December and April; Weavers 1983).

Diet

Lace monitors are opportunistic carnivores, taking a wide variety of prey and carrion, both native and exotic (Weavers 1989). The 50 animals successfully sampled in southeastern Australia by Weavers (1989) ranged in mass from 750 to 9800 g. I found arthropods in more than half (56%), macropods (kangaroos and wallabies) in 32%, the exotic rabbit in 28%, birds in 16%, and reptiles and reptile eggs in 12%. Although exotic mammals were found altogether in 38% of the samples, this surely represents opportunity rather than a preference for exotics. Vast changes in the Australian environment throughout the range of *V. varius* over the past 200 years since European settlement means that the overall habitat for this species and its natural prey has been greatly reduced. Fortunately for the lace monitor, some of the exotics now present have been able to provide satisfactory substitute food.

Guarino (2001) also found *Varanus varius* to be a generalist carnivorous scavenger and predator, with carrion dominating the diet. He found 18 taxa of animals represented. In the central west of New South Wales where he conducted the study, Guarino found similar categories of food taken but different percentages compared with Weavers (1989). In particular, the percentage of birds was up to 59% in spring and early summer, whereas I found an overall average of just 6%. Apart from differences in habitat and seasonal conditions, a couple of reasons might account for this. Guarino points out that Weavers (1989) was conducted before the release of rabbit calicivirus disease, and Guarino thinks that the lower availability of rabbits at his site may have shifted the balance of the lacies' diet. Also, Guarino's results were estimates of the percentage volumes of the taxa, whereas I simply measured the frequency of occurrence in the samples (very different measures).

Although one might imagine that smaller lace monitors would take predominantly smaller prey, this was not statistically obvious in my sample of 50 (Weavers 1989). Basically, a lace monitor will eat any meat that it can get into its stomach, dead or alive. I have observed quite large adults waddling along tracks, snapping from side to side, feeding on minor plague locusts. Conversely, a juvenile animal of only 335 mm SVL had grey kangaroo (*Macropus giganteus*) in its gut (one presumes that this was carrion!). Weavers (1989) also failed to find that any individuals had particular preferences in food, or that there were any obvious seasonal choices.

Varanids have a kinetic skull (Rieppel 1979), well illustrated by an observation I made of a relatively small lace monitor. At a sedate gathering around a barbecue in rural Victoria to celebrate the 45th wedding anniversary of a friend's parents, I was asked to show a relatively small, but fat, goanna that I had caught a short time beforehand. As soon as I produced this little lacy out of its bag, it proceeded to produce a rabbit out of its mouth (it wasn't wearing a hat)—it regurgitated the carcass of a complete, near-perfect 500-g rabbit. Empty, the goanna was just 1200 g, which means that this single mouthful amounted to 42% of the predator's body mass. Quite some kinetics must have been present in its skull to allow such a swallowing feat!

Lacies may also become prey. Although hatchlings and small juveniles usually become prey, perhaps to birds, adults are also occasionally at risk. Webb (1994) observed three dingoes (*Canis familiaris dingo*) capture and kill a large lace monitor in about 2 minutes in Morton National Park (southeast New South Wales).

Reproduction

Mating occurs between mid-November and early January in the field in southeastern Australia (Carter 1990), although mating of one pair at Taronga Zoo was observed in March (Boylan 1995). Animals communicate through scent, touch, and sight. Females may mate with several males, including subordinates. Males tend to use their hemipenes alternately in repeated copulation with the same female (Carter 1990).

Lacies mate frequently over quite long periods (up to 16 times over 3 hours), something not reported about many other varanids (Carter 1990). However, King and Green (1999a, p. 27) note that intense mating activities may occur over a period of days in *V. rosenbergi*.

In southeastern Australia, lacies lay their eggs in termite mounds in the middle of summer. Typically the mounds are those of *Nasutitermes exitiosus*. Carter (1999b) could find no confirmed records of lacies using nest sites anywhere other than in termite mounds, although, as Tasoulis (1992) records, the termite mound may be in a tree and not necessarily only on the ground. Eggs overwinter in the field and hatch after about 290 days (Carter 1999b), although artificially incubated eggs may take between 70 and 317 days to hatch (Carter 1999a).

Around the time of hatching in the field, adults dig into the mound and release the young. In captivity at Taronga Zoo, a female was observed burrowing into a termite mound containing freshly hatched young (Boylan 1995), and Horn (1991) observed a female exhibiting protective behavior toward her nest site. Although Carter believes the mother digs the nest and later releases the young, I have observed known males (telemetered or marked animals) digging holes in termite mounds too.

A 3-year-old captive female in Taronga Zoo (Sydney) was recorded laying five eggs, although it was not known whether the eggs were fertile (Boylan 1995).

Cowles (1930) was probably the first to record varanids using termite mounds for their eggs (*V. niloticus*), although he believed that in this species, the fluid from ruptured eggs allowed hatchlings to dig themselves out of the softened mound.

Cogger (1967) first speculated that female lace monitors dug the tunnel to release the hatchlings. Carter (1999b) thinks that female lacies are quite capable of relocating the termite mound in which they laid their eggs, especially if they use the same mound over several seasons. Hatchlings can survive at least 2 to 3 weeks within the mound, allowing the mother to take up to this long to get there and dig them out. An alternative possibility is that the release of hatchlings by an adult developed as an incidental outcome of other behavior.

Lacies in southeastern Australia seem to be digging into termite mounds from the beginning of spring (September)—virtually as soon as their active season begins. This is several months before egg laying (between late December and late January). At my study site in southeastern New South Wales, some *Nasutitermes exitiosus* termite mounds had evidence of recent digging by lacies between late August and February (and virtually every mound had one or more holes during December and January). Carter (1999b) observed mostly females doing the digging, but I saw a number of animals that I knew were males (from radio transmitters or previous capture and marking). With all this digging centering on termite mounds, mounds make a good place for socializing and meeting mates (the goanna "day-club scene"). Early spring coincides with males' readiness to mate. Perhaps the digging activity forms some sort of display to potential partners? Perhaps both males and females engage in this display toward each other? Perhaps a good hole digger represents a good prospect for a mate? The gathering around the mound and the digging there at this time may have originally developed as a mating ritual. Any hatchlings inside the mound released as a result of this social activity may actually be incidental. Most goanna tunnels that both Carter and I have observed tend to aim more or less at the core of the mound, and so whatever the original motivation for the tunneling, the effect of a deep-enough hole will be to release any hatchlings present. Originally the egg laying and ritual digging need not have been very deep into the mounds, but some advantage may still have been gained by the animals practicing this procedure. Indeed, it seems unlikely that the first use of mounds went straight to the core, but probably the depth of burrowing would have increased over generations (with increased advantages of temperature regulation and protection for hatching the eggs, but also an increased reliance on adults to dig out hatchlings).

There is a good chance that individual lacies will regularly visit some mounds within their home range. This familiarity would not be surprising because several researchers, including me, have observed lacies repeatedly using particular roost sites. For a lacy, therefore, digging at a regular mound gives you a good chance of releasing your own offspring (and perhaps mating again with your partner from last season), even though neither outcome need be deliberate nor the behavior that is directly selected for. My scenario requires lacies to be able to remember a place (a capability already accepted regarding roost sites) and to be able to remember an activity (one that brings the immediate reward of mating—probably the easiest thing to remember), but does not require a lacy to remember the correct time to go and release young (young that the parent has never seen and is not aware of). Digging in a termite mound is just a routine part of life for a lacy as soon as the active season starts.

Another possible selective advantage of a prolonged digging ritual is the chance to discover (and abandon) dead termite mounds that are not particularly useful for hatching eggs. Dead mounds cannot fill a goanna hole. A dead mound will not provide thermoregulatory advantage for the eggs and gives no protection from predators or fungal attack. However, living *N. exitiosus* termite colonies in good condition will partially fill a goanna burrow each night (unless already filled by a goanna spending the night there). On a wet night, termites completely filled a 1.3-m-deep hole that I had watched a male lacy dig that day. Digging and redigging a hole in the same place over a prolonged period confirms the fact that the mound is still living, and that the mound is therefore able to provide all the best nursery conditions (stable and elevated temperature and humidity over an extended season, disposal of fungus and other possible infections of the egg, and protection from predators). A dead mound will not do these things. If the mound is dead and the hole is still there when the lacy revisits, and no more digging is required, then the lacy would presumably go somewhere else to perform his or her digging ritual.

Movement

Lacies do seem to move around a regular patch that can be termed a home range. For males (with a mean mass of 5.1 kg), this area is about 65 ha in southeastern Australia (Weavers 1993), and Carter (thesis 1992, reported in Carter 1999b) measured a home range of 25 ha for females.

Animals in Weavers' (1993) study were radiotracked for between 2 and 19 months, so it is not known whether the range of a particular animal is relatively fixed over its lifetime, or whether it may shift and change. It seems likely, though, that hatchlings would start with a relatively small range, probably mostly arboreal, and move to the ground and expand this range progressively over the following few years until the animal's maturity at least. Perhaps interactions with other lacies would cause some individuals to shift ranges too.

Massive variations in seasonal conditions in southeastern Australia can mean that some parts of any particular home range may be suitable in some years and unsuitable in others. Lacies do not seem to move very far into completely open ground without cover. Some areas without trees or understory may nevertheless have very dense ground cover (e.g., long grass or herbs) in "good" years, which would encourage lacies to use that place. But in other years, the same place may be almost bare earth. When bare ground, these places would not be an effective part of the animal's home range.

Stebbins and Barwick (1968), Weavers (1983, 1993), and Carter (thesis 1992, reported in Carter 1999b) have all observed that lacies seem quite familiar with at least the core of their home ranges. I have often noted lacies reusing roosting sites, sometimes after long intervals.

Physiology

The use of water is a critical factor for most animals in the variable climate of temperate Australia. Water efflux in lacies is strongly seasonal, ranging from a mean ± SD of 24.7 ± 7.3 mL^{-1}·kg^{-1}·d^{-1} in summer, to 5.1 ± 2.4 mL^{-1}·kg^{-1}·d^{-1} over winter. Efflux is not correlated with body mass (Weavers 1983). Total ± SD body water content of lacies in the field during summer is 72.7 ± 4.3% (Weavers 1983).

In a laboratory metabolism chamber, Bartholomew and Tucker (1964) found heating rates faster than cooling rates for *Varanus varius* (heating 0.19–0.53°C, cooling 0.17–0.39°C, between 20 and 40°C).

In the field, I measured mean heating rates for lacies basking in ideal conditions (full sun) of 0.29°C·min^{-1} in summer and 0.21°C·min^{-1} in winter (Weavers 1983).

Seebacher (2000) reports that heart rates in free-ranging *Varanus varius* are significantly higher during body heating than during body cooling. Resting heart rates also tend to decrease with increasing body mass (and rose to over 50 beats per minute in less than 5% of the heating and cooling episodes that Seebacher observed for his six lizards, with masses of 4 to 5.6 kg).

In free-ranging animals in the field, the lowest heart rate that I measured was 4 beats per minute in a resting animal with a body temperature of 12.4°C at the beginning of an autumn day. The maximum heart rate I recorded was 86 beats per minute when the animal in question had a body temperature of 37°C while feeding at a carcass in full sun. All the heart rates that I recorded from lacies at rest during the operating periods of the day were between 38 and 58 beats per minute (three lacies over 8 days; Weavers 1983).

Fat Bodies, Testicular Cycles

Lacies in good condition have two large abdominal fat bodies, and in a large adult, they can be up to about the size of a human hand each. However, there has been no seasonal analysis of trends in the deposition and use of these fat bodies by lacies (which is understandable, given the difficulty of doing this from preserved specimens and the enormous extravagance of killing enough animals for the purpose).

Some lizards in a lace monitor population seem to produce sperm throughout the year (Weavers 1983), although the peak sperm abundance (indicating peak readiness to mate) seems to be from early spring through to the first half of summer (September to January).

Parasites

Parasites that I have found incidentally in stomach, scat, and blood samples of lacies from southeastern Australia include nematodes (order Nematoda: *Abbreviata* sp., family Physalopteridae; an unidentified species of the suborder Ascaridata), and cestodes (order Cestoda: tapeworm) (identified by W. L. Nicholas, Department of Zoology, Australian National University, Canberra). A blood parasite *Haemoproteus* sp. (family Haemoproteidae) was also found (identified by M. J. Howell, Department of Zoology, Australian National University, Canberra).

Externally, one species of tick (*Aponomma undatum*) occurs in very high numbers on virtually all lacies I have handled (identified by

M. Bull, School of Biological Sciences, Flinders University). Ticks are typically found around the cloaca and in the more protected areas of the axilla and groin.

References

Anonymous. 2000. *Hydrosaurus gouldii* Gray, 1838 (currently *Varanus gouldii*) and *Varanus panoptes* Storr, 1980 (Reptilia, Squamata): Specific names conserved by the designation of a neotype for *H. gouldii*. *Bull. Zoo. Nomenclature* 57(1):63–65.

Ast, J. C. 2001. Mitochondrial DNA evidence and evolution in Varanoidea (Squamata). *Cladistics* 17:211–226.

Australia Zoo. 2000. Canopy goanna (*Varanus keithhornei*) breeding program. Records and data, 1992–2000.

Barnett, B. 1977. Additional notes on centralian bluetongues (*Tiliqua multifasciata*). *Newsl. Vic. Herpetol. Soc.* 1:10.

———. 1979. Incubation of sand goanna (*Varanus gouldii*) eggs. *Herpetofauna* 11(1):21–22.

Bartholomew, G. A., and V. A. Tucker. 1963. Control of changes in body temperature, metabolism, and circulation by the agamid lizard, *Amphibolurus barbatus*. *Physiol. Zool.* 36:199–218.

Baverstock, P. R., D. King, M. King, J. Birrell, and M. Krieg. 1993. The evolution of species of the Varanidae: Microcomplement fixation analysis of serum albumins. *Austr. J. Zool.* 41:621–638.

Becker, H. O., W. Böhme, and S. F. Perry. 1989. Die Lungenmorphologie der Warane (Reptilia: Varanidae) und ihre stammesgeschichtliche Bedeutung. *Bonn. Zool. Beitr.* 40:27–56.

Bedford, G. S., and K. A. Christian. 1996. Tail morphology related to habitat of varanid lizards and some other reptiles. *Amphibia-Reptilia* 17:131–140.

Bennett, A. F. 1972. A comparison of activities of metabolic enzymes of lizards and rats. *Comp. Biochem. Physiol.* 42B:637–547.

———. 1973. Ventilation in two species of lizards during rest and activity. *Comp. Biochem. Physiol.* 46A:653–671.

Bennett, D. 1995. *A little book of monitor lizards*. Aberdeen: Viper Press.

———. 1998. *Monitor lizards: Natural history, biology and husbandry*. Frankfurt am Main: Chimaira.

Bennett, A. F., and P. Licht. 1972. Anaerobic metabolism during activity in lizards. *J. Comp. Physiol.* 81: 277–288.

Berney, F. L. 1936. Gould's monitor (*Varanus gouldii*). *Mem. Queensl. Naturalist* 10(1):12–14.

Bickler, P. E., and R. A. Anderson. 1986. Ventilation, gas exchange and aerobic scope in a small monitor lizard, *Varanus gilleni*. *Physiol. Zool.* 59(1):76–83.

Blamires, S. J. 1999. A note on the reproductive seasonality of *Varanus panoptes* in the wet-dry tropics of Australia. *Hamadryad* 24:49–51.

Blamires, S. J., and M. Nobbs. 2000. Observations of mangrove habitation by the monitor lizard *Varanus panoptes*. *N. Terr. Naturalist* 16:21–23.

Bligh, J., and K. G. Johnson. 1973. Glossary of terms for thermal physiology. *J. Appl. Physiol.* 35:941–961.

Böhme, W. 1988a. Der Arguswaran (*Varanus panoptes* Storr, 1980) auf Neuguinea: *V. panoptes horni* ssp. n. (Sauria: Varanidae). *Salamandra* 24:87–101.

———. 1988b. Zur Genitalmorphologie der Sauria: Funktionelle und Stammesgeschichtliche Aspekte. *Bonn. Zool. Monogr.* 271–176.

———. 1991. The identity of *Varanus gouldii* (Gray, 1838), and the nomenclature of the *V. gouldii*–species complex. *Mertensiella* 2:38–41.

Böhme, W., and T. Ziegler. 1998. Comments on the proposed conservation of the names *Hydrosaurus gouldii* Gray, 1838 and *Varanus panoptes* Storr, 1980 (Reptilia, Squamata) by the designation of a neotype for *Hydrosaurus gouldii*. *Bull. Zoo. Nomenclature* 55(3):173–174.

Boulenger, G. A. 1885. *Catalogue of lizards in the British Museum (Natural History)*. Vol. 2. London: British Museum.

———. 1898. Third report on additions to the collection of lizards in the British Museum. *Proc. Zool. Soc. Lond.* 1898:912–923.

———. 1906. Description of a new lizard and a new snake from Australia. *Ann. Mag. Nat. Hist.* 18(7):440–441.

Boylan, T. 1995. Field observations, captive breeding and growth rates of the lace monitor, *Varanus varius*. *Herpetofauna* 25:10–14.

Branch, W. R. 1982. Hemipenal morphology of platynotan lizards. *J. Herpetol.* 16:16–38.

Braysher, M., and B. Green. 1970. Absorption of water and electrolytes from the cloaca of an Australian lizard *Varanus gouldii* (Gray). *Comp. Biochem. Physiol.* 35:607–614.

Bredl, J. 1987. First captive breeding of the perentie. *Thylacinus* 12:2–3.

Bredl, J., and H.-G. Horn. 1987. Über die Nachzucht des Australischen Riesenwarans *Varanus giganteus* (Gray 1845). *Salamandra* 23:90–96.

Broer, W., and H.-G. Horn. 1985. Erfahrungen bei Verwendung eines Motorbrüters zur Zeitigung von Reptilieneiern. *Salamandra* 21(4): 304–310.

Brooker, M. G., and J. C. Wombey. 1978. Some notes on the herpetofauna of the western Nullarbor Plain, Western Australia. *W. Austr. Naturalist* 14(2):36–41.

Browne-Cooper, R. 1998. Predation on a ridge-tailed monitor (*Varanus acanthurus*) by a pygmy python (*Antaresia perthensis*). *Herpetofauna* 28(2):49.

Brunn, W. 1978. A trip to Alice Springs. *S. Austr. Herpetologist* 1978 (April):1–3.

———. 1980. Beyond the banana curtain. *S. Austr. Herpetol. Newsl.* 1980(April):2–7.

———. 1981. Goannas caught on trips. *S. Austr. Herpetol. Group* 1981 (October):4.

———. 1982. Goannas caught on trips. *S. Austr. Herpetol. Group* 1982 (March):6–9.

Burnett, S. 1997. Colonizing cane toads cause population declines in native predators: Reliable anecdotal information and management implications. *Pacific Conserv. Biol.* 3:65–72.

Bustard, H. R. 1970. *Australian lizards*. Sydney: Collins.

Butler, W. H. 1970. A summary of the vertebrate fauna of Barrow Island, W.A. *W. Austr. Naturalist* 11:149–160.

Card, W. 1994. A reproductive history of monitors at the Dallas Zoo. *Vivarium* 6(1):26–27, 44, 45, 47.

———. 1995. Double clutching Gould's monitors (*Varanus gouldii*) and Gray's monitors (*Varanus olivaceus*) at the Dallas Zoo. *Herpetologica* 25:111–114.

Card, W., and A. G. Kluge. 1995. Hemipeneal skeleton and varanid lizard systematics. *J. Herpetol.* 29:275–280.

Carpenter, C. C., J. C. Gillingham, J. B. Murphy, and L. A. Mitchell. 1976. A further analysis of the combat ritual of the pygmy mulga monitor, *Varanus gilleni* (Reptilia, Varanidae). *Herpetologica* 32:35–40.

Carter, D. B. 1990. Courtship and mating in wild *Varanus varius* (Varanidae: Australia). *Mem. Queensl. Mus.* 29:333–338.

———. 1999a. Reproductive cycle of the lace monitor (*Varanus varius*). *Mertensiella* 11:129–135.

———. 1999b. Nesting and evidence of parental care by the lace monitor *Varanus varius*. *Mertensiella* 11:137–147.

Case, T. J., and T. D. Schwaner. 1993. Island/mainland body size differences in Australian varanid lizards. *Oecologia* 94:12–19.

Christian, T. 1977. Notes on *Varanus glebopalma*. *Vic. Herpetol. Soc. Newsl.* 6:11–13.

———. 1979. Notes on Spencer's monitor (*Varanus spenceri*). *Vic. Herpetol. Soc. Newsl.* 14:13–14.

Christian, K. A. 1995. *Varanus panoptes* and *Varanus gouldii* diet and predation. *Herpetol. Rev.* 26:146.

Christian, K. A., and C. R. Tracy. 1982. Reproductive behavior of the Galapagos land iguana *Conolophus pallidus* on Isla Santa Fe, Galapagos. In *Iguanas of the world: Behavior, ecology and conservation*, eds. G. M. Burghardt and A. S. Rand, 366–379. Park Ridge, N.J.: Noyes.

Christian, A., H.-G. Horn, and H. Preuschoft. 1994. Bipedie bei rezenten Reptilien. *Natur. Mus.* 124:45–56.

Christian, K. A., and K. E. Conley. 1994. Activity and resting metabolism of varanid lizards as compared to "typical" lizards. *Austr. J. Zool.* 42:185–193.

Christian, K. A., and B. W. Weavers. 1994. Analysis of the activity and energetics of the lizard *Varanus rosenbergi*. *Copeia* 1994:289–295.

———. 1996. Thermoregulation of monitor lizards in Australia: An evaluation of methods in thermal biology. *Ecol. Monogr.* 66:139–157.

Christian, K. A., L. Corbett, B. Green, and B. Weavers. 1995. Seasonal activity and energetics of two species of varanid lizards in tropical Australia. *Oecologia* 103:349–357.

Christian, K. A., and G. Bedford. 1996. Thermoregulation by the spotted tree monitor, *Varanus scalaris*, in the seasonal tropics in Australia. *J. Thermal Biol.* 21:67–73.

Christian, A., and T. Garland. 1996. Scaling of limb proportions in monitor lizards (Squamata: Varanidae). *J. Herpetol.* 30:219–230.

Christian, K. A., G. S. Bedford, and S. R. Shannahan. 1996a. The solar absorptance of some Australian lizards and its relationship to temperature. *Austr. J. Zool.* 44:59–67.

Christian, K. A., B. Green, G. Bedford, and K. Newgrain. 1996b. Seasonal metabolism of a small, arboreal monitor lizard, *Varanus scalaris*, in tropical Australia. *J. Zool.* 240:383–396.

Christian, K. A., B. Weavers, B. Green, and G. Bedford. 1996c. Energetics and water flux in a semi-aquatic lizard *Varanus mertensi*. *Copeia* 1996:354–362.

Christian, K. A., R. V. Baudinette, and Y. Pamula. 1997. Energetic costs of activity by lizards in the field. *Funct. Ecol.* 11:392–397.

Christian, K., G. Bedford, and T. Schultz. 1999. Energetic consequences of metabolic depression in tropical and temperate-zone lizards. *Austr. J. Zool.* 47:133–141.

Cloudsley-Thompson, J. L. 1972. Temperature regulation in desert reptiles. *Symp. Zool. Soc. Lond.* 31:39–59.

———. 1994. *Predation and defence amongst reptiles*. Somerset: R&A.

Cogger, H. G. 1967. *Australian reptiles in colour*. Sydney: Reed.

————. 1986. *Reptiles and amphibians of Australia*. 3rd ed. Sydney: Reed.

————. 1992. *Reptiles and amphibians of Australia*. 5th ed. Sydney: Reed.

————. 2000. *Reptiles and amphibians of Australia*. New rev. ed. Sydney: Reed.

Cogger, H. G., E. E. Cameron, and H. M. Cogger. 1983. *Zoological catalogue of Australia 1. Amphibia and Reptilia*. Canberra: Bureau of Flora and Fauna, Australian Government Publishing Service.

Cogger, H. G., E. E. Cameron, R. A. Sadlier, and P. Eggler. 1999. *The action plan for Australian reptiles*. Australian Nature Conservation Agency Endangered Species Program Project 124.

Covacevich, J., and P. Couper. 1994. Type specimens of frog and reptile species, Queensland Museum: Recent additions and new information. *Mem. Queensl. Mus.* 37:53–65.

Cowles, R. B. 1930. The life history of *Varanus niloticus* (L.) as observed in Natal, South Africa. *J. Entomol. Zool.* 22:3–31.

Czechura, G. 1980. The emerald monitor *Varanus prasinus* (Schlegel): An addition to the Australian mainland herpetofauna. *Mem. Queensl. Mus.* 20(1):103–109.

Delean, S. 1981. Notes on aggressive behaviour by Gould's goanna (*Varanus gouldii*) in captivity. *Herpetofauna* 12:31.

De Lisle, H. F. 1996. *The natural history of monitor lizards*. Malabar, Fla.: Krieger.

Doles, M., and W. Card. 1995. Delayed fertilization in the monitor lizard *Varanus gouldii*. *Herpetol. Rev.* 26(4):196.

Dryden, G. L., B. Green, D. King, and J. Losos. 1990. Water and energy turnover in a small monitor lizard, *Varanus acanthurus*. *Aust. Wildl. Res.* 17:641–646.

Dunson, W. A. 1974. Salt gland secretion in a mangrove monitor lizard. *Comp Biochem. Physiol.* 47A:1245–1255.

Ehmann, H. 1992. *Encyclopedia of Australian animals: Reptiles*. Pymble. Australia: Angus and Robertson.

Ehmann, H., G. Swan, and G. S. B. Swan. 1991. Nesting, egg incubation and hatching by the heath monitor (*Varanus rosenbergi*) in a termite mound. *Herpetofauna* 21:17–24.

Eidenmüller, B. 1994. Bemerkungen zur Haltung und Zucht von *Varanus acanthurus* Boulenger, 1885, *V. storri* Mertens, 1966 und *V. gilleni* Lucas and Frost, 1895. *Herpetofauna* 16(88):6–12.

————. 1997. *Warane—Lebensweise*, Pflege, Zucht. Offenbach: Herpeton Verlag Elke Köhler.

————. 2001. Between the rocks: Tips on breeding and keeping King's rock monitor. *Reptiles* 9(5):79–80.

Eidenmüller, B., and H.-G. Horn. 1985. Eigene Nachzuchten und der gegenwärtige Stand der Nachzucht von *Varanus* (*Odatria*) *storri* Mertens, 1966. *Salamandra* 21(1):55–61.

Eidenmüller, B., and R. Wicker. 1997. The breeding of Gillen's pygmy monitor, *Varanus gilleni*, Lucas and Frost 1895. *Herpetofauna* 27(1):2–6.

Eidenmüller, B., and C. Langner. 1998. Bemerkungen zu Haltung und Zicht des Pilbara-Felsen Warans *Varanus pilbarensis* Storr 1980. *Herpetofauna* 20:5–10.

Estes, R. 1983. *Handbuch Paläoherpetologie/Encyclopedia Paleoherpetologie. Sauria terrestria, Amphisbaenia*. Stuttgart: Fischer.

Farlow, J. O., and E. R. Pianka. 2000. Body form and trackway pattern in Australian desert monitors (Squamata: Varanidae): Comparing zoological and ichnological diversity. *Palaios* 15:235–247.

Firth, B. T., and J. S. Turner. 1982. Sensory, neural and hormonal aspects of thermoregulation. In *Biology of the Reptilia*, ed. Carl Gans and F. Harvey Pough, 12:213–274. London: Academic Press.

Flower, S. S. 1937. Further notes on the duration of life in animals. III. Reptiles. *Proc. Zool. Soc. Lond.* 1937A:1–39.

Frappell, P. B., T. J. Schultz, and K. A. Christian. 2002a. Oxygen transfer during aerobic exercise in a varanid lizard, *Varanus mertensi,* is limited by circulation. *J. Exp. Biol.* 205:2725–2736.

Frappell, P. B., T. J. Schultz, and K. A. Christian. 2002b. The respiratory system in varanid lizards: Determinants of O_2 transfer. *Comp. Biochem. Physiol. A* 133:239–258.

Frauca, H. 1973. *Australian reptile wonders.* Rigby.

Fuller, S., P. Baverstock, and D. King. 1998. Biogeographic origins of goannas (Varanidae): A molecular perspective. *Mol. Phylogenet. Evol.* 9:294–307.

Garrett, C. M., and W. C. Card. 1993. Chemical discrimination of prey by naive neonate Gould's monitors *Varanus gouldii. J. Chem. Ecol.* 19:2599–2604.

Glauert, L. 1951. A new *Varanus* from East Kimberley, *Varanus mertensi* sp. n. *Western Aust. Nat.* 3:14–16.

Glazebrook, R. 1977. Old man goanna. *N. Queensl. Naturalist* 44(170): 4–6.

Gow, G. F. 1982. Notes on the reproductive biology of the pygmy mulga goanna *Varanus gilleni* Lucas and Frost 1895. *N. Terr. Naturalist* 5:4–5.

Gray, J. E. 1838. A catalogue of the slender tongued saurians, with descriptions of many new genera and species. *Ann. Mag. Nat. Hist.* 1:388–394.

———. 1845. *The zoology of the voyage of H.M.S. Erebus and Terror, under the command of Captain Sir James Clark Ross, during the years 1839 to 1843.* Vol. 2. London: Longman, Brown, Green and Longmans.

Green, B. 1969. Water and electrolyte balance in the sand goanna *Varanus gouldii* (Gray). Ph.D. thesis, University of Adelaide.

———. 1972a. Aspects of renal function in the lizard *Varanus gouldii. Comp. Biochem. Physiol.* 43A:747–756.

———. 1972b. Water losses of the sand goanna (*Varanus gouldii*) in its natural environments. *Ecology* 53:452–457.

Green, B., and D. King. 1978. Home range and activity patterns of the sand goanna, *Varanus gouldii* (Reptilia: Varanidae). *Austr. Wildl. Res.* 5:417–424.

Green, B., D. King, and H. Butler. 1986. Water, sodium and energy turnover in free-living perenties, *Varanus giganteus. Austr. Wildl. Res.* 13:589–595.

Green, B., G. Dryden, and K. Dryden. 1991. Field energetics of a large carnivorous lizard, *Varanus rosenbergi. Oecologia* 88:547–551.

Green, B., and K. Christian. 1997. Allometry in the metabolic rates of varanid lizards? Presented at the Third World Congress of Herpetology.

Green, B., M. McKelvey, and P. Rismiller. 1999. The behaviour and energetics of hatchling *Varanus rosenbergi. Mertensiella* 11:105–112.

Greer, A. E. 1989. *The biology and evolution of Australian lizards.* Chipping Norton, Australia: Surrey Beatty and Sons.

Guarino, F. 2001. Diet of a large carnivorous lizard, *Varanus varius. Wildl. Res.* 28:627–630.

Heatwole, H. 1976. *Reptile ecology.* St. Lucia: University of Queensland Press.

Heatwole, H., B. T. Firth, and G. J. W. Webb. 1973. Panting thresholds of lizards—I. Some methodological and internal influences on the panting threshold of an agamid, *Amphibolurus muricatus*. *Comp. Biochem. Physiol.* 46A:799–826.

Heatwole, H. F., and J. Taylor. 1987. *Ecology of reptiles*. 2nd ed. Chipping Norton, Australia: Surrey Beatty and Sons.

Heger, N. A. 2000. The impact of size on thermal efficiency: Size related costs and benefits in *Varanus giganteus*. Ph.D. diss., University of Texas, Austin.

Heger, N. A., and T. G. Heger. 1992. Investigation of potential census and survey methods and conservation related information for perenties (*Varanus giganteus*) in Cape Range National Park. Final report to the Department of Conservation and Land Management (CALM), Perth.

Hermes, N. 1981. Merten's water monitor feeding on trapped fish. *Herpetofauna* 13:34.

Holmes, R. S., M. King, and D. King. 1975. Phenetic relationships among varanid lizards based upon comparative electrophoretic data and karyotypic analysis. *Biochem. System. Ecol.* 3:257–262.

Horn, H.-G. 1978. Nachzucht von *Varanus gilleni*. (Reptilia: Sauria: Varanidae). *Salamandra* 14(1):29–32.

———. 1981. *Varanus spenceri*, nicht *Varanus giganteus*: Eine Richtigstellung (Reptilia: Sauria: Varanidae). *Salamandra* 17(1/2):78–81.

———. 1985. Beiträge zum Verhalten von Waranen: Die Ritualkämpfe von *Varanus komodoensis* Owens, 1912 und *Varanus semiremex* Peters, 1867 sowie die Imponierphasen der Ritualkämpfe von *V. timorensis timorensis* (Gray, 1831) und *V. t. similis* Mertens, 1958. *Salamandra* 21(2/3):169–179.

———. 1991. Breeding of the lace monitor (*Varanus varius*) for the first time outside of Australia. *Mertensiella* 2:168–175.

———. 1999. Evolutionary efficiency and success in monitors: A survey on behavior and behavioral strategies and some comments. *Mertensiella* 11:167–180.

Horn, H.-G., and U. Schurer. 1978. Bemerkungen zu *Varanus* (*Odatria*) *glebopalma* Mitchell 1955 (Reptilia: Sauria: Varanidae). *Salamandra* 14:105–116.

Horn, H.-G., and G. J. Visser. 1988. Freilandbeobachtungen und einige morphometrische Angaben zu *Varanus giganteus* (Gray, 1845). *Salamandra* 24(2/3):102–118.

———. 1990a. Australien, ein herpetologisches Traumland. *D. Aqu. Terr. Zeitschr. (DATZ)* 43(4):230–234 (Tl.2)

———. 1990b. Australien, ein herpetologisches Traumland. *D. Aqu. Terr. Zeitschr. (DATZ)* 43(5):296–299 (Tl. 3).

———. 1991. Basic data on the biology of monitors. *Mertensiella* 2:176–187.

———. 1997. Review of reproduction of monitor lizards *Varanus* spp. in captivity II. *Int. Zoo Yrbk.* 35:227–246.

———. 1989. Review of reproduction of monitor lizards (*Varanus* spp.) in captivity. *Int. Zoo Yrbk.* 29:140–150.

Horn, H.-G., M. Gaulke, and W. Böhme. 1994. New data on ritualized combats in monitor lizards (Sauria: Varanidae) with remarks on their function and phylogenetic implications. *Zool. Garten N.F.* 64(5):265–280.

Hoser, R. T. 1989. *Australian reptiles and frogs*. Sydney: Pierson.

———. 1999. Comments on the proposed conservation of the names

Hydrosaurus gouldii Gray, 1838 and *Varanus panoptes* Storr, 1980 (Reptilia, Squamata) by the designation of a neotype for *H. gouldii*. *Bull. Zoo. Nomenclature* 56(1):66–71.

Houston, T. F. 1978. *Dragon lizards and goannas of South Australia*. Special Educational Bulletin Series. Adelaide: South Australian Museum.

Houston, T., and M. Hutchinson. 1998. *Dragon lizards and goannas of South Australia*. Adelaide: South Australian Museum.

Husband, G. A. 1980. Notes on a nest and hatchlings of *Varanus acanthurus*. *Herpetofauna* 11(1):29–30.

———. 2001. Natural history and captive maintenance of the Northern bluntsoined monitor *Varanus primordius*. *Herpetofauna* 31:126–131.

Husband, G., and M. Vincent. 1999. Pygmy mulga monitor (*Varanus gilleni*). *Reptiles* 7(5):10–18.

Ingram, G. J., and R. J. Raven. 1991. *An atlas of Queensland's frogs, reptiles, birds and mammals*. Brisbane: Board of Trustees, Queensland Museum.

Irwin, B. 1986. Captive breeding of two species of monitor. *Thylacinus* 11(2):4–5.

Irwin, S. 1996a. Capture, field observations and husbandry of the rare canopy goanna. *Thylacinus* 21(2):12–19.

———. 1996b. Courtship, mating, and egg deposition by the captive perentie *Varanus giganteus* at Queensland Reptile and Fauna Park. *Thylacinus* 21:8–10.

———. 1997. Courtship, mating and egg deposition by the captive perentie *Varanus giganteus* (Gray, 1845). *Vivarium* 8(4):26–31

Irwin, S., K. Engle, and B. Mackness. 1996. Nocturnal nesting by captive varanid lizards. *Herpetol. Rev.* 27:192–194.

James, C. 1994. Spatial and temporal variation in structure of a diverse lizard assemblage in arid Australia. In *Lizard ecology: Historical and experimental perspectives*, ed. L. J. Vitt and E. R. Pianka, 287–317. Princeton, N.J.: Princeton University Press.

———. 1996. Ecology of the pygmy goanna (*Varanus brevicauda*) in spinifex grasslands of central Australia. *Austr. J. Zool.* 44:177–192.

James, C. D., J. B. Losos, and D. R. King. 1992. Reproductive biology and diets of goannas (Reptilia: Varanidae) from Australia. *J. Herpetol.* 26:128–136.

Johnson, C. R. 1976. Some behavioural observations on wild and captive sand monitors, *Varanus gouldii* (Sauria: Varanidae). *Zool. J. Linn. Soc.* 59:377–380.

Johnstone, R. E. 1983. Herpetofauna of the Hamersley Range National Park Western Australia. In *A fauna of the Hamersley Range National Park Western Australia*, ed. B. G. Muir, 7–11. Nedlands: W. A. National Parks Authority of Western Anstralia, 1980.

Jones, H. I. 1983a. Abbreviata (Nematoda: Physalopteroidea) in lizards of the *Varanus gouldii* complex (Varanidae) in Western Australia. *Austr. J. Zool.* 31:285–298.

———. 1983b. Prevalence and intensity of *Abbreviata travassos* (Nematoda: Physalopteridae) in the ridge-tailed monitor *Varanus acanthurus* Boulenger in Northern Australia. *Rec. W. Austr. Mus.* 11(1):1–9.

———. 1985. Gastrointestinal nematodes of the perentie, *Varanus giganteus* in Western Australia, with description of a new species of *Abbreviata travessos* (Nematoda: Physalopteridae). *Rec. W. Austr. Mus.* 12:379–387.

————. 1988. Nematodes from nine species of *Varanus* (Reptilia) from tropical northern Australia, with particular reference to the genus *Abbreviata* (Physalopteridae). *Austr. J. Zool.* 36:691–708.

————. 1991. Speciation, distribution and host-specificity of gastric nematodes in Australian varanid lizards. *Mertensiella* 2:195–203.

————. 1992. The gastric nematodes of *Varanus caudolineatus* (Reptilia: Varanidae) in Western Australia. *Rec. W. Aust. Mus.* 16(1):19–112.

————. 1995. Gastric nematode communities in lizards from the Great Victoria Desert, and an hypothesis for their evolution. *Austr. J. Zool.* 43:141–164.

Kennerson, K. J. 1979. Remarks on the longevity of *Varanus varius. Herpetofauna* 10:32.

Kennett, R., K. Christian, and D. Pritchard. 1993. Underwater nesting by the tropical chelid turtle *Chelodina rugosa* from northern Australia. *Austr. J. Zool.* 41:47–52.

Kierans, J. E., D. R. King, and R. D. Sharrad. 1994. *Aponomma* (*Bothriocroton*) *glebopalma*, n. subgen., n. sp., and *Amblyomma glauerti*, n. sp. (Acari: Ixodida: Ixodidae), parasites of monitor lizards (Varanidae) in Australia. *J. Med. Entomol.* 31:132–147.

King, D. R. 1977. Temperature regulation in the sand goanna *Varanus gouldii* (Gray). Ph.D. thesis, University of Adelaide.

————. 1980. The thermal biology of free-living sand goannas (*Varanus gouldii*) in Southern Australia. *Copeia* 1980:755–767.

————. 1981. The thermal biology of sand goannas (*Varanus gouldii*) in southern Australia. *Proc. Melbourne Herpetol. Symp.* 22–23.

————. 1997. Comment on the proposed conservation of the specific names of *Hydrosaurus gouldii* Gray, 1838 (currently *Varanus gouldii*) and *Varanus panoptes* Storr, 1980 (Reptilia, Squamata) by the designation of a neotype for *H. gouldii. Bull. Zoo. Nomenclature* 54:249–250.

————. 1999. The land of the lizards. *Geo (Australasia)* 21(3):22–31.

King, M. 1990. Chromosomal and immunogenetic data: A new perspective on the origin of Australia's reptiles. In *Cytogenetics of amphibians and reptiles*, ed. E. Olne, 153–180. Basel: Birkhäuser.

King, M., and D. R. King. 1975. Chromosomal evolution in the lizard genus *Varanus* (Reptilia). *Austr. J. Biol. Sci.* 28:89–108.

King, D., and B. Green. 1979. Notes on diet and reproduction of the sand goanna, *Varanus gouldii rosenbergi. Copeia* 1979:63–70.

————. 1993a. Dimensions of the perentie (*Varanus giganteus*) and other large varanids. *W. Austr. Naturalist* 19(3):195–200.

————. 1993b. Family Varanidae. In *Fauna of Australia*, ed. C. J. Glasby, G. B. J. Ross, and L. P. Beesly, 2A:255–260. Canberra: Australian Government Publishing Service.

————. 1993c. *Goanna: The biology of varanid lizards.* Kensington: New South Wales University Press.

————. 1999. *Goannas: The biology of varanid lizards.* 2nd ed. Sydney: New South Wales University Press. Published in the United States as *Monitors: The biology of varanid lizards.* Malabar, Fla.: Krieger.

King, D. R., and L. Rhodes. 1982. Sex ratio and breeding season of *Varanus acanthurus. Copeia* 1982:784–787.

King, M., and P. Horner. 1987. A new species of monitor (Platynota: Reptilia) from northern Australia and a note on the status of *Varanus insulanicus* Mertens. *Beagle* 4(1):73–79.

King, D. R., B. Green, and H. Butler. 1989. The activity pattern, temperature regulation and diet of *Varanus giganteus* on Barrow Island, Western Australia. *Austr. Wildl. Res.* 16:41–47.

King, D. R., M. King, and P. Baverstock. 1991. A new phylogeny of the Varanidae. *Mertensiella* 2:211–219.

King, D. R., S. Fuller, and P. Baverstock. 1999. The biogeographic origins of varanid lizards. *Mertensiella* 11:43–49.

Krebs, U. 1999. Experimental variation of breeding season and incubation time in the spiny-tailed monitor (*Varanus acanthurus*). *Mertensiella* 11:227–237.

Licht, P., W. R. Dawson, and V. H. Shoemaker. 1966. Observation on the thermal relations of Western Australian lizards. *Copeia* 1966:97–11.

Limpus, D. J. 1995. Observations of *Varanus gouldii* (Varanidae) at Mon Repos Beach, Bundaberg. *Herpetofauna* 25(2):14–16.

Losos, J. B., and H. W. Greene. 1988. Ecological and evolutionary implications of diet in monitor lizards. *Biol. J. Linn. Soc.* 35:379–407.

Loveridge, A. 1934. Australian reptiles in the Museum of Comparative Zoology, Cambridge, Massachusetts. *Bull. Mus. Comp. Zool.* 77:243–383.

Lucas, A. H. S., and C. Frost. 1895. Preliminary notice of certain new species of lizards from central Australia. *Proc. R. Soc. Vic.* 7:264–269.

———. 1896. Reptilia. In *The Horn scientific expedition to Central Australia. II. Zoology*, 112–151. London: Dulau.

———. 1903. Description of two new Australian lizards, *Varanus spenceri* and *Diplodactylus bilineatus*. *Proc. R. Soc. Vic.* 15(1):145–147.

Lydekker, R. 1888. *Catalogue of the fossil Reptilia and Amphibia in the British Museum (Natural History), Cromwell Road, S. W. Pt. 1. The Orders Ornithosauria, Crocodilia, Dinosauria, Squamata, Rhychocephalia, and Proterosauria*. London: The Trustees.

Magnuson, J. J., L. B. Crowder, and P. A. Medvick. 1979. Temperature as an ecological resource. *Am. Zool.* 19:331–343.

McPhee, D. R. 1979. *Snakes and lizards of Australia*. Sydney: Methuen.

Mertens, R. 1941. Zwei neue Warane des Australischen Faunengebietes. *Senck. Biol.* 23:266–272.

Mertens, R. 1942a. Die Familie der Warane (Varanidae). *Abh. Senck. Naturf. Ges.* 462, 466 467:1–391.

———. 1942b. Ein uciterer neuer Waran aus Australien. *Zool. Anz.* 137:41–44.

———. 1957a. Ein neuer melanistischer Waran aus dem südlichen Australien. *Zool. Anz.* 159:17–20

———. 1957b. Two new goannas from Australia. *W. Austr. Naturalist* 5:183–185.

———. 1958. Bemerkungen über die Warane Australiens. *Senck. Biol.* 39(5/6):229–264.

———. 1963. Liste der rezenten Amphibien and Reptilien. Helodermatidae, Varanidae. Lanthanotidae. *Tierreich. Lief.* 79: i–x, 1–26.

———. 1966. Ein neuer Zwergwaran von Australien. *Senck. Biol.* 47(6):437–441.

Minton, S. A., and M. R. Minton. 1973. *Giant reptiles*. New York: Scribner.

Mitchell, F. J. 1955. Preliminary account of the Reptilia and Amphibia collected by the National Geographic Society—Commonwealth Government—Smithsonian Institution expedition to Arnhem Land (April to November 1948). *Rec. S. Austr. Mus.* 11:373–408.

Mitchell, L. A. 1989. Reproduction of Gould's monitors (*Varanus gouldii*) at Dallas Zoo. *Bull. Chicago Herpetol. Soc.* 25(1):8–9.

Murphy, J. B., and L. A. Mitchell. 1974. Ritualized combat behavior of the pygmy mulga monitor lizard *Varanus gilleni* (Sauria, Varanidae). *Herpetologica* 30:251–260.

Overton, B. M. 1987. An observation of apparent Goanna (*Varanus rosenbergi*) predation on a juvenile echidna (*Tachyglossus aculatus* var.). *S. Austr. Naturalist* 62:84–85.

Owerkowicz, T., C. G. Farmer, J. W. Hicks, and E. L. Brainerd. 1999. Contribution of gular pumping to lung ventilation in monitor lizards. *Science* 284:1661–1663.

Pengilley, R. 1981. Notes on the biology of *Varanus spenceri* and *V. gouldii*, Barkly Tablelands, Northern Territory. *Aust. J. Herpetol.* 1(1):23–26.

Peters, W. 1869. Über neue Gattungen und Arten von Eidechsen. *Mber. K. Preuss. Akad. Wiss. Berl.* 1869:57–66.

Peters, U. 1968. *Moloch horridus, Varanus spenceri, V. mitchelli, Egernia bungana,* und *Heteronotia binoei* im Taronga-Zoo, Sydney. *Aquar. Terr. Zeit.* 21(8):252–254.

———. 1969. Zum ersten Mal in Gefangenschaft: Eiablage und Schlupf von *Varanus spenceri. Aquar. Terrar.* 16(9):306–307.

———. 1971. The first hatching of *Varanus spenceri* in captivity. *Bull. Zoo. Man.* 3(2):17–18

———. 1973. A contribution to the ecology of *Varanus (Odatria) storri. Koolewong* 2:12–13.

Pianka, E. R. 1968. Notes on the biology of *Varanus eremius. W. Austr. Naturalist* 11:39–44.

———. 1969. Notes on the biology of *Varanus caudolineatus* and *Varanus gilleni. W. Austr. Naturalist* 11:76–82.

———. 1970a. Notes on the biology of *Varanus brevicauda. W. Austr. Naturalist* 11:113–116.

———. 1970b. Notes on the biology of *Varanus gouldii flavirufus. W. Austr. Nauralist* 11:141–144.

———. 1971. Notes on the biology of *Varanus tristis. W. Austr. Naturalist* 11:180–183.

———. 1982. Observations on the ecology of *Varanus* in the Great Victoria Desert. *W. Austr. Naturalist* 15:37–44.

———. 1986. *Ecology and natural history of desert lizards: Analyses of the ecological niche and community structure.* Princeton, N.J.: Princeton University Press.

———. 1994a. Comparative ecology of *Varanus* in the Great Victoria Desert. *Aust. J. Ecol.* 19:395–408.

———. 1994b. *The lizard man speaks.* Austin: University of Texas Press.

———. 1995. Evolution of body size: Varanid lizards as a model system. *Am. Naturalist* 146:398–414.

———. 1996. Long-term changes in lizard assemblages in the Great Victoria Desert: Dynamic habitat mosaics in response to wildfires. In *Long-term studies of vertebrate communities,* ed. M. L. Cody and J. A. Smallwood, 191–215. New York: Academic Press.

Polleck, R. 1980. Ein temperamentvoller Zwerg im Terrarium: Der Gillenwaran (*Varanus gilleni*) Lucas and Frost 1895. *Herpetofauna* 2(6):19–20.

———. 2001. Haltung und Zucht vom Gillenwaran, *Varanus gilleni* Lucas and Frost, 1895. *Herpetofauna* 23(131):15–18.

Pough, F. H., and C. Gans. 1982. The vocabulary of reptilian thermoregulation. In *Biology of the Reptilia,* ed. C. Gans and F. H. Pough, 12:17–23. London: Academic Press.

Regal, P. J. 1978. Behavioral differences between reptiles and mammals: An analysis of activity and mental capabilities. In *Behavior and neurology of lizards,* ed. N. Greenberg and P. D. MacLean, 183–202. Rockville, Md.: National Institute of Medical Health.

Rieppel, O. 1979. A functional interpretation of the varanid dentition (Reptilia, Lacertilia, Varanidae). *Gegenbaurs Morphol. Jahrb.* 125: 797–817.

Rismiller, P., M. McKelvey, and B. Green. 1997. Body temperature and activity patterns of *Varanus rosenbergi* during the breeding season. Presented at the Third World Congress of Herpetology.

Robinson, A. C. 1992. Perenties, predators and prey. *S. Austr. Naturalist* 67:30–34.

Rokylle, G. 1989. An addition to the ranges of two rock monitors: *Varanus glauerti* and *Varanus glebopalma. Herpetofauna* 7:4–8.

Ruegg, R. 1974. Nachzucht beim Timor-Baumwaran, *Varanus timorensis similis* Mertens 1958. *Aquarium* 8:360–363.

Saint Girons, H., G. E. Rice, and S. D. Bradshaw. 1981. Histologie comparée et ultrastructure de la glande nasale externe de quelques Varanidae (Reptilia: Lacertilia). *Ann. Sci. Nat. Zool. Paris* 12(3):15–21.

Schlegel, H. 1839. *Abbildungen neuer oder unvollstandig bekannter Amphibien, nach der Natur der dem Leben entworfen.* Dusseldorf: Arnz.

Schmida, G. E. 1974. Die Kurzschwanzwaren (*Varanus brevicauda*). *Aquar. Terr. Zeit.* 27:390–394.

———. 1975. Däumlinge aus der Familie der Riesen, Kurzschwanz-, Gillen- und Schwanzstrichwaran. *Aquarrien-Mag.* 9(1):8–11.

———. 1985. *The cold blooded Australians.* Sydney: Doubleday.

Schultz, T. 2002. Oxygen transport in varanid lizards during exercise. Ph.D. thesis, Northern Territory University, Darwin.

Seebacher, F. 2000. Heat transfer in a microvascular network: The effect of heart rate on heating and cooling in reptiles (*Pogona barbata* and *Varanus varius*). *J. Theor. Biol.* 203:97–109.

Sharrad, R. D., and D. R. King. 1981. The geographic distribution of reptile ticks in Western Australia. *Austr. J. Zool.* 29:861–873.

Shea, G. M., and G. L. Reddacliff. 1986. Ossifications in the hemipenis of varanids. *J. Herpetol.* 20(4):566–568.

Shea, G. M., and H. G. Cogger. 1998. Comment on the proposed conservation of the names *Hydrosaurus gouldii* Gray, 1838 and *Varanus panoptes* Storr, 1980 (Reptilia, Squamata) by the designation of a neotype for *Hydrosaurus gouldii. Bull. Zoo. Nomenclature* 55:106–111.

Shea, G. M., and R. A. Sadlier. 2001. An ovigerous argus monitor, *Varanus panoptes panoptes. Herpetofauna* 31:132–133.

Shine, R. 1986. Food habit, habitats and reproductive biology of four sympatric species of varanid lizards in tropical Australia. *Herpetologica* 42(3):346–336.

Smith, M. J. 1976. Small fossil vertebrates from Victoria Cave, Naracoorte, South Australia. IV. Reptiles. *Trans. R. Soc. S. Austr.* 100:39–51.

Smith, L. A. 1988. Notes on a clutch of monitor (*Varanus caudolineatus*) eggs. *W. Austr. Naturalist* 17(5):96.

Sprackland, R. G. 1991. Taxonomic review of the *Varanus prasinus* group with descriptions of two new species. *Mem. Queensl. Mus.* 30:561–576.

Sprackland, R. G., H. M. Smith, and P. D. Strimple. 1997. *Hydrosaurus gouldii* Gray, 1838 (currently *Varanus gouldii*) and *Varanus panoptes*

Storr, 1980 (Reptilia, Squamata): Proposed conservation of the specific names by designation of a neotype for *H. gouldii*. *Bull. Zoo. Nomenclature* 54(2):95–99.

Stammer, D. 1970. Goannas. *Wildl. Austr.* 7:118–120.

Stebbins, R. C., and R. E. Barwick. 1968. Radiotelemetric study of thermoregulation in a lace monitor. *Copeia* 1968:541–547.

Stirling, E. C. 1912. Observations on the habits of the large central Australian monitor (*Varanus giganteus*), with a note on the "fat bodies" of this species. *Trans. R. Soc. S. Austr.* 36:26–33.

Stirnberg, E., and H.-G. Horn. 1981. Eine Unerwartete Nachzucht im Terrarium: *Varanus* (*Odatria*) *storri*. *Salamandra* 17:55–62.

Storr, G. M. 1966. Rediscovery and taxonomic status of the Australian lizard *Varanus primordius*. *Copeia* 1966:583–584.

———. 1980. The monitor lizards (genus *Varanus* Merrem, 1820) of Western Australia. *Rec. W. Austr. Mus.* 8:237–293.

Storr, G. M., L. A. Smith, and R. E. Johnstone. 1983. *Lizards of Western Australia. II. Dragons and monitors.* Perth: Western Australia Museum.

Strimple, P. D. 1988. *Varanus giganteus* (Gray, 1845). *Forked Tongue* 13(2):5–15.

Strimple, P. D., and J. L. Strimple. 1996. Australia's largest goanna, the perentie. *Reptiles* 1996(February):76–84.

Sutherst, R. W., R. B. Floyd, and G. F. Maywold. 1996. The potential geographic distribution of the cane toad, *Bufo marinus* L. in Australia. *Conserv. Biol.* 9:294–299.

Swanson, S. 1979. Some rock-dwelling reptiles of the Arnhem Land escarpment. *N. Terr. Naturalist* 1:14–18.

———. 1987. *Lizards of Australia.* North Ryde, Australia: Angus and Robertson.

Sweet, S. S. 1999. Spatial ecology of *Varanus glauerti* and *V. glebopalma* in Northern Australia. *Mertensiella* 11:317–366.

Tasoulis, T. 1992. Nesting observations on the lace monitor *Varanus varius*. *Herpetofauna* 22:46.

Templeton, J. R. 1970. Reptiles. In *Comparative physiology of thermoregulation*, vol. 1, *Invertebrates and nonmammalian vertebrates,* ed. G. C. Whittow, 167–221. New York: Academic Press.

Thissen, R. 1992. Breeding the spiny-tail monitor (*Varanus acanthurus* Boulenger). *Vivarium* 3(5):32–34.

Thompson, G. G. 1992. Daily distance travelled and foraging areas of *Varanus gouldii* (Reptilia: Varanidae) in an urban environment. *Wildl. Res.* 19:743–753.

———. 1993. The behavioural ecology of *Varanus caudolineatus* (Reptilia: Varanidae). *Wildl. Res.* 20:227–231.

———. 1994. Activity area during the breeding season of *Varanus gouldii* (Reptilia: Varanidae) in an urban environment. *Wildl. Res.* 21:633–641.

———. 1995. Foraging patterns and behaviours, body postures and movement speed for goannas, *Varanus gouldii* (Reptilia: Varanidae), in a semi-urban environment. *J. R. Soc. W. Austr.* 78:107–114.

———. 1996. Notes on the diet of *Varanus gouldii* in a semi-urban environment. *W. Austr. Naturalist* 21(1):49–54.

———. 1997. Do training and captivity affect maximal metabolic rate of *Varanus gouldii* (Squamata: Varanidae)? *Amphibia-Reptilia* 18:112–116.

————. 1999. Goanna metabolism: Different to other lizards and if so, what are the ecological consequences? *Mertensiella* 11:79–90.

Thompson, G. G., and P. C. Withers. 1992. Effects of body mass and temperature on standard metabolic rates for two Australian varanid lizards (*Varanus gouldii* and *V. panoptes*). *Copeia* 1992:343–350.

————. 1994. Standard metabolic rates of two small Australian varanid lizards (*Varanus caudolineatus* and *V. acanthurus*). *Herpetologica* 50(4):494–502.

————. 1997a. Evaporative water loss of Australian goannas (Squamata: Varanidae). *Amphibia-Reptilia* 18:177–190.

————. 1997b. Standard and maximal metabolic rates of goannas (Squamata: Varanidae). *Physiol. Zool.* 70:307–323.

————. 1997c. Comparative morphology of Western Australian varanid lizards (Squamata: Varanidae). *J. Morphol.* 233:127–152.

————. 1997d. Patterns of gas exchange and extended non-ventilatory periods in small goannas (Squamata: Varanidae). *Comp. Biochem. Physiol.* 118A(4):1411–1417.

————. 1998. Standard evaporative water loss and metabolism of juvenile *Varanus mertensi* (Squamata: Varanidae). *Copeia* 1998:1054–1059.

Thompson, G. G., P. C. Withers, and S. A. Thompson. 1992. The combat ritual of two monitor lizards, *Varanus caudolineatus* and *Varanus gouldii*. *W. Austr. Naturalist* 19(1):21–25.

Thompson, G., and D. King. 1995. Diet of *Varanus caudolineatus* (Reptilia: Varanidae). *W. Austr. Naturalist* 20(4):199–204.

Thompson, G. G., N. A. Heger, T. G. Heger, and P. C. Withers. 1995. Standard metabolic rate of the largest Australian lizard, *Varanus giganteus*. *Comp. Biochem. Physiol.* 111A:603–608.

Thompson, G. G., S. D. Bradshaw, and P. C. Withers. 1997. Energy and water turnover rates of a free-living and captive goanna, *Varanus caudo-lineatus* (Lacertilia: Varanidae). *Comp. Biochem. Physiol.* 116A(2):105–111.

Thompson, G. G., and E. R. Pianka. 1999. Reproductive ecology of the black-headed goanna *Varanus tristis* (Squamata: Varanidae). *J. R. Soc. W. Austr.* 82:27–31.

————. 2001. Allometry of clutch and neonate sizes in monitor lizards (Varanidae: *Varanus*). *Copeia* 2001:443–458.

Thompson, G. G., M. de Boer, and E. R. Pianka. 1999a. Activity areas and daily movements of an arboreal monitor lizard, *Varanus tristis* (Squamata: Varanidae) during the breeding season. *Austr. J. Ecol.* 24:117–122.

Thompson, G. G., E. R. Pianka, and M. de Boer. 1999b. Thermoregulation of an arboreal monitor lizard, *Varanus tristis* (Squamata: Varanidae) during the breeding season. *Amphibia-Reptilia* 20:82–88.

Thompson, G. G., and S. A. Thompson. 2002. Clutch of *Varanus caudolineatus* (Varanidae). *W. Austr. Naturalist* 23(3):228.

Thompson, S. A., P. C. Withers, G. G. Thompson, and D. Robinson. 2003. Range extension for the perentie, *Varanus gigantus*. *W. Austr. Naturalist* 23 (in press).

Thomson, D. F., and W. Hosmer. 1963. A preliminary account of the herpetology of the Great Sandy Desert of central Western Australia. *Proc. R. Soc. Vic.* 77:217–237.

Valentic, R. A. 1995. Further instances of nocturnal activity in agamids and varanids. *Herpetofauna* 25:49–50.

———. 1997. Diet and reproductive status on a roadkill Spencer's monitor, *Varanus spenceri*. *Herpetofauna* 27(2):43–45.

Vincent, M., and S. Wilson. 1999. *Australian goannas*. Sydney: New Holland.

Visser, G., and H.-G. Horn. 1989. Bijzondere dieren van Australie: De reuzen varaan. *Dieren* 5:176–178.

Waite, E. R. 1929. *The reptiles and amphibians of South Australia*. Adelaide: British Science Guild.

Weavers, B. W. 1983. Thermal ecology of *Varanus varius* (Shaw), the lace monitor. Ph.D. thesis, Australian National University, Canberra.

———. 1988. Vital statistics of the lace monitor lizard (*Varanus varius*) in south-eastern Australia. *Vic. Naturalist* 105:142–5.

———. 1989. Diet of the lace monitor lizard (*Varanus varius*) in south-eastern Australia. *Austr. Zool.* 25:83–85.

———. 1993. Home range of male lace monitors, *Varanus varius* (Reptilia: Varanidae), in south-eastern Australia. *Wildl. Res.* 20:303–13.

Webb, J. K. 1994. Observation of three dingoes killing a large lace monitor (*Varanus varius*). *Austr. Mammal.* 19:55–56.

Wells, R. W., R. C. and Wellington. 1985. A classification of the Amphibia and Reptilia in Australia. *Austr. J. Herpetol. Suppl. Ser.* 1:1–61.

Werner, Y. L. 1988. Are hemipeneal "ossifications" of Gekkonidae and *Varanus* ossified? *Isr. J. Zool.* 35:99–100.

Whitford, W. G. 1998. Contribution of pits dug by goannas (*Varanus gouldii*) to the dynamics of banded mulga landscapes in eastern Australia. *J. Arid Environ.* 40:453–457.

Wilson, S. K., and D. G. Knowles. 1988. *Australia's reptiles: A photographic reference to the terrestrial reptiles of Australia*. Sydney: Collins.

Worrell, E. 1963. *Reptiles of Australia*. Sydney: Angus and Robertson.

Ziegler, T., and W. Böhme. 1997. Genitalstrukturen und Paarungsbiologie bei squamaten Reptilien, speziell der Platynota, mit Bemerkungen zur Systematik. *Mertensiella* 8:3–207.

8. Other Varanoids

Three other varanoids still exist today, although all three are endangered as a result of habitat destruction. Two species of helodermatids occur in the southwestern deserts of the United States and western Mexico. Helodermatidae is the sister group to a clade consisting of the little-known *Lanthanotus* from Borneo. Varanids, in turn, are the sister group to the *Heloderma* and *Lanthanotus* clade. A species account is also included for the Mongolian fossil monstersaurian *Estesia*.

8.1. Overview of the Family Helodermatidae for Varanophiles

DANIEL D. BECK

No extant varanid lizards are found in the New World. This unfortunate situation is partly remedied by the occurrence of their sister group, the Helodermatidae, in North and Central America. Helodermatids are the New World equivalent of a sluggish, venomous monitor lizard. They are similar to *Varanus* in many respects, such as several skeletal features, their spectacular ritualized male combat behavior, osteoderms (a diagnostic trait of *Heloderma* and also present in *Lanthanotus* as well as many species of *Varanus*), and their specialized chemosensory systems associated with forked tongues.

As a member of the Varanoidea, *Heloderma* shares a more recent common ancestor with the Varanidae than with any other extant lizard group. Helodermatid lizards, especially the Mexican beaded lizard, *Heloderma horridum*, have a resemblance to varanid lizards that is more than merely superficial. Their demeanor, gait, and general body form are sufficiently reminiscent of *Varanus* that many familiar with *H. horridum* call it the "Mexican monitor." For this work, the Helodermatidae might thus be informally referred to as the "Helodermonitor" lizards.

Comparisons with Varanid Lizards

Upon closer examination, however, many traits distinguish helodermatid from varanid lizards. With only two extant species (*Heloderma horridum* and *H. suspectum*), the Helodermatidae is a depauperate family when compared to the 50-plus species found within the Varanidae.

Helodermatids have a number of skeletal characteristics that set them apart from the varanids, including a steep nasal process of the maxilla, which gives them a more rounded muzzle and shorter face than *Varanus* (Estes et al. 1988). Helodermatid lizards have shorter tails than varanids, as well as a shorter, fatter tongue that includes a villose basal portion and no sheath (Bogert and Martín del Campo 1956; Schwenk 1988). Helodermatid lizards are unique in having grooved teeth and venom glands, although the precursor of *Heloderma*'s venom apparatus, the gland of Gabe, is also possessed by varanid lizards (Kochva 1974; Gabe and Saint Girons 1976; Pregill et al. 1986).

In addition to morphology, helodermatids differ from varanids in a number of physiological and behavioral characteristics. Although varanids are well known for their speed, high activity, and relatively high metabolic rates (Bennett 1983), helodermatids are slow moving and characterized by low levels of activity and low rates of metabolism during rest (Beck 1990; Beck and Lowe 1991, 1994; Beck et al. 1995). Interestingly, however, helodermatid lizards, like varanids, have a high capacity for sustained aerobic activity, a trait that may be advantageous during their strenuous male combat rituals (see below) (Beck 1990; Beck and Ramírez-Bautista 1991; Beck et al. 1995).

Helodermatid lizards do not tolerate the high body temperatures sustained by varanid lizards. Their preferred temperature during activity and basking is around 30°C. They begin seeking refuge at temperatures approaching 38°C and become paralyzed at 42°C, which is near their critical thermal maximum. Varanid lizards take a much wider variety of prey items than *Heloderma* but do not show the same capacity to ingest relatively large prey (Pregill et al. 1986). Even predation by varanids on large vertebrates, for example by *Varanus komodoensis* (Auffenberg 1981), involves tearing the flesh with serrated teeth and ingesting pieces, rather than large, intact morsels (Pregill et al. 1986). Helodermatids, on the other hand, show dietary specialization: they prefer the eggs and young of vertebrates, which they take from nests, and have the ability to swallow relatively large prey, like nestling cottontail rabbits. It is thought that their skull architecture (and perhaps even their venom system) set *Heloderma* on a track separate from other varanoids early in their evolutionary history toward a more sedentary lifestyle and a diet on immobile, larger, intact prey items (Pregill et al. 1986).

Evolution and the Fossil Record

The Helodermatidae has a rich and diverse evolutionary history that extends well into the Cretaceous across Europe, Asia, and North America. The fossil record shows that the two remaining species of helodermatid lizards are relics of a more diverse lineage that included at least six other genera (*Lowesaurus, Eurheloderma, Paraderma, Gobiderma, Estesia, Primaderma*) inhabiting subtropical desert, forest, and savanna habitats (Gilmore 1928; Hoffstetter 1957; Estes 1964; Yatkola 1976; Borsuk-Białynicka 1984; Pregill et al. 1986; Cifelli and Nydam 1995; Norell and Gao 1997; Nydam 2000). The genus *Heloderma* has existed since at least the early Miocene (~23 mya). Fossils of *Heloderma* include *H. texana* from the early Miocene near Big Bend, Texas (Stevens 1977), and *H. suspectum* from late Pleistocene deposits (8000 to 10,000 years old) near Las Vegas, Nevada (Brattstrom 1954).

The helodermatid clan, a well-supported lizard clade recently named the Monstersauria, dates back 98 million years to a time well before many dinosaurs had appeared (Norell and Gao 1997; Nydam 2000). Family members somehow managed to survive the great Cretaceous extinctions, which vanquished the dinosaurs 65 mya. Helodermatid lizards have undergone relatively little gross morphological change over this time and many regard them as living fossils (Pregill et al. 1986; Nydam 2000).

The Venom System

Helodermatids are the only lizards known to be venomous. The venom system consists of paired venom glands that empty through ducts at the base of venom-conducting teeth. The venom glands of helodermatid lizards are visible externally as conspicuous swellings below the lower lips. The venom glands, unlike those of venomous snakes, are multilobed, located in the lower jaw, and drain through ducts associated with each of the lobes (Loeb et al. 1913). Snake venom glands are situated behind the eye above the upper jaw and drain through a single duct that leads to openings at the base of the fangs (Greene 1997). The venom glands of *Heloderma* are not surrounded by the compressor musculature possessed by most venomous snakes. Instead, tension within the glands produced by jaw movements propels venom into the vicinity of the venom-conducting teeth, and capillary action carries the venom from grooved teeth into the wound. Each specialized tooth normally has two grooves, one anteriorly and another (sometimes absent) posteriorly. Each groove is flanked by a cutting flange, which makes the tooth better adapted for piercing flesh than a merely conical tooth. The largest, most deeply grooved teeth are the fourth to seventh pair of dentaries (in the lower jaw, counting from the front), which can be up to 6 mm long in *H. horridum* and 5 mm long in *H. suspectum*. The maxillary teeth (in the upper jaw) are shorter and less strongly grooved.

Helodermatids cannot quickly sprint away from a potential threat, as can most other lizards (Beck et al. 1995), which necessitates an effective defense during their infrequent, although occasionally extensive, above-ground forays. An elaborate venom system is not needed to subdue defenseless young or eggs in vertebrate nests, but it does serve an important function for defense (Beck 1990). The venomous bite of helodermatid lizards serves to deter and incapacitate an adversary by means of a diversity of proteins and peptides that manipulate the adversary's physiology in a variety of ways. Major symptoms include excruciating pain, edema, and weakness often accompanied by a rapid drop in blood pressure. Over a dozen proteins and peptides have been isolated from the venom of *Heloderma*. These include hyaluronidase (a spreading factor), serotonin, phospholipase A_2, several important bioactive peptides, and several kallikrein-like glycoproteins (Raufman 1996; Tu 2000). Kallikreins are hypotensive enzymes that release bradykinins—powerful local hormones that produce pain and inflammation (Greger 1996). The kallikrein-like glycoproteins are largely responsible for the savage pain and edema resulting from bites by *Heloderma*.

Four lethal toxins have been isolated from *Heloderma* venom, some of which are also kallikrein-like glycoproteins. One of these, horridum toxin, causes hemorrhage in internal organs and bulging of the eyes (Nikai et al. 1988; Datta and Tu 1997); another, helothermine, causes lethargy, partial paralysis of the limbs, and hypothermia in rats (Mochca-Morales et al. 1990).

The most fascinating constituents of *Heloderma* venom are the bioactive peptides: helodermin, helospectin, exendin-3, and exendin-4. Most are similar in structure and action to vasoactive intestinal peptide (VIP), a mammalian hormone secreted by nerves found throughout the gastrointestinal tract. VIP is a powerful relaxant of smooth muscle (hence its name) and mediates the secretion of water and electrolytes by the small and large intestines. Many of the bioactive peptides in *Helo-*

derma venom bind to VIP receptors in a number of human tissues in the gastrointestinal tract, lungs, and even on human breast cancer cells (Raufman 1996). Helodermin has even been shown to inhibit growth and multiplication of lung cancer cells (Maruno and Said 1993).

The best-known constituent of *Heloderma* venom, exendin 4 (isolated only from venom of the Gila monster) has enjoyed notoriety as a promising new treatment for type 2 diabetes. Exendin-4 shows strong structural homology to human glucagon-like peptide-1 (GLP-1), a hormone that stimulates the release of insulin and helps to moderate blood glucose levels. Exendin-4 is even more effective at inducing insulin release than is GLP-1 and has a much longer biological action (Doyle and Egan 2001). These traits have jettisoned exendin-4 into the forefront of pharmacological research on the treatment of diabetes (Seppa 2001).

Physiological Ecology

Helodermatid lizards are well known for their ability to store fat in their tails, which in well-fed individuals become quite rotund. They show a preference for a relatively low body temperature during activity and spend a good deal of the year at temperatures below 25°C (for *H. horridum*) or 20°C (for *H. suspectum*). During rest, helodermatids have among the lowest metabolic rates of any lizard measured (Beck and Lowe 1994). With a Q_{10} of 3, metabolic savings are even greater at reduced temperatures (a Q_{10} of 3 suggests that metabolic rate decreases threefold with a 10°C drop in temperature). Helodermatid lizards are therefore well adapted for inhabiting harsh, seasonal environments (such as deserts and tropical dry forests), showing a presence on the surface only during relatively rare bouts of surface activity. Their low metabolic rates, low activity levels, ability to ingest large meals, and great capacity for fat storage enable helodermatid lizards to subsist long periods without feeding, and to exploit a resource, the contents of vertebrate nests, upon which few other vertebrates specialize.

Helodermatid lizards also have a high capacity to sustain aerobic activity (John-Alder et al. 1983; Beck et al. 1995). This, coupled with their low resting metabolic rates, gives *Heloderma* among the highest factorial aerobic scope (30.4) of any lizard measured (Beck et al. 1995). A high aerobic capacity may seem surprising for such a sedentary lizard, but it may be advantageous during the intensive and physically exhausting combat rituals of male *Heloderma*. Males with superior strength and endurance more frequently emerge as winners during combat encounters and may thereby enjoy greater reproductive success. Male *Heloderma* have also been shown to have higher aerobic capacities than do females (Beck et al. 1995), which suggests that sexual selection may have played a role in shaping the high aerobic capacities of Helodermonitor lizards.

Although contemporary *Heloderma* and *Varanus* still show many similarities, the two seem to have diverged early in their evolutionary history. The ecology and behavior of the two species of fascinating Helodermonitor lizards are outlined in the following species accounts.

8.2. *Heloderma horridum* (Wiegmann 1829)

DANIEL D. BECK

Figure 8.1. The Mexican beaded lizard, Heloderma horridum *(photo by Dan Beck).*

Synonomy

Type

Trachyderma horridum (Wiegmann 1829). Although there is some uncertainty about the actual location of the type, it apparently came from the state of Morelos, Mexico, near the town of Huajintlán (Bogert and Martín del Campo 1956) (Fig. 8.1).

Nomenclature

Heloderma horridum (Wiegmann 1829). Mexican beaded lizard or *escorpión*. Noting that *Trachyderma horridum* was preoccupied, Wiegmann proposed the name *Heloderma* for the genus (Bogert and Martín del Campo 1956).

Heloderma horridum exasperatum (Bogert and Martín del Campo 1956). Río Fuerte beaded lizard, or *escorpión*. Type locality: "near village of Guirocoba, Sonora," Mexico.

Heloderma horridum alvarezi (Bogert and Martín del Campo 1956). Río Grijalva, Chiapan, or black beaded lizard or *escorpión*. Type locality: "immediate vicinity of Tuxtla Gutiérrez, Chiapas," Mexico.

Heloderma horridum charlesbogertii (Campbell and Vannini 1988). Motagua Valley beaded lizard or *escorpión*. Type locality: "Espíritu Santo, 17 km E El Rancho, Departmento de El Progreso, Guatemala, elevation 300 m."

Etymology

Heloderma is derived from the Greek words *"helos"* for a nail stud and *"derma"* for skin; *"horridum"* is derived from Latin *"horridus"* for horrible; hence, "the horrible one with studded skin."

Common Names

The standard common name for *Heloderma horridum* is the Mexican beaded lizard. In Mexico, it is commonly called *"escorpión"* as well as *"heloderma negro"* and *"lagarto enchaquirado"* (Spanish for beaded lizard). Numerous local names are also used by regional Indians and villagers.

Closely Related Species

Heloderma horridum is sympatric with *H. suspectum* in the northern part of its range in southern Sonora, Mexico (Beck 1993). However, differences in habitat preferences between the two species probably keep them from occupying the same microhabitat. *Heloderma horridum* is a longer, lankier, more arboreal lizard than is *H. suspectum*. The two species are readily distinguished from each other by the proportionately longer tail of *H. horridum* (at least 65% of the snout–vent length [SVL]; no more than 55% in *H. suspectum*) and by conspicuously enlarged postanal scales (usually two) in *H. suspectum*. *Heloderma horridum* normally have no enlarged postanal scales, although some individuals from Guatemala are an exception (Campbell and Vannini 1988). The two species also show differences in number of infralabials that are in contact with the chin shields: one pair in *H. suspectum*, two pairs in *H. horridum* (Bogert and Martín del Campo 1956).

Geographic Distribution

Heloderma horridum occurs from sea level to about 1600 m along the Pacific foothills of Mexico from southern Sonora to Chiapas, along Pacific drainages in southern Guatemala, and along two Atlantic drainages in Chiapas and eastern Guatemala (Bogert and Martín del Campo 1956; Campbell and Vannini 1988; Campbell and Lamar 1989, 2003). Four subspecies are recognized: *H. h. alvarezi* from the Rio Grijalva valley of central Chiapas, Mexico, to extreme western Guatemala; *H. h. charlesbogerti* from the Río Motagua Valley and adjacent foothills of eastern Guatemala; *H. h. exasperatum* in southern Sonora and northern Sinoloa, Mexico; and *H. h. horridum* from the remainder of the species range in western Mexico (Fig. 8.2).

Habitat

Heloderma horridum inhabits primarily tropical deciduous forest (=dry forest) and tropical thornscrub, occurring less frequently in lower pine–oak woodland. It frequents relatively open sandy and rocky ar-

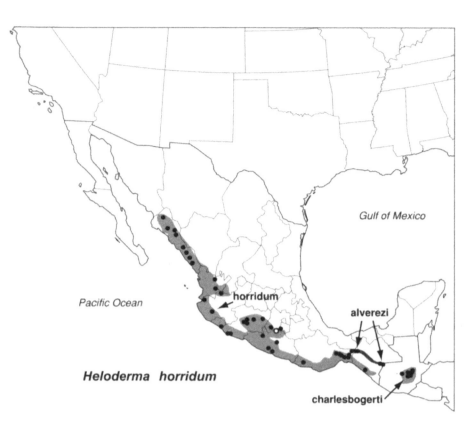

Pacific Ocean

Gulf of Mexico

horridum

alverezi

Heloderma horridum

charlesbogerti

Figure 8.2. Geographic distribution of beaded lizards (based on a map from Campbell and Lamar 2003; used with permission).

royos as well as densely vegetated upland hilly terrain and plateaus. Within tropical dry forest, it frequents both deciduous upland forest and semievergreen forests of arroyos and canyon bottoms, although upland forest seems more commonly preferred (Beck and Lowe 1991). Like Gila monsters, beaded lizards spend the vast majority of their time hidden in shelters; therefore, shelters are very important habitat features. Arroyos, rocky slopes, and the forest floor, including both dense and more open understory vegetation, all provide suitable substrata for beaded lizard shelters. Hollow trunks and branches within trees are also used as shelters, especially during rainy periods.

Fossil Record

See the summary of fossil history in the introductory section on helodermatid lizards. No fossils of *H. horridum* have been reported.

Description

Heloderma horridum is a large, stout, ground-dwelling, and semi-arboreal lizard. Juveniles have a background color of black or dark brown with spots and bars or blotches of yellow. Head, neck, and limbs are mostly dark; the tail is distinctly banded with alternating dark and yellow. Adult pattern varies from reticulated yellow crossbands to mostly dark brown or slate gray. No differences between males and females are known in coloration or patterns. Males have slightly bulkier heads than females, but this difference has not yet been quantified. The thick, red tongue is forked distally; the black snout is bluntly rounded. The genus can easily be recognized by the presence of beadlike osteoderms on dorsal surfaces of the head, limbs, body, and tail. The limbs are relatively short; the fourth toe of the hind foot is not enlarged. Fat reserves are stored in the tail, which can be thick in well-fed individuals but is often quite thin in wild-caught lizards. A robust skull architecture, large jaw muscles, and venom glands of the lower jaw give the head a bulky appearance.

Size

Adults reach a SVL of 330 to 470 mm, an overall length of 570 to 800 mm, and a mean adult body mass of about 900 g (Beck and Lowe 1991). Individuals may rarely approach 1 m in overall length and weigh over 2 kg.

Natural History

Heloderma horridum is largely terrestrial, although it occasionally ascends trees in search of vertebrate nests, the contents of which it feeds on. It also commonly uses shelters in hollow branches or trunks of trees, especially during the wet season.

Timing of Activity

In Jalisco, Mexico, *Heloderma horridum* exhibits a bimodal, diurnal activity pattern with a strong peak of activity in the afternoon between 1600 and 2000 hours and a smaller morning peak between 0700 and 1000 hours. This bimodal pattern becomes more widely separated during the rainy season, with an earlier morning activity peak, and a second peak later in the afternoon and early evening (Beck and Lowe 1991). Nocturnal activity also occurs occasionally, predominantly during rainy periods. *Heloderma horridum* appears diurnal in other parts of its range, with a similar bimodal pattern. In Chiapas, *H. h. alvarezi* are more commonly observed active toward dusk and shortly thereafter, with less activity earlier in the day, except during rainy periods (Alvarez del Toro 1982).

Thermoregulation

In Jalisco, mean monthly body temperatures of *H. horridum* while resting in shelters ranged from 20°C in December to 28.5°C in July (Beck and Lowe 1991). Extremes in body temperatures ranged from 17.7°C (below ground in January) to 37.2°C (thermal inertia after activity in June).

Beaded lizards tend to be thermoconformers while resting in shelters, which can be for weeks at a time during the activity season and longer at other times of the year. Those using subsurface refugia show little daily variation in body temperature, whereas body temperatures of lizards that use shelters in trees fluctuate considerably in relation to daily variation in the temperature of their surroundings (Beck and Lowe 1991). Lower body temperatures of winter-dormant lizards within shelters, and associated reduced metabolic rates, may represent a significant energy savings for Heloderma horridum.

When foraging on the surface or basking, Heloderma horridum appears to actively thermoregulate. Body temperatures of surface-active lizards in Jalisco range from 22.5°C to 36.0°C, averaging 29.5°C (±3.2°C, n = 96), which is significantly higher than mean body temperature when at rest (inside shelters) during the activity season (Beck and Lowe 1991) (all values are expressed as ±standard error of the mean [SEM] unless otherwise indicated). Activity temperatures during the wet season tend to be lower than during the dry season, probably because beaded lizards frequently leave their shelters during and shortly after rainstorms when they are exposed to lower surface and air temperatures. I have observed beaded lizards sun basking after heavy rains, and in January. One lizard, monitored while basking in the sun in a tree after a heavy September rain in Jalisco, maintained its body temperature between 28.2°C and 29.4°C ($\overline{X}$ = 28.8°C, ± 0.48°C, n = 12) between 1415 and 1620 hours (body temperatures were sampled every 10 minutes). Air temperature fell from 28.4°C to 26.8°C during the same period (Beck and Lowe 1991).

Foraging Behavior and Diet

Heloderma horridum is a widely searching forager with a dietary specialization on the contents of vertebrate nests: reptilian and avian eggs, young birds, and juvenile mammals. The diet of H. horridum appears to be more varied than that of its congener, H. suspectum, largely because its dry tropical habitats harbor a greater diversity of potential prey items than do the desert ecosystems inhabited by Gila monsters. Heloderma horridum has well-developed chemosensory and auditory systems it uses to search for nests both on the ground and in trees. In Chamela, Jalisco, rate of travel during surface activity ranged from 0.7 to 7.0 m/min (±153 m/min; $\overline{X}$ = 236 m/min; n = 116) (Beck and Lowe 1991). Assuming a travel rate of 3.5 m/min, mean duration of an above-ground foray was approximately 67 minutes.

In 14 fecal samples collected from 10 beaded lizards in coastal Jalisco, Beck and Lowe (1991) found remains of reptilian eggshells (10 of 14), feathers (8 of 14), and insects (5 of 14). In Jalisco, a major food source for Heloderma is eggs of the spiny-tailed iguana, Ctenosaura pectinata, whose nesting season in May and June coincides with a peak activity period for Heloderma (Beck and Lowe 1991). Eggs of other lizards and snakes have also been observed to be excavated by H. horridum or found in their stomachs, both in Jalisco and in other regions. Some avian prey (eggs or nestlings) observed to be taken by H. horridum include Beechey's jay (Cissilopha beecheyi) in Nayarit, Mexico (Pregill et al. 1986), white-tipped dove (Leptotila verrauxi) in Jalisco, and possibly nestlings of chachalacas (Ortalis poliocephlus), trogons (Trogon citreolus), quail, and other birds in Jalisco and other regions (McDiarmid 1963; Campbell and Vannini 1988; Beck and

Lowe 1991). Throughout its range, *Heloderma horridum* feeds on juvenile mammals in nests (Bogert and Martín del Campo 1956). Insects, especially large beetle larvae, may also be an important component of the diet of *H. horridum*. Large insects (exclusive of those incidentally ingested with vertebrate prey) have been found in the stomachs of specimens from Guatemala, and from Jalisco, Nayarit, and Sonora, Mexico (McDiarmid 1963; Pregill et al. 1986; Beck and Lowe 1991; Campbell and Lamar 2003).

Reproduction

Heloderma horridum exhibit spectacular ritualized male–male combat behaviors that are strikingly similar to those of *Varanus gilleni, Varanus caudolineatus,* and other species of *Varanus* (Murphy and Mitchell 1974; Ramírez-Velázques and Guichard-Romero 1989; Beck and Ramírez-Bautista 1991; Thompson et al. 1992; Horn et al. 1994). Combat consists of a sequence of ritualized wrestling positions that eventually result in formation of a high arch posture, with venters adpressed and snouts, forelimbs, and tail tips forming contact points on the ground. Further pressure exerted by the combatants eventually collapses the arch, with the dominant lizard ending up on top. Lizards may repeat the sequence many times over several hours, each bout usually being initiated by the subordinate lizard. The objective during each bout is apparently to force the opponent onto his back, collapsing the arch. Attempts are made by the dominant lizard to keep the subordinate on his back after the arch has collapsed. Size and tail strength are important factors in the ability to arch higher than an opponent and end up in a superior position after the arch collapses (Beck and Ramírez-Bautista 1991). A typical combat session demands considerable physical effort and apparently leaves both participants exhausted.

Combat in *H. horridum* takes place in September and October and coincides with courtship, mating, and spermiogenesis (Alvarez del Toro 1982; Ramírez-Velázques and Guichard-Romero 1989; Beck and Ramírez-Bautista 1991; Goldberg and Beck 2001). Eggs are laid between October and December, and hatchlings appear in June or July with the onset of the wet season, suggesting an incubation period of 6 to 8 months (Alvarez del Toro 1982; Beck and Ramírez-Bautista 1991; Ramírez-Bautista 1994). In captivity, incubation periods range from 154 to 226 days at incubation temperatures ranging from 21°C to 29°C, and clutch size ranges from 2 to 22 eggs (mean, 7–9 eggs; Perry 1996; Johnson and Ivanyi 2001). In captive individuals, egg size varies from 28 to 32 mm wide by 34 to 65 mm long, and egg mass varies from 25 to 40 g (Perry 1996). Wild hatchlings have a SVL of 115 to 127 mm and weigh 23 to 27 g (Ramírez-Bautista 1994). Captive-raised *H. horridum* hatchlings range in mass from 38 to 47 g (Conners 1987; Applegate 1991; Gonzalez-Ruiz et al. 1996).

Home Range

In a population of *H. horridum* radio tracked in Jalisco, mean home range size was 21.6 ha, and there was substantial overlap in home ranges of neighboring individuals (Beck and Lowe 1991). In regions of home range overlap, some shelters are used by more than one lizard, either at different times or concurrently.

Seasonal Patterns in Activity

In Jalisco, surface activity of *Heloderma horridum* peaks near the end of the dry season. Animals are most frequently sighted in early May, although considerable activity continues throughout June and July, gradually decreasing in September and October. Throughout its range, *Heloderma horridum* is also frequently encountered on the surface after rains, especially at the onset of the rainy season in July or August.

On the basis of results of a field study using radiotelemetry in coastal Jalisco (Beck and Lowe 1991), *Heloderma horridum* were found to be active on the surface on 47% of the days they were monitored during May through July, traveling an average of 252 m for each day active (m/d active). By September and October, radio-equipped lizards were active on only 33% of the days they were monitored, traveling an average of 242 m/d active. In January, they were active only 17% of the days they were monitored and traveled an average of 45 m/d active. Some surface activity in the local population most likely occurs during every month of the year, although it is greatly reduced during the winter months of December through February. In Jalisco, *H. horridum* were estimated to travel 25.3 km annually during only 121 hours of surface activity, spending over 95% of their time hidden in shelters (Beck and Lowe 1991).

Physiology

Heloderma horridum, like *H. suspectum,* has a low resting metabolic rate and a high capacity for sustained aerobic activity (John-Alder et al. 1983; Beck and Lowe 1994; Beck et al. 1995). Metabolic rates of the two species do not differ significantly from each other (Beck and Lowe 1994; Beck et al. 1995). (See the introductory section on helodermatid lizards for an overview of their physiological ecology).

Testicular Cycles

Testicular material examined from road-killed specimens collected in Sonora and Jalisco suggests that sperm formation (spermiogenesis) begins in late August and continues through at least early October (Goldberg and Beck 2001).

Parasites

A number of helminth parasites (Goldberg and Bursey 1990, 1991; Upton et al. 1993) as well as ectoparasites (Smith 1910) have been identified from *Heloderma suspectum,* but little published information exists on parasites of *Heloderma horridum.*

8.3. *Heloderma suspectum* (Cope 1869)

Daniel D. Beck

Figure 8.3. Heloderma suspectum
(photo by Dan Beck).

Synonomy

Type

Heloderma suspectum (Cope 1869). Locality "on the International Boundary, at Monument 146, or Sierra de Moreno" (Bogert and Martín del Campo 1956) (Fig. 8.3).

Nomenclature

Heloderma suspectum suspectum (Bogert and Martín del Campo 1956), reticulate Gila monster.

Heloderma suspectum cinctum (Bogert and Martín del Campo 1956). Banded Gila monster. Type locality: "Las Vegas, Clark County, Nevada . . . at an elevation of 2033 feet in the Las Vegas Valley" (Bogert and Martín del Campo 1956).

Etymology

Heloderma is derived from the Greek words "*helos*" for a nail stud and "*derma*" for skin; and "*suspectum*" from Cope's assertion that, on the basis of its grooved teeth and sinister reputation, he "suspected" the lizard was venomous.

Pacific Ocean

Heloderma suspectum

Figure 8.4. Geographic
distribution of Gila monsters
(based on a map from Campbell
1989; used with permission).

Common Names

The standard common name for *Heloderma suspectum* is the Gila
monster. In Mexico, it is commonly called "*el monstro de Gila*" or
"*escorpión Pintada.*"

Closely Related Species

Heloderma suspectum is sympatric with *H. horridum* in the southern part of its range in southern Sonora, Mexico (Beck 1993). However,
differences in habitat preferences between the two species probably
keep them from occupying the same microhabitat. *Heloderma suspectum* is a shorter, chunkier, less arboreal lizard than *H. horridum*.
The two species are readily distinguished from each other by the
proportionately longer tail of *H. horridum* (at least 65% of the SVL; no
more than 55% in *H. suspectum*) and by conspicuously enlarged
postanal scales (usually two) in *H. suspectum*. *Heloderma horridum*
normally have no enlarged postanal scales, although some individuals
from Guatemala are an exception (Campbell and Vannini 1988). The
two species also show differences in number of infralabials that are in
contact with the chin shields: one pair in *H. suspectum,* two pairs in *H.
horridum* (Bogert and Martín del Campo 1956). The thick, forked

tongue of a Gila monster is black, rather than pink, as it is in most individuals of *H. horridum.*

Geographic Distribution

Heloderma suspectum occurs from the Mojave Desert in extreme southern Nevada, southwestern Utah, southeastern California, and northwestern Arizona, throughout the Sonoran Desert region in Arizona and Sonora, Mexico (exclusive of Baja California), and into a small part of the Chihuahuan desert in southeastern Arizona and southwestern New Mexico. It ranges in elevation from near sea level in desert-scrub habitats up to about 5100 ft (1600 m) in grassland/oak–juniper–woodland communities. Two subspecies are recognized. *H. s. suspectum,* the reticulate Gila monster, occurs from northwestern Sinaloa/southern Sonora, Mexico into southern Arizona and extreme southwestern New Mexico. The banded Gila monster (*H. suspectum cinctum)* occurs primarily in the Mojave Desert, from northwestern Arizona into California, Nevada, and Utah (Fig. 8.4)

Habitat

Heloderma suspectum is primarily a desert dweller, although it also inhabits semidesert grassland and woodland communities along mountain foothills. In the Sonoran desert, it is more commonly found in the Arizona Upland subdivision than in the drier and sandier Lower Colorado Valley subdivision (Lowe et al. 1986). In the Mojave Desert, it prefers canyons or adjacent rocky slopes (more rarely in open valleys) in regions having at least 25% of annual precipitation falling as summer rain. Occurrence is strongly influenced by availability of suitable microenvironments (e.g., boulders, burrows, pack rat middens) used as shelters, where Gila monsters spend most of their time.

Fossil Record

Fragments of skin and osteoderms of *Heloderma suspectum* have been found in 8000- to 10,000-year-old late Pleistocene deposits in southern Nevada (Brattstrom 1954). See summary of fossil history in introductory section on helodermatid lizards.

Description

Heloderma suspectum is a stout, bulky, ground-dwelling lizard with striking patterns and coloration. The color pattern varies geographically and changes with age. Juveniles have five black saddlelike crossbands on a pale yellow or pale orange background and a tail with four to five black bands. In *H. s. suspectum,* the bandlike juvenile pattern often breaks up to form an irregular "reticulated" pattern of black mottles or blotches on a rose, orange, or yellow background. However, some adults within the range of *H. s. suspectum* (e.g., southwestern New Mexico) largely retain the banded pattern of juveniles.

Adults of the northern subspecies, *H. s. cinctum,* tend to be lighter in color and retain the juvenile pattern of wide black crossbands on a normally yellow background. Exceptions include dark individuals (without crossbands) found to inhabit black basaltic lava flows in southwestern Utah (Beck 1985). Males have wider heads than females, but this difference has not been quantified. The thick, black tongue is forked distally; the black snout is bluntly rounded. The genus can easily be recognized by the presence of beadlike osteoderms on dorsal surfaces of the head, limbs, body, and tail. See the species account for *H. horridum* for additional characteristics of the genus.

Size

Adults reach a SVL of 300 to 360 mm, an overall length of 350 to 500 mm, and a mean adult body mass of about 600 g (Beck, unpublished data). Wild individuals may rarely approach 550 mm in overall length and weigh up to 1 kg. Males become reproductive at a SVL of 221 mm, females at a SVL of 239 mm (Goldberg and Lowe 1997).

Natural History

Heloderma suspectum is largely terrestrial, although it has been observed to climb up to 2.5 m into desert subtrees (Cross and Rand 1979).

Timing of Activity

Heloderma suspectum can be active at any time of day but is largely diurnal, especially during its seasonal period of peak activity in April and May. In the spring, peak activity time is in the morning, with a smaller peak in the late afternoon (Beck 1990). Timing of activity becomes more variable with the arrival of higher summer temperatures. Nocturnal activity seem to occur more frequently after summer rains. Gila monsters emerge to bask near the entrance to shelters occasionally during the winter (especially after rainy periods) and more frequently in the spring (Beck 1990).

Thermoregulation

In southwestern Utah, mean monthly body temperatures of *H. suspectum cinctum* taken during 1983–1984 while resting in shelters ranged from 12°C in December to 28.0°C in July (Beck 1990) and in southern New Mexico (during 1992–1993) 11.0°C in December to 27.6°C in July (Beck, unpublished data). Extremes in body temperatures ranged from 9.1°C (below ground in December) to 36.8°C (thermal inertia after activity in June).

Like beaded lizards, Gila monsters tend to be thermoconformers while they are resting in shelters, which can be for weeks at a time during the activity season and longer at other times of the year. Lower body temperatures of winter-dormant lizards within shelters, and associated reduced metabolic rates, may permit a significant energy saving for *Heloderma suspectum.*

Mean body temperature during activity (29.3°C) is similar to that of *H. horridum* (29.5°C) but is more variable (range, 17.4–36.8°C) (Beck 1990). When foraging on the surface or basking, *Heloderma suspectum* appears to actively thermoregulate. One Utah lizard, monitored for several days while basking in April, frequently exited and reentered its shelter throughout each day, maintaining a body temperature around 30°C despite considerably cooler air and substrate temperatures (Beck 1990).

Foraging Behavior and Diet

Like *Heloderma horridum*, the Gila monster feeds on the contents of vertebrate nests, primarily reptilian and avian eggs, and juvenile mammals. In southwestern Utah, a typical above-ground foray consists of considerable searching and lasts an average of 50 minutes, covering approximately 215 m (Beck 1990). Important prey observed to be eaten by Gila monsters include juvenile cottontail rabbits (*Sylvilagus auduboni*), rock squirrels (*Spermophilus variegatus*), cotton rats (*Sigmodon*), and other juvenile mammals; eggs from Gambel's quail (*Lophortyx gambeli*), mourning doves (*Zenaida macroura*), desert tortoises (*Gopherus agassizi*), mud turtles (*Kinosternon*), and other reptiles; and nestling birds (Arnberger 1948; Hensley 1949; Stahnke 1950; Zweifel and Norris 1955; Martín del Campo 1956; Jones 1983; Beck 1990).

Encountering such widely distributed prey requires a widely searching foraging strategy. Widely searching foragers typically must shunt a relatively large portion (~25%–35%) of their annual maintenance energy budget to activity (Anderson and Karasov 1981; Huey and Pianka 1981; Nagy et al. 1984). Interestingly, this is not the case for *Heloderma suspectum*, which allocates less than 13% of its annual maintenance energy budget to surface activity, despite its occasionally extensive foraging bouts. Its relatively large size, low resting metabolic rate, and ability to take large meals make frequent foraging activity unnecessary for the Gila monster and greatly reduces its activity costs (Beck 1990).

Reproduction

Like *Heloderma horridum*, the Gila monster exhibits spectacular ritualized male–male combat behaviors (Demeter 1986; Beck 1990). Combat in *Heloderma* suspectum also consists of a series of bouts, reminiscent of ritualized wrestling matches, whereby combatants straddle each other, then perform a body twist in an effort to gain the superior position (Beck 1990). Gila monsters do not form the arching postures performed by *H. horridum* and some varanid lizards, probably because their tails are too short. Each bout ends when pressure exerted by the body twist causes the lizards to separate, but bouts can be repeated many times over several hours. Two fighting males observed in southwestern Utah performed at least 13 individual bouts over nearly 3 hours of continuous exertion (Beck 1990).

Combat in *H. suspectum* takes place in late April through early

June and coincides with courtship, mating, and spermiogenesis (Goldberg and Lowe 1997). Gila monsters often cohabit shelters at this time. Analyses of reproductive tissues (Goldberg and Lowe 1997; see testicular cycles, below) and field observations reveal that eggs are laid in July and August, which coincides with the onset of summer rains in the Sonoran and Chihuahuan deserts. Interestingly, hatchling Gila monsters are not observed until at least mid-April, suggesting that either eggs overwinter before hatching, or (less likely) that eggs hatch in the fall and hatchlings remain hidden until the following spring. In captivity, incubation periods range from 120 to 145 days at incubation temperatures of 27 to 31°C (Wagner et al. 1976; Grow and Branham 1996; Strimple 1995), temperatures that are considerably higher than those likely experienced by eggs under natural conditions. Clutch size of 17 specimens collected near Tucson, Arizona, varied from 2 to 12 eggs, with a mean of 5.7 eggs (Goldberg and Lowe 1997). At hatching, newborns average 165 mm total length and 33 g (Lowe et al. 1986).

Home Range

Home range size, estimated from a population in southwestern Utah, ranged from 5.6 to 66.2 ha (Beck 1990). Mean home range size from a population in southwestern New Mexico was 58.1 ha (±12.4 ha; range, 6.2–104.8 ha; n = 7) (Beck and Jennings 2003). Individuals show considerable overlap in their home ranges and commonly share the same shelters. Gila monsters also commonly reuse shelters, often returning to the same shelters year after year.

Seasonal Patterns in Activity

Gila monsters typically emerge from winter dormancy in March, when they often emerge to bask near shelter entrances. The peak activity season is April through early June. This period coincides with courtship and mating, as well as emergence of many of the food resources (eggs, juvenile mammals, and birds) used by *Heloderma suspectum*. With the hotter summer temperatures, *Heloderma suspectum* becomes less active on the surface and may remain within shelters for weeks at a time (Lowe et al. 1986; Beck 1990). Summer activity usually occurs on relatively cooler, cloudy days or on warm nights. By mid-October, surface activity decreases abruptly and Gila monsters gradually return to overwintering shelters, which they may reuse year after year. By late November, a period of winter dormancy ensues that continues until the following spring. Gila monsters will occasionally emerge on warm, sunny winter days to bask at the shelter entrance, especially after rainy periods. In a Utah population, 64% of annual activity occurred from late April to early July (Beck 1990); similar patterns have been observed in Arizona (Lardner 1969). In Utah, lizards were active only an average of 9 days per month during the activity season, spending less than 70 hours per year in surface activity (Beck 1990).

In New Mexico, Gila monsters show seasonal changes in their

habitat preferences. Shelters on rocky south-facing slopes are selected in winter. During the hot, dry summer before the onset of summer rain, Gila monsters select shelters that are more variable in orientation, more soil-like in composition, and cooler and more humid than other available shelters. During the spring, when they are foraging, basking, searching for mates, and pairing, Gila monsters select shelters more strongly oriented toward the east (Beck and Jennings, unpublished data).

Physiology

Heloderma suspectum has among the lowest standard metabolic rates of any lizard measured (Beck and Lowe 1994) and a high capacity for sustained aerobic activity (John-Alder et al. 1983; Beck and Lowe 1994; Beck et al. 1995). Metabolic rates of *H. suspectum* do not differ significantly from *H. horridum* (Beck and Lowe 1994; Beck et al. 1995). (See introductory section on helodermatid lizards for an overview of their physiological ecology.)

Testicular Cycles

The reproductive cycle of *H. suspectum* in southern Arizona has been investigated by Goldberg and Lowe (1997). They found that on emergence from hibernation in March, testes are in a state of recrudescence, wherein germinal epithelium contain primary and/or secondary spermatocytes for the upcoming reproductive season. By May and into June, testes are largest and epididymides contain masses of sperm, suggesting that Gila monsters come into breeding condition in May and remain so through at least part of June. By July–August, testes have shrunken and epididymides are devoid of sperm. Recrudescence begins in September and continues through October–November. Female *H. suspectum* emerge from hibernation with yolk deposition in progress, and follicles become enlarged during late March through May. Oviductal eggs appear by mid-June through at least early August (Goldberg and Lowe 1997). Average clutch size is about 5.6 eggs. In Red Rock, New Mexico, Gila monsters contain shelled, oviductal eggs by late June and have been observed to lay eggs in mid-July (Beck and Jennings, unpublished data).

Parasites

A number of helminth parasites (Goldberg and Bursey 1990, 1991; Upton et al. 1993) as well as ectoparasites (Smith 1910) have been identified from *Heloderma suspectum*.

8.4. *Lanthanotus borneensis*

Eric R. Pianka

Nomenclature

Commonly called the earless monitor lizard, *Lanthanotus borneensis* was named in 1878 by Franz Steindachner. The type specimen remains well preserved in a museum in Vienna (Fig. 8.5).

Geographic Distribution

These very secretive lizards, sometimes given their own familial level status, are found only in riverine regions of Sarawak on northern Borneo, Malaysia; there are unsubstantiated reports of their existence in nearby Kalimantan, Malaysia. *Lanthanotus* is considered to be the sister group to varanids and may be the closest relatives of snakes. Only about 100 of these lizards have ever been collected. Figure 8.6 shows their approximate known geographic range.

Fossil Record

No fossils of *Lanthanotus borneensis* are known. A related fossil lanthanotid from the Cretaceous (75 my BP) of the Gobi Desert was described and named *Cherminotus* (Borsuk-Białynicka 1984). This creature is reputed to be similar to *Lanthanotus*, suggesting that lanthanotids may have changed relatively little over a vast time period. However, Gao and Norell (2000) have questioned the putative affinity

Figure 8.6. Geographic range of Lanthanotus borneensis.

with *Lanthanotus. Lanthanotus* could be a "living fossil" that might provide valuable insights into the ancestor of all anguimorphan lizards (anguids, xenosaurids, helodermatids, lanthanotids, and varanids) and perhaps snakes as well.

Diagnostic Characteristics

- Medium-sized brown lizards, about 42 to 55 cm in total length.
- Body and tail cylindrical.
- Scales small with six longitudinal rows of enlarged scales running from head down the back and two central rows run on to the long prehensile tail.
- Short legs; long, curved, sharp claws.
- No external ear opening.

Description

One of the strangest and least known of all lizards, *Lanthanotus* are medium-sized lizards, adult size about 420 to 550 mm total length with a relatively long cylindrical body, long neck, and long tail (Harrisson 1966).

Viewed from below, males have blunt, rectangular jaws, whereas jaws of females are more pointed. Because of the hemipenes in males, the base of the tail is broader than in females. They have short legs but long, curved, sharp claws. They can wrap their muscular bodies and prehensile tails around a branch in a manner that suggests that they might climb (Proud 1978). Most of their scales are small, but six longitudinal rows of enlarged scales run from the head down the back, and two central rows run out onto the tail. *Lanthanotus* tails are prehensile and do not regenerate. They shed their skin in one piece, as do some other anguimorphans and snakes (J. Arnett, personal communication). The braincase is more solidly encased than in varanids, more similar to that of snakes. The upper temporal arch has been lost, and there is a hinge joint in the middle of the lower jaw. *Lanthanotus* have sharp recurved teeth on their premaxillaries, maxillaries, palatines, pterygoids, and dentaries. *Lanthanotus* is the only species among anguimorphans with translucent windows in its lower eyelids, which could be a precursor to the "spectacle" covering the eye of snakes. Like snakes, *Lanthanotus* have no external ear openings, and they have forked tongues. Indeed, *Lanthanotus* could well be closely related to snakes (McDowell and Bogert 1954; Underwood 1957).

Size

Total length of adults is about 420 to 550 mm. SVL averages about 200 mm, with a maximum of 400 mm SVL (Das 2003).

Habitat and Natural History

In captivity, *Lanthanotus* are sluggish lizards that spend most of their time lying in water, seldom moving (Mertens 1966; Proud 1978; J. Arnett, personal communication). Captives shed their skins very infrequently, less than once per year (Mertens 1966; Proud 1978). Such observations could be mere artifacts of the unusual environmental conditions in captivity and could be largely irrelevant to behaviors of free-ranging wild lizards.

One was captured "while hiding in a cave" at 8 A.M. The only eyewitness account of *Lanthanotus* in the wild states, "Sometime in the month of July 1961, while burning the area for the farming season, I suddenly saw two lizards come out from the earth, walking through the hot ashes. The bigger lizard followed about two feet behind the smaller. I quickly caught it: it struggled and coiled its tail. But it soon died on the spot. I caught the smaller lizard and took them both home to our long-house" (Harrisson 1961a, 1961b). *Lanthanotus* dig burrows in banks along watercourses and retreat into the water when threatened (Robert Murphy, personal communication). They are sometimes captured by local people seining for fishes. Das (2003) reports that they are nocturnal, foraging on land and in the water at night. They swim by lateral undulation (Das 2003).

Unfortunately, virtually nothing is known about the natural history of these important and fascinating lizards (Harrisson 1963a). A field study of the natural history and ecology of *Lanthanotus* is greatly needed.

Terrestrial, Arboreal, Aquatic

By some accounts, *Lanthanotus* is aquatic (they have been captured in fish seines); by other accounts, it is a burrower. Very likely, it is both. It does seem to prefer cool, moist habitats. A number of these unusual lizards were collected after severe flooding in Sarawak in 1963 (Harrisson 1963b; Sprackland 1970, 1972). They could have been inactive in underground retreats and emerged when it flooded. They also use their prehensile tails to climb (Proud 1978).

Time of Activity (Daily and Seasonal)

Some reports of captive lizards suggest that *Lanthanotus* could be nocturnal. Das (2003) asserts that wild lizards are nocturnal.

Thermoregulation

Little is known about the thermal biology of *Lanthanotus*. In captivity, *Lanthanotus* appear to prefer relatively low ambient temperatures of about 24 to 28°C.

Foraging Behavior

Little is known about foraging behavior of *Lanthanotus borneensis*

Diet

Citing Losos and Greene (1988) and Pregill et al. (1986), Das (2003) says that the diet in the wild is composed of earthworms and crustaceans. Captives have eaten squid, small bits of fish, earthworms, liver, and even beaten eggs. Earthworm setae were present in the stomach of a specimen in the Museum of Comparative Zoology at Harvard (Pregill et al. 1986). Their natural diet remains largely unknown.

Reproduction

Little is known about reproduction in *Lanthanotus*. Proud (1978) found one lizard with six large eggs in September 1976, but stated that these eggs may not have been *Lanthanotus* eggs. Sprackland (1999) asserts that eggs have never been laid in captivity, and that clutch size from dissected females is three to four. Eggs are large, about 30 mm long. Das (2003) says that mating has been observed in early February (Harrisson 1963b) and asserts that clutches of two to five oval, leathery-shelled eggs are produced at a time.

Movement

Nothing is known about movements, physiology, fat bodies, testicular cycles, or parasites of *Lanthanotus*.

8.5. *Estesia mongoliensis*

Mark A. Norell

Figure 8.7. Fossilized skull of Estesia mongoliensis *from the Gobi Desert (about 77 my BP).*

Nomenclature

Norell et al. (1992) described and illustrated the species on the basis of a single specimen (M 3/14) housed in the paleontological collections of the Institute of Geology, Ulaanbaatar, Mongolia. Subsequently, the braincase was described and illustrated in more detail by Norell and Gao (1997) in a phylogenetic treatment of the taxon (Fig. 8.7).

Geographic Distribution

Estesia mongoliensis is known from four localities in Mongolia's Gobi Desert (Gao and Norell 2000); the holotype locality Khulsan, Bayn Dzak (the Flaming Cliffs), Kheerman Tsav, and Ukhaa Tolgod. These localities are roughly equivalent in age and are thought to be Late Cretaceous, about 80 million years old (Loope et al., personal communication; Gao and Norell 2000).

Fossil Record

Estesia mongoliensis is known exclusively from fossils. A list of specimens identified to this taxon is provided in Gao and Norell (2000).

Other specimens are in the collections of the Institute of Geology, Mongolia.

Diagnostic Characteristics

The diagnosis of *Estesia mongoliensis* is dependent on the analysis of a large matrix of anguimorphs, which places it in a group with the extant *Heloderma* (Norell and Gao 1997). It can be "Distinguished from *Heloderma* and its fossil relatives in having the following autapomorphies: presence of a distinct single, median palatal trough formed by vomers; long median contact of vomers; presence of extensive convex surface of quadrate; no distinct articular; foramen ovale located far anterior to sphenooccipital tubercle; sphenoccipital tubercle strongly elongate and posteriorly oriented" (Norell and Gao 1997, p. 2).

Description

Initially *Estesia mongoliensis* was considered to be the sister group to a group composed of the Varanidae and *Lanthanotus* (Norell et al. 1992). However, in more extensive studies, Norell and Gao (1997) and Gao and Norell (1998) showed that *Estesia mongoliensis* is part of a group of largely extinct taxa that includes the extant *Heloderma* called the Monstersauria.

Probably the most interesting thing about this taxon is the presence of "poison grooves" on the teeth, similar to those of extant helodermatids. This indicates that the poison delivery system found in modern helodermatids has a long history—extending back into the Mesozoic and distributed across a number of species. The osteodermal covering of *Estesia* is not as extensive as most other monstersaurians. *Estesia* also shows the retracted nares found in the Varanidae. However, this feature appears to have arisen convergently.

Size

The skull of the holotype is about 14 cm in length. Although complete animals have not been collected, the animal undoubtedly exceeded a meter in length.

Habitat and Natural History

Because *Estesia mongoliensis* is known only from fossil remains, little can be determined about its natural history. Some habitat information is, however, available through geologic investigations at the localities where the specimens have been collected (Loope et al. 1998; Loope and Dingus 1999; Loope et al. 1999). At the best-studied locality, Ukhaa Tolgod (Dashzeveg et al. 1995), fossils are preserved in alluvial fans that represent the runoff from collapsing destabilized sand dunes (Loope et al. 1998). Ancillary evidence indicates that these dunes were vegetated and that interdunal water was available at least seasonally. The presence of caliche in paleosols at several of these localities suggests an arid climate (Loope et al., personal communication).

Terrestrial, Arboreal, Aquatic

Although it is impossible to determine with any precision, the geological and taphonomic evidence (Loope et al. 1998) strongly suggest that *Estesia mongoliensis* is terrestrial.

Diet

Although direct dietary evidence in the way of stomach contents is unknown, *Estesia mongoliensis* doubtlessly preyed on small animals and may have eaten dinosaur eggs. Remains of nests and fossil eggs and eggshells are very common in areas where *Estesia mongoliensis* remains are found.

References

Alvarez del Toro, M. 1982. *Los reptiles de Chiapas: Coleccion libros de Chiapas,* Serie Espanol. Chiapas, Mexico: Instituto de Historia Natural, Tuxtla Gutierrez.

Anderson, R. A., and W. H. Karasov. 1981. Contrasts in energy intake and expenditure of sit-and-wait and widely foraging lizards. *Oecologia (Berlin)* 49:67–72.

Applegate, R. W. 1991. Tails of Gila monsters and beaded lizards. In *Proceedings of the 1991 Northern California Herpetological Society Conference on Captive Propagation and Husbandry of Reptiles and Amphibians,* pp. 39–44.

Arnberger, L. P. 1948. Gila monster swallows quail eggs whole. *Herpetologica* 4:209–210.

Auffenberg, W. 1981. *The behavioral ecology of the Komodo monitor.* Gainesville: University Press of Florida.

Beck, D. D. 1985. *Heloderma suspectum cinctum* (banded Gila monster): Pattern/coloration. *Herpetol. Rev.* 16:53.

———. 1990. Ecology and behavior of the Gila monster in southwestern Utah. *J. Herpetol.* 24:54–68.

———. 1993. A retrospective of *The Gila monster and its allies.* In *The Gila monster and its allies: the relationships, habits, and behavior of the lizards of the family Helodermatidae,* by C. M. Bogert and R. Martin del Campo, 1956. Reprinted by Oxford, Ohio: Society for the Study of Amphibians and Reptiles.

Beck, D. D., and A. Ramírez-Bautista. 1991. Combat behavior of the beaded lizard *Heloderma h. horridum* in Jalisco, Mexico. *J. Herpetol.* 25:481–484.

Beck, D. D., and C. H. Lowe. 1991. Ecology of the beaded lizard, *Heloderma horridum,* in a tropical dry forest in Jalisco, México. *J. Herpetol.* 25:395–406.

———. 1994. Resting metabolism of helodermatid lizards: Allometric and ecological relationships. *J. Comp. Physiol. B Biochem. Syst. Environ. Physiol.* 164:124–129.

Beck, D. D., M. R. Dohm, T. Garland Jr., A. Ramírez-Bautista, and C. H. Lowe. 1995. Locomotor performance and activity energetics of helodermatid lizards. *Copeia* 1995:577–585.

Beck, D. D., and R. D. Jennings. 2003. Habitat use by Gila monsters: The importance of shelters. *Herpetol. Monogr.* 17 (in press).

Bennett, A. F. 1983. Ecological consequences of activity metabolism. In *Lizard ecology: Studies of a model organism,* ed. R. B. Huey, E. R. Pianka, and T. W. Shoener, 11–23. Cambridge: Harvard University Press.

Bogert, C. M., and R. Martín del Campo. 1956. The Gila monster and its allies: The relationships, habits, and behavior of the lizards of the family Helodermatidae. *Bull. Am. Mus. Nat. Hist.* 109:1–238.

Borsuk-Białynicka, M. 1984. Anguimorphans and related lizards from the late Cretaceous of the Gobi Desert, Mongolia. *Palaeontol. Polonica* 46:5–105.

Brattstrom, B. H. 1954. Amphibians and reptiles from Gypsum Cave, Nevada. *Bull. Southern Acad. Sci.* 53:8–12.

Campbell, J. A., and J. P. Vannini. 1988. A new subspecies of beaded lizard, *Heloderma horridum,* from the Motagua Valley of Guatemala. *J. Herpetol.* 22:457–468.

Campbell, J. A., and W. L. Lamar. 1989. *The venomous reptiles of Latin America.* Ithaca, N.Y.: Cornell University Press.

———. 2003. *The venomous reptiles of the Western Hemisphere.* 2 vols. Ithaca, N.Y.: Cornell University Press.

Cifelli, R. L., and R. L. Nydam. 1995. Primitive, helodermatid-like platynotan from the early cretaceous of Utah. *Herpetologica* 51:286–291.

Conners, J. S. 1987. Captive breeding of beaded lizards at the Detroit Zoo. *Proc. Ann. Conf. Am. Assoc. Zool. Parks Aquar.* 514–516.

Cope, E. D. 1869. Diagnosis of *Heloderma suspectum. Proc. Acad. Nat. Sci. Philadelphia* 21:4–5.

Cross, J. K., and M. S. Rand. 1979. Climbing activity in wild-ranging Gila monsters, *Heloderma suspectum* (Helodermatidae). *Southwest. Naturalist* 24:703–705.

Das, I. 2003. *The lizards of Borneo.* Borneo: Natural History Publications.

Dashzeveg, D., M. J. Novacek, M. A. Norell, J. M. Clark, L. M. Chiappe, A. Davidson, M. C. McKenna, L. Dingus, C. Swisher, and A. Perle. 1995. Unusual preservation in a new vertebrate assemblage from the Late Cretaceous of Mongolia. *Nature* 374:446–449.

Datta, G., and A. T. Tu. 1997. Structure and other chemical characterizations of Gila toxin, a lethal toxin from lizard venom. *J. Peptide Res.* 50:443–450

Demeter, B. J. 1986. Combat behavior in the Gila monster (*Heloderma suspectum cinctum*). *Herpetol. Rev.* 17:9–11.

Doyle, M. E., and J. M. Egan. 2001. Glucagon-like peptide-1. *Recent Progr. Horm. Res.* 2001:56377–56399.

Estes, R. 1964. Fossil vertebrates from the Late Cretaceous Lance Formation, eastern Wyoming. *Univ. Calif. Publ. Geol. Sci.* 49:1–180.

Estes, R., K. de Queiroz, and J. Gauthier. 1988. Phylogenetic relationships within Squamata. In *Phylogenetic relationships of the lizard families,* ed. R. Estes and G. Pregill, pp. 119–281. Stanford, Calif.: Stanford University Press.

Gabe, M., and H. Saint Girons. 1976. Contribution a la morphologie comparée des fosses nasales et de leurs annexes chez les lepidosauriens. *Mem. Mus. Natl. Hist. Nat. Paris* A98:1–106.

Gao, K., and M. A. Norell. 1998. Taxonomic revision of *Carusia* (Reptilia: Squamata) from the Late Cretaceous of the Gobi Desert and phylogenetic relationships of anguimorphan lizards. *Am. Mus. Novitates* 3230:1–51.

———. 2000. Taxonomic composition and systematics of Late Cretaceous lizard assemblages from Ukhaa Tolgod and adjacent localities, Mongolian Gobi Desert. *Am. Mus. Nat. Hist. Bull.* 249:1–118.

Gilmore, C. W. 1928. Fossil lizards of North America. *Mem. Natl. Acad. Sci.* 22:1–201.

Goldberg, S. R., and C. R. Bursey. 1990. Redescription of the microfilaria *Piratuba mitchelli* (Smith) (Onchocercidae) from the Gila monster, *Heloderma suspectum* Cope (Helodermatidae). *Southwest. Naturalist* 35:458–460.

———. 1991. Gastrointestinal helminths of the reticulate Gila monster, *Heloderma suspectum suspectum* (Sauria: Helodermatidae). *J. Helminthol. Soc. Wash.* 58:146–149.

Goldberg, S. R., and C. H. Lowe. 1997. Reproductive cycle of the Gila monster, *Heloderma suspectum* in southern Arizona. *J. Herpetol.* 31:161–166.

Goldberg, S. R., and D. D. Beck. 2001. *Heloderma horridum* (beaded lizard) reproduction. *Herpetol. Rev.* 32:255–256.

Gonzalez-Ruiz, A., E. Godinez-Cano, and I. Rojas-Gonzalez. 1996. Captive reproduction of the Mexican Acaltetepon, *Heloderma horridum*. *Herpetol. Rev.* 27:192–193.

Greene, H. W. 1997. *Snakes: The evolution of mystery in nature.* Berkeley: University of California Press.

Greger, R. 1996. Cell communication by autacoids and paracrine hormones. In: *Comprehensive human physiology: From cellular mechanisms to integration,* ed. R. Greger and U. Windhorst, 1:115–137. Berlin: Springer-Verlag.

Grow, D. T., and J. Branham. 1996. Reproductive husbandry of the Gila monster (*Heloderma suspectum*). *Adv. Herpetoculture* 57–64.

Harrisson, T. 1961a. *Lanthanotus borneensis*—Habits and observations. *Sarawak Mus. J.* 10:286–292.

———. 1961b. The earless monitor lizard, *Lanthanotus borneensis*. *Discovery* 22(July 1961):290–293.

———. 1963a. Earless monitor lizards in Borneo. *Nature* 198(4878):407–408.

———. 1963b. *Lanthanotus borneensis*—The first 30 live ones. *Sarawak Mus. J.* 11:299–301.

———. 1966. A record size *Lanthanotus* alive. *Sarawak Mus. J.* 14:323–334.

Hensley, M. M. 1949. Mammal diet of *Heloderma*. *Herpetologica* 5:152.

Hoffstetter, R. 1957. Un Saurien hélodermatidé (*Eurheloderma gallicum* nov. gen. et sp.) dans la faune fossile des Phosphorites du Quercy. *Bull. Géol. Soc. France* 7:775–786.

Horn, H.-G., M. Gaulke, and W. Böhme. 1994. New data on ritualized combats in monitor lizards (Sauria: Varanidae), with remarks on their function and phylogenetic implications. *Zool. Garten N.F.* 64:265–280.

Huey, R. B., and E. R. Pianka. 1981. Ecological consequences of foraging mode. *Ecology* 62:991–999.

John-Alder, H. B., C. H. Lowe, and A. F. Bennett. 1983. Thermal dependence of locomotory energetics and aerobic capacity of the Gila monster (*Heloderma suspectum*). *J. Comp. Physiol.* 151:119–126.

Johnson, J. P., and C. I. Ivanyi. 2001. *Beaded lizard (Heloderma horridum) North American regional studbook*. 3rd ed. Tucson: Arizona-Sonoran Desert Museum.

Jones, K. B. 1983. Movement patterns and foraging ecology of Gila monsters (*Heloderma suspectum* Cope) in northwestern Arizona. *Herpetologica* 39:247–253.

Kochva, E. 1974. Glandes specialisées de la machoire inferieure chez les Anguimorphes. *Récherches biologiques contaemporaines: Ouvrage dédiée a la Memoire du Dr. Manfred Gabe Vagner*, ed. L. Arvy, 281–286.

Lardner, P. J. 1969. Diurnal and seasonal locomotory activity in the Gila monster, *Heloderma suspectum* Cope. Ph.D. diss., University of Arizona, Tucson.

Loeb, L., C. L. Alsberg, E. Cooke, E. P. Corson-White, M. S. Fleisher, H. Fox, T. S. Githens, S. Leopold, M. K. Meyers, M. E. Rehfuss, D. Rivas, and L. Tuttle. 1913. The venom of *Heloderma*. *Publ. Carnegie Inst. Wash.* 177:1–244.

Loope, D. B., L. Dingus, C. C. Swisher III, and C. Minjin. 1998. Life and death in a Cretaceous dunefield, Nemegt Basin, Mongolia. *Geology* 26:27–30.

Loope, D. B., and L. Dingus. 1999. Mud filled *Ophiomorpha* from Upper Cretaceous continental redbeds of Southern Mongolia: An ichnologic clue to the origin of detrital, grain-coating clays. *Palaios* 14:452–459.

Loope, D. B., J. A. Mason, and L. Dingus. 1999. Lethal sandslides from eolian dunes. *J. Geol.* 107:707–713.

Losos, J. B., and H. W. Greene. 1988. Ecological and evolutionary implications of diet in monitor lizards. *Biol. J. Linn. Soc.* 35:379–407.

Lowe C. H., C. R. Schwalbe, and T. B. Johnson. 1986. *The venomous reptiles of Arizona*. Phoenix: Arizona Game and Fish Department.

Maruno, K., and S. I. Said. 1993. Small-cell lung carcinoma: Inhibition of proliferation by vasoactive intestinal peptide and helodermin and enhancement of inhibition by anti-bombesin antibody. *Life Sci.* 52: PL267–PL271.

McDiarmid, R. W. 1963. A collection of reptiles and amphibians from the highland faunal assemblage of western Mexico. *Contrib. Sci. Nat. Hist. Mus. Los Angeles County* 68:1–15.

McDowell, S., and C. Bogert. 1954. The systematic position of *Lanthanotus* and the affinities of the anguimorphan lizards. *Bull. Am. Mus. Nat. Hist.* 105:1–142.

Mertens, R. 1966. The keeping of Borneo earless monitors (*Lanthanotus*). *Sarawak Mus. J.* 14:320–322.

Mochca-Morales, J., B. M. Martin, and L. D. Possani. 1990. Isolation and characterization of helothermine, a novel toxin from *Heloderma horridum horridum* (Mexican beaded lizard) venom. *Toxicon* 28:299–309.

Murphy, J. B., and L. A. Mitchell. 1974. Ritualized combat behavior of the pygmy mulga monitor, *Varanus gilleni* (Sauria: Varanidae). *Herpetologica* 30:90–97.

Nagy, K. A., R. B. Huey, and A. F. Bennett. 1984. Field energetics and foraging mode of Kalahari lacertid lizards. *Ecology* 65:588–596.

Nikai, K. I., H. Sugihara, and A. T. Tu. 1988. Isolation and characterization of horridum toxin with arginine ester hydrolase activity from *Heloderma horridum* (beaded lizard) venom. *Arch. Biochem. Biophys.* 264:270–280.

Norell, M. A., M. C. McKenna, and M. J. Novacek. 1992. *Estesia mongoliensis,* a new fossil varanoid from the Cretaceous Barun Goyot Formation of Mongolia. *Am. Mus. Novitates* 3045:1–24.

Norell, M. A., and K. Gao. 1997. Braincase and phylogenetic relationships of *Estesia mongoliensis* from the late Cretaceous of the Gobi Desert and the recognition of a new clade of lizards. *Am. Mus. Novitates* 3211:1–25.

Nydam, R. L. 2000. A new taxon of helodermatid-like lizard from the Albian-Cenomanian of Utah. *J. Vertebr. Paleontol.* 20:285–294.

Perry, J. J. 1996. *Beaded lizard/Escorpion, Heloderma horridum ssp.* Silver Spring, Md.: Lizard Advisory Group, Taxon Management Account, American Zoo and Aquarium Association.

Pregill, G. K., J. A. Gauthier, and H. W. Greene. 1986. The evolution of helodermatid squamates, with description of a new taxon and an overview of Varanoidea. *Trans. San Diego Soc. Nat. Hist.* 21:167–202.

Proud, K. R. S. 1978. Some notes on a captive earless monitor lizard, *Lanthanotus borneensis. Sarawak Mus. J.* 24:235–242.

Ramírez-Bautista, A. 1994. *Los Reptiles y Anfibios de Chamela.* México, D.F.: Instituto de Biología Universidad Nacional Autónoma de México.

Ramírez-Velázques, A., and C. A. Guichard-Romero. 1989. *El escorpion negro: Combates ritualizados.* Chiapas, Mexico: Instituto de Historia Natural Tuxtla Gutierrez.

Raufman, J. P. 1996. Bioactive peptides from lizard venoms. *Regul. Peptides* 61:1–18.

Rieppel, O. 1983. A comparison of the skull of *Lanthanotus borneensis* (Reptilia: Varanoidea) with the skull of primitive snakes. *Zeit. Zool. Syst. Evol.-Forsch.* 21:142–153.

Schwenk, K. 1988. Comparative morphology of the lepidosaur tongue and its relevance to squamate phylogeny. In *Phylogenetic relationships of the lizard families,* ed. R. Estes and G. Pregill, 569–598. Stanford, Calif.: Stanford University Press.

Seppa, N. 2001. Reptilian drug may help treat diabetes. *Sci. News* 160(3):47.

Smith, A. J. 1910. A new filarial species (*F. mitchelli* n.s.) found in *Heloderma suspectum,* and its larvae in a tick parasitic upon the Gila monster. *Univ. Pennsylvania Med. Bull.* 23:487–497.

Sprackland, R. G. 1970. Further notes on *Lanthanotus. Sarawak Mus. J.* 18:412–413.

———. 1972. A summary of observations of the earless monitor. *Sarawak Mus. J.* 40–41:323–328.

———. 1999. Sarawak's earless monitor lizard. *Reptiles* 7:72–80.

Stahnke, H. L. 1950. The food of the Gila monster. *Herpetologica* 6:103–106.

Steindachner, F. 1878. Über zwei neue Eidechsen-Arten aus Sud-Amerika und Borneo. *Denkschr. Adad. Wiss., Vienna* 38:93–96.

Stevens, M. S. 1977. Further study of the Castolon local fauna (Arikareean: Early Miocene) Big Bend National Park, Texas. *Pearce-Sellards Ser. Texas Mem. Mus.* 28:1–69.

Strimple, P. 1995. Captive reproduction of the Gila monsters: A review. *Reptiles* 3(7):16–24

Thompson, G. G., P. C. Withers, and S. A. Thompson. 1992. The combat ritual of two monitor lizards, *Varanus caudolineatus* and *Varanus gouldii. Western Australian Naturalist* 19: 21–25.

Tu, A. T. 2000. Lethal toxins of lizard venoms that possess kallikrein-like activity. Natural and Selected Synthetic Toxins ACS Symposium Series 745:283–301. Washington, D.C.: American Chemical Society.

Underwood, G. 1957. *Lanthanotus* and the anguinomorphan lizards: A critical review. *Copeia* 1957:20–30.

Upton, S. J., C. T. McAllister, and C. M. Garrett. 1993. Description of a new species of *Eimeria* (Apicomplexa: Eimeriidae) from *Heloderma suspectum* (Sauria: Helodermatidae). *Texas J. Sci.* 45:155–159.

Wagner, E., R. Smith, and F. Slavens. 1976. Breeding the Gila monster *Heloderma suspectum* in captivity. *Int. Zoo Yrbk.* 16:74–78.

Wiegmann, A. F. A. 1829. Über das Acaltetepan oder Temaculcahua des Hernandez, eine neue Gattung der Saurer, *Heloderma*. *Isis von Oken* 22:624–629.

Yatkola, D. A. 1976. Fossil *Heloderma* (Reptilia, Helodermatidae). *Occ. Pap. Mus. Nat. Hist. Univ. Kansas* 51:1–14.

Zweifel, R. G., and K. S. Norris. 1955. Contributions to the herpetology of Sonora, Mexico: Descriptions of new subspecies of snakes (*Micruroides euryxanthus* and *Lampropeltis getulus*) and miscellaneous collecting notes. *Am. Midl. Nat.* 54:230–249.

Part III

9. Evolution of Body Size and Reproductive Tactics

Eric R. Pianka

Variation in size among species of varanoids, especially among *Varanus*, is impressive, ranging from only 20 cm in total length (weight, 8–10 g) in the diminutive *V. brevicauda* to a total length of over 3 m (weight, more than 150 kg) in the giant *V. komodoensis* (Pianka 1995). Even larger fossils are known, such as the enormous *Varanus* (formerly *Megalania*) *prisca* from Australia, which reached a massive size of 6 m and weighed over 600 kg.

The probable course of evolution of body size can be traced using the most recent and well-resolved phylogenetic tree proposed for extant varanoids (Ast 2001). Body sizes of hypothetical ancestors can be estimated as the average of those of their living descendants and plotted on the phylogeny (Fig. 9.1). Sizes at deeper nodes can be estimated from averages of shallower nodes. The sister group to *Varanus, Lanthanotus,* is a moderate-sized lizard about 55 cm in total length. *Heloderma* are larger, reaching about 80 cm in total length. Figure 9.1 shows that small body size has evolved three times among varanids, in the Australian *Odatria* clade, and, in the Asian clade, in *V. flavescens* and in the *prasinus* species complex. Most African monitors are similar in size to their ancestors, although *V. niloticus* has evolved a larger size. Large body size also evolved in *V. bengalensis* and *V. salvator* in the Asian clade and independently in the common ancestor to *V. salvadorii, V. komodoensis,* and *V. varius,* as well as in the Australian perentie *V. giganteus* (the gigantic *V. prisca* [formerly *Megalania*] presumably belongs to this clade as well).

Thompson and Pianka (2001) examined and reviewed various aspects of the evolution of reproductive tactics among monitor lizards. Body size influences reproductive tactics more strongly than phylogeny.

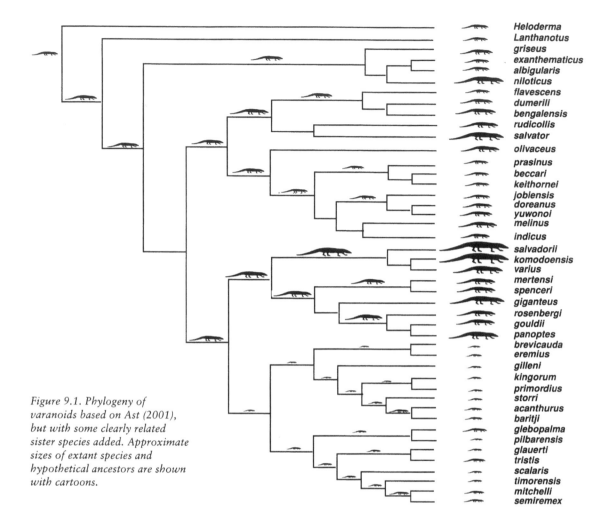

Figure 9.1. Phylogeny of varanoids based on Ast (2001), but with some clearly related sister species added. Approximate sizes of extant species and hypothetical ancestors are shown with cartoons.

Maximum snout–vent length (SVL) is positively correlated with egg mass, clutch size, clutch mass, neonate SVL, and neonate body mass. Incubation time is positively correlated with egg mass, neonate SVL, and maximum adult SVL. Incubation period of heavier eggs is proportionally less than for smaller eggs. Eggs of small species are laid in the spring and hatch in about 100 days, typically in the summer. Eggs of larger species are laid later, often overwinter, and hatch the next year. Although neonates of larger species are absolutely larger, smaller species have relatively larger hatchlings compared with adult size than do larger species. Clutch sizes are larger and more variable among larger species (Fig. 9.2).

However, relative to maximum SVL, clutch sizes for larger species are smaller than for smaller species. Maternal SVL influences clutch size much more strongly within a species than it does between species (Fig. 9.3).

The most recent phylogenetic tree proposed for extant varanoids (Ast 2001) can be exploited to trace the probable course of evolution of

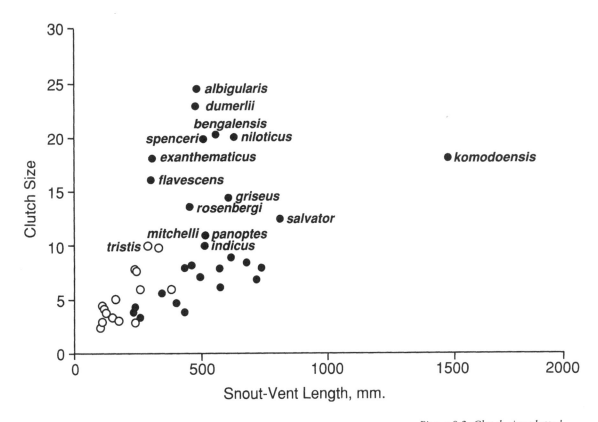

clutch size among varanoids. For example, clutch size could be plotted on the tree, and ancestral clutches sizes can be estimated from those of their descendants. However, such data points are not independent because of phylogenetic relatedness. Related species do not constitute independent observations because of shared common ancestry. A method, known as independent contrasts, developed by Felsenstein (1985, 1988), removes the effects of relatedness. Phenotypic data for N tip species of a monophyletic group are transformed into N – 1 independent "contrasts" or differences. Phylogenetically independent contrasts make comparisons within evolutionary radiations—immediate descendants of a common ancestor are compared with one another. By use of the computer program CAIC (Purvis and Rambaut 1995), independent contrasts were calculated for body size and for clutch size. Figure 9.4 plots these contrasts against each other and shows that after the effects of phylogenetic relatedness are removed, differences in clutch size remain correlated with differences in body size. Several very high independent contrasts suggest rapid evolution of clutch size. For example, V. tristis has an average clutch size of 10 eggs, whereas its smaller sister species V. glauerti has a clutch of only 3 eggs. Interestingly, relative clutch mass of V. tristis is similar to that of other sympatric desert varanids with smaller clutch sizes (Pianka 1994), suggesting that its larger clutch is achieved at the expense of relative neonate size (hatchling V. tristis are relatively small compared to adults). Another

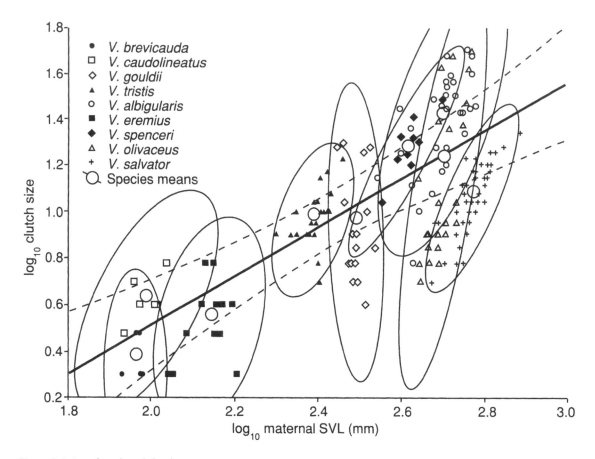

Figure 9.3. Log–log plot of clutch size versus SVL for nine species of varanids. Ellipses enclose data points for individuals within species (small symbols). Species means are shown with large open circles in the center of each ellipse (from Thompson and Pianka 1999).

example is *V. spenceri,* which has much larger clutch size (20 eggs) than does its similar-sized sister species, *V. mertensi,* which lays only about 8 eggs. *V. salvator* are considerably larger than their sister species, *V. rudicollis,* and the former species lay 12 eggs, whereas the latter lays 8 eggs.

Varanus species known to deposit their eggs in termitaria include African, Asian, and Australian species: *bengalensis, niloticus, prasinus, rosenbergi, salvator,* and *varius.* Large South American teiids in the genus *Tupinambis* also lay their eggs in termitaria (Pianka and Vitt 2003), an example of evolutionary convergence. Because the above six monitor species are scattered across Ast's phylogenetic tree, the habit of laying eggs in termite mounds may well have arisen multiple times within varanids. Of course, other monitor species could also use termitaria as nest sites, but it might not yet have been recorded.

Most large monitor species engage in ritual combat, with males standing erect on their hind legs and tail, chests pressed together, grappling with their forelegs wrapped around each other. Called the "clinch phase," these displays appear to be wrestling matches, with the two contenders trying to throw one another off balance. Sometimes both fall to the ground and continue wrestling while rolling over and over (Thompson et al. 1992). When this happens, the winner some-

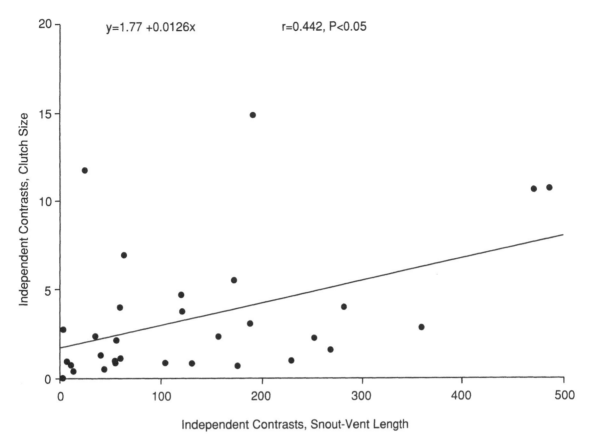

y=1.77 +0.0126x   r=0.442, P<0.05

Independent Contrasts, Clutch Size

Independent Contrasts, Snout-Vent Length

Figure 9.4. Independent contrasts in clutch size plotted against independent contrasts in SVL (see text).

times bites the loser (not in all species, however). Such combat rituals have been observed in most species of large monitors, including *V. albigularis, V. bengalensis, V. dumerilii, V. giganteus, V. gouldii, V. indicus, V. komodoensis, V. mertensi, V. niloticus, V. olivaceus, V. panoptes, V. salvadorii, V. salvator, V. spenceri,* and *V. varius* (King and Green 1999). Interestingly, *Heloderma* males engage in similar grappling matches (Beck and Ramírez-Bautista 1991), so combat is most likely a primitive trait, passed down from varanoid ancestors.

Male *Heloderma* have higher aerobic capacities than females (Beck et al. 1995), suggesting that sexual selection could have been a factor in the evolution of high aerobic capacities of varanoid lizards. However, female varanids occasionally engage in such wrestling, but the function of this remains unstudied. Smaller varanid species in the Australian subgenus *Odatria* have dispensed with the clinch phase and evolved a slightly different ritual: they do not stand, but roll around sideways on the ground in a similar grappling position, but with bellies pressed tightly together, both forelegs and hind legs wrapped around each other, and tails intertwined (Murphy and Mitchell 1974; Thompson et al. 1992). Sometimes they bite as well.

Snakes share many features with varanoids, including teeth and tooth replacement, skull structure, forked tongues, and behavioral

Evolution of Body Size and Reproductive Tactics • 553

traits such as their male combat dance (Estes et al. 1988; Beck and Ramírez-Bautista 1991; Lee 1997). A phylogenetic connection between snakes and varanoids has long been suspected (Camp 1923; McDowell and Bogert 1954) and has recently been validated in sophisticated cladistic studies (Lee 1997; Norell and Kequin 1997; Kequin and Norell 1998, 2000). Male rattlesnakes rear up, face off, press their bellies together, and weave from side to side in a combat dance that is essentially like that of large male varanids (however, they have no forelegs to wrap around each other).

References

Ast, J. C. 2001. Mitochondrial DNA evidence and evolution in Varanoidea (Squamata). *Cladistics* 17:211–226.

Beck, D. D., and A. Ramírez-Bautista. 1991. Combat behavior of the beaded lizard, *Heloderma h. horridum*, in Jalisco, Mexico. *J. Herpetol.* 25:481–484.

Beck, D. D., M. R. Dohm, J. T. Garland, A. Ramírez-Bautista, and C. H. Lowe. 1995. Locomotor performance and activity energetics of helodermatid lizards. *Copeia* 1995:577–585.

Camp, C. L. 1923. Classification of lizards. *Bull. Am. Mus. Nat. Hist.* 48:289–481.

Felsenstein, J. 1985. Phylogenies and the comparative method. *Am. Naturalist* 125:1–15.

———. 1988. Phylogenies and quantitative charactgers. *Ann. Rev. Ecol. System.* 19:445–471.

Estes, R., K. de Queiroz, and J. Gauthier. 1988. Phylogenetic relationships within Squamata. *Phylogenetic relationships of the lizard families: Essays commemorating Charles L. Camp*, ed. R. Estes and G. Pregill, 119–281. Stanford, Calif.: Stanford University Press.

Kequin, G., and M. Norell. 1998. Taxonomic revision of *Carusia intermedia* (Reptilia: Squamata) from the upper Cretaceous of Gobi Desert and phylogenetic relationships of anguimorphan lizards. *Am. Mus. Novitiates* 3230:1–51.

———. 2000. Taxonomic composition and systematics of Late Cretaceous lizard assemblages from Ukhaa Tolgod and adjacent localities, Mongolian Gobi Desert. *Bull. Am. Mus. Nat. Hist.* 249:1–118.

King, D., and B. Green. 1999. *Monitors: The biology of varanid lizards.* Malabar, Fla.: Krieger.

Lee, M. S. Y. 1997. The phylogeny of varanoid lizards and the affinities of snakes. *Phil. Trans. R. Soc. Lond.* B 352:53–91.

McDowell, S., and C. Bogert. 1954. The systematic position of *Lanthanotus* and the affinities of the anguimorphan lizards. *Bull. Am. Mus. Nat. Hist.* 5:1–142.

Murphy, J. B., and L. A. Mitchell. 1974. Ritualized combat behavior of the pygmy mulga monitor lizard, *Varanus gilleni* (Sauria: Varanidae). *Herpetologica* 3:90–97.

Norell, M. A., and G. Kequin. 1997. Braincase and phylogenetic relationships of *Estesia mongoliensis* from the Late Cretaceous of the Gobi Desert and the recognition of a new clade of lizards. *Am. Mus. Novitiates* 3211:1–25.

Pianka, E. R. 1994. Comparative ecology of *Varanus* in the Great Victoria Desert. *Austr. J. Ecol.* 19:395–408.

———. 1995. Evolution of body size: Varanid lizards as a model system. *Am. Naturalist* 146:398–414.

Pianka, E. R., and L. J. Vitt. 2003. *Lizards: Windows to the evolution of diversity.* Berkeley: University of California Press.

Purvis A., and A. Rambaut. 1995. Comparative analysis by independent contrasts (CAIC): An Apple Macintosh application for analysing comparative data. *Comput. Appl. Biosci.* 11:247–251.

Thompson, G. G., P. C. Withers, and S. A. Thompson. 1992. The combat ritual of two monitor lizards, *Varanus caudolineatus* and *Varanus gouldii. W. Austr. Naturalist* 19:21–25.

Thompson, G. G., and E. R. Pianka. 1999. Reproductive ecology of the black-headed goanna *Varanus tristis* (Squamata: Varanidae). *J. R. Soc. W. Austr.* 62:27–31.

———. 2001. Allometry of clutch and neonate sizes in monitor lizards (Varanidae: *Varanus*). *Copeia* 2001:443–458.

10. Keeping Monitors in Captivity: A Biological, Technical, and Legislative Problem

HANS-GEORG HORN

Preliminary Remarks

Before coming to the main point of this contribution, I feel it's necessary to say a few words about my attitude toward monitors. This can be done best citing an Austrian herpetologist, Franz Werner: "Varanids form a small and well-characterized family containing only one genus *Varanus*. They belong to the proudest, well-proportioned, most powerful and most intelligent group of lizards" (Werner 1904). This striking statement needs no further comment.

Most people, professional herpetologists as well as keepers (both private keepers and zookeepers) treat monitors as simple, primitive reptiles. But this is far from correct. Two statements, both by physiologists investigating oxygen consumption, heart efficiency, and energy output of monitors established the following: "*Varanus gouldii* is shown in this study to possess a standard metabolic rate equal to that of comparably sized lizards. However, its capacities for transporting oxygen during activity are high and maximal oxygen consumption *exceeds* levels of *homeotherms* of equal size" (Bennett 1972). And "Varanid lizards have a higher oxygen requirement than other reptiles. This is reflected in the control of ventilation, the specialized lung morphology, the high arterial saturation due to low intracardiac shunting, pH regulation and other *mammal-like* features of *Varanus*" (Wood et al. 1977).

Furthermore, ecological investigations of lizard communities in Australia lead to the conclusion that monitors occupy the same ecological niche as certain mammals in other countries. Pianka (1969) emphasized, "but it is fairly safe to assert that the large monitor lizard *V. gouldii* is an ecological analogue of the North American kit fox (*Vulpes* spec.)."

The unusual staying power, which monitors may demonstrate by swimming through bays or straits or by running down prey, is only comprehensible on the basis of oxygen consumption and heart efficiency in these reptiles, and the unique mechanism of oxygen uptake in running monitors (investigated in *Varanus exanthematicus*, but results can be assumed to be valid for all varanids). As well as the usual method of oxygen uptake as in any reptile (bilaterally operating muscles), monitors pump additional oxygen into their lungs by means of gular pumping (Owerkowicz et al. 1999).

Additionally, monitors exhibit a number of unusual behavioral strategies as compared with other reptiles. For example, methods of reproduction in some species may include breeding in termite mounds; strategies of hunting and overpowering prey in most species may involve hurling tactics (a rapid tossing of the head and prey to and fro while eating); and *Varanus niloticus* cooperates in raiding nests of crocodiles for eggs. A comprehensive overview of such behavioral strategies that led to evolutionary efficiency and success in monitors is described in Horn (1999).

Thus, we should emphatically keep in mind that monitors cannot be treated as simple, primitive reptiles.

Introduction

Why should monitors be kept in captivity? One reason may be scientific interest. Fields of investigation worthy of study include oxygen consumption and heart efficiency (which I discussed above), biomechanical measurements, or biochemical research (e.g., for the hormonal level depending on season). Obviously, such investigations cannot be carried out in the wild, because the investigator needs a laboratory. Another scientific field may be ethology, which may require filming of, say, ritualized movements.

In exhibitions such as zoos and aquaria, monitors are displayed, thus becoming material of biological education for the public. For several decades, both private keepers and zookeepers have tried to investigate the biological basis of reproduction of these reptiles. There are several reasons why keepers want to be able to breed monitors in captivity. First, a vanishing nature, with less common flora and fauna, does not allow capturing wildlife without limit. Second, a need exists to substitute younger animals for old ones. It soon may become even more important to conserve monitors through captive propagation because of the increasing hole in the ozone layer over Australia. Monitors are better protected by their skin than humans. But monitors must bask every day, and therefore increasing ultraviolet (UV) radiation may damage varanid genes in the wild. Thus, a gene pool of self-sustaining and flourishing populations in captivity is needed for reintroduction into the wild.

Here, I will discuss the factors that limit reproductive efforts (which of course include the simple keeping of these animals), particularly the biological, technical, and legislative aspects thereof. Because of lack of space, most problems will be indicated only briefly.

Biological Problems

It is a widely accepted view in modern breeding biology that individuals of very distant places of origin should not be interbred and species and subspecies should not be hybridized. One of the most limiting factors of reproductive efforts in monitors is our restricted knowledge of the taxonomy in this family—and further, dependent on this, the distribution of taxa. Different scientific methods often lead to different taxonomic categories. For example, Robert Mertens put *V. exanthematicus* and *V. flavescens* into the same subgenus *Empagusia*. This was based on external morphology, and Mertens did not consider convergence. It seems that keepers of the yellow monitor have been led astray by this because they often kept *V. flavescens* like a Savannah monitor. Only recently has Auffenberg et al. (1989) demonstrated *V. flavescens* to be a dweller of water-rich areas, although Carleyle had called *V. flavescens* an aquatic species as early as 1869! King and King (1975), Holmes et al. (1975), King et al. (1991), and Baverstock et al. (1993), by means of karyotypic analysis, electrophoretic properties of (four) blood proteins, and microcomplement fixation of serum albumins; Branch (1981), Böhme (1988), and Ziegler and Böhme (1997), by means of hemipeneal morphology; and Becker et al. (1989) and Becker (1991), by means of lung morphology, demonstrated that these two species do not belong to the same subgenus.

According to these researchers, species of monitors are now arranged in species clusters. This leads to a good overview on taxonomic and evolutionary relationships, but it is not suitable for combining sexes of species. If we leave the subgeneric level and go down the hierarchy to the species or subspecies level, difficulties increase drastically in combining animals belonging to the same species for reproductive purposes. The species/subspecies concept is still based mainly on morphology. Let me discuss this with two examples: the *gouldii-panoptes* and the *yemenensis-exanthematicus-albigularis* groups.

If we compare the appearances of *V. p. horni* from New Guinea, *V. p. panoptes* from northern Australia, and *V. p. rubidus* from, say, Wiluna in Western Australia, it is not difficult to identify and to name these species/subspecies correctly, although they look relatively similar. Most of the remaining distribution area of all these members of the *gouldii-panoptes* group is occupied by *V. g. flavirufus*. While traveling through Central Australia, I frequently encountered *V. g. flavirufus* (the Central Sand goanna), and I could easily identify this orange-eyed and characteristically colored and patterned subspecies of *gouldii*. But then, near Wallara Ranch in Central Australia, which lies in the center of the distribution area of *V. g. flavirufus*, I encountered a totally dissimilar-looking specimen. Unfortunately, this specimen could not be caught to count scales, so it remained unclear whether it was a different species or subspecies. Nevertheless, this specimen and *V. g. flavirufus* occur in the same distribution area, and so we could try to breed these. However, the three subspecies of *V. panoptes* ssp., mentioned above, although they look relatively similar, would not be allowed to be used for breeding purposes.

The African monitors of the *exanthematicus-albigularis* group and the recently described *yemenensis* of Saudi Arabia present even greater difficulties for identification. This is because the distribution area of all these forms (in part) is not clearly defined and because zones of overlap between species or subspecies may exist. But we do not know their border lines, nor do we know whether any are hybrids. An additional difficulty arises if we don't know the accurate place of capture of a specimen.

Although it is relatively simple to distinguish *V. exanthematicus* on one hand and *V. yemenensis* and the forms of *V. albigularis* on the other, it is difficult to distinguish between, say, *V. yemenensis* and *V. a. microstictus*. Both species have a pale yellow band on their snout; their size and general appearance are similar; and their distribution area is cut only by the Red Sea. This means that if one intends to set up a breeding pair of one of those species, one has to count scales and carry out electrophoretic investigations of blood proteins, then compare results with original descriptions to ensure accurate identification. If we go down the African east coast and then up the west coast, we might get into trouble with the white-throated and black-throated *V. a. albigularis*. Then the question arises: what is a *V. a. angolensis*? Is this an *albigularis* or an *exanthematicus* (in northern Angola)? We only may assume the dense and huge rain forests of middle Africa act as a fauna barrier that stops immigration of *V. exanthematicus* to the south. On the other hand, in the possible southern distribution area of *V. a. microstictus*, a form of monitor has been encountered in eastern part of Tsavo National Park in Kenya that looks totally different from a "typical" *V. a. microstictus* or a *V. a. albigularis*. For a series of color plates of these different species and subspecies of monitors, see Röder and Horn (1994).

Similar problems exist in the *salvator* and the *indicus* groups, even in well-described species such as the emerald or lace monitor. Thus, difficulties arise in combining the sexes to make a breeding pair for similar taxonomic reasons.

Ethological problems may occur because of uncertainties in taxonomy. For instance, *V. p. panoptes* of northern Australia often assumes an upright bipedal position; this is rarely seen in *V. g. flavirufus*. There may be other, minor, remarkable behavioral characteristics that may be important in reproduction of those species. If one tries to combine sexes of different species, reproduction will never occur, because the species simply do not "understand" each other. Furthermore, members of the same species inhabiting a wide area of distribution may have evolved different times of sexual activities, depending on slightly different seasons.

What can be done in practice to avoid such problems? One point has already been mentioned above: we can compare original descriptions of scale numbers and apply biochemical methods to unknown specimens. Or we have to personally catch a dozen or so individuals occurring in a local area (but see the section on legislative problems with this practice!). Or if we intend to buy such species from a dealer, we would have to buy all the individuals he offers, hoping they have been caught from a single collection area.

With live animals, there is no chance that a dealer in the field or in the shop will carry out karyotyping or evert hemipenes looking for their structure. Of course, this last point is especially true with females. The user of taxonomy needs simple, easily identifiable, external morphological characteristics—for example, the Australian central sand goanna, *V. g. flavirufus*, is a yellow- to orange-red-eyed species. All other forms of *gouldii* or *panoptes* seem to be black- or brown-eyed.

It could appear to be a trivial assessment that a male plus a female are required for reproduction of monitors. But it is not as trivial as it seems because discrimination of the sexes in these reptiles is difficult. Varanids are (nearly) monomorphic, which means the sexes cannot be distinguished easily from each other. Therefore, some sexing techniques applied to monitors are detailed below, along with some comments.

- Head proportions. Reptile keepers have tried to distinguish sexes in monitors by head proportions (especially width). This is not very reliable and can be applied only if a great number of specimens of roughly the same total size can be compared.

- View of bulges posterior to cloaca. This is not a very reliable technique because not every adult male bulges out clearly, and juveniles do not or rarely bulge out.

- Pre- and postanal scalation. Auffenberg (1981) demonstrated a statistical relationship between sex and postanal scalation in male *V. bengalensis*. But results are somewhat disappointing: only 96% of males show such scalation.

- Lateral postanal scales. Male *V. tristis* have special lateral postanal scales near the base of the tail, as does *V. storri*. In *V. storri,* this scalation can be seen only in males if inspection takes place by vertical view from above to the base of the tail (Flugi 1990). This is not a reliable way to determine sex because of individual differences, and this characteristic is restricted to a few species.

- Depth of hemipeneal and hemicliteral pockets. Use of a probe to determine the depth of hemipeneal and hemiclitoral pockets (Honegger 1978) is dangerous (penetration of bottom of pockets) and not very reliable (Card 1995). Horn and Visser (1988) found two different depths (7.4 and 3.6 cm) in the two hemipeneal pockets of a large male *V. giganteus*.

- Ethological observations. Only very experienced keepers of monitors may use this technique. In wild-caught, relatively large species (e.g., *V. bengalensis, V. salvator, V. varius*), females tend to be shyer in general, and if food is offered, males tend to accept it (e.g., live mice or rats) far earlier. Successful rate of sex determination by ethological observation may reach 80%.

• Hormone analysis. This method—determination of testosterone and/ or estrogen level—has been applied to *V. komodoensis* (Judd et al. 1977) and *V. albigularis* (Phillips 1994). Testosterone level is extremely high in males as compared with females and increases dramatically during the breeding season in both species. Estrogen level also increases in female *V. albigularis* during reproduction. This technique needs a well-experienced investigator and a very well equipped laboratory. It is never applied to dwarf monitors and medium-sized species.

• Ultrasonography or ultrasound. Abdominal ultrasonographic imaging has been used widely for sex determination (and of course to examine other organs) in monitors. Females can be identified on the basis of follicle dimensions down to 1 to 2 mm, depending on the resolution of the instrument. Growth of eggs can be followed too, during reproductive time. This method has been applied to *V. indicus* (Schildger 1992), *V. albigularis* (Morris and Alberts 1996), *V. komodoensis* (Hildebrandt et al. 1996; Spelman 1998), *V. gouldii,* and *V. indicus* (Schildger et al. 1999, 2000; Schildger 1999/2000). The advantage of this method is that it is noninvasive and therefore can be applied many times; but sex determination in males seems to be less successful.

• Endoscopy. Recently, endoscopy has frequently been used for sex determination in birds and has been transferred with excellent success to reptiles, including monitors (Schildger and Wicker 1992; Schildger 1998, 1999/2000). A total of 252 specimens comprising *V. acanthurus, V. gilleni, V. prasinus, V. storri, V. timorensis, V. tristis, V. bengalensis, V. doreanus, V. dumerilii, V. exanthematicus, V. gouldii* ssp., *V. griseus* ssp., *V. indicus* ssp., *V. mertensi, V. niloticus, V. rudicollis, V. salvadorii, V. salvator cumingi,* and *V. varius,* have been investigated endoscopically and sexed successfully (Schildger 1999/ 2000; Schildger and Wicker 1992). This method needs a highly experienced veterinary surgeon and a well-equipped laboratory. Its disadvantage is that it is an invasive method, which means a specimen has to be anesthetized with a modern anesthetic such as isoflurane. The mortality rate in 700 investigated reptiles (including monitors) was zero (Schildger 1998, 1999/2000).

• Radiography. This is an excellent, noninvasive X-ray method for sex determination in semiadult to adult male specimens of monitors. The observer looks for mineralized hemipeneal bones in male monitors. This technique has been successfully applied to *V. prasinus* (Juschka, Löbbecke Museum, personal communication), *V. komdoensis* (2.5 years old; Meier, Honolulu Zoo, personal communication), *V. doreanus* (adult), and *V. varius* (semiadult) (Horn, unpublished data). A well-equipped laboratory and an experienced investigator are needed. Disadvantages include the fact that females cannot be sexed this way;

neither can juveniles, because mineralization takes place at a certain age in monitors. So far as is known, adult males of all species have these mineralized bones.

From the above list of sexing methods in monitors, I conclude that combination of sexes in monitors remains a difficult problem, and only three methods (ultrasonography, endoscopy, and radiography), with some minor limitations, produce reliable results.

In this context, another fascinating hypothesis concerning use, sense, and reproductive efficiency of the hemipenes in snakes and varanids has to be mentioned. Three species of snakes—*Chondropython viridis, Epicrates c. cenchria,* and *Corallus caninus*—differ significantly in use of the two hemipenes. During copulation, the male *Chondropython viridis* changes from the right to the left side of the female, and vice versa, many times. The same is true for *Corallus caninus.* Such reptiles may be called multiple or two-side copulators. In contrast with this, copulation in *Epicratis c. cenchria* takes place in most cases from one side only, and hence this species may be called a mono- or one-side copulator. The difference has been explained by the structure of the hemipenes. Two-side copulators possess a simple, straight, even hemipenes structure, whereas one-side copulators have deeply bifurcated hemipenes, which allow a better fixing during copulation. If the keeper separates a two-side copulator from the female, allowing copulation only from one side, reproductive success is reduced dramatically, to less than 50% (Böhme and Sieling 1993).

According to observations in the wild, the lace monitor, *V. varius,* is a two-side copulator. This is easily demonstrated by its hemipenes structure. In contrast to *V. varius,* the hemipenes structure of *V. komodoensis* and *V. bengalensis* is deeply bifurcated, as is the case in *V. exanthematicus.* Although observations in *V. komodoensis* and *bengalensis* are still lacking, we observed a one-side copulation in *V. exanthematicus.* All eggs deposited were fertile. Some time ago, D. King (personal communication) informed me that *V. rosenbergi* is also a two-side copulator.

Again, I have to remind the reader of the taxonomic problems mentioned before. Another problem not yet understood are the differences of mean activity temperatures in the wild and in captivity (King 1991; King and Green 1993). As we can learn from these data, the mean activity temperature of several species, taken in the wild, is significantly higher than those taken in captivity. At present, no explanation can be given, but one can imagine the consequences—for example, digestion is not as efficient in captivity, compared with in nature, and difficulties in reproduction may arise as a result. For example, development of eggs in the ovaries or sperm production may be inhibited.

Another problem arises from diets supplied to captive lizards. Recently published results of determination of stomach contents (James et al. 1992) demonstrated that most smaller species are insectivorous, and larger species prey on vertebrates, mainly on lizards. But in captivity, mainly rats, mice, hamsters, chicks, and canned food are offered to larger species; smaller species are fed with baby mice and a few species of insects, e.g., crickets, cockroaches, locusts, mealworms, and a few species of moth. This may lead to wrong amino acid combinations for protein synthesis in the animal fed in this way. Unfortunately, only one investigation on amino acid compositions of varanid proteins has been carried out, but results of similar investigations in dendrobatid frogs (Birkhahn 1991) can be taken for comparison. These data will illuminate the problem. This author investigated the amino acid content of proteins by use of an amino acid analyzer and dried material of different species of Dendrobatidae (e.g., *Dendrobates leucomelas, auratus,* and *tinctorius*). Studied were wild-caught specimens and captive-bred but deformed offspring of the same species. The amino acid content of *Drosophila melanogaster* and of crickets (*Acheta domestica*) was investigated in the same manner. As can be expected, there were differences in amino acid contents in different species of dendrobatid frogs. But it was astonishing that there was a clear difference in amino acid content of normal (wild-caught) species of Dendrobatidae and of the captive-bred but deformed offspring of the same species. Obviously, there is a clear difference in amino acid content between insects used for diet and their consumers. But if feeding is based on a biased diet in the parental generation, this could lead to amino acid deficiencies in the filial generation, thus generating deformed offspring. Wrong and insufficient amino acid combinations in the diet disturb protein synthesis.

What can be done to feed monitors properly? First, the diet should contain as many different species of vertebrates and invertebrates as possible. Second, the diet needs to be rich in trace elements and vitamins; the habit monitors have of rubbing their prey through the soil appears to be related to this need. Third, if similar investigations in monitors as in frogs have been carried out, it might be necessary to feed captive animals artificial amino acid combinations, such as those available for space pilots, in addition to their usual diet.

It is beyond the scope of this contribution to discuss such details as cage size, length of artificial daylight, and intervals of feeding. Such details (e.g., cage sizes depend on legislative prescriptions created in Germany and may be different in other countries) may be taken from textbooks (e.g. Eidenmüller 2003; Vincent and Wilson 1999) or published articles (e.g. Horn and Visser 1989, 1997).

Technical Problems

Keeping and breeding fishes is a comparatively simple procedure compared with keeping reptiles, because industry provides aquarists with a lot of technical equipment—e.g., heating filaments, numerous types of filters and pumping systems, specialized electric bulbs and other electric equipment, standardized diets. Even field investigations are simplified by offering complete sets of chemicals and different instruments for pH and conductivity measurements. Similar equipment for terraria and reptile keepers is not available. If one intends to construct terraria, he is forced to construct cages on his own. He has to make use of electric equipment of different types (e.g., electric bulbs, heat lamps, fluorescent tubes with daylight spectrum) manufactured for different purposes. Bigger heating filaments and similar heating elements are available, but they have been produced for other purposes—for example, the heating of retorts in chemical industry. Thermometers and hygrometers are available but are often of low stability over the long term. The same is true for wire mesh panels made from steel, copper, or aluminum. All of these materials mentioned are expensive because they were developed for special purposes, different from those required by herpetoculturists. Thus, these technical problems are also limiting factors for keeping monitors in captivity.

A more detailed discussion seems to be necessary on light requirements in terraria. Of course, these considerations may be of some importance to many other species of reptiles.

Ultimately, the sun is the energy source of all biological processes of this planet. Therefore, the sun, along with a few characteristics of sunlight, have to be taken as a model for all lighting appliances used in terraria.

Sun emits a continuous spectrum comprising UV-C, UV-B (280–315 nm), UV-A (315–400 nm), visible light, and infrared (IR), with wavelengths of about 200 to 800 nm as the most interesting part. Further: In physics, a black body is defined as a body with an absorptivity of 1, which means the quotient of incident and absorbed radiation energy equals 1. Sun can be named approximately a black body radiator with a color temperature of approximately 6000 K. Color temperature is defined as that temperature of any radiator where the color of the emitted radiation is the same as that of a black body radiator. For a more detailed discussion on the color temperature of the sun, see Horn (2004).

Let me give a few examples to illustrate this term. An area of 1 cm^2 behaving like the surface of a black body with color temperature of 6000 K emits 7.35 kW. If the color temperature is reduced to 3000 K (comparable to a light bulb), the emitted radiation energy is reduced to 460 W (=6%), and a further reduction to 1500 K (comparable to a candle) results in a reduced radiation energy of 13 W (= 0.2%) (Horn 2004).

A number of commercially available lighting appliances emit a continuous spectrum between 400 to 700 nm; some emit between 300 to 750 nm, and all of these are additionally characterized by a high color temperature. For instance, Osram produces fluorescent tubes named Biolux with a color temperature of 6500 K and a nearly continuous spectrum between 425 to 700 nm; UV and IR are of relative low intensity. If these lamps are combined with special switch panels, they produce a light flux of 1100 to 3700 lm and thus are suitable for keeping reptiles and amphibians.

Far more suitable, but depending on their spectral power and especially the height and general dimensions of the terraria in which they are to be used, are metal-halide lamps (Dobrusskin 1971; Reinhold 1976) with special rare-earth additives and/or xenon-vapor pressure lamps. Such lamps (HQI Powerstars, HMI lamps, XBO lamps; all manufactured by Osram) emit a spectral power distribution similar to that of sunlight, which reaches from UV to IR, and they show a color temperature of 6000 K. Thus, they are suitable for all species of monitors (Horn 2004). A comparison of the spectral power distributions of various metal halide lamps with different rare-earth halide additives and with that of daylight 55 (color temperature 5500 K) can be found elsewhere (Dobrusskin 1971; Horn and Visser 1997). Disadvantages of these lights are their (relatively) high cost and their high power, which make them unsuitable for small terraria.

Apart from—as it seems—a never-ending discussion on usefulness or disadvantages of UV light in keeping and rearing of reptiles, including monitors, a serious technical problem does exist. Most keepers, both private and professional, are convinced of the benefit of UV (Horn 2004). But what part of the UV spectrum? UV-A, UV-B, or both?

Numerous investigators have revealed that vitamin D_3 synthesis from a precursor takes place in the skin under exclusive influence of UV-B radiation. For a detailed discussion on this point, see Horn (2004). But for the point under discussion, here are the problems involved in generation of UV light in general and in UV-B especially.

Gehrmann (1987) has determined the contributions of UV-B, UV-A, and visible light in a number of various tubular lamps commercially available in the United States. Only two sunlamps (Westinghouse FS 20 and FS 40) irradiated UV-A and UV-B with relatively high intensity; all the others investigated either irradiated with relatively low intensity or showed no UV-A and UV-B. This means that most tubular lamps except these two are of no value in synthesis of vitamin D_3, vital for bone production, which thus does not occur in animals reared under these lamps. Related to these more general considerations are some practical measurements on the skin of monitors.

V. spenceri may be taken as an example. This species has a series of gray and white dorsal bands. With the support of Phillips Research Laboratories, Aachen, I obtained transmission and reflectivity spectra of shed skin of *V. spenceri* and several other monitors. The range of wavelength covered 200 to 2300 nm, which means from UV-C to far IR. The most interesting part of this spectrum for practical use concerns the range from 290 to 750 nm, taken separately for gray and bright parts of the dorsal pattern of *V. spenceri*. It was unexpected that there was nearly no or only a slight transmission or reflectivity in the UV-B plus UV-A range (~280 to 400 nm). But this makes sense. Because this is the most energetic part of the (visible) spectrum, the skin should be protective.

These measurements seem to coincide with another observation of my way of rearing young varanids: I never irradiated these with UV light. There were no visible, rickety deformations of the pelvic girdle in these animals. But I used a lot of mineral salts and some vitamins to avoid trace element deficiencies. These observations are in contrast to several published articles that deal with UV light and reptile keeping.

Legislative Problems

As I already have pointed out, keeping of monitors in captivity can be necessary for scientific reasons or of breeding purposes—e.g., to diminish capture from the wild, or for reintroduction to the wild because of diminishing populations depending on, for example, exploitation for the leather trade.

But legislation, beginning with the Washington Convention (CITES; all species of monitors are listed in appendix I and II of this convention) and in most cases, the more restrictive national regulation impedes more and more the keeping of all wild animals and of monitors too. This is especially true for Germany. Let me give a few examples. *V. salvadorii,* the Papuan monitor, is listed in appendix II of the Washington Convention. If, for instance, one wants to import *V. salvadorii,* our federal office in one case said no, initially because *V. salvadorii* is believed to be a rare species. Some time later, import of this monitor was not allowed because it allegedly grows to 4.5 m total length, making it a "dangerous monster." But no census of this species has been undertaken (even its exact distribution is not known in New Guinea), and there is no reliable report of animals of this length, although the cited length has been reported in a few books. Two or three years later, a smuggler tried to pass through Germany with his luggage filled with dozens of monitors, 10 *V. salvadorii* among them (Horn 1995). The same federal institutions were happy that the above questioner took over several of these "dangerous monsters." Sometimes it happens that if one applies for an import permit of other species of reptiles the administration initially says no, but if your attorney applies for that permit, the officers agree (Lambertz 1993). If you successfully breed an endangered species with, say, an illegal male and a legal female, officers have already suggested castration of the illegal part of the parents' generation or suggested that they be put into different cages.

Of course, these restrictions, along with arbitrary actions, impede, and even make impossible, continuous ecological, zoological, and taxonomic research and work. But none of the legislation acts really stopped the importation of leather skins in huge numbers either to Europe or to the United States. Let me give one example. Between 1975 and 1980, a total of 2,559,008 skins of monitors, including *V. bengalensis, V. flavescens* (a relatively rare species), and *V. indicus* (also a relatively rare species), were exported from their countries of origins (Anonymous 1983). The author commented on export of this huge number of monitor skins: "The trade in live specimens is relatively small and insignificant in terms of the exploitation of wild populations." The newest data on the export of skins of *V. niloticus* from only one African state (The Gambia) have been reported by Lenz (1995): within 3 years (1990, 1991, 1992), 119,405 skins plus 67,279 watch straps and numerous belts and handbags were exported to Germany. In contrast to these huge numbers of skins, only 522 live specimens reached Germany during the same time (Lenz 1995).

In this context, a thoroughly prepared contribution on exploitation of the Australian herpetofauna by Ehmann and Cogger (1985) has to be mentioned. They defined collecting of reptiles and frogs as mortality and estimated the annual mortality rate of collecting for research to be 0.005%; they found the same amount for the collection of sea snake skins plus amateur collecting. This means collecting of both types does not contribute significantly to mortality defined as above and does not lead to a decline of Australian reptiles. Monitors do not play an important role in this census, and this can be assumed to be true for all other countries of origin for monitors.

I suspect politicians produce these types of laws on nature and conservation to convince an uninformed public that something is being done, which means that these laws exist mostly for public relations. The world's first environmental conference was held in Rio de Janeiro in 1992. Did this stop killing of the whales, destruction of the rain forests, or the leather trade of wildlife? No! Not at all! Further, I suspect biologists, who promote these activities, are only interested in the ecological niche they specialize in. Michael Tyler (Adelaide), one of the leading frog biologists of the world, commented on legislation: "Wildlife laws are easier to establish in ignorance than repeal in intelligence." The last chance for many species to survive by captive propagation will be lost because of restrictive international and national laws.

In particular, keepers of monitors and giant snakes—private people as well as zoo personnel—have to suffer under the bureaucracy of such regulations. Therefore, in Germany an umbrella organization called BNA, which means "umbrella society of nature and protection of species," was founded in 1985. Wildlife-keeping organizations, clubs, and sanctuaries form this society. BNA now has more than 150,000 members through its suborganizations and has cooperating groups in Denmark, the Netherlands, Switzerland, and other European countries. I was elected president and held this position from 1985 to 1991. The Society of German Zoo Directors cooperated with me. We tried to have some significant and intelligent influence on the legislation of fauna and flora in Germany. But success remained moderate. One of the big political parties, the Social Democrats, supports an extremely restrictive policy, saying that animal keeping does not fit into modern times, whereas the other big party, the Christian Democrats, follows a less restrictive kind of policy. Of course, not all difficulties and impediments in keeping wildlife—the numerous vexations and the complex bureaucracy—can be discussed here, because of lack of space. But one can imagine that these problems based on legislation nearly stop all activities, or make them difficult for both for private and professional keepers of wildlife as well as for scientists.

Epilogue

Now, if you did a good job with your animals, despite the biological, technical, and legislative problems involved in undertaking such care, you may be happy to breed a few species of monitors. Recent results on breeding of numerous species of monitors, including *V. s. salvator, V. s. cumingi, V. salvadorii,* and *V. varius,* can be found in comprehensive overviews, along with details on incubation methods, diets applied by different keepers, terraria used, climatic considerations, and so on (Broer and Horn 1985; Horn and Visser 1989, 1997; Horn 1991).

Acknowledgments

I acknowledge scientific support by Dr. Hörster and Dipl. Phys. G. Frank, Phillips Research Laboratories, Aachen, and several friends for their longtime support. I would especially like to mention Prof. Dr. W. Böhme, Bonn; Wolfgang Broer, Dortmund; PD Dr. B. Schildger, Berne; and Eduard Stirnberg, director of the Bochum Zoo.

References

Anonymous. 1983. International trade in skins of monitor and Teju lizards 1975–1980. *IUCN Traffic Bull.* IV 6:71–79.

Auffenberg, W. 1981. Combat behaviour in *Varanus bengalensis* (Sauria: Varanidae). *J. Bombay Nat. Hist. Soc.* 78:54–72.

Auffenberg, W., H. Raman, F. Iffy, and Z. Peruvian. 1989. A study of *Varanus flavescens* (Hardwicke and Gray) (Sauria: Varanidae). *J. Bombay Nat. Hist. Soc.* 86:286–307.

Baverstock, P. R., D. King, M. King, J. Birell, and M. Krieg. 1993. The evolution of species, of the Varanidae: Microcomplement fixation analysis of serum albumins. *Aust. J. Zool.* 41:621–638.

Becker, H. O. 1991. The lung morphology of *Varanus yemenensis* Böhme, Joger and Schätti, 1989, and its bearing on the systematics of the Afro-Asian monitor radiation. *Mertensiella* 2:29–37.

Becker, H. O., W. Böhme, and S. F. Perry. 1989. Die Lungenmorphologie der Warane (Reptilia: Varanidae) und ihre systematisch stammesgeschichtliche Bedeutung. *Bonner Zool. Beitr.* 40:27–56.

Bennett, A. 1972. The effect of activity on oxygen consumption, oxygen debt, and heart rate in the lizards *Varanus gouldii* and *Sauromalus hispidus. J. Comp. Physiol.* 79:259–280.

Birkhahn, H. 1991. Neue Erkenntnisse über die Aminosäureversorgung bei Dendrobatiden. *herpetofauna* 13(74):23–28.

Böhme, W. 1988. Zur Genitalmorphologie Der Sauria: Funktionelle und Stammesgeschichtliche Aspekte. *Bonner Zool. Monogr.* 27:1–176.

Böhme, W., and U. Sieling. 1993. Zum Zusammenhang zwischen Genitalstruktur. Paarungsverhalten und Fortpflanzungserfolg bei squamaten Reptilien: Erste Ergebnisse. *herpetofauna* 15(82):15–23.

Branch, W. R. 1981. Hemipeneal morphology of Platynotan lizards. *J. Herpetol.* 16:16–38.

Broer, W., and H.-G. Horn. 1985. Erfahrungen bei Verwendung eines Motorbrüters zur Zeitigung von Reptilieneiern. *Salamandra* 21:304–310.

Card, W. 1995. Monitor lizard husbandry. *Bull. Assoc. Rep. Amph. Vets.* 5(3):9–14.

Carlleyle, A. C. 1869. Description of two new species, belonging to the genera *Varanus*, and *Feranioides* respectively, from near Agra. *J. Asiat. Soc. Bengal* 33, Part 2, Nos. 1–4, 192–200.

Dobrusskin, A. 1971. Metal halide lamps with rare earth additives. *Lighting Res. Techn.* 3(2):125–132.

Ehmann, H. , and H. Cogger. 1985. Australia's endangered herpetofauna: A review of criteria and policies. In *The biology of Australasian frogs and reptiles,* ed. W. Grigg, R. Shine, and H. Ehmann, 435–447. Surrey Beatty.

Eidenmüller, B. 2003. *Warane.* Offenbach: Herpeton.

Flugi, U. 1990. Bericht über die Haltung und Nachzucht des Storr'schen

Zwergwarans (*Varanus storri* Mertens, 1966). *herpetofauna* 12(67): 31–34.

Gehrmann, W. H. 1987. Ultraviolet irradiances of various lamps used in animal husbandry. *Zoo Biol.* 6:117–127.

Hildebrandt, T. B., F. Göritz, C. Pitra, L. H. Spelman, T. A. Walsh, R. Roscoe, and N. C. Pratt. 1996. Sonomorphological sex determination in subadult Komodo dragons. *Proc. Ann. Mtng. Am. Assoc. Zoo Vet.* 1996:251–253.

Holmes, R. S., M. King, and D. King. 1975. Phenetic relationships among varanid lizards based upon comparative electrophoretic data and karyotypic analysis. *Biochem. System. Ecol.* 3:257–262.

Honegger, R. E. 1978. Geschlechtsbestimmung bei Reptilien. *Salamandra* 14(2):69–79.

Horn, H.-G. 1991. Breeding of the lace monitor (*Varanus varius*) for the first time outside of Australia (Reptilia: Sauria: Varanidae). *Mertensiella* 2:168–175.

———. 1995. Odyssee und Rettungsversuch geschmuggelter Reptilien mit Anmerkungen zu Sinn und Unsinn praktischen Naturschutzes. Int. Symp. Vivaristik, 22–27 Sept., Dokum. 35–37.

———. 1999. Evolutionary efficiency and success in monitors: A survey on behavioral strategies and some comments. *Mertensiella* 11:167–180.

———. 2004. Beleuchtung im Vivarium (Paludarium, Terrarium). In *Praxis Ratgeber: Vivariumbeleuchtung,* ed. Kh. Sauer, B. Steck, H. Schuchardt, and H.-G. Horn. Frankfurt am Main: Edition Chimaira.

Horn, H.-G., and G. J. Visser. 1988. Freilandbeobachtungen und einige morphometrische Angaben zu *Varanus giganteus* (Gray 1845). *Salamandra* 24(2/3):102–118.

———. 1989. Review of reproduction of monitor lizards *Varanus* spp. in captivity. *Int. Zoo Yrbk.* 28:140–150.

———. 1997. Review of reproduction of monitor lizards *Varanus* spp. in captivity II. *Int. Zoo Yrbk.* 35:227–246.

James, C. D., J. B. Losos, and D. R. King. 1992. Reproductive biology and diets of goannas (Reptilia: Varanidae) from Australia. *J. Herpetol.* 26:128–136.

Judd, H. L., J. P. Bacon, D. Rüedi, J. Girard, and K. Benirschke. 1977. Determination of sex in the Komodo dragon, *Varanus komodoensis.* *Int. Zoo Yrbk.* 17:208–209.

King, D. 1991. The effect of body size on the ecology of varanid lizards. *Mertensiella* 2:204–210.

King, M., and D. King. 1975. Chromosomal evolution in the lizard genus *Varanus* (Reptilia). *Aust. J. Biol. Sci.* 28:89–108.

King, D., M. King, and P. Baverstock. 1991. A new phylogeny of the Varanidae. *Mertensiella* 2:211–219.

King, D., and B. Green, D. 1993. *Goanna.* New South Wales University Press.

Lambertz, K. 1993. Praktizierter Artenschutz in der BRD: Eigenimport von WA-Tieren. *Elaphe N.F.* 1(2):24.

Lenz, S. 1995. Zur Biologie und Ökologie des Nilwarans, *Varanus niloticus* (Linnaeus 1766) in Gambia, Westafrika. *Mertensiella* 5:3–256.

Morris, P. J., and A. C. Alberts. 1996. Determination of sex in white-throated monitors (*Varanus albigularis*), Gila monsters (*Heloderma suspectum*), and beaded lizards (*H. horridum*) using two-dimensional ultrasound imaging. *J. Zoo Wildl. Med.* 27:371–377.

Owerkowicz, T., C. G. Farmer, J. W. Hicks, and E. L. Brainerd. 1999.

Contribution of gular pumping to lung ventilation in monitor lizards. *Science* 284:1661–1663.

Phillips, A. 1994. Recommendations for captive breeding of medium to large-sized monitor lizards. In *Conserv. Assess. Manage. Plan for Iguanidae and Varanidae*, ed. R. Hudson, A. Aberts, S. Ellis, and O. Byers. Publ. IUCN/SSC. Conserv. Breed. Spec. Group.

Pianka, E. R. 1969. Habitat specificity, speciation, and species density in Australian lizards. *Ecology* 50:498–502.

Reinhold, H. 1976. Zinnhalogenid-Entladungslampen—Funktion und Eigenschaften. *Fernseh Kinotechn.* 30:343–346.

Röder, A., and H.-G. Horn. 1994. Über zwei Nachzuchten des Steppenwarans (*Varanus exanthematicus*). *Salamandra* 30(2):97–108.

Schildger, B.-J. 1992. Zur Ultraschalldiagnostik bei Reptilien. *Monitor* 1(1):33–37.

———. 1998. Endoscopy in reptiles. In *Endoscopy in birds, reptiles, amphibians and fish*, ed. M. J. Murray, B. Schildger, and M. Taylor, 31–56. Tuttlingen: (Endo-) Press.

———. 1999/2000. *Endoskopie bei Reptilien*. Vlg. Büchse Der Pandora.

Schildger, B.-J., and R. Wicker. 1992. Endoskopie bei Reptilien und Amphibien—Indikationen, Methoden, Befunde. *Prakt. Tierarzt* 73:516, 518, 523–526.

Schildger, B.-J., H. Tenhu, M. Kramer, M. Gerwing, G. Kuchling, G. Thompson, and R. Wicker. 1999. Comparative diagnostic imaging of the reproductive tract in monitors: Radiology–ultrasonography–coelioscopy. *Mertensiella* 11:193–211.

Schildger, B.-J., W. Häfeli, M. Kramer, H. Tenhu, and R. Wicker. 2000. Die Anwendung bild-gebender Verfahren zur Geschlechterbestimmung bei Reptilien. *Prakt. Tierarzt.* 81:150–156.

Spelman, L. H. 1998. Medical management of Komodo dragons (*Varanus komodoenis*) at the Nation Zoological Park. In *Program and abstracts of the European Assoc. Zoo Wildl. Vets. scientific meeting*.

Vincent, M. and S. Wilson. 1999. *Australian lizards*. Sydney: New Holland Publ.

Werner, F. 1904. Die Warane. *Bl. Aqu. Terr. K.* 15:84–87.

Wood, S. C., M. L. Glass, K. Johansen. 1977. Effects of temperature on respiration and acid–base balance in a monitor lizard. *J. Comp. Physiol.* 116:287–296.

Ziegler, T. and W. Böhme. 1997. Genitalstrukturen und Paarungsbiologie bei squamaten Reptilien, speziell der Platynota, mit Bemerkungen zur Systematik. *Mertensiella* 8:7–207.

adaptation. Conformity between an organism and its environment, or all the ways that a given organism copes with its physical and biotic environments. Adaptations result from natural selection and enhance fertility or survival.

adaptive radiation. Diversification into a large group of ecologically diverse species.

aerobic. Metabolic activities that use oxygen and rely on oxidation to provide energy.

allometry. Differential growth rates of body parts with age.

allopatry. The occurrence of two species in different, nonoverlapping geographic areas.

anaerobic. Metabolic energy obtained without the use of oxygen.

Anguimorpha. A large monophyletic clade that includes anguids, xenosaurids, beaded lizards (*Heloderma*), *Lanthanotus*, varanids, mosasaurs, and snakes, the sister group to Scincomorpha, together comprising the Autarchoglossa.

arboreal. Living above ground in trees or shrubs.

Autarchoglossa. A large monophyletic clade that includes Scincomorpha and Anguimorpha; the sister group to Gekkota.

autotomy. "Self-loss": in lizards, referring to separation of a piece of a lizard's tail.

brummate. Becoming inactive during cold weather, the ectothermic equivalent of hibernation.

buccal. Oral cavity of the mouth.

chemoreception. Detection of chemicals.

clade. A group of descendant species that share a common ancestor.

cloaca. The vent at the end of the digestive tract, which also includes orifices from the urinary and reproductive tracts.

congener (*adj.* **congeneric**). A member of the same genus.

conspecific. A member of the same species.

crepuscular. Active at dusk, dawn, or both.

crypsis (*adj.* **cryptic**). Anatomical and behavioral traits that camouflage an animal, rendering it nearly indistinguishable from its background.

diploid. A fertilized egg or somatic body cell that contains both maternal and paternal genetic material (two full sets of chromosomes).

distal. A part of an appendage that is away from the body.

diurnal. Active during the day.

dorsal. The top side (upper surface) of an animal.

ecomorph. An animal's body plan that can be directly related to its ecology.

ecotone. An edge community at the boundary of two distinct biomes.

ectotherm. An animal that relies on heat sources and sinks in the external environment to gain or lose heat.

endemic. A taxon restricted to a specific defined geographical region.

endotherm. An animal that generates body heat metabolically to control its body temperature.

epithelium. The outer layer of cells, such as those lining the intestine or the surface of the skin.

estivate. To enter a dormant, low metabolic state during warm weather; the process of allowing body temperature to fall and becoming inactive during warm weather.

extant. Surviving to the present day.

fecundity. The number of eggs or offspring produced.

femoral pores. Glands found on the underside of the base of the hind legs, which produce lipid-based substances used to deposit scent trails.

fitness. An organism's ability to perpetuate its genes in a population gene pool.

gamete. A haploid cell that carries only half of a parental genome (one set of chromosomes), such as a sperm or an egg.

gene pool. All the genes present in a given population at a particular moment in time.

gular. The throat region of the lower neck.

gustatory. Chemical signals that stimulate taste buds.

haploid. A sperm cell or an egg cell carrying only half of a parent's genes.

heliothermic. Deriving bodily heat from the sun, as in basking.

hemibaculum. A bone in the hemipenes of some varanid lizards.

hemipenes. Paired copulatory organs that evert from the base of the tail of a male squamate.

heterogamety. The situation in which sex cells or gametes (sperm and eggs) have different sex chromosomes, as in XY heterogamety.

heterozygosity. Having two different alleles at a particular locus on homologous chromosomes.

holocrine gland. A gland that secretes chemical products.

homeothermy. Maintenance of a stable internal body temperature.

hygroscopic. Taking up of moisture from the external environment.

Iguania. A large monophyletic clade of relatively primitive lizards, consisting of iguanids, agamids, and chameleons; the sister group to all other lizards (Scleroglossa).

incertae sedis. The situation in which a taxon is known to have descended from within a particular clade but its exact position in that clade remains undetermined.

inguinal. Referring to the groin region.

integumentary. Referring to the skin.

intromission. Insertion of a male's hemipenis into a female's cloaca.

Jacobson's organ. Paired organs in the roof of the mouth used in analyzing scents; part of the vomeronasal system.

maxithermy. Maintenance of a high active body temperature close to the upper thermal limit to provide high performance ability and maximal fitness.

meiosis. The reduction division whereby homologous chromosomes align and separate into haploid sex cells, or gametes.

mesic. Relatively moist.

mesokinesis. A skull condition whereby joints allow the muzzle to move with respect to the braincase, rising upward when the mouth is opened and clamping downward when it is closed.

monophyletic. Referring to a natural group whose members have all evolved from a single common ancestor.

nares. Nostrils.

nocturnal. Active at night.

olfactory. Referring to the sense of smell.

osteoderm. Bones embedded within scales.

oviductal egg. A shelled egg remaining in the oviduct.

oviparous. Egg laying.

oviposition site. A place where eggs are laid.

phenotype. An organism's external appearance.

pheromone. A chemical used in communication.

phylogenetic. Having to do with evolutionary history.

Platynota. A clade of lizards that includes helodermatids, lanthanotids, varanids, and mosasaurids (and perhaps snakes).

pleurodont. A type of dentition in which teeth are set in sockets in the jawbones.

poikilothermy. The condition in which body temperature fluctuates with ambient temperature.

quadrate. A bone that connects the upper jaw with the mandible in squamates.

riparian. Living along the banks of creeks or rivers.

rupicolous. Rock dwelling.

saxicolous. Rock dwelling.

Scincomorpha. A large monophyletic clade that includes lacertids, teiids, gymnophthalmids, skinks, and xantusiids; the sister group to Anguimorpha.

Scleroglossa. A large monophyletic clade consisting of Autarchoglossa and Gekkota; the sister group to Iguania.

sexual dimorphism. Sex-specific differences in anatomy or behavior.

squamate. A lizard or snake.

streptostyly. A hanging jaw setup whereby the lower jaw is attached to the quadrate bone, which is free to rotate on the skull at the base of the upper jaw.

sympatry. The co-occurrence of two (or more) species at the same place.

synapomorphy. A shared, derived trait.

taxon (*pl.* taxa). A named taxonomic unit, such as a species, genus, family, or higher unit of classification. Any clade is a taxon.

tetrapod. A chordate with four legs (amphibians, reptiles, and mammals) or one derived from an ancestor with four legs (e.g., snakes).

thermoconformer. An animal that does not actively thermoregulate but allows its body temperature to mirror ambient environmental temperature.

thermoregulator. An animal that maintains a narrow range of body temperature in the face of variable external environmental temperatures.

trophic level. A functional classification of organisms according to their feeding relationships.

Varanoidea. A clade that includes beaded lizards (*Heloderma*), *Lanthanotus,* varanids, mosasaurs, and snakes.

ventral. Referring to the bottom surface of an animal.

ventrolateral. Referring to the bottom sides of an animal.

vomerolfaction. The ability to discriminate chemicals based on a sensitive chemosensory system in the roof of the mouth, which can detect and analyze nonairborne chemicals picked up with the tongue.

vomeronasal. Having to do with the chemosensory system that perceives vomodors; also called the vomerolfactory system, or Jacobson's organ.

xeric. Relatively arid; a dry environment.

Page numbers in italics refer to illustrations.

ERIC R. PIANKA,

a world-renowned ecologist, has spent his life studying the evolutionary ecology of the second-largest group of terrestrial vertebrates, lizards. Pianka is the Denton A. Cooley Centennial Professor of Zoology at the University of Texas at Austin. His past and present research covers a broad range of topics pertaining to the ecology, biology, and evolution of reptiles. During his more than 35-year academic career, Pianka has published more than a hundred scientific papers, four of which have become Citation Classics. His intercontinental comparisons of desert lizard ecology have become a standard textbook example. His text *Evolutionary Ecology,* first published in 1974, went through six editions and was translated into Japanese, Polish, Russian, and Spanish. It is currently being translated into Chinese, Italian, and Greek. In 1986, he published a synthesis of his life's research, a small but important book entitled *Ecology and Natural History of Desert Lizards.* He was a Guggenheim Fellow in 1978–1979 and a Fulbright Senior Research Scholar during 1990–1991 (both these were spent doing fieldwork in Australia). In 1994, he published an autobiographical account of his adventures in Australia, *The Lizard Man Speaks* (University of Texas Press). In 2003, with coauthor Laurie J. Vitt of the University of Oklahoma, he published *Lizards: Windows to the Evolution of Diversity* (University of California Press).

DENNIS R. KING (1942–2002)

was one of the world's leading experts on varanid lizards and was also considered a pioneer in several aspects of animal–plant interaction. Dr. King was born in Canada. His early research was on blue grouse before he traveled to Australia, where he began his lifelong interest in herpetology—in particular, varanid lizards. This interest took him to all parts of Australia, to South America, and to Southeast Asia. He studied many aspects of the physiology and ecology and phylogeny of varanids and other lizards. For many years before his retirement in 1996, he was employed as a research scientist by the Agriculture Protection Board of Western Australia. Dr. King was one of the key researchers who discovered, and then clearly demonstrated, that many native Australian animals have developed a tolerance to the toxic substance fluoroacetate (also known as compound 1080) through their long-term association with fluoroacetate-bearing native vegetation. This finding has become the classic worldwide example of plant–animal interaction involving vertebrates. It led to the use of compound 1080 to control introduced vertebrates in Australia and New Zealand, providing the basis for reducing the effects of predation by introduced predators on Australian wildlife. This achievement has been vital to many wildlife conservation programs in Australia, including "Western Shield," the statewide conservation project of the Department of Conservation and Land Management, Western Australia. A successful program for reducing exotic predators, Western Shield has ultimately enabled several native marsupials to be removed from the rare and endangered species list. Dr. King was an honorary associate of the Western Australian Museum from 1986 until his death in early 2002. He is author of more than 100 scientific articles on Australian wildlife, including varanids, mammals, and vertebrate pests, and on conservation and education. In 1993, with coauthor Brian Green, he published *Goannas: The Biology of Varanid Lizards,* which went into its second edition in 1999. This book was published in the United States as *Monitors: The Biology of Varanid Lizards.* He also lectured to undergraduate students and co-supervised several postgraduate students, giving freely of his time and knowledge.